D0491869

Mathematics in Action Group

55

Thomas Nelson and Sons Ltd
Nelson House Mayfield Road
Walton-on-Thames Surrey
KT12 5PL UK

51 York Place
Edinburgh
EH1 3JD UK

Nelson Blackie
Westercleddens Road
Bishopbriggs
Glasgow
G64 2NZ UK

Thomas Nelson (Hong Kong) Ltd
Toppan Building 10/F
22A Westlands Road
Quarry Bay Hong Kong

Thomas Nelson Australia
102 Dodds Street
South Melbourne
Victoria 3205 Australia

Nelson Canada
1120 Birchmount Road
Scarborough Ontario
M1K 5G4 Canada

© Mathematics in Action Group 1989

First published by Blackie and Son Ltd 1989
ISBN 0-216-92349-2

This edition published by Thomas Nelson and Sons Ltd 1992

ISBN 0-17-431400-0
NPN 9 8 7 6 5 4 3

All rights reserved. No paragraph of this publication may be reproduced, copied or transmitted save with written permission or in accordance with the provisions of the Copyright, Design and Patents Act 1988, or under the terms of any licence permitting limited copying issued by the Copyright Licensing Agency, 90 Tottenham Court Road, London, W1P 9HE.

Any person who does any unauthorised act in relation to this publication may be liable to criminal prosecution and civil claims for damages.

Printed in Great Britain.

Robin D. Howat, Mathematics Adviser, Ayrshire
Edward C. K. Mullan, Galashiels Academy
Ken Nisbet, Madras College, St Andrews, Fife
Doug Brown, St Anne's High School, Heaton Chapel, Stockport, Cheshire

with
J. L. Hodge, **J. Hunter,** **A. G. Robertson,** **P. Whyte**

MATHEMATICS IN ACTION GROUP

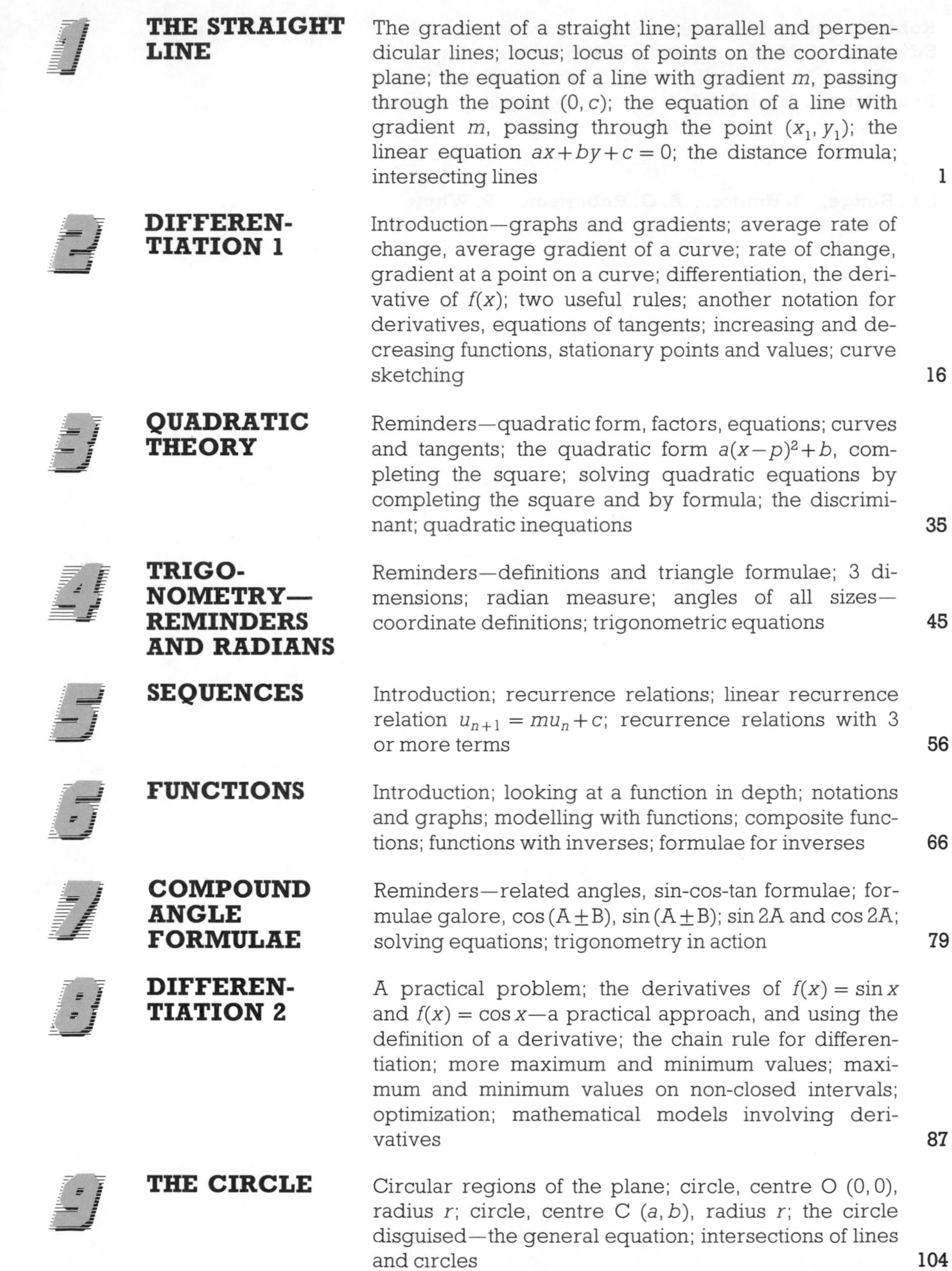

CONTENTS

INVESTIGATIONS will always be in a right-handed box

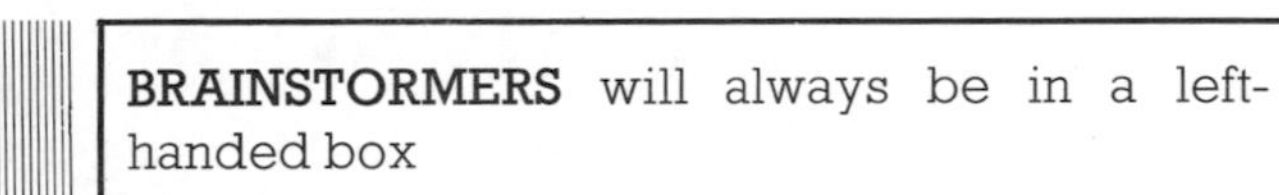

BRAINSTORMERS will always be in a left-handed box

INTRODUCTION

Mathematics is in action all around you. You need mathematics in your daily life. **Mathematics in Action** has been written to help you to understand and to use mathematics sensibly and well—to save you time and effort. Some parts of mathematics are needed in other subjects and some parts will be studied in greater detail later on.

Mathematics in Action follows the latest thinking in *what* mathematics should be studied and how hard, or easy, it should be. So you are taken forward, stage by stage, as far as you can go.

Exercises for practice, Puzzles and Games for fun, Brainstormers to make you think, Investigations to explore, Practical Activities, even Check-ups (to see how you're doing)—all are here.

Enjoy maths with **Mathematics in Action!** Let's hope that you will find a lot that is worthwhile, interesting, and above all, useful.

MiAG March 1989

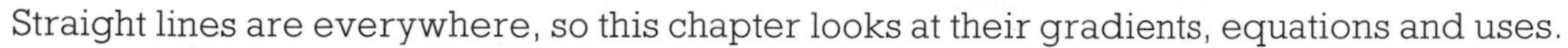

Straight lines are everywhere, so this chapter looks at their gradients, equations and uses.

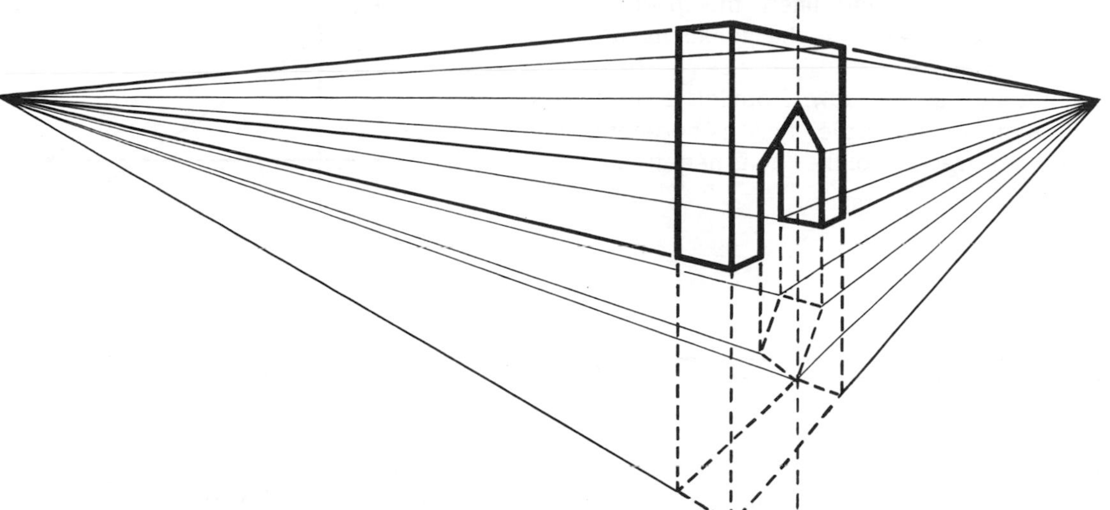

THE GRADIENT OF A STRAIGHT LINE

In Book 4 the gradient of AB, m_{AB}, was defined as $\dfrac{\text{change in } y \text{ from A to B}}{\text{change in } x \text{ from A to B}}$

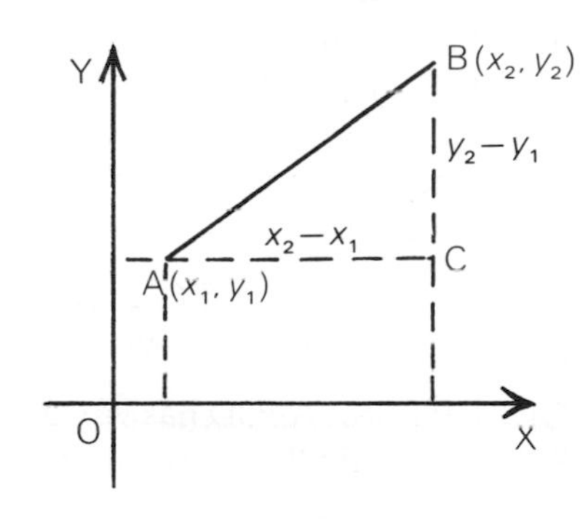

So $m_{AB} = \dfrac{y_2 - y_1}{x_2 - x_1}$.

Also, $m_{AB} = \tan\theta°$, where $\theta°$ is the anticlockwise angle from OX to AB (or AB produced).

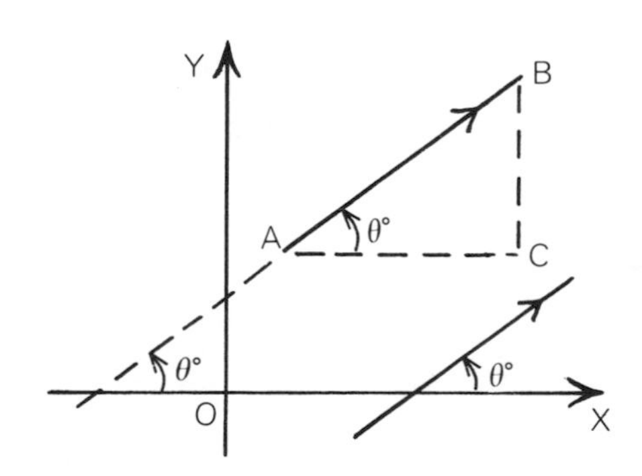

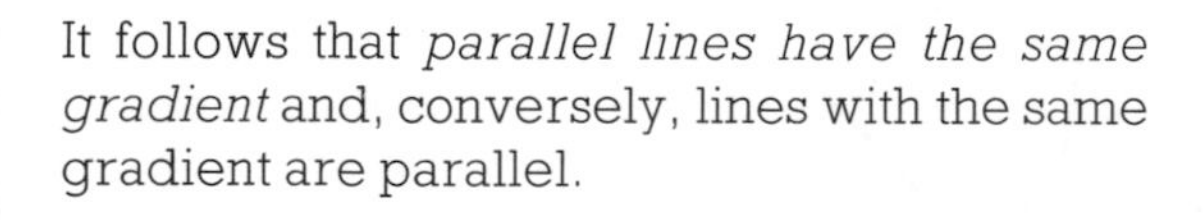

It follows that *parallel lines have the same gradient* and, conversely, lines with the same gradient are parallel.

Notes

(i) Since $\tan 90^\circ$ is undefined, the gradient of a line parallel to the y-axis is undefined.

(ii) For a line sloping *down* from left to right, θ° is an obtuse angle, $\tan\theta^\circ$ is negative, and the gradient of the line is negative.

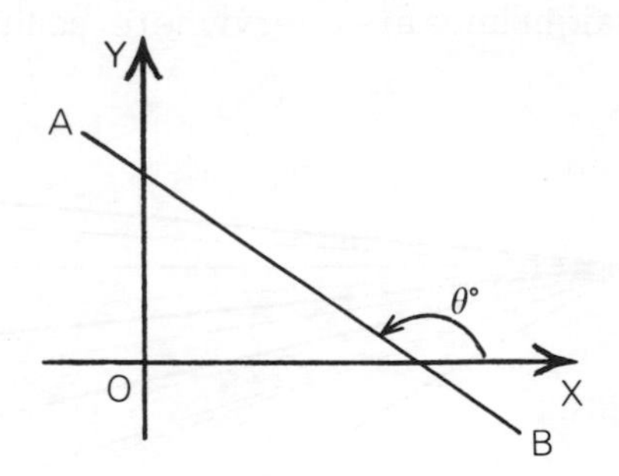

Exercise 1

1 a Calculate the gradient of the line joining each pair of points:

(i) A(5, 3), B(8, 4) (ii) O(0, 0), C(3, 6) (iii) D(−3, −3), E(3, 3)
(iv) F(−2, 0), G(1, 1) (v) H(−1, 5), K(2, 0) (vi) M(3, −5), N(8, −5)

b Which two lines are parallel?

c Which line is parallel to the x-axis?

d Which lines slope: (1) up from left to right (2) down from left to right?

2 a Calculate the gradients of these lines:

(i) P(2, 1) Q(4, 7) (ii) R(−3, −5) S(−7, 7) (iii) T(−1, 4) U(3, 8) (iv) V(4, −2) W(−8, 7)

b (i) Use the definition $m = \tan\theta^\circ$ to calculate, to the nearest degree, the anti-clockwise angle which each line makes with OX ($0 < \theta < 180$).

(ii) Show the lines in a sketch.

3 D is the point (2, 2), E(−3, 2), F(1, −4), G(4, −1). Prove that two pairs of these points can be joined to give parallel lines.

4 Quadrilateral ABCD has vertices A(2, 3), B(8, 9), C(9, 3), D(3, −3).
Prove that ABCD is a parallelogram.

5 Triangles PQS and RQT share a common vertex. P is the point (−10, −1), Q(−4, 1), R(5, 4), S(−6, 3), T(−1, −2). Prove that:

a P, Q, R are collinear (in the same straight line)

b S, Q, T are collinear

c triangles PQS and RQT are similar.

6 Calculate the gradient (correct to 1 decimal place) of:

a AB

b AC

c BC.

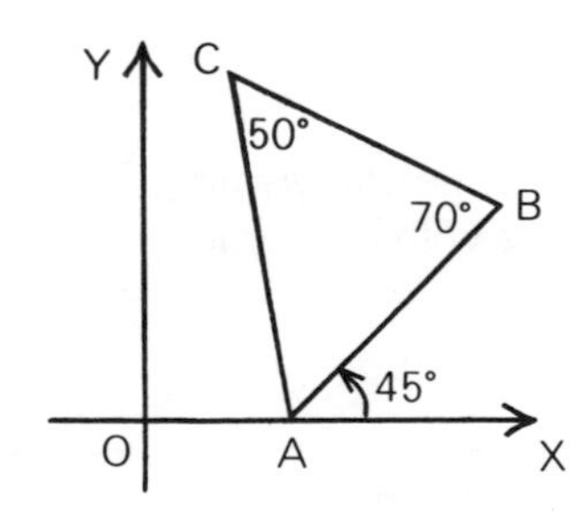

7 Geographers use 6-figure references to pinpoint places. For example, A is 505615. Mathematicians would interpret this as co-ordinates (50·5, 61·5). Ben Nevis Road forms an angle ABC, where A is 505615, B is 514618 and C is 518631.

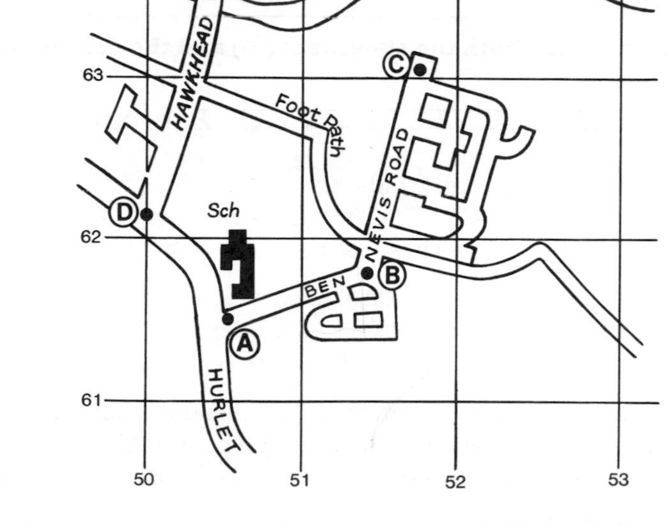

a Taking the grid as a co-ordinate system, find the gradients of AB and BC, correct to 1 decimal place.

b The vertical lines run north-south. Find the bearing of: (i) B from A (ii) C from B.

c Calculate the size of ∠ABC.

d Hawkhead Road runs parallel to part BC of Ben Nevis Road from D 500622 to E 504***. Find the full 6 figure reference for E.

8 The acute angle between two intersecting lines is 40°. One of the lines has a gradient of 3. Make a sketch, and find the possible gradients of the other line, correct to 1 decimal place.

9

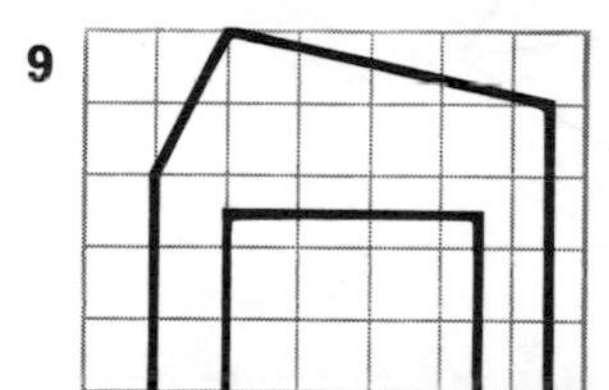

On an architect's blueprints for a factory, one side of the roof has a gradient of 2. Calculate the gradient, correct to 1 decimal place, of the other side of the roof if the angle at the apex is:

a 100° **b** 110° **c** 90°.

PERPENDICULAR LINES

Under a rotation about O of 90°, $A(a, b) \rightarrow B(-b, a)$.

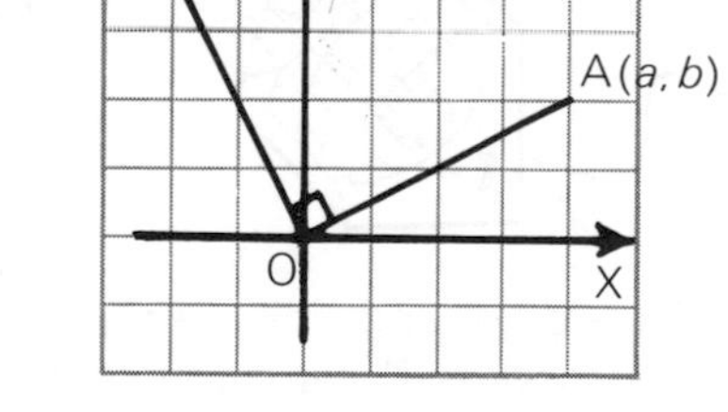

$$m_{OA} = \frac{b-0}{a-0} = \frac{b}{a}, \text{ and}$$

$$m_{OB} = \frac{a-0}{-b-0} = -\frac{a}{b}.$$

$$\text{So } m_{OA} \times m_{OB} = \frac{b}{a} \times \left(-\frac{a}{b}\right) = -1.$$

The converse is also true.
Suppose OC is perpendicular to OA.
Then $m_{OA} \times m_{OC} = -1$.
But $m_{OA} \times m_{OB} = -1$.
So $m_{OC} = m_{OB}$, and OB is parallel to OC.
But OC is perpendicular to OA, so OB is perpendicular to OA.

Summary. If two straight lines have gradients m_1 and m_2 then:
a $m_1 = m_2 \Leftrightarrow$ the lines are parallel
b $m_1 m_2 = -1 \Leftrightarrow$ the lines are perpendicular.

$\Leftrightarrow$ means that both the statement ($\Rightarrow$) and the converse ($\Leftarrow$) are true.

Exercise 2

1 Some lines and their gradients:
(i) AB, $\frac{1}{2}$ (ii) CD, 2 (iii) EF, $-\frac{1}{2}$ (iv) GH, -2 (v) IJ, $\frac{1}{2}$ (vi) KL, $-\frac{1}{2}$
Which pairs of lines are: **a** parallel **b** perpendicular?

2 Write down the gradients of lines which are:
a parallel **b** perpendicular to lines with gradients:
(i) 4 (ii) -1 (iii) $\frac{1}{3}$ (iv) $\frac{3}{4}$ (v) $-2\frac{1}{2}$

3 $y = mx$ is the equation of a line through the origin, with gradient m. Write down the equations of lines through the origin perpendicular to:
a $y = 3x$ **b** $y = -5x$ **c** $y = \frac{1}{10}x$ **d** $y = \frac{2}{3}x$ **e** $y = -\frac{1}{2}x$

4 In which of these pairs are the lines: **a** parallel **b** perpendicular?
(i) $y = 2x, y = -\frac{1}{2}x$ (ii) $x + y = 0, x - y = 0$ (iii) $y = 4x + 3, y = 4x - 3$
(iv) $y = \frac{3}{2}x, 3x - 2y = 1$.

5 Prove that the triangles in the diagram are right-angled.

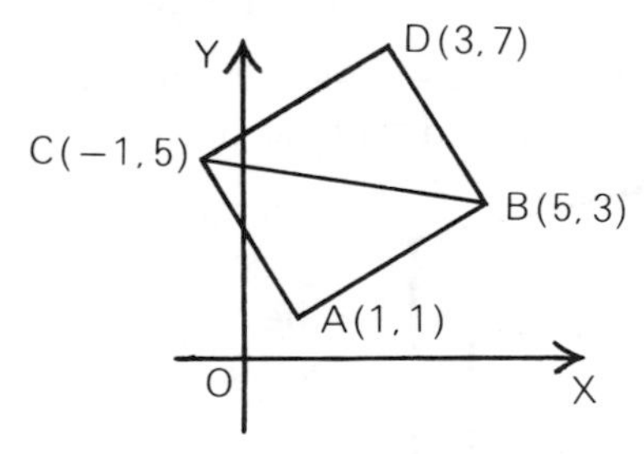

6 a Prove that the quadrilateral with vertices P(-1, 0), Q(5, -2), R(6, 1), S(0, 3), in order, is a rectangle.
b What angle does each side make with OX?

LOCUS

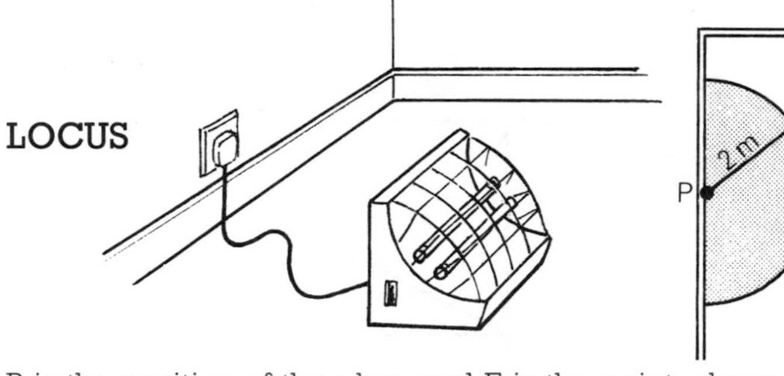

The electric fire has a flex 2 m long from the plug to the point where it enters the fire. Look at the right-hand diagram.

P is the position of the plug, and F is the point where the flex enters the fire. The fire can be placed anywhere as long as F remains in the shaded region bounded by the semi-circle, centre P and radius PF. The shaded region is called *the locus of F*, locus meaning 'place' or 'location'.

A locus can be thought of as a well-defined set of points, and can be described by means of 'set builder' notation. In the example above, the locus, L, of P is given by $L = \{P: PF \leqslant 2\}$, that is 'the set of points P such that PF is less than or equal to 2'.

Example
Workmen have arrived to connect the mains water to the supply pipe for this block of flats. They know that the junction is 5 m from the base of the building, which is 20 m square. The locus of J, the junction, is shown by the dotted line. Why are there quarter circles at each corner? What would the locus be if the junction is *at most* 5 m from the building?

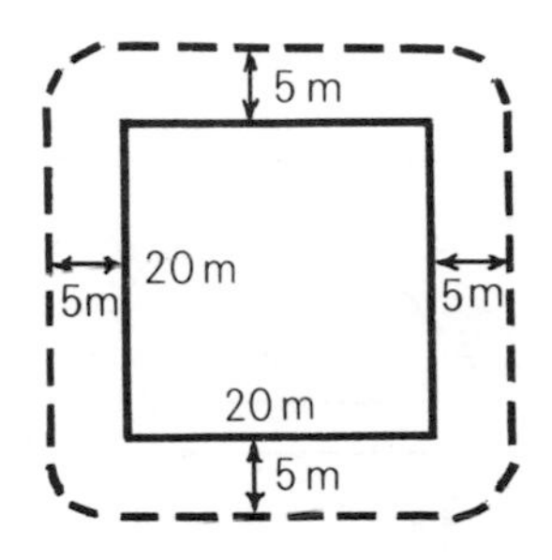

Exercise 3

1 The C.B. radio in Mike's cab has a range of 10 miles.
Show in a sketch the locus of points where his signal could have been picked up if:
a he was parked in the taxi rank
b he travelled in a straight line for 10 miles.

2

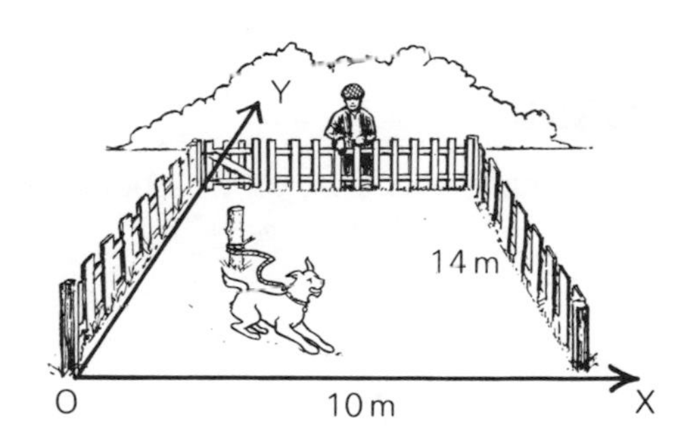

Samson the guard dog is tied to a rope 6 m long, in a rectangular yard 10 m by 14 m. The rope is tied to a post at point (3, 7) with reference to the axes.

a Show by shading in a diagram the locus of S, representing Samson's position.
b Dot the 'complementary' locus of points where you'd be safe from Samson.

3 A rotary clothes drier is to be set up in a patio which is a 10 metre square. The centre post of the drier, C, has to be at least 6 m from each corner of the patio. Show in a diagram the locus of C.

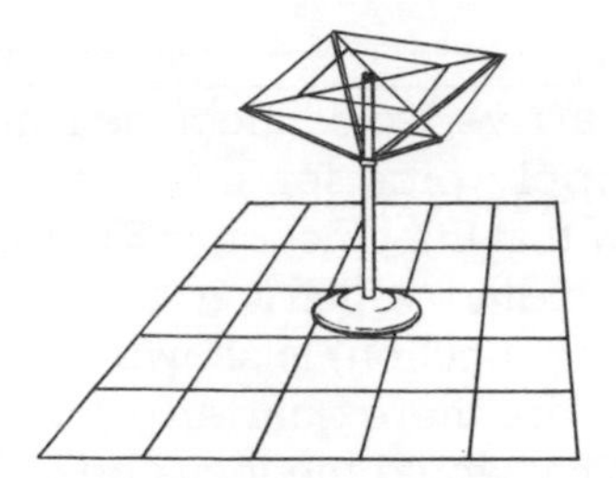

4 At the local fête, coins are thrown onto a board. To win a prize the whole coin must land inside a square of side 4 cm. The coin has a diameter of 1 cm.

a (i) Draw the locus of the centre, C, of a coin in a winning position.
(ii) What fraction of the area of the square is covered by this locus?

b Repeat **a** when the target area is:
(i) a circle of diameter 4 cm (ii) an equilateral triangle of side 4 cm.

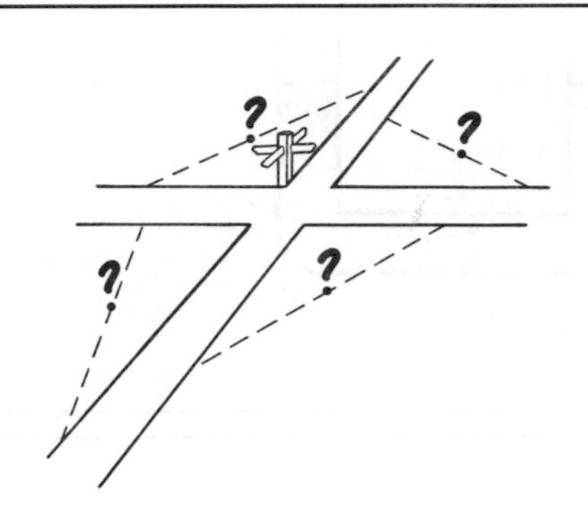

There are two roads across the moor, at right angles to each other. Colin left one of the roads and cut across the moor for 1 mile before reaching another road. Halfway across he stopped for a snack, and left his pack, P, there by mistake.

a Investigate the locus of P. (Use single lines at right angles for the roads.)
b Investigate the locus if Colin had stopped one quarter of the way across the moor.

LOCUS OF POINTS ON THE COORDINATE PLANE

P(x, y) can be anywhere on the XOY plane. But if you put conditions on x or y or both, then you'll restrict P's position.

Example 1
If $x = 0$, the locus of P is $\{P(x, y): x = 0\}$.
$x = 0$ is the equation of the locus, which is the y-axis.

Example 2
If $x > 0$ and $y > 0$, then P is restricted to the first quadrant. Its locus is:
$L = \{P(x, y): x > 0\} \cap \{P(x, y): y > 0\}$
$= \{P(x, y): x > 0 \text{ and } y > 0\}$.

Note The locus L is the clear, or unshaded region.

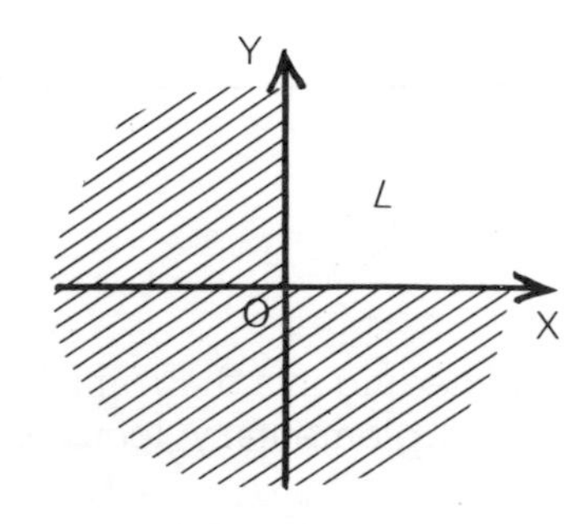

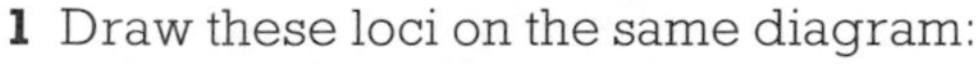

Exercise 4

1 Draw these loci on the same diagram:
a $\{P(x, y): x = 6\}$ **b** $\{P(x, y): x = -2\}$ **c** $\{P(x, y): y = 4\}$ **d** $\{P(x, y): y = -3\}$.

2 Show by clear regions (by shading the unwanted regions):
a $A = \{(x, y): 0 < x < 4\}$ **b** $B = \{(x, y): -2 < y < 2\}$ **c** $C = A \cap B$.

3 a Show the locus of points which are less than 3 units from the line $y = 4$.
b Express the locus in set-builder notation.

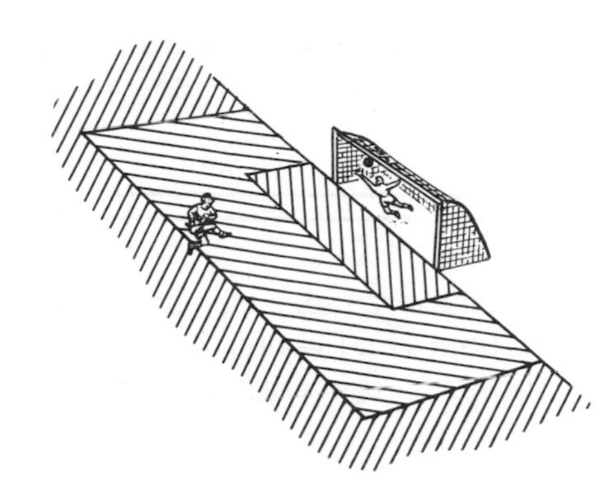

4 a (i) Draw the locus of all points in the first and third quadrants which are equidistant from the x and y-axes.
(ii) Write down the equation of the locus, including the origin.
b Repeat **a** for points in the second and fourth quadrants.

5 Draw diagrams showing these loci:
a $\{P(x, y): OP = 4\}$ (the set of points 4 units from the origin)
b $\{P(x, y): OP = 7\}$ **c** $\{P(x, y): AP = 3\}$, where A is the point (3, 4)
d $\{P(x, y): OP < 5\}$ **e** $\{P(x, y): 2 < OP \leqslant 6\}$
f $\{P(x, y): OP < 5\} \cap \{P(x, y): AP < 5\}$, where A is the point (2, 1).

6 a (i) Plot some points for which $y = x + 4$.
(ii) Draw the locus of all such points.
b (i) Plot 3 points for which $y \leqslant x + 4$.
(ii) Show the locus of all such points (a clear region).
c Repeat **a** and **b** to identify the loci:
(i) $y < x + 3$ (ii) $y > 2x$ (iii) $y > 2x + 4$ (iv) $2x < y < 3x$.

7 On a map, OX is drawn due east and OY due north. A new motorway is represented by $\{(x, y): y = x - 3\}$, where 1 unit = 1 mile. A grant is to be made to people living within 1 mile north or south of the motorway so that they can sound-proof their homes. Make a sketch to show the region within which the grant can be claimed, and describe it using set notation. What difference would it make if the grant were given to people living within 1 mile of the motorway?

THE LINE WITH GRADIENT m, THROUGH THE POINT $(0, c)$

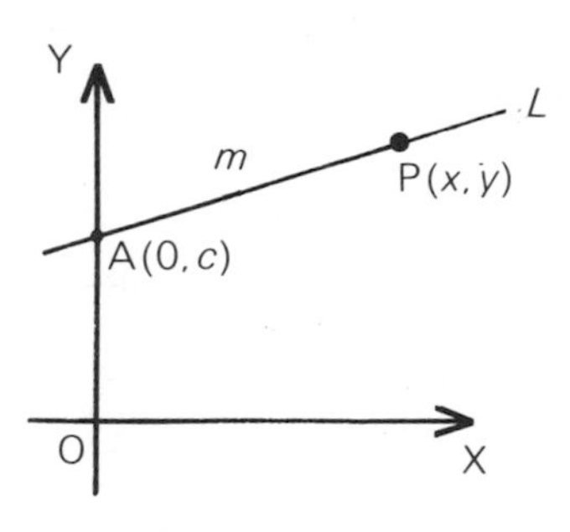

The line L consists of the point A and the set of points P such that $m_{AP} = m$.

$L = \{A\} \cup \{P(x, y): m_{AP} = m\}$

$$= \{A\} \cup \left\{P(x, y): \frac{y-c}{x-0} = m,\ x \neq 0\right\}$$

$$= \{P(x, y): y - c = mx\}$$

$$= \{P(x, y): y = mx + c\}$$

The equation of the line is $y = mx + c$.	c is the *intercept* on the y-axis.

Example

a Write down the equation of the line with gradient 3, and y-intercept 5.
b Which of these points are above, below or on the line? (i) $(1, 7)$ (ii) $(-2, -1)$ (iii) $(-1, 3)$.

a Using $y = mx + c$, the equation of the line is $y = 3x + 5$.
b (i) If $x = 1$, $y = 3 + 5 = 8$. So $(1, 7)$ is below the line.
(ii) If $x = -2$, $y = -6 + 5 = -1$. So $(-2, -1)$ is on the line.
(iii) If $x = -1$, $y = -3 + 5 = 2$. So $(-1, 3)$ is above the line.

Exercise 5

1 Write down the equation of the line:
a with gradient 4 and y-intercept: (i) 2 (ii) -1 (iii) 0
b with gradient -1, passing through the point: (i) $(0, 2)$ (ii) $(0, -5)$ (iii) $(0, 0)$.

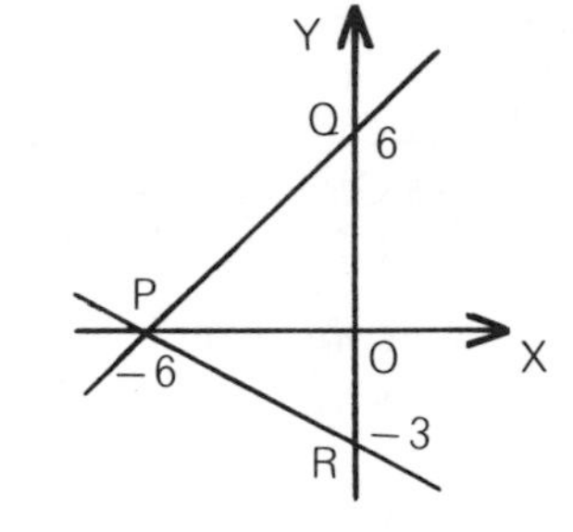

2 Find the gradients, y-intercepts and equations of: **a** PQ **b** PR.

3 Find the gradient, then the equation of the line joining:
a A(0, 4) B(2, 6) **b** C(0, 2) D(3, 8)
c E(0, 3) F(6, -1) **d** G(0, -2) H(-2, 6).

4 Write down the equation of each line, and find whether the given point is above, below or on the line.

	a	**b**	**c**	**d**	**e**
m	1	-2	$\frac{1}{3}$	$-\frac{2}{3}$	$\frac{4}{3}$
c	2	-1	-4	2	-1
Point	$(1, 3)$	$(-2, 4)$	$(3, -4)$	$(6, -2)$	$(-3, -6)$

5 Express each equation in the form $y = mx + c$, and hence give its gradient and y-intercept:
a $y - 2x = 4$ **b** $2y + 6x = 12$ **c** $3y - 2x = 6$ **d** $x + y = 5$ **e** $3x + 4y - 12 = 0$.

6 On squared paper draw the lines in question **5**, either by using the form $y = mx + c$, or by finding where the lines cut the x and y-axes.

7 **a** A is the point $(-2, 4)$ and B is $(6, 6)$. Calculate the gradient of AB.
b Find the equation of the line through C(0, 5): (i) parallel to AB (ii) perpendicular to AB.

8 Find the equation of the line through the first pair of points, and then check whether the third point is above, on or below the line:
a $(-4, 3)$, $(4, 7)$; $(8, 9)$ **b** $(-1, -5)$, $(1, 3)$; $(3, 10)$.

THE LINE WITH GRADIENT m, THROUGH THE POINT (x_1, y_1)

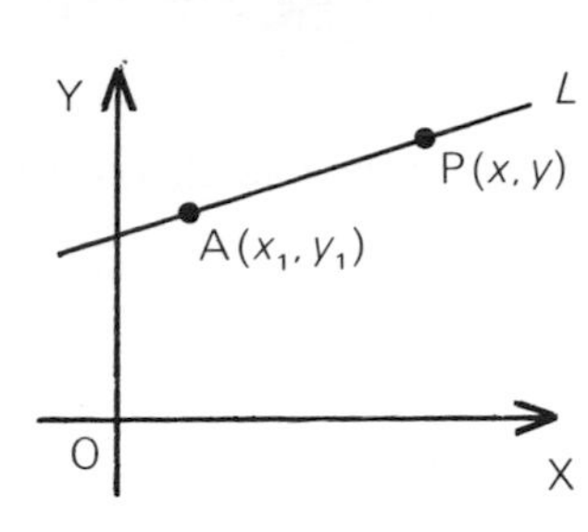

$$L = \{A\} \cup \{P(x, y): m_{AP} = m\}$$

$$= \{A\} \cup \left\{P(x, y): \frac{y-y_1}{x-x_1} = m, x \neq x_1\right\}$$

$$= \{P(x, y): y-y_1 = m(x-x_1)\}$$

The equation of the line is $y-y_1 = m(x-x_1)$

Example

Find the equation of the line through A$(-1, 2)$, perpendicular to the line through B$(2, -3)$ and C$(4, 3)$.

$$m_{BC} = \frac{3+3}{4-2} = \frac{6}{2} = 3.$$

So the gradient of the perpendicular line is $-\frac{1}{3}$.

The required equation is $y-2 = -\frac{1}{3}(x+1)$

$3y-6 = -x-1$

$x+3y = 5.$

Exercise 6

1 Find the equation of the line through the given point, with the given gradient:
a $(4, 6)$, $m = 2$ **b** $(-2, -3)$, $m = 3$ **c** $(3, 4)$, $m = \frac{1}{3}$ **d** $(3, -1)$, $m = -\frac{2}{5}$.

2 Find the gradient, and hence the equation, of the line through each pair of points:
a $(-2, 1), (4, 5)$ **b** $(-2, 5), (6, 3)$ **c** $(-1, 8), (5, -1)$ **d** $(-2, -1), (4, -3)$.

3 Find the equation of the line parallel to the given line, passing through the given point:
a $y = x+4$, $(5, 6)$ **b** $y = 3x+2$, $(4, -1)$ **c** $x+y-4 = 0$, $(-5, 0)$
d $2y-3x+1 = 0$, $(1, -3)$.

4 Find the equation of the line perpendicular to the given line, passing through the given point:
a $y = 2x-3$, $(6, -1)$ **b** $y-4x = 1$, $(9, 3)$ **c** $2y-x-5 = 0$, $(3, 4)$
d $2x+3y+1 = 0$, $(5, 0)$.

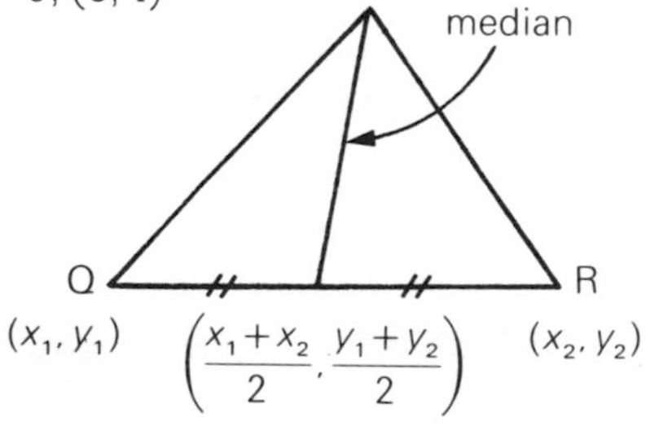

5 $\triangle$PQR has vertices P$(3, 4)$, Q$(-6, 6)$, R$(-4, -2)$.
a Find the equation of:
(i) the median through P
(ii) the altitude through P.
b What kind of triangle is PQR? Give a reason for your answer.

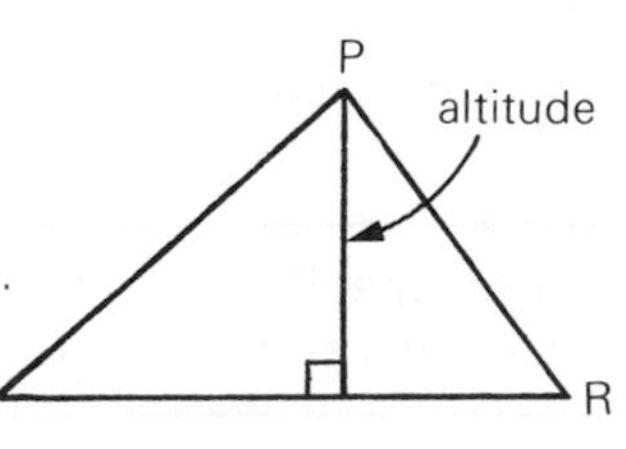

6 **a** Prove that S(−3, 2), T(5, 6), U(2, −1), V(−6, −5) are the vertices, in order, of a parallelogram.
b Find the equations of its sides.

7 A(2, 4), B(9, 8), C(8, 0) are vertices of quadrilateral ABCD which has an axis of symmetry BD.
a Calculate the gradient of: (i) AC (ii) BD.
b Find the equation of BD.
c Write down the coordinates of the midpoint of AC.
d Describe the locus of D if ABCD is:
(i) a V-kite (ii) a kite (not a V-kite) (iii) a rhombus.

8 In a CAD (Computer Aided Design) program a drive chain passes round two cog wheels.

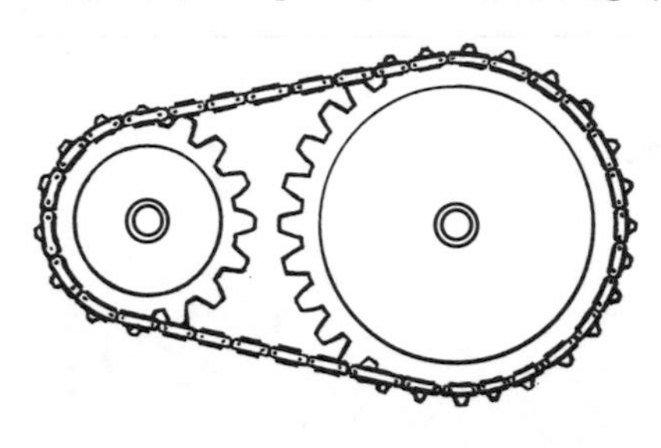

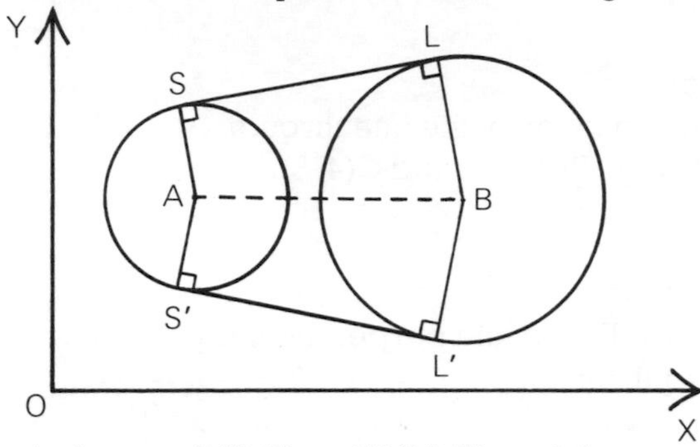

In the coordinate 'model', the centres of the circles are A(5, 7) and B(15, 7), and the tangent LS touches one circle at S(4, 10). Find the equations of:
a the radii AS and BL **b** the tangent SL
c the axis of symmetry AB **d** the tangent S′L′.

THE LINEAR EQUATION $ax+by+c=0$

Let $L = \{(x, y): ax+by+c=0\}$.

(i) If $a = 0$ and $b \neq 0$, $L = \{(x, y): by+c=0\}$

$$= \left\{(x, y): y = -\frac{c}{b}\right\}$$

which is a line parallel to the x-axis.

(ii) If $a \neq 0$ and $b = 0$, $L = \{(x, y): ax+c=0\}$

$$= \left\{(x, y): x = -\frac{c}{a}\right\}$$

which is a line parallel to the y-axis.

(iii) If $a \neq 0$ and $b \neq 0$, $L = \{(x, y): ax+by+c=0\}$

$$= \left\{(x, y): y = -\frac{a}{b}x-\frac{c}{b}\right\}$$

which is the line with gradient $-\frac{a}{b}$ and y-intercept $-\frac{c}{b}$.

> $ax+by+c=0$ represents a straight line if a and b are not both zero. It is called a linear equation; it is of the first degree in x and y.

Exercise 7

1 Arrange each equation in the form $ax+by+c=0$, where a, b and c are integers:

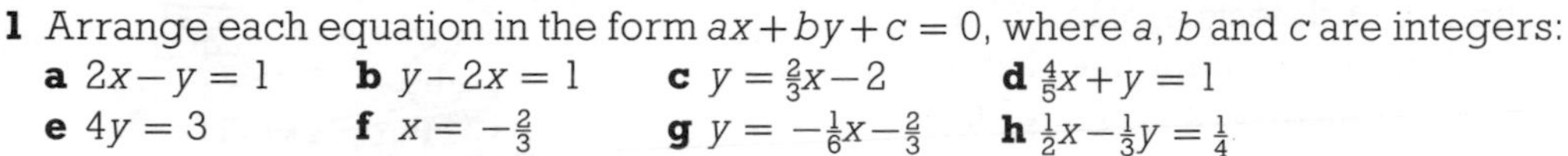

a $2x-y=1$ **b** $y-2x=1$ **c** $y=\frac{2}{3}x-2$ **d** $\frac{4}{5}x+y=1$
e $4y=3$ **f** $x=-\frac{2}{3}$ **g** $y=-\frac{1}{6}x-\frac{2}{3}$ **h** $\frac{1}{2}x-\frac{1}{3}y=\frac{1}{4}$

2 Sketch the lines which have equations:
a $2x+y=4$ **b** $4x-3y=12$ **c** $2x+y-2=0$ **d** $3x-y=0$

3 Find, in the form $ax+by+c=0$, the equation of the line through:
a $(0, 5)$, with gradient -2 **b** $(0, -3)$, with gradient $-\frac{3}{4}$ **c** $(3, 5)$, with gradient $\frac{1}{2}$
d $(0, 0)$ and $(-4, 2)$ **e** $(-2, 7)$ and $(4, 4)$ **f** $(-2, -4)$ and $(2, -1)$.

4 AB has equation $3x+4y-8=0$. Find the equation of the line through $(7, 3)$:
a parallel to AB **b** perpendicular to AB.

5 P is $(-4, 0)$ and Q is $(2, 6)$. Find:
a (i) the gradient of PQ (ii) the equation of PQ
b the equations of the lines through P and Q, perpendicular to PQ.

6 Diagonal AC is the axis of symmetry of kite ABCD, and has equation $3x-y-3=0$.
a A is $(3, k)$ and C is $(h, -3)$. Find the values of h and k.
b BD intersects AC at $(2, 3)$. Find the equation of diagonal BD.
c The x-coordinate of B is -1. Calculate the coordinates of B and D.

THE DISTANCE FORMULA

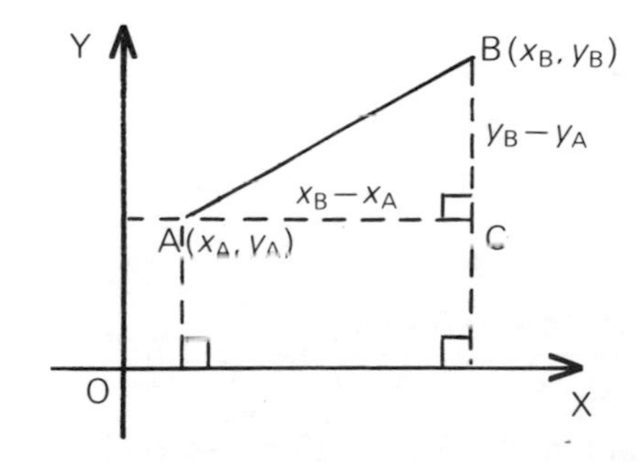

By Pythagoras' Theorem,

$$AB^2 = AC^2 + CB^2$$
$$= (x_B - x_A)^2 + (y_B - y_A)^2$$

So $AB = \sqrt{(x_B - x_A)^2 + (y_B - y_A)^2}$

Exercise 8

1 Calculate the distance between each pair of points:
a $(-2, -2)$ and $(6, 4)$ **b** $(1, 6)$ and $(6, -6)$ **c** $(1, 2)$ and $(8, 26)$.

2 $\triangle ABC$ has vertices A$(4, 6)$, B$(5, -1)$, C$(10, 4)$. Prove that the triangle is isosceles.

3 P is the point $(-6, -4)$, Q is $(2, 2)$ and R$(-2, 4)$. Prove that $\triangle PQR$ is right-angled, using:
a gradients **b** the distance formula.

4 Quadrilateral ABCD has vertices A$(-3, -1)$, B$(5, -3)$, C$(7, 5)$ and D$(-1, 7)$. Prove that:
a its opposite sides are parallel
b its sides are all equal
c its angles are all right angles.
What type of quadrilateral is it?

5 A radar station at O detects a plane 5 km north, at A, flying in a straight line. Later the plane is noted at B, 10 km east of O. Taking OE and ON as x and y-axes:

a find the equation of: (i) AB
(ii) OD, which is perpendicular to AB

b find the coordinates of the point where the plane was nearest O

c calculate the bearing and distance of D from O.

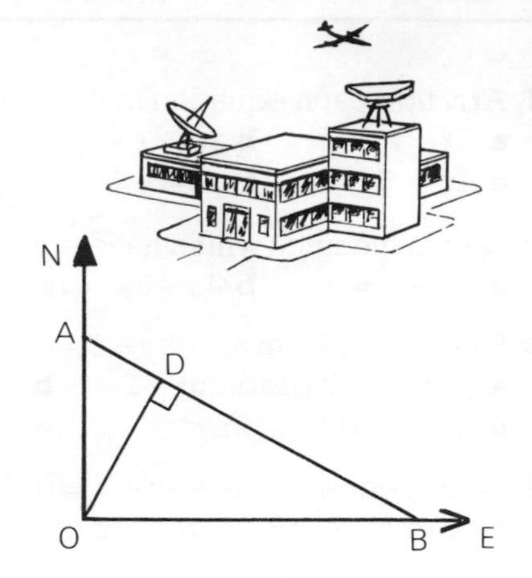

INTERSECTING LINES

Example
Show that the perpendicular bisectors of the sides of $\triangle$ A$(-4, 8)$, B$(6, 12)$, C$(6, -2)$ are concurrent.

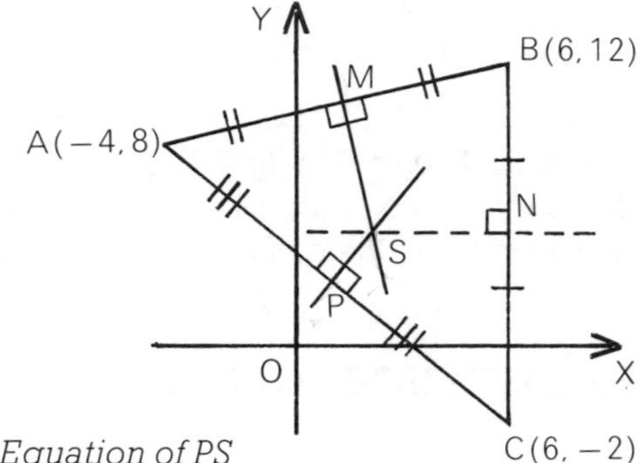

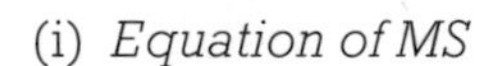

(i) *Equation of MS*

$$m_{AB} = \frac{12-8}{6+4} = \frac{2}{5}$$

So $m_{MS} = -\frac{5}{2}$

M is $\left(\frac{-4+6}{2}, \frac{8+12}{2}\right)$, i.e. $(1, 10)$

Equation of MS is $y - 10 = -\frac{5}{2}(x-1)$

i.e. $\underline{5x + 2y = 25}$

(ii) *Equation of PS*

$$m_{AC} = \frac{-2-8}{6+4} = -1$$

So $m_{PS} = 1$

P is $\left(\frac{-4+6}{2}, \frac{8-2}{2}\right)$, i.e. $(1, 3)$

Equation of PS is $y - 3 = 1(x-1)$

i.e. $\underline{x - y = -2}$

To find S, solve $\begin{cases} 5x + 2y = \ \ 25 \ldots \times 1 \\ \ \ x - y = -2 \ldots \times 2 \end{cases}$ $\begin{aligned} 5x + 2y &= \ \ 25 \\ 2x - 2y &= -4 \end{aligned}$

Add $7x = 21$
$x = 3$
$y = x + 2 = 5$
$\underline{\text{S is the point } (3, 5)}$

(iii) *Equation of NS*
NS is parallel to the x-axis
N is $(6, 5)$.
The equation of NS is $y = 5$.
S$(3, 5)$ lies on NS, so the three lines are concurrent.

Exercise 9

1 Find the point of intersection of the lines in each pair:

a $3x+y=7$, $x-y=1$ **b** $2x+3y=10$, $x+y=5$ **c** $3x-4y=13$, $4x+2y=10$ **d** $4x+3y=1$, $3x-2y=5$

2 a Find the equation of:
(i) median PS
(ii) altitude QT, in $\triangle PQR$.

b Find the coordinates of V, the point of intersection of PS and QT.

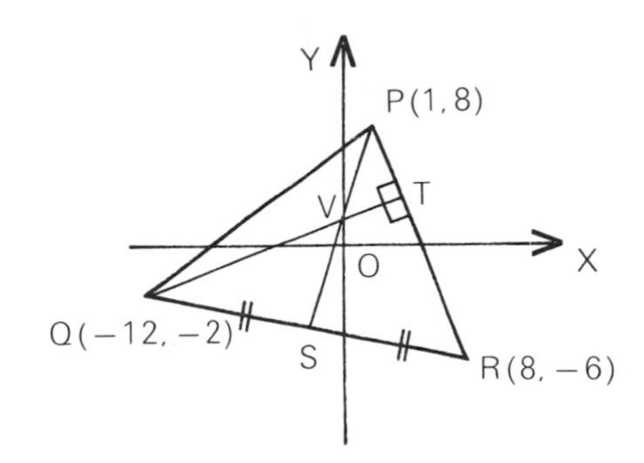

3

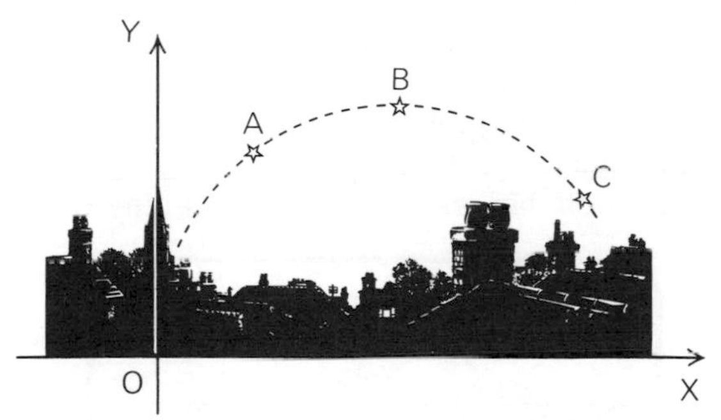

Ann and Liz are keen astronomers. They observe the movement of a star as it circles the Pole Star. With reference to their chosen axes they record the star's position at three different times as A(4, 12), B(10, 14) and C(18, 10).

a Calculate the gradients of chords AB and BC.
b Find the equations of the perpendicular bisectors of AB and BC.
c By calculating the point of intersection of the perpendicular bisectors of AB and BC find the coordinates of the Pole Star.

4 The equations of the sides of $\triangle ABC$ are:
AB, $2x-5y+25=0$; BC, $3x-y-8=0$; CA, $4x+3y+11=0$.
a Find: (i) the coordinates of A, B and C
(ii) the equations of medians AP, BQ and CS of $\triangle ABC$.
b Show that the point of intersection of AP and BQ lies on CS. What does this tell you about the medians of the triangle?

5

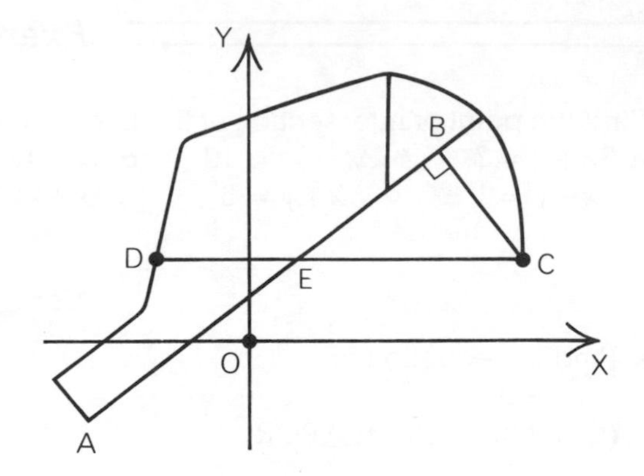

In the coordinate 'model' the origin is taken at the control tower of the airport. A is the point $(-6, -2)$ and B is $(6, 6)$.

a Find the equation of: (i) Runway 1 (AB) (ii) the taxi-strip (BC).

b Runway 2 points W–E. D is the point $(-2, 3)$. Find:
(i) the equation of Runway 2 (DC) (ii) the coordinates of the point of intersection of the runways (E), and of point C (iii) the length of BC.

6 The vertices of a triangle are $A(5, 5)$, $B(-10, 0)$ and $C(0, -10)$.

a Find: (i) the equations of the altitudes of the triangle through A and B.
(ii) the point of intersection, K, of these altitudes.

b Show that K lies on the altitude through C. What does this tell you about the altitudes of the triangle?

7 A computer can represent solid objects using perspective drawing and coordinate geometry. In a program, all lines going to the right converge on a vanishing point $R(16, 10)$, and all lines to the left converge on $L(0, 10)$. $A(4, 7)$, $E(4, 5)$, $C(12, 7)$ and $G(12, 5)$ are four vertices of a cuboid.

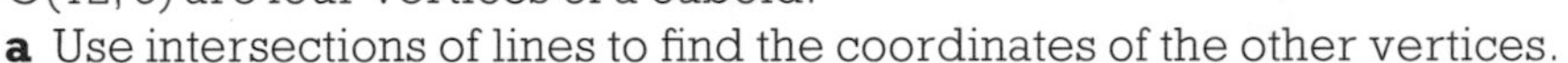

a Use intersections of lines to find the coordinates of the other vertices.

b Write down the coordinates of the centre of the cuboid.

8

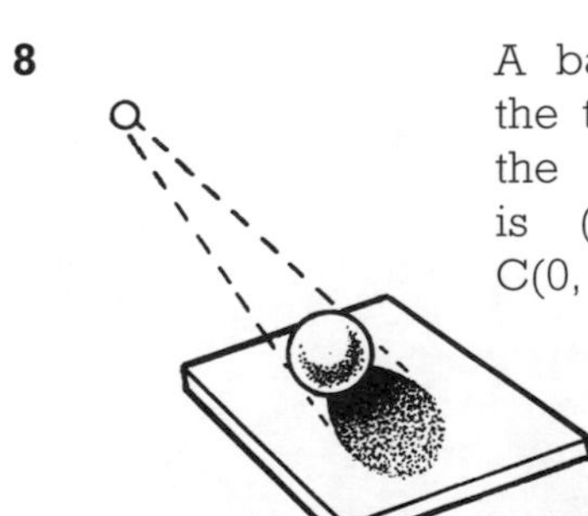

A ball casts a shadow on the table. With origin O at the centre of the ball, A is $(-3, 4)$, $B(4, -3)$ and $C(0, -5)$.

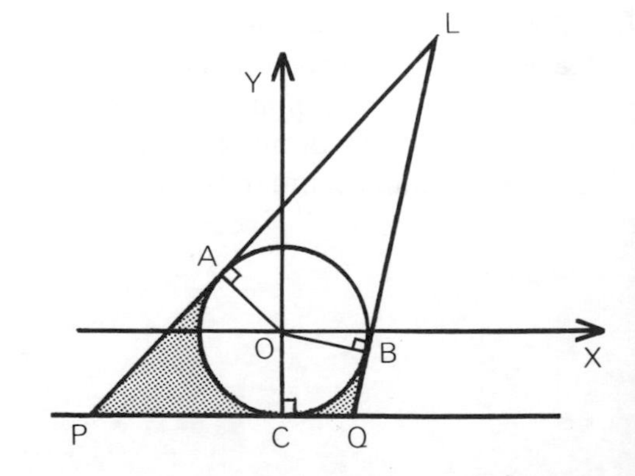

Calculate:

a the gradients of OA and OB
b the equations of AL and LB
c the coordinates of L
d the distance of the light, L, from the centre of the ball
e the coordinates of P and Q
f the width of the shadow (PQ).

CHECK-UP ON **THE STRAIGHT LINE**

1 A is the point $(-5, -2)$ and B is $(7, 3)$. Find:
a the midpoint of AB **b** the length of AB **c** the gradient of AB
d the gradients of CD, parallel to AB, and EF, perpendicular to AB.

2 Draw the locus of all points which are 3 cm from the circumference of a circle with radius:
a 3 cm **b** 4 cm.

3 Calculate the size of the anti-clockwise angle, correct to 1 decimal place, which the line joining the points $(-2, 6)$ and $(4, 7)$ makes with the positive direction of the x-axis.

4 Find the gradient and y-intercept of the line with equation:
a $y = 3x+4$ **b** $3y = 2x-6$ **c** $4x+5y+3 = 0$.

5 **a** Find the equation of the line:
(i) with gradient $\frac{3}{4}$, passing through the point $(-4, 1)$
(ii) passing through the points $(7, 3)$ and $(10, -1)$.
b Prove that the two lines in **a** are perpendicular, and find their point of intersection.

6 Quadrilateral ABCD has vertices A$(-4, 4)$, B$(6, 4)$, C$(-2, -2)$ and D$(3, 3)$.
a Find the point of intersection of its diagonals.
b Show that BD is an axis of symmetry.
c Use the distance formula to check that two pairs of sides are equal.
d What type of quadrilateral is ABCD?

7 **a** Find the point of intersection of the lines $2x-3y = 13$ and $3x+5y = -9$.
b Find also the sizes of the angles, correct to $0{\cdot}1°$, of the triangle they form with the x-axis.

8 A is the point $(0, 2)$, B$(8, 2)$ and C$(8, 6)$.
a Find the equation of:
(i) median CM (ii) altitude BP, in $\triangle$ABC.
b Find the coordinates of S, the point of intersection of CM and BP.

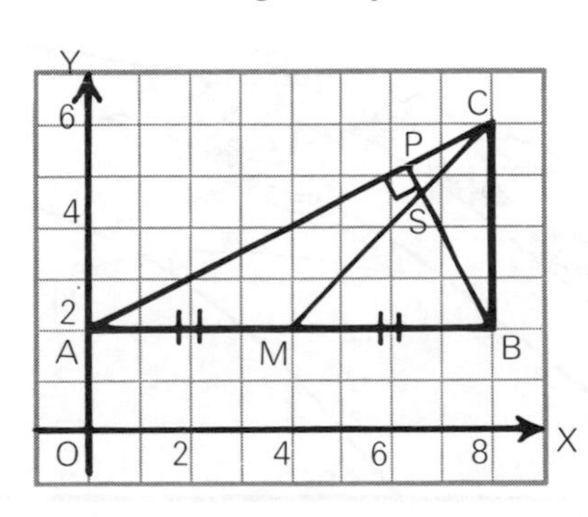

INTRODUCTION: GRAPHS AND GRADIENTS

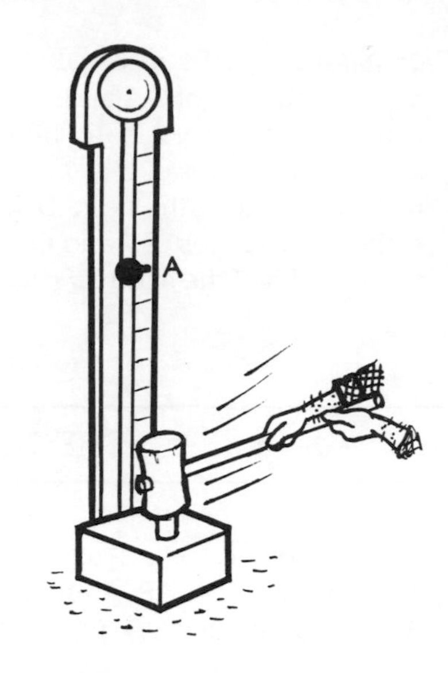

Example 1 Having a go at the GIANT TEST YOUR STRENGTH machine, Bill must hit the button, sending a striker, A, up to try to ring the bell. The motion of the striker is modelled by the equation $h(t) = 20t - 5t^2$, where $h(t)$ metres is its height after t seconds.

a Copy and complete:

t	0	0·5	1	1·5	2	2·5	3	3·5	4
$h(t)$	0	8·75	15	18·75					

b Plot the points (t, h) and sketch the smooth graph $h = 20t - 5t^2$, as shown below.

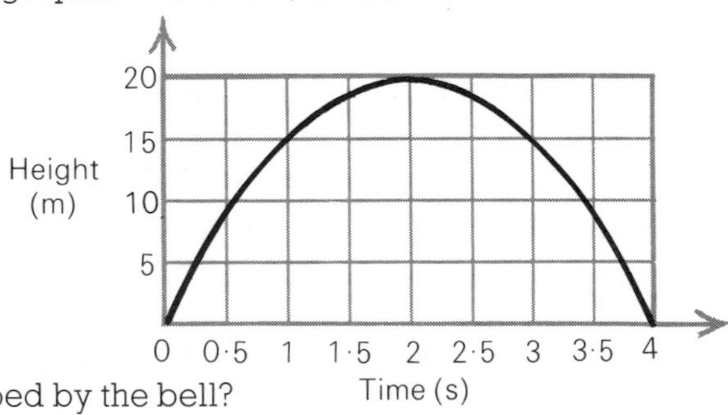

c Estimate the speed at which the striker is going up after 1 second (i.e. the gradient of the graph at $t = 1$).

d What is the initial speed (i.e. the speed at $t = 0$)?

e How high would the striker go, if not stopped by the bell?

Example 2 A ski run down a hill-side is modelled by the graph of the function

$$f(x) = 6x^2 - x^3 \text{ for } 0 \leqslant x \leqslant 4.$$

a Construct a table of values for $x = 0, 1, 2, 3, 4$.

b Sketch the graph of f over this interval.

c At which point is the slope of the ski run steepest (i.e. where the gradient is greatest)?

Discuss how you could solve these two practical problems:

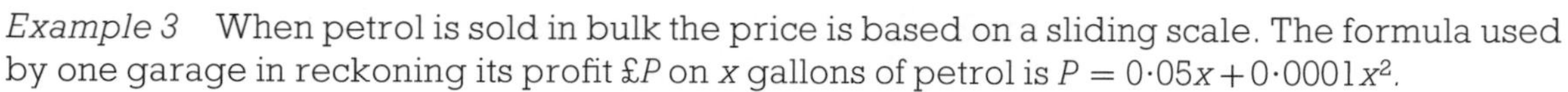

Example 3 When petrol is sold in bulk the price is based on a sliding scale. The formula used by one garage in reckoning its profit £P on x gallons of petrol is $P = 0{\cdot}05x + 0{\cdot}0001x^2$.
a How does the profit change as more and more petrol is sold?
b How can you find the rate of change of P at $x = 1000$?

Example 4 A bridge CD has to be built across the canal so that the walking distance from Sunnyside Cottage (A) to Canal View Villa (B) is as short as possible. Where would the bridge have to be constructed?

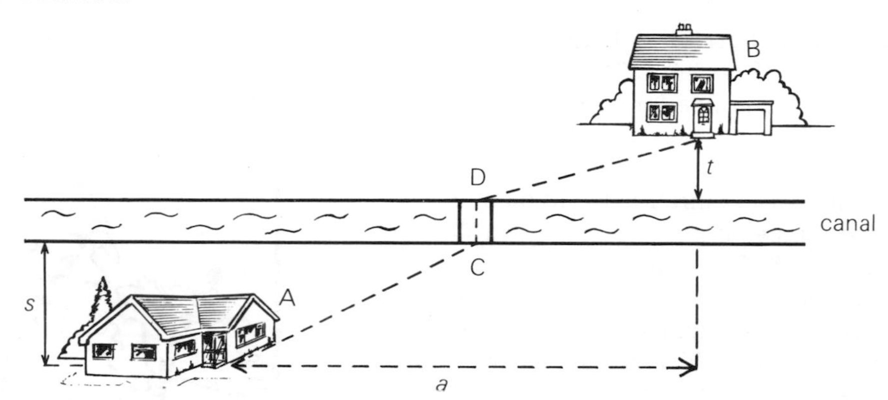

The solutions to all of these problems depend upon making a mathematical model, graphing the model and then reading off a maximum (or minimum) value, or, in the case of a rate, estimating the gradient of a tangent. This method can take a long time and is often inaccurate. In this chapter you will start to develop much more powerful methods.

AVERAGE RATE OF CHANGE; AVERAGE GRADIENT OF A CURVE

Exercise 1

1 James Good is driving along a straight section of road where the speed limit is 40 mph. Glancing at the clock (it's 2 pm) he notices to his dismay that his speedometer is not working, although the mileometer and clock are ticking away as normal. The reading on his mileometer is 5321·2 miles. 30 minutes later the reading is 5345·6.

a Calculate his *average speed* (i.e. *average rate of change of distance*) from 2.00 pm to 2.30 pm.

b James' mileometer readings, taken at 5-minute intervals are:

Time (pm)	*Mileometer reading*
2.00	5321·2
2.05	5324·7
2.10	5328·5
2.15	5332·2
2.20	5336·3
2.25	5341·0

(i) Calculate (correct to 1 decimal place) his average speeds over the following time intervals: (2.00, 2.25), (2.00, 2.20), (2.00, 2.15), (2.00, 2.10) and (2.00, 2.05).
(ii) Is a pattern appearing in your answers? Can you estimate his probable speed at 2 pm?

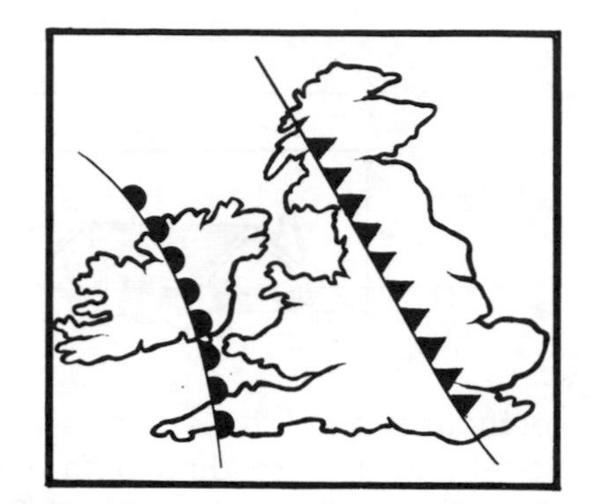

2 A cold front passing through the Midlands makes the temperature drop from 11·1°C at 7.00 pm to −5·0°C at 1.00 am.

a Calculate the *average rate of change of temperature* over this period, correct to 1 decimal place. Why is your answer negative?

b A table of temperatures was recorded:

Time (pm)	8.00	9.00	10.00	11.00	12.00
Temperature (°C)	10·0	8·4	4·9	−0·2	−2·8

(i) Calculate the average rate of fall in temperature at hourly intervals (7.00 to 8.00, 8.00 to 9.00, etc).
(ii) Between which hours was the rate of fall greatest?

3 *Average gradient of a curve*
The points P(1, 1) and Q(2, 4) lie on the parabola $y = x^2$. As Q moves along the curve towards P the gradient of the curve changes. The *average gradient of the parabola* from P to Q is taken as the gradient of the chord PQ(m_{PQ}).

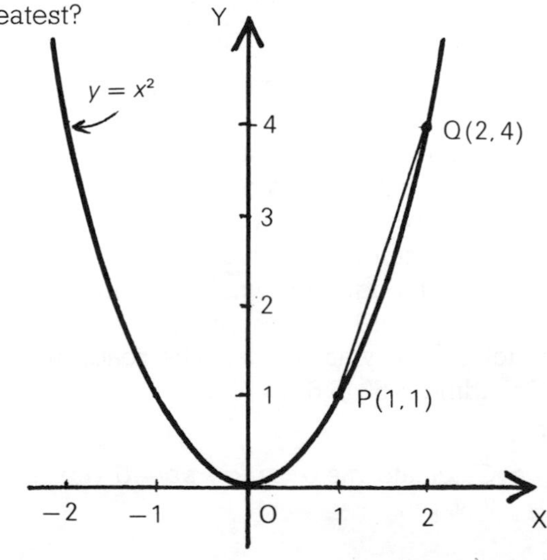

a Copy and complete:
The average gradient from $x = 1$ to $x = 2$ is

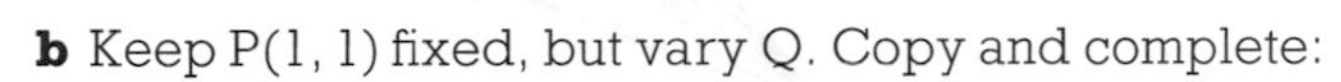

$$m_{PQ} = \frac{\text{change in } y}{\text{change in } x} = \frac{y_Q - y_P}{x_Q - x_P} = \frac{\dots}{\dots} = \dots.$$

b Keep P(1, 1) fixed, but vary Q. Copy and complete:

x_Q	y_Q	$x_Q - x_P$	$y_Q - y_P$	Average gradient m_{PQ}
1·8	3·24	0·8	2·24	2·8
1·6	2·56	0·6	1·56	2·6
1·4	1·96	. . .	. . .	. . .
1·3	. . .	. . .	. . .	. . .
1·2	. . .	. . .	. . .	. . .
1·1	. . .	. . .	. . .	. . .

(i) As Q gets closer and closer to P, what happens to the gradients of the chords?
(ii) Can you predict the gradient of the parabola at P?

RATE OF CHANGE; GRADIENT AT A POINT ON A CURVE

Exercise 2

1 Look back at James Good driving along the road. How could he find out his speed at 2.00 pm? If he took mileometer readings at times closer and closer to 2.00 pm:

Time (pm)	Mileometer reading	Average speed from 2.00 pm (mph)
2.00	5321·2	. . .
2.05	5324·7	42·0
2.04	5323·93	41·1
2.03	5323·22	40·4
2.02	5322·54	40·2
2.01	5321·87	40·2

This suggests that his *speed at 2.00 pm* was approximately 40 mph.
Do you see that you are using the same technique of calculating gradients of chords (on the graph of miles against time) as you did for $y = x^2$, and that these gradients approach a definite value?

2 *Gradient of tangent PT at P(1, 1) on parabola $y = x^2$*
Draw lots of chords through P with Q getting closer and closer to P. m_{PQ} gets closer and closer to the gradient of the tangent at P, the gradient of the parabola at P.

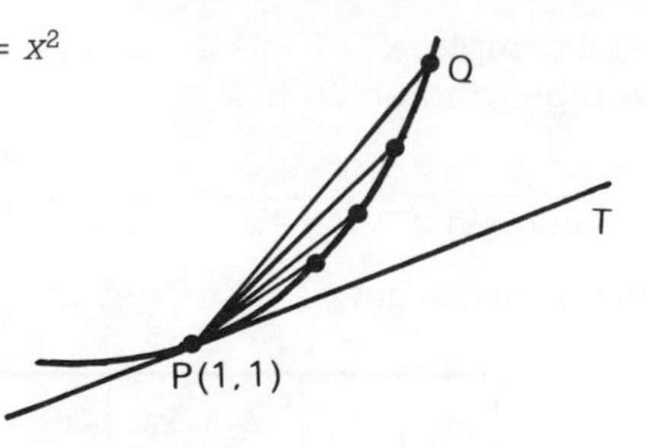

Continue the table you started in question **3b** in Exercise **1**, as far as your calculator will allow.

Copy and complete:

x_Q	y_Q	$x_Q - x_P$	$y_Q - y_P$	Average gradient of PQ
1·09				
1·08				
...				

Do your results suggest that *the gradient of the tangent at P(1, 1) is 2*?

3 You can *prove* that the gradient of the tangent at P is 2, using a 'limit' method.

Start at P(1, 1). Take Q on $y = x^2$ with x-coordinate h along from P($h \neq 0$).
$x_Q = 1 + h$, $y_Q = (1 + h)^2$.

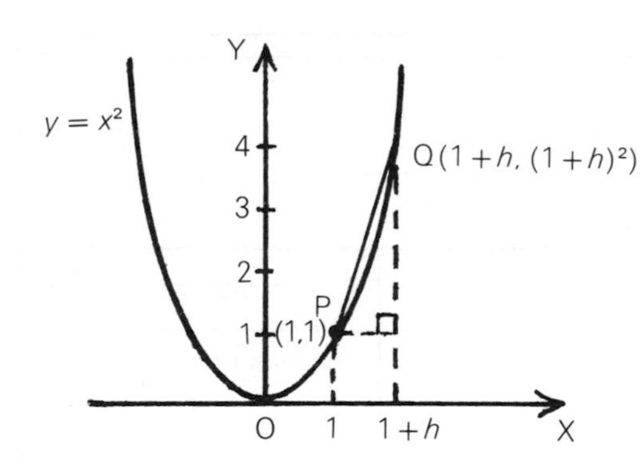

$$m_{PQ} = \frac{y_Q - y_P}{x_Q - x_P} = \frac{(1+h)^2 - 1}{(1+h) - \ldots} = \frac{1 + h^2 + \ldots - 1}{h} = \frac{h}{h}(\ldots + \ldots)$$

$$= 2 + h.$$

As Q gets closer and closer to P (Q→P),
h gets closer and closer to 0 ($h \to 0$),
and m_{PQ} gets closer and closer to 2 ($m_{PQ} \to 2$).
Throughout this 'limit' process m_{PQ} is a gradient. As Q 'disappears' at P, m_{PQ} 'becomes the gradient of the tangent at P'.

As $h \to 0$, $m_{PQ} \to 2$. So the gradient of the tangent at P(1, 1) is 2.

4 Use this 'h' method to calculate the gradient of the tangent to the parabola $y = x^2$ at P(2, 4).
[Take $Q(2+h, (2+h)^2)$.

$$m_{PQ} = \frac{y_Q - y_P}{x_Q - x_P} = \frac{(2+h)^2 - 4}{(2+h) - 2} = \frac{4 + h^2 + \ldots - \ldots}{\ldots} = \frac{h(\ldots)}{\ldots} = \ldots.$$

As $h \to 0$, $m_{PQ} \to \ldots$. So the gradient of the tangent at P is . . . .]

5 Repeat question **4** for the gradient of the tangent to $y = x^2$ at:
a (3, 9) **b** (a, a^2); take $Q(a+h, (a+h)^2)$.

6 In the same way, calculate the gradient of the tangent to $y = 3x^2$ at:
a (1, 3) **b** $(a, 3a^2)$; here Q is $(a+h, 3(a+h)^2)$.

7 Use the 'h' method to calculate the gradient of the tangent to $y = x^2 + x$ at P(3, 12). Here $x_Q = 3+h$ and $y_Q = (3+h)^2 + (3+h) = \ldots = 12 + 7h + h^2$, etc.

8 Calculate the gradient of the tangent to:
a $y = x^2 + 3$ at P(2, 7) **b** $y = x^2 + 3x$ at P(1, 4) **c** $y = x^2 - x$ at P(3, 6).

---

1 Carry out your 'h' method to find the gradient of the tangent at $x = 5$ on:
a $y = 2x$ **b** $y = x^2$ **c** $y = x^2 + 2x$
Is there any relationship between **a**, **b** and **c**?

2 Repeat for: **a** $y = 3x$ **b** $y = 5x^2$ **c** $y = 5x^2 + 3x$ at $x = 1$.
Are you reaching any conclusion? If so, write it down.

You have now made a start on an important part of mathematics, CALCULUS. The subject was developed mainly by Isaac Newton (1642–1727) and Gottfried Wilhelm von Leibniz (1646–1716), working quite independently. It can be divided into two parts, *differentiation* and *integration*. This chapter deals only with differentiation.

DIFFERENTIATION, THE DERIVATIVE OF $f(x)$

Take points $P(x, f(x))$ and $Q(x+h, f(x+h))$ on graph $y = f(x)$.

The gradient of the tangent at P

$$= \lim_{Q \to P} m_{PQ} = \lim_{Q \to P} \frac{y_Q - y_P}{x_Q - x_P} = \lim_{h \to 0} \frac{f(x+h) - f(x)}{(x+h) - x} = \lim_{h \to 0} \frac{f(x+h) - f(x)}{h}.$$

The gradient is denoted by $f'(x)$ ('f dash x'). So

$$f'(x) = \lim_{h \to 0} \frac{f(x+h)-f(x)}{h}.$$

Since the number $f'(x)$ is derived from $f(x)$, it is called the derivative of f at x. The process of finding $f'(x)$ is called *differentiation*. When $f'(x)$ exists, f is *differentiable* at x. The formula for $f'(x)$ defines a new function f', the *derived function* of f. Its value at $x = a$ is $f'(a) = \lim_{h \to 0} \frac{f(a+h)-f(a)}{h}$.

Rate of change

Let's go back with Bill to the GIANT TEST YOUR STRENGTH machine on Page 16. There you had a function, $f(t) = 20t - 5t^2$ of time t. One of the problems was to find the speed (in m/s) of the striker at $t = 1$, i.e. the rate of change of f at $t = 1$. This is simply the gradient of the tangent at P(1, 15) on the graph $y = 20t - 5t^2$.
So the speed is $f'(1)$, i.e. the value of $f'(t)$ at $t = 1$.

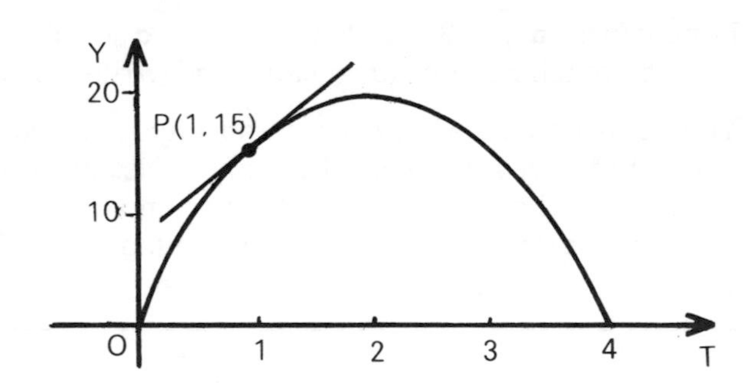

Formulae save time

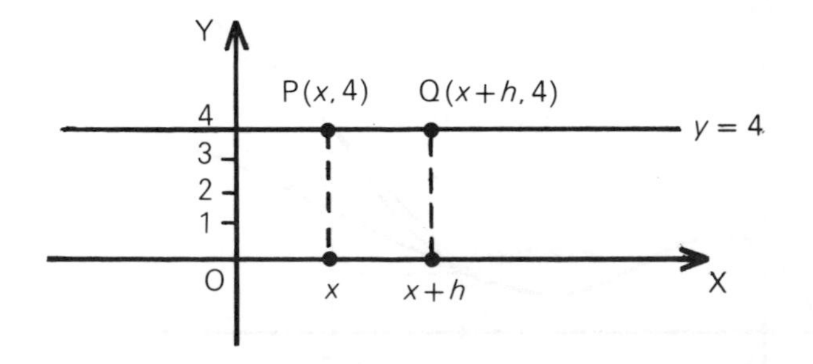

a *Derivative of f(x) = 4.*
Graph is the line $y = 4$.

$$m_{PQ} = \frac{f(x+h)-f(x)}{h} = \frac{4-4}{h} = 0.$$

So $f'(x) = \lim_{h \to 0} m_{PQ} = 0$.

For $f(x) = c$, where c is a constant, $f'(x) = 0$.

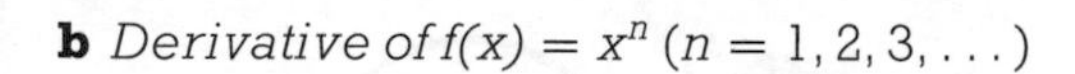

b *Derivative of $f(x) = x^n$ ($n = 1, 2, 3, \ldots$)*

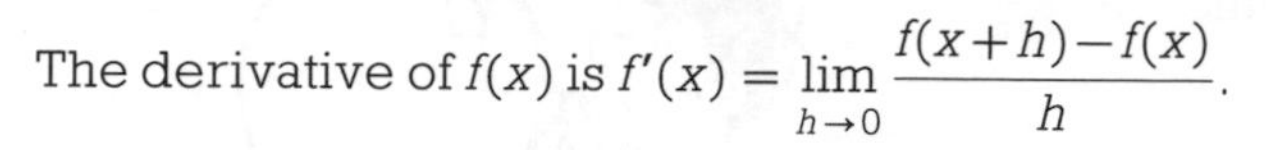

The derivative of $f(x)$ is $f'(x) = \lim_{h \to 0} \dfrac{f(x+h) - f(x)}{h}$.

(i) $f(x) = x$. $\quad \dfrac{f(x+h)-f(x)}{h} = \dfrac{(x+h)-x}{h} = \dfrac{h}{h} = 1.$

So $f'(x) = \lim_{h \to 0} 1 = 1.$

(ii) $f(x) = x^2$. $\quad \dfrac{f(x+h)-f(x)}{h} = \dfrac{(x+h)^2-x^2}{h} = \dfrac{x^2+h^2+2xh-x^2}{h} = \dfrac{h(h+2x)}{h} = h+2x.$

So $f'(x) = \lim_{h \to 0} (h+2x) = 2x.$

(iii) $f(x) = x^3$. $\quad \dfrac{f(x+h)-f(x)}{h} = \dfrac{(x+h)^3-x^3}{h} = \dfrac{(x+h)(x^2+h^2+2xh)-x^3}{h}$

$$= \frac{(x^3+3x^2h+3xh^2+h^3)-x^3}{h}$$

$$= \frac{h(3x^2+3xh+h^2)}{h} = 3x^2+3xh+h^2.$$

So $f'(x) = \lim_{h \to 0} (3x^2+3xh+h^2) = 3x^2.$

Do you see the pattern for these derivatives? Copy and complete:

$f(x)$	x	x^2	x^3	x^4	x^5	x^6	x^{10}
$f'(x)$	1	$2x$	$3x^2$				

In fact for all rational numbers n there is a powerful rule:

> For $f(x) = n^n$ ($n \in Q$), $f'(x) = nx^{n-1}$.

Examples

1 Given $f(x) = x^7$, $f'(x) = 7x^{7-1} = 7x^6$.

2 Given $f(x) = \dfrac{1}{x^2} = x^{-2}$, $f'(x) = (-2)x^{-2-1} = -2x^{-3} = -\dfrac{2}{x^3}$.

3 A curve has equation $y = \sqrt{x}$. Calculate the gradient of the tangent at $x = 9$.
Let $f(x) = \sqrt{x} = x^{1/2}$.

$f'(x) = \dfrac{1}{2}x^{1/2-1} = \dfrac{1}{2}x^{-1/2} = \dfrac{1}{2\sqrt{x}}$, so $f'(9) = \dfrac{1}{2\sqrt{9}} = \dfrac{1}{6}$.

Gradient of tangent at $x = 9$ is $\frac{1}{6}$.

4 A balloon expands so that its volume $V(t)\,\text{cm}^3$ at time t seconds is $V(t) = t^3$. Calculate the rate of change at $t = 2$.
$V'(t) = 3t^2$, so $V'(2) = 3 \,.\, 2^2 = 12$.
The rate of change is $12\,\text{cm}^3/\text{s}$.

Exercise 3

1 Write down the derivatives of:
a 2 **b** x **c** x^2 **d** x^3 **e** x^4 **f** x^5 **g** x^{50} **h** x^{100}.

2 Differentiate:
a x^{-2} **b** x^{-10} **c** x^{-15} **d** $\frac{1}{x}$ **e** $\frac{1}{x^3}$ **f** $\frac{1}{x^5}$ **g** $\frac{1}{x^8}$ **h** $\frac{1}{x^{20}}$.

3 Differentiate:
a $x^{3/2}$ **b** $x^{5/2}$ **c** $x^{1/3}$ **d** $x^{-1/2}$ **e** $x^{-5/2}$ **f** $\frac{1}{\sqrt{x}}$ **g** $\frac{1}{x^{2/3}}$ **h** $x^{-3/4}$.

4

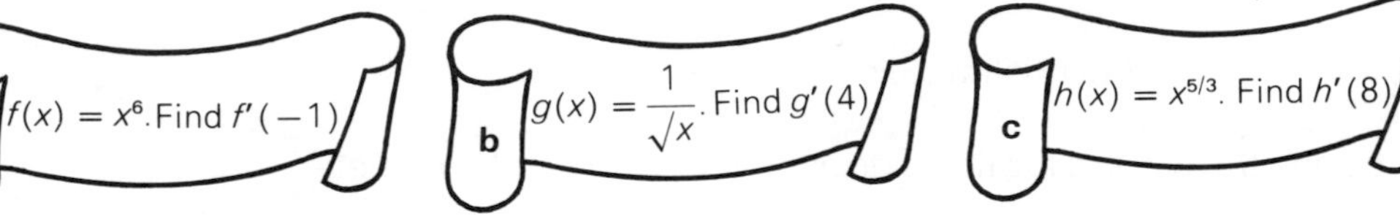

a
f(x) = x⁶. Find f′(−1)
b
g(x) = 1/√x. Find g′(4)

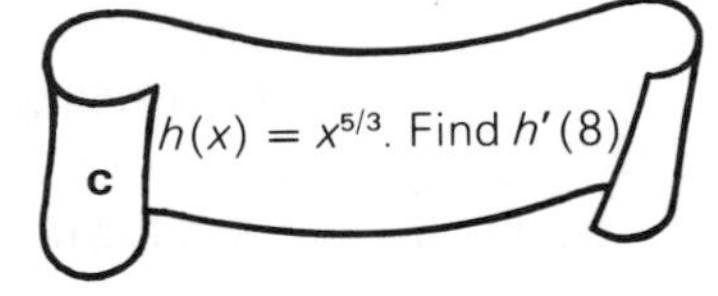

5 Find the gradient of the graph $y = f(x)$ at $x = a$ for $f(x) =$:
a x^3 **b** x^{12} **c** $\frac{1}{x^3}$ **d** x^{-4} **e** $x^{1/3}$ **f** $x^{-2/5}$ **g** $\frac{1}{(\sqrt[4]{x})^3}$ **h** $\sqrt[5]{x^4}$.

6 For $f(x) = x^3$ find the value of:
a $f'(1)$ **b** $f'(2)$ **c** $f'(0)$ **d** $f'(-\frac{1}{2})$ **e** the derivative of f at $x = -1$.

7 For $f(x) = \sqrt{x}$ find:
a $f'(x)$ **b** $f'(\frac{1}{9})$ **c** a value of x at which $f'(x) = \frac{1}{4}$.

8 Calculate the rate of change of:
a $x^{1/2}$ at $x = 4$ **b** x^5 at $x = 1$ **c** $x^{5/3}$ at $x = 27$ **d** x^{100} at $x = 1$
e x^3 at $x = -2$ **f** $\frac{1}{x^{1/3}}$ at $x = -8$ **g** $x^{-1/4}$ at $x = 16$ **h** $x^{-2/5}$ at $x = -32$.

9 Sketch the graph of $f'(x)$, given: **a** $f(x) = x^2$ **b** $f(x) = x^3$.

10 A bacterial culture has a mass of t^2 grams after t seconds of growth. Calculate its rate of growth (in g/s) when $t = 2$.

11 A land-yacht travels $\sqrt{t^5}$ km in t hours. Calculate its speed after:
a 1 hour **b** 15 minutes.

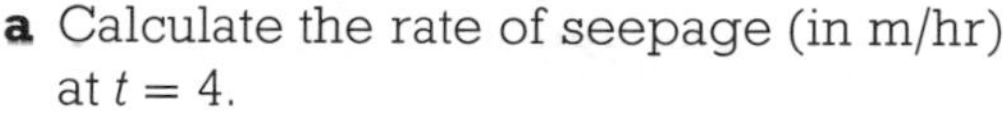

12 Water seeps $\sqrt{t}$ metres into soil in t hours.

a Calculate the rate of seepage (in m/hr) at $t = 4$.

b What happens to the rate of seepage as t increases? Is the answer surprising?

P is the point $\left(a, \frac{1}{a}\right)$ on the curve $y = \frac{1}{x}\ (a > 0)$.

a Calculate the gradient of the tangent at P.

b Show that the equation of the tangent is $x + a^2y = 2a$.

c If the tangent meets the axes at A and B, find out what you can about the area of triangle OAB as P varies on the curve.

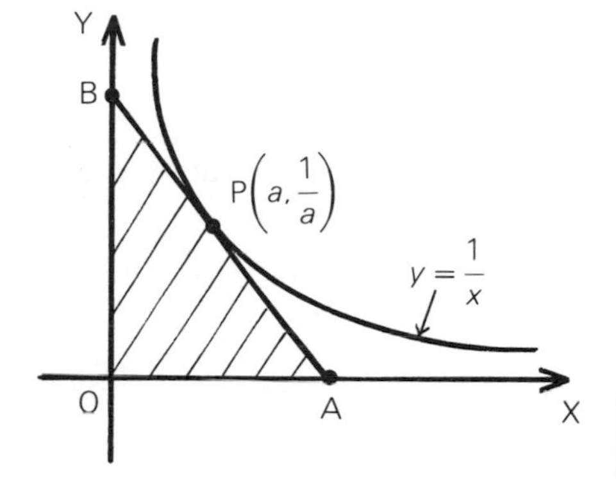

TWO USEFUL RULES

1 If $f(x) = x^2$, then $f'(x) = 2x$. What about $g(x) = 3x^2$?

$$g'(x) = \lim_{h \to 0} \frac{g(x+h) - g(x)}{h} = \lim_{h \to 0} \frac{3(x+h)^2 - 3x^2}{h} = 3 \lim_{h \to 0} \frac{(x+h)^2 - x^2}{h}$$

$$= 3f'(x) = 3 \,.\, 2x = 6x.$$

> If $g(x) = kf(x)$, k a constant, then $g'(x) = kf'(x)$.

2 In the Investigation after Exercise **2** you discovered that the derivative of $x^2 + 2x$ = (derivative of x^2) + (derivative of $2x$).

> If $h(x) = f(x) + g(x)$, then $h'(x) = f'(x) + g'(x)$.

Examples

1 If $f(x) = 2x^6 - 7x - 5$, then
$f'(x) = 2 \,.\, 6x^5 - 7 \,.\, 1 - 0 = 12x^5 - 7.$

2 It costs a firm £C(x), where

$$C(x) = 1000 + \frac{x}{100}(2500 + x)$$

to produce x radios.

a C(0) = 1000. Can you explain this?

b Calculate the *marginal cost*, C′(x).
C(x) is not a sum of powers of x, but don't despair, look at it closely. Can you express it as a sum of powers?

$$C(x) = 1000 + 25x + \frac{1}{100}x^2, \text{ so } C'(x) = 0 + 25 \,.\, 1 + \frac{1}{100} \,.\, 2x = 25 + \frac{1}{50}x.$$

3 Calculate the rate of change of $t\left(1 + \frac{2}{\sqrt{t}}\right)$ at $t = 4$.

Let $f(t) = t\left(1 + \frac{2}{\sqrt{t}}\right) = t + t \,.\, 2 \,.\, t^{-1/2} = t + 2t^{1/2}$.

$f'(t) = 1 + 2 \,.\, \frac{1}{2} t^{-1/2} = 1 + \frac{1}{\sqrt{t}}$, so $f'(4) = 1 + \frac{1}{2} = \frac{3}{2}$.

Rate of change of $f(t)$ at $t = 4$ is $\frac{3}{2}$.

Exercise 4

1 Differentiate each of the following with respect to the relevant variable:

a $1 + x^2$ **b** $2x^3 - 3$ **c** $3x^4 - 2x$ **d** $t^2 + 2t$ **e** $2t^3 - 5t + 3$
f $3p^2 - 7p + 5$ **g** $p^3 - 2p^6$ **h** $2x^5 + 11x - 1$ **i** $27x^3 - 9x$ **j** $3x^4 - 2x^3 + 5x - 8$.

2 Find $f'(x)$, given $f(x) =$

a $x^2 + \frac{1}{x}$ **b** $2x^3 + \frac{1}{x^2}$ **c** $x\left(1 - \frac{1}{x}\right)$ **d** $x^4(3 - 2x)$ **e** $x(2 - x)^2$

f $\frac{1}{3x}$ **g** $\frac{3}{5x^2}$ **h** $2x^3 + \frac{1}{2x^3}$ **i** $3\sqrt{x} + \frac{1}{3\sqrt{x}}$ **j** $\frac{2x - 1}{x^{1/2}}$.

3 For each curve find the gradient of the tangent at the point given by the value of x:

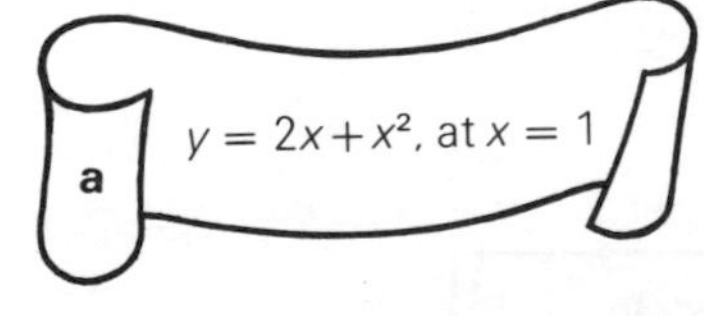

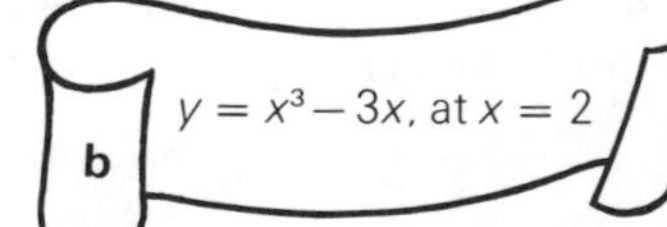

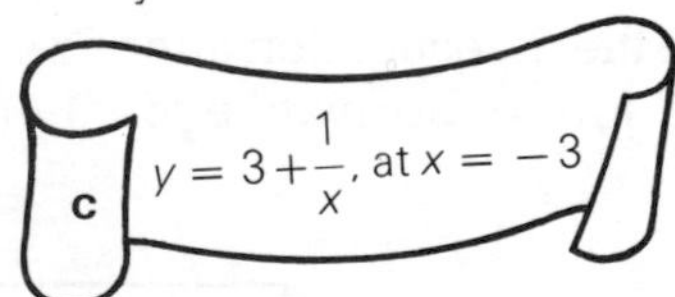

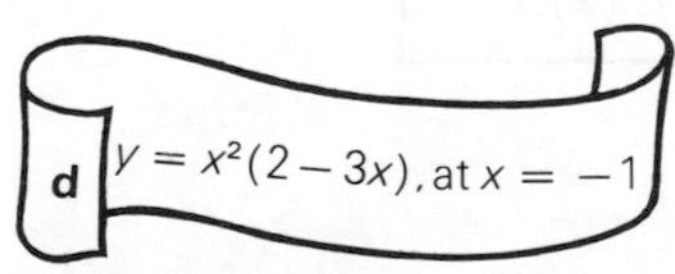

e $y = \frac{5 - 2x}{x^2}$, at $x = 2$

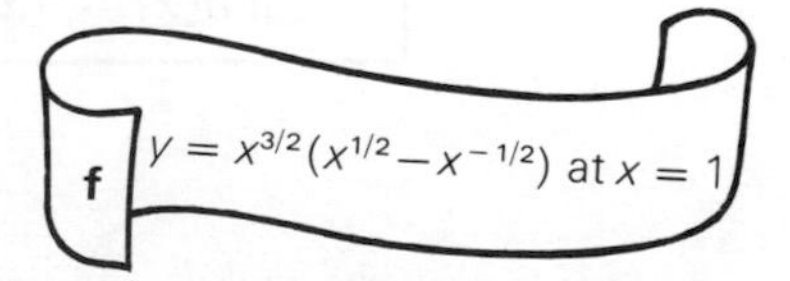

4 Calculate the rate of change of:

a $f(x) = x^4 - 2x^3 + 5x - 6$ at $x = -2$ **b** $g(t) = 2t^2 + \dfrac{3}{t}$ at $t = 1$

c $h(u) = (u^2+2)(u+2)$ at $u = -1$ **d** $k(x) = \dfrac{3-x^2}{x^4}$ at $x = 1$.

5 Calculate the rate of change of πr^2 at $r = 1$.
Can you think of a geometrical meaning for this?

6 A balloon is being blown up steadily.
a Find the rate of change of its volume V with respect to radius r. [$V = \frac{4}{3}\pi r^3$]
b Explain geometrically.

7 The number N of bacteria in a culture varies with time t (seconds) according to the formula $N = 2000 + 300t + 18t^2$.
How fast is the population growing when $t = 10^2$?

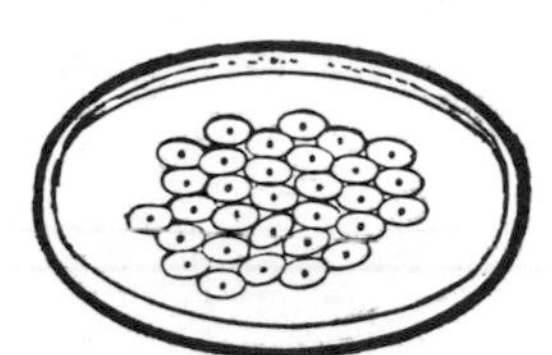

8 A psychologist carrying out memory tests finds the number N of words a person can memorise in t minutes is (approximately) $N = 10\sqrt{t}$.
Calculate the rate at which N is changing when $t = 16$.

9 Doug's model of the 'Forties Flier' runs along a straight track. Its displacement OP metres from the signal at O after t seconds is given by: $x(t) = 1 + 4t - t^2 (t \geqslant 0)$.

a Find its velocity at time t (i.e. $x'(t)$).
b At what time is the velocity zero? What does this mean?
c Calculate $x'(3)$. What does the negative sign tell you?

ANOTHER NOTATION FOR DERIVATIVES; EQUATIONS OF TANGENTS

$P(x_P, y_P)$ and $Q(x_Q, y_Q)$ are points on the graph $y = f(x)$.
(i) Replace $y_Q - y_P$, the change in y, by Δy. (Read it as 'delta y'.)
(ii) Replace $x_Q - x_P$ by Δx.

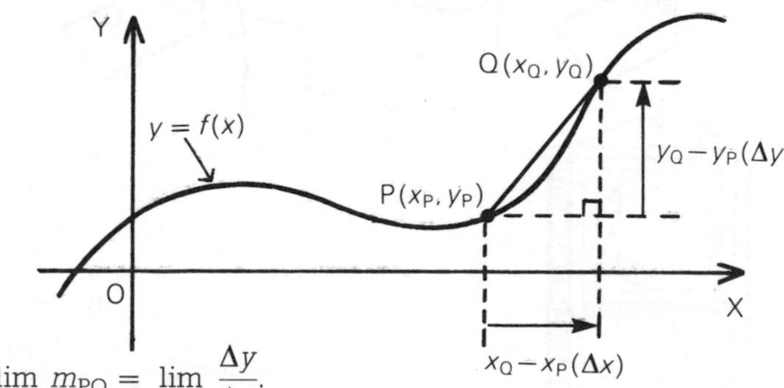

Then:

$$m_{PQ} = \frac{y_Q - y_P}{x_Q - x_P} = \frac{\Delta y}{\Delta x}, \text{ and } f'(x) = \lim_{Q \to P} m_{PQ} = \lim_{\Delta x \to 0} \frac{\Delta y}{\Delta x}.$$

$\lim\limits_{\Delta x \to 0} \dfrac{\Delta y}{\Delta x}$ is denoted by $\dfrac{dy}{dx}$. (Read it as 'dee y by dee x'.)

For the graph $y = f(x)$, $\dfrac{dy}{dx} = f'(x)$.

Equivalent notations: $f'(x)$, $\dfrac{df}{dx}$, $\dfrac{d}{dx}(f(x))$, $\dfrac{dy}{dx}$, y', $y'(x)$.

Example A parabola has equation $y = x^2 - 4x + 1$.
Calculate the gradient of the tangent at $x = 3$, and then find the equation of this tangent.

$\dfrac{dy}{dx} = 2x - 4$, so the gradient of the tangent at $x = 3$ is $2 \cdot 3 - 4 = 2$.

The point has coordinates $(3, 3^2 - 4 \cdot 3 + 1)$, i.e. $(3, -2)$.
The tangent has equation $y - (-2) = 2(x - 3)$, i.e. $y = 2x - 8$.

Exercise 5

Find the derivatives in questions **1** and **2**:

1 a $\dfrac{d}{dx}(2x^2 + 1)$ **b** $\dfrac{d}{dx}(3x^3 - 2x)$ **c** $\dfrac{d}{dt}(t^4 - 2t^2)$ **d** $\dfrac{d}{dp}(2p^2 - 4p + 7)$

2 a $\dfrac{d}{dx}(2x^4 - 3x^3 + 4x)$ **b** $\dfrac{d}{dx}\left(3x + \dfrac{1}{3x}\right)$ **c** $\dfrac{d}{dt}\left(12t^{2/3} + \dfrac{4}{t}\right)$ **d** $\dfrac{d}{du}\left(2u^3 - \dfrac{1}{2u^3}\right)$

3 Find the equation of the tangent to:
a $y = 3x^2$ at $x = 2$ **b** $y = x^2 + 2x$ at $x = 1$ **c** $y = x^4$ at $x = 1$

d $y = \sqrt{x}$ at $x = 4$ **e** $y = x^{3/2}$ at $x = 9$ **f** $y = \dfrac{1}{x^3}$ at $x = -2$.

4

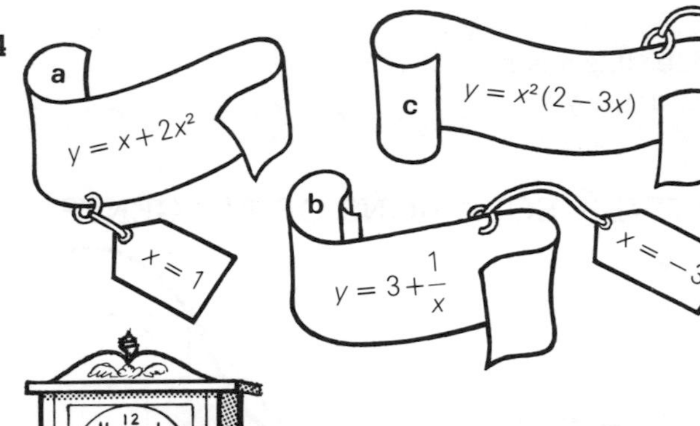

For each curve find the equation of the tangent at the point given by the value of x on the label.

5 The curve $y = (x + 1)(x^2 + 1)$ meets the x-axis at A and the y-axis at B. Find the equations of the tangents at A and B.

6 The period T of the swing of a pendulum depends on its length L and on g, the 'constant of gravity'. The formula is $T = 2\pi\sqrt{\dfrac{L}{g}}$. Find $\dfrac{dT}{dL}$, and calculate the rate of change of T when $L = 16$.

7 Differentiate with respect to the relevant variable:

a $(2x-3)^2$ **b** $x^3\left(3-\dfrac{2}{x^4}\right)$ **c** $\dfrac{1+\sqrt{u}}{2u}$ **d** $t^2(2-t)^2$ **e** $\dfrac{2-3v}{v^{1/3}}$

8 At time t seconds a ball thrown vertically upwards has height $h(t) = 2+10t-5t^2$ metres.

a What is its initial height?
b When does it return to its initial height?
c Calculate its speed at $t = 1$. Explain.
d Find its initial speed, and its speed when it returns to its starting point.

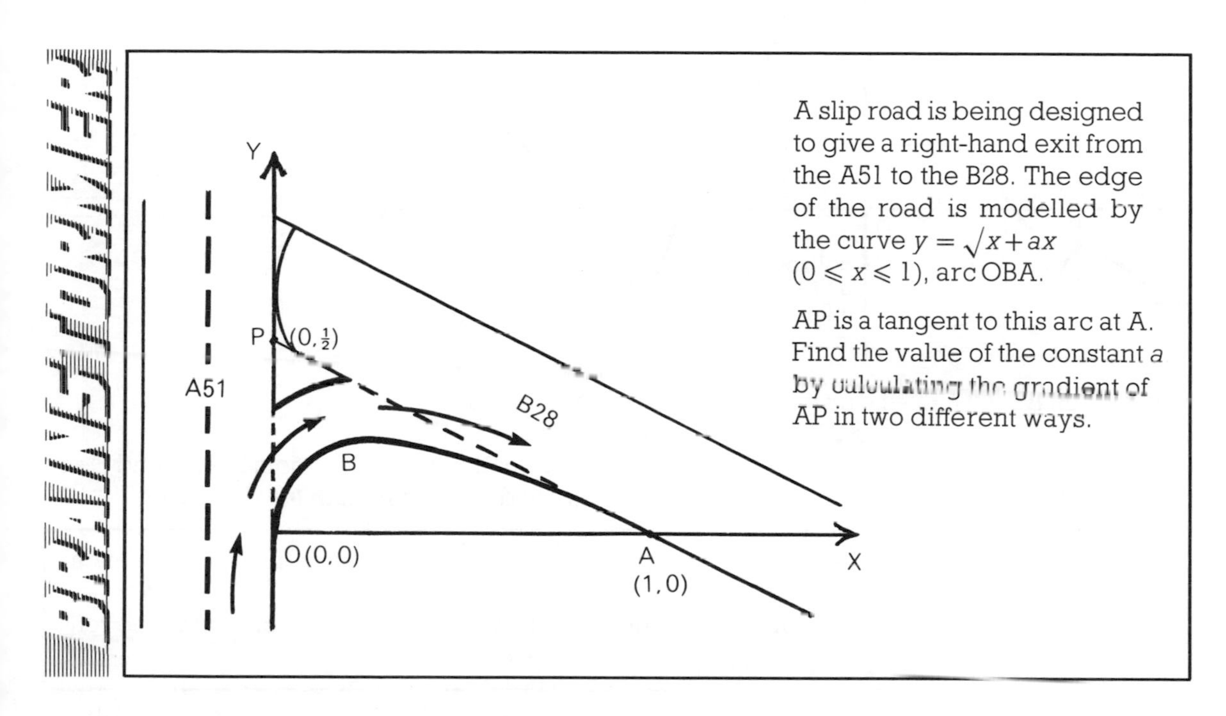

A slip road is being designed to give a right-hand exit from the A51 to the B28. The edge of the road is modelled by the curve $y = \sqrt{x}+ax$ $(0 \leqslant x \leqslant 1)$, arc OBA.

AP is a tangent to this arc at A. Find the value of the constant a by calculating the gradient of AP in two different ways.

INCREASING AND DECREASING FUNCTIONS, STATIONARY POINTS AND VALUES

1 The modulus function is

$$f(x) = |x| = \begin{cases} x \text{ when } x \geqslant 0 \\ -x \text{ when } x < 0. \end{cases}$$

a Its graph is shown.

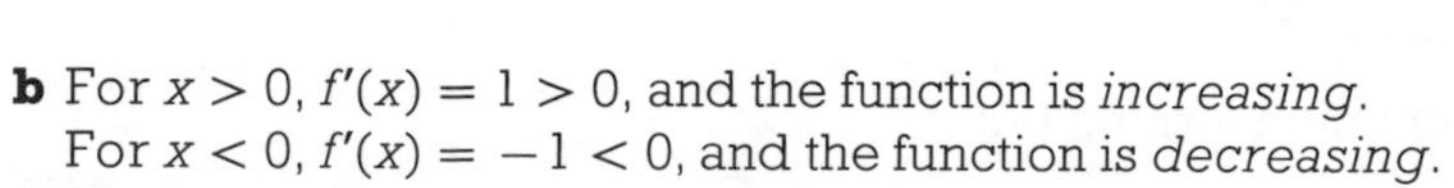

b For $x > 0$, $f'(x) = 1 > 0$, and the function is *increasing*.
For $x < 0$, $f'(x) = -1 < 0$, and the function is *decreasing*.

c There is no tangent at $x = 0$, so $f'(0)$ does not exist.

2 Examine the 'helter-skelter' curve below.

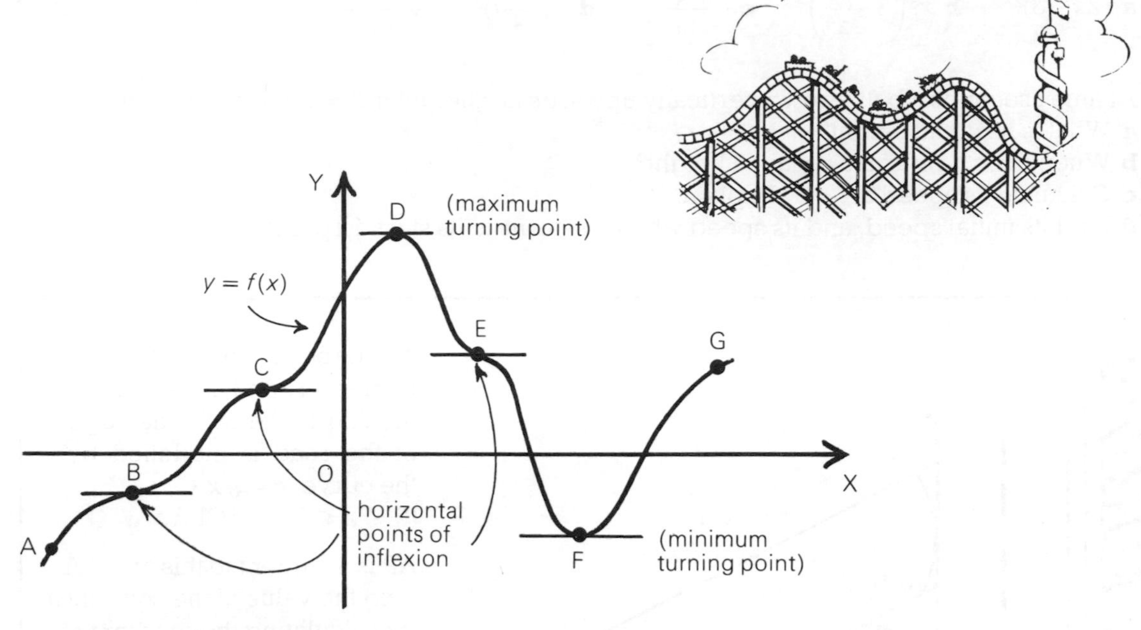

a On AB, BC, CD, FG the graph is rising and f is increasing.
b On DE, EF, the graph is falling and f is decreasing.
c At B, C, D, E and F the gradient is 0, and these are called *stationary points*. B, C and E are *points of inflexion*, D is a *maximum turning point* and F is a *minimum turning point*.

> If $(a, f(a))$ is a stationary point, $f'(a) = 0$, so *a is a root of the equation* $f'(x) = 0$. $f(a)$ is a *stationary value* of f.
> The nature of a stationary point can be determined by finding the sign of $f'(x)$ to the left and to the right of $x = a$.

Example 1 Find the stationary point of the parabola $y = x^2 - 6x + 1$, and the nature of the point.

a $\frac{dy}{dx} = 2x - 6$. If $\frac{dy}{dx} = 0$, $2x - 6 = 0$, and $x = 3$.

When $x = 3$, $y = -8$, so the only stationary point is $(3, -8)$.
b Nature of $(3, -8)$? Here is a good technique for deciding:

x:	$\rightarrow$	**3**	$\rightarrow$
$\frac{dy}{dx} = 2(x-3)$:	$-$	**0**	$+$
Shape of graph:	╲	―	╱

(The arrow$\rightarrow$3 means 'x approaching 3 from the left', while 3$\rightarrow$means 'x going away from 3 to the right'.)

$(3, -8)$ is a minimum turning point.

Note It is useful to use SP for stationary point, SV for stationary value, TP for turning point and PI for point of inflexion.

Example 2 A function f is given by $f(x) = 5x^3 - 3x^5$. Find:
a the coordinates of its SPs
b the nature of each SP
c the intervals on which f is increasing or decreasing.

a $f'(x) = 15x^2 - 15x^4 = 15x^2(1-x^2) = 15x^2(1+x)(1-x)$.
So f has three SPs, $(-1, -2)$, $(0, 0)$ and $(1, 2)$, (and three SVs, $f(-1) = -2$, $f(0) = 0$ and $f(1) = 2$).
b Use a *table of sign* for $f'(x)$ to find the nature of the SPs and the intervals of increase and decrease of f.

x	$\longrightarrow$	-1	$\longrightarrow$	0	$\longrightarrow$	1	$\longrightarrow$
$15x^2$ $1+x$ $1-x$	$+$ $-$ $+$	$+$ 0 $+$	$+$ $+$ $+$	0 $+$ $+$	$+$ $+$ $+$	$+$ $+$ 0	$+$ $+$ $-$
$f'(x)$	$-$	0	$+$	0	$+$	0	$-$

Shape of graph:

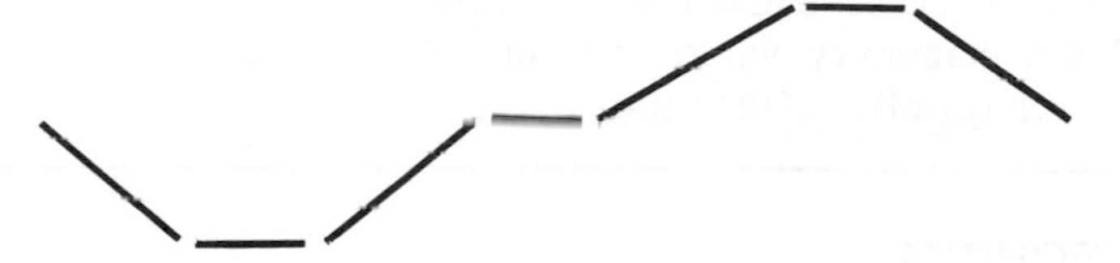

$(-1, -2)$ is a minimum TP, $(0, 0)$ is a PI, and $(1, 2)$ is a maximum TP.

c f is increasing for $-1 < x < 0$ and $0 < x < 1$, and
f is decreasing for $x < -1$ and $x > 1$.

Exercise 6

1 Find the SPs on the following curves and decide their nature:
a $y = 2x^2 - 4x + 1$ **b** $y = 5 + 4x - x^2$ **c** $y = 4 - x^3$

2 For each function find:
(i) the coordinates of its SPs
(ii) the nature of each SP
(iii) the intervals on which f is increasing or decreasing.
a $f(x) = x^2 + 1$ **b** $f(x) = 3 - 2x^2$ **c** $f(x) = x^3$
d $f(x) = 3x - x^3$ **e** $f(x) = 3x^4 - 4x^3$ **f** $f(x) = x^4 - 2x^2 + 5$
g $f(x) = x^3 + 3x$ **h** $f(x) = 3x^5 - 5x^3 + 2$ **i** $f(x) = 1 + 2x^3 - 3x^4$

3 Back with Bill again to the GIANT TEST YOUR STRENGTH machine on Page 16. The height $h(t)$ of the striker at time t is $h(t) = 20t - 5t^2$. Determine the intervals on which the striker is rising and falling.

4 During the first two months of growth, the weight w grams of a laboratory mouse t days after birth is given by the formula $w(t) = 15 + \frac{7}{3}t - \frac{1}{30}t^2$ $(0 \leqslant t \leqslant 60)$. Find the time interval on which w is: (i) increasing (ii) decreasing.

5 A manufacturer finds that the annual cost £C for maintaining his stock depends on warehouse capacity x (in tens of cubic metres) according to the formula $C = 500 + 4x + \frac{400}{x}$ $(x > 0)$. Discuss the increase and decrease of C.

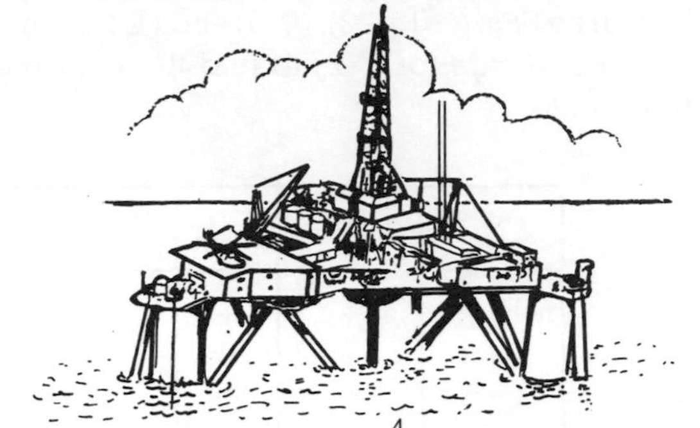

The cost £C of extracting a million barrels of oil from a North Sea Field depends on the cost £C_1 of pipe laying and the cost £C_2 of pumping. These are related to the radius, x feet, of the pipe by the formulae $C_1 = x + a$ and $C_2 = \frac{4}{x}$, where a is a positive constant.

a Discuss the increase or decrease of C_1 and C_2 with respect to x.

b Similarly discuss the total cost $C = C_1 + C_2$.

c Find the stationary value of C and deduce the minimum cost, and the corresponding radius of the pipe.

CURVE SKETCHING

A useful guide:

Determine: (i) stationary points and their nature
(ii) points of intersection with the axes, and other useful points
(iii) the behaviour for x numerically large.

Example Sketch the graph $y = 8x^3 - 3x^4$

(i) For SPs, $\frac{dy}{dx} = 0$.

Here $\frac{dy}{dx} = 24x^2 - 12x^3 = 12x^2(2 - x)$.

$\frac{dy}{dx} = 0$ has two roots, $x = 0$ and $x = 2$, so there are two SPs, (0, 0) and (2, 16).

Check the table of sign for $\frac{dy}{dx}$:

x	$\longrightarrow$	0	$\longrightarrow$	2	$\longrightarrow$
$12x^2$ $2-x$	+ +	0 +	+ +	+ 0	+ −
$\frac{dy}{dx}$	+	0	+	0	−

Shape of graph:

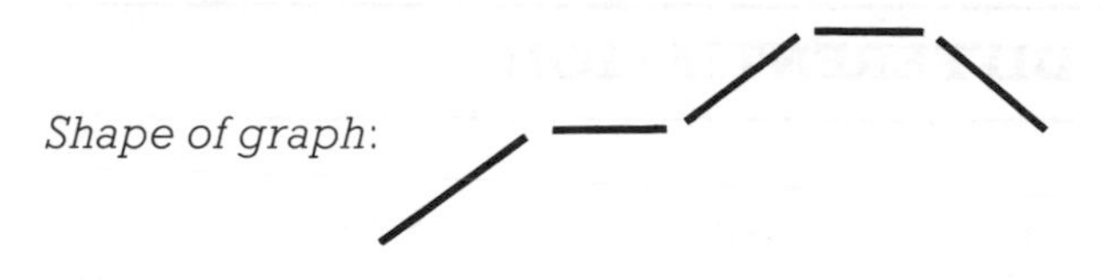

(0, 0) is a PI, and (2, 16) is a maximum TP.

(ii) The graph meets the x-axis where $y = 0$.
If $y = 0$, $8x^3 - 3x^4 = 0$, i.e. $x^3(8-3x) = 0$, and $x = 0$ or $x = \frac{8}{3}$.
The points of intersection with the x-axis are (0, 0) and ($2\frac{2}{3}$, 0).
The graph meets the y-axis only at the origin (0, 0).
A few extra points on the curve are useful, e.g. (1, 5) and (−1, −11).

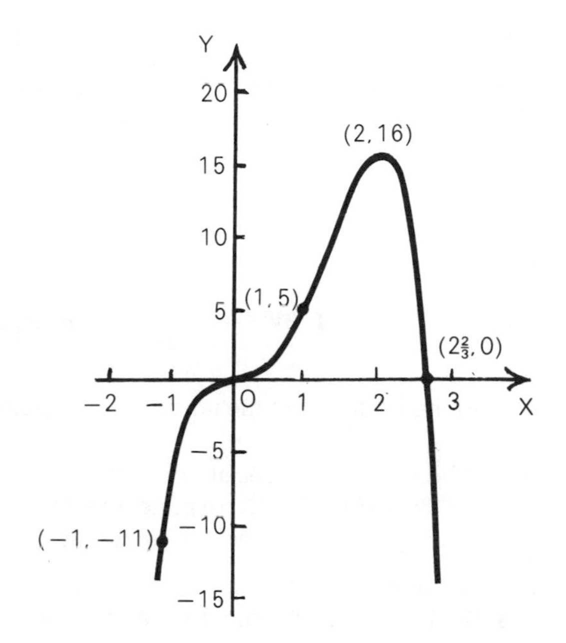

(iii) $y = 8x^3 - 3x^4$ behaves like $y = -3x^4$ for $|x|$ large.
For x large and positive, y is large and negative.
For x large and negative, y is large and negative.

Exercise 7

1 Follow the guide on Page 32 to sketch the graph of:
a $y = x^2 - 4x + 5$ **b** $y = 8 + 2x - x^2$ **c** $y = x^2 + 2$

In questions **2–4**, sketch the graphs $y = f(x)$, where $f(x) =$

2 a $4 - x^2$ **b** x^3 **c** $x^3 - 3x$ **d** $3x - x^3$

3 a $3x^5 - 5x^3$ **b** $4x^3 + 6x^2$ **c** $x^6 - 6x^4$ **d** $6x^4 - x^6$

4 a $x(x+2)$ **b** $x^2(3-2x)$ **c** $x(x-3)^2$ **d** $(4-x^2)(2+x^2)$

5 The height $h(t)$ metres at time t seconds of a rocket shot upwards is given by the formula $h(t) = 5 + 30t - 5t^2$.
a Sketch the graph of $h(t) = 5 + 30t - 5t^2 (0 \leqslant t \leqslant 6)$.
b What is the maximum height reached by the rocket?

6 A small wine-making firm finds that its profit £P in selling x litres of its special wine is given by $P = 100 - 12x + x^3$.
a Sketch the graph for $0 \leqslant x \leqslant 10$.
b What is the least profit that the firm can make, and what is the maximum profit?

CHECK-UP ON **DIFFERENTIATION—1**

1 a Calculate the average gradient of the curve $y = 3x + x^2$ from $x = 1$ to $x = 3$.

b A car travels 25·4 miles in 23 minutes. Calculate, correct to 1 decimal place, its average speed in mph over this time interval.

2 a Using $f'(x) = \lim_{h \to 0} \dfrac{f(x+h) - f(x)}{h}$, find the derivative of $f(x) = 3x^2$.

b Using this h-method, find: (i) the gradient at the point P(2, 5) on the curve $y = x^2 + 1$ (ii) the rate of change of $f(x) = x^3$ at $x = -1$.

3 Differentiate each of these with respect to the relevant variable:

a $x^3 + 5x$ **b** $2x^7 - 4x^3 + 2$ **c** $t^{1/2} - t - \dfrac{1}{2}$ **d** $2t - \dfrac{1}{2t^2}$

e $u^{3/2} + \dfrac{1}{\sqrt{u}}$ **f** $3v^{1/3} + 1 - \dfrac{1}{3v^{1/3}}$ **g** $x^3\left(2x + \dfrac{5}{x^4}\right)$ **h** $\dfrac{(2-u)^2}{u^5}$.

4 Find the equation of the tangent at the point given by $x = -1$ on the curve $y = 3x^2 - \dfrac{1}{x}$.

5 The height $h(t)$ metres at time t seconds of a projectile shot upwards is given by the formula $h(t) = 50 + 40t - 5t^2$. Determine the intervals on which the projectile is rising and falling, assuming that it returns to its initial height.

6 A function f is defined by the formula $f(x) = 3x^4 - 4x^3$.

a Find a formula for the derived function f'.

b Find the intervals of increase and decrease for f.

c Determine the stationary points on the graph of f, and the nature of each.

d Sketch the graph $y = 3x^4 - 4x^3$.

INTRODUCTION

1 The sky rocket rises s metres vertically in t seconds, where $s = 80t - 5t^2$.
What is the maximum height it reaches?

2 Is the line $y = 2x - 1$ a tangent to the parabola $y = x^2$?

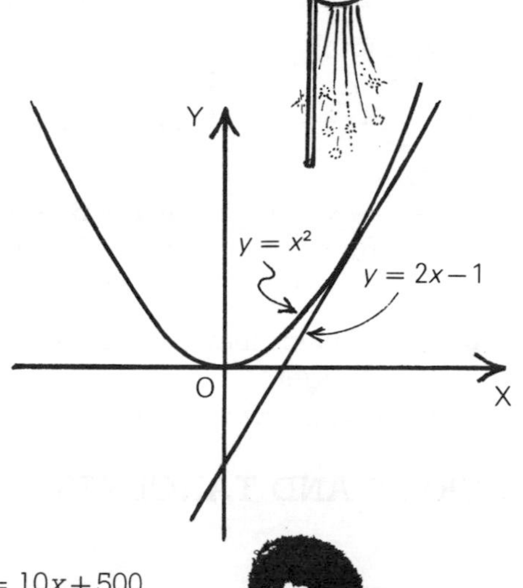

3 The cost to a small firm of making x guitars is $C = 10x + 500$ (in £). The income (in £) arising from the sale of these guitars is $I = 15x + x^2$.
How many guitars does the firm need to sell to break even?

The solution to these and many other questions can be found by using 'quadratic theory'.

REMINDERS

1 Quadratic form: $ax^2 + bx + c\ (a \neq 0)$
(i) $4x^2 + 2x - 1$ (ii) $1 - 4x^2$ (iii) $6t^2 - 2t$, are in quadratic form.
In (i), $a = 4$, $b = 2$, $c = -1$; in (ii), $a = -4$, $b = 0$, $c = 1$; in (iii), $a = 6$, $b = -2$, $c = 0$.

2 Factorising: $ax^2 + bx + c$
Example 1 $2x^2 + 5x - 3 = (2x - 1)(x + 3)$.
Example 2 $16 - 9t^2 = (4 - 3t)(4 + 3t)$.

3 Solving a quadratic equation: $ax^2 + bx + c = 0$
Example Solve these equations:

a $3x^2 + 2x - 1 = 0$
$(3x - 1)(x + 1) = 0$
$3x - 1 = 0$ or $x + 1 = 0$
$x = \frac{1}{3}$ or $x = -1$
Roots are $\frac{1}{3}$ and -1.

b $4y^2 - 12y + 9 = 0$
$(2y - 3)(2y - 3) = 0$
$2y - 3 = 0$ or $2y - 3 = 0$
$y = \frac{3}{2}$ (equal, or repeated, roots).

Exercise 1 (Revision)

1 Write down the values of a, b, c in:

a $6x^2-7x+1$ **b** $4+2x-x^2$ **c** $7-3x^2$ **d** $2x^2$ **e** $4x-10x^2$.

2 Factorise:

a t^2+t-2 **b** p^2-6p+8 **c** n^2-2n+1 **d** $2x^2+x-3$

e $x^2-14x+49$ **f** $2y^2-7y+6$ **g** m^2-1 **h** $16-25r^2$.

3 Solve:

a $t^2+2t-3=0$ **b** $x^2+5x+6=0$ **c** $p^2+16p+15=0$ **d** $y^2-2y-8=0$

e $9x^2-6x+1=0$ **f** $y^2-9=0$ **g** $2r^2-r=0$ **h** $2q^2+4q=0$

i $2x^2-5x-3=0$ **j** $4d^2-12d+9=0$ **k** $x(x-1)=12$ **l** $y(y+2)=15$.

4 **a** Ali maintains that there is only one quadratic equation with roots 1 and 2. Peter disagrees. Who is right?

b Write down two quadratic equations with roots 2 and -3.

CURVES AND TANGENTS

Let's answer question **2** in the Introduction (Is $y=2x-1$ a tangent to the parabola $y=x^2$?), and also look at the line $y=2x$.

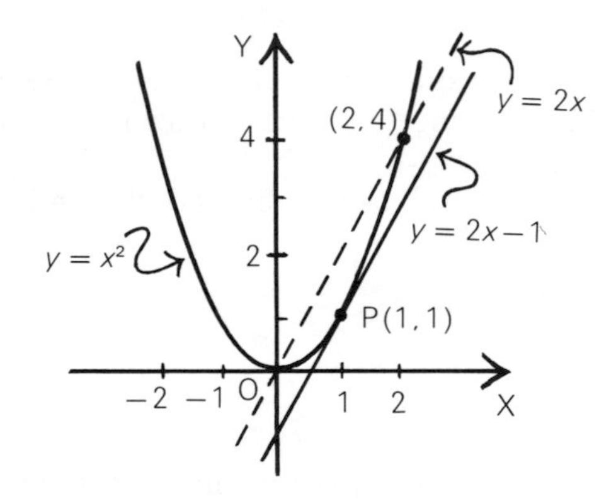

$y=2x$ meets $y=x^2$ where

$$x^2=2x$$
$$x^2-2x=0$$
$$x(x-2)=0$$
$$x=0 \text{ or } x=2$$

There are two intersections: (0, 0) and (2, 4).

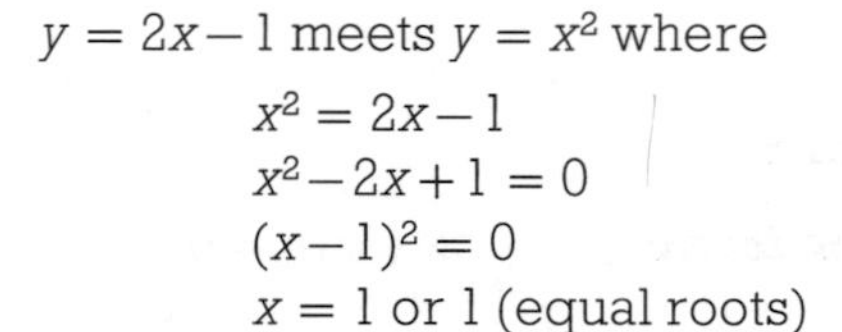

$y=2x-1$ meets $y=x^2$ where

$$x^2=2x-1$$
$$x^2-2x+1=0$$
$$(x-1)^2=0$$
$$x=1 \text{ or } 1 \text{ (equal roots)}$$

There is only one intersection: (1, 1). The line is a tangent.

> The line is a tangent if equal roots appear.

Exercise 2

1 Show that $y=4x-4$ is a tangent to $y=x^2$, and find the point of contact.

2 Either $y=x+2$ or $y=6x-9$ is a tangent to $y=x^2$. Decide which one is the tangent and find its point of contact.

3 Prove that $4x+y+4=0$ is a tangent to $y=x^2$ and find the point of contact.

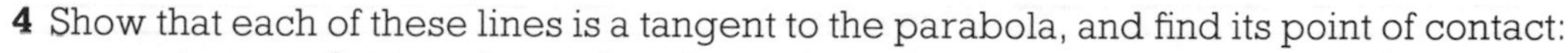

4 Show that each of these lines is a tangent to the parabola, and find its point of contact:
a $y = 4x$, $y = x^2+4$ **b** $y = 2x+1$, $y = 4x-x^2$ **c** $y = 6+2x$, $y = 5-x^2$.

5 A hyperbola has equation $xy = 1$. Prove that $y = 4(1-x)$ is a tangent to the hyperbola and find the point of contact.

6 Either $y = 4-2x$ or $y = 3+2x$ is a tangent to the hyperbola $y = \frac{2}{x}$.

a Which one is it?
b Find the points of intersection (or contact) in each case.

7 Is the line $x+y+3 = 0$ a tangent to the parabola $y = x^2+x-2$? Find the answer and the point of contact or points of intersection.

8 Judy makes two claims:
a $x^2+y^2 = 2$ is the equation of a circle.
b $y = 2-x$ is a tangent to the circle.
Investigate her claims.

It's possible to draw a line that cuts across a curve but which still is a tangent.
a Make a sketch of such a possibility.
b Test this out for the line $y = -x$ and curve $y = x^3-2x^2$.

THE QUADRATIC FORM $a(x-p)^2+b$; COMPLETING THE SQUARE

In question **1** of the Introduction the height reached by the rocket at t seconds was $s = 80t-5t^2$.
$s = -5(t^2-16t) = -5[(t-8)^2-8^2] = 320-5(t-8)^2$.
The least value of $(t-8)^2$ is zero, when $t = 8$. So the greatest value of s is 320.
This means that the rocket's maximum height is 320 metres after 8 seconds.
To solve the problem we had to put $80t-5t^2$ into the form $a(x-p)^2+b$, by 'completing the square'.

Reminder $(x\pm k)^2 = x^2\pm 2kx+k^2$.
Notice that k^2 is ($\frac{1}{2}$ coefficient of x)^2.

Example Express these in the form $a(x-p)^2+b$ by completing the square:

1 $x^2+6x = x^2+6x+3^2-3^2 = (x+3)^2-9$

2 $x^2-4x = x^2-4x+(-2)^2-(-2)^2 = (x-2)^2-4$

3 $y^2+3y+5 = y^2+3y+(\frac{3}{2})^2-(\frac{3}{2})^2+5 = (y+\frac{3}{2})^2+2\frac{3}{4}$

4 $4x^2+8x+3 = 4(x^2+2x)+3 = 4(x^2+2x+1^2-1^2)+3 = 4(x+1)^2-1$

5 $7+8y-2y^2 = 7-2(y^2-4y) = 7-2(y^2-4y+2^2-2^2) = 15-2(y-2)^2$.

Note In **1**, the minimum value is -9, when $x = -3$.
In **5**, the maximum value is 15, when $y = 2$.

Exercise 3

1 Express each of these in the form $a(x-p)^2+b$ by completing the square:

a x^2+2x+5 **b** y^2+6y-1 **c** $t^2-10t+3$ **d** k^2-2k-5
e $2x^2+8x+1$ **f** $3y^2-6y-2$ **g** $2t^2+4t+4$ **h** $5p^2-20p+7$
i u^2+3u-1 **j** n^2-n+5 **k** $2c^2+2c+3$ **l** $4r^2-12r+3$
m $7-2x-x^2$ **n** $5-6x-3x^2$ **o** $12-4t-2t^2$ **p** $6+6x-2x^2$

2 **a** Show that $x^2-4x+5=(x-2)^2+1$.
b Check the graphs of this family of parabolas:

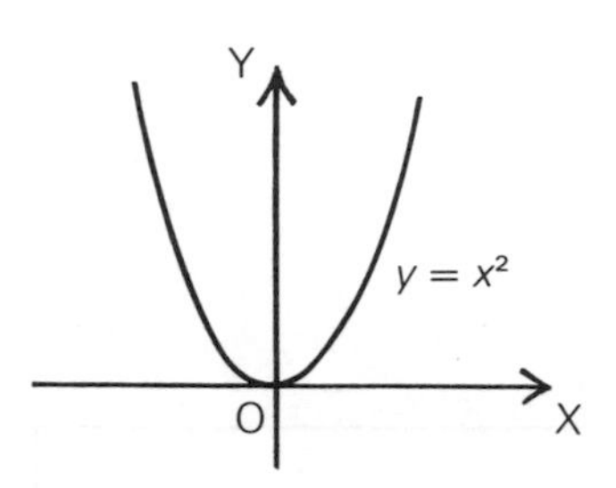

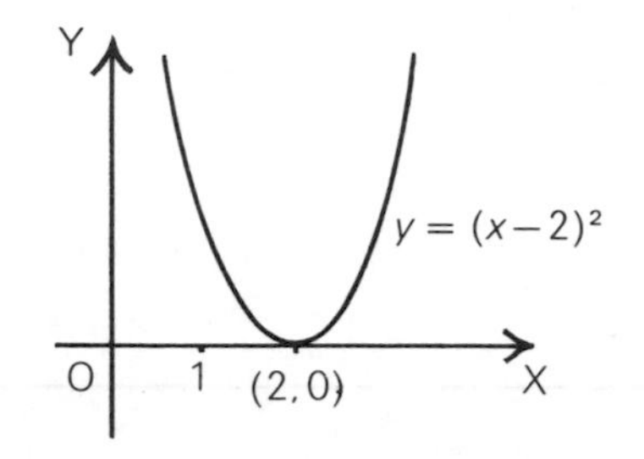

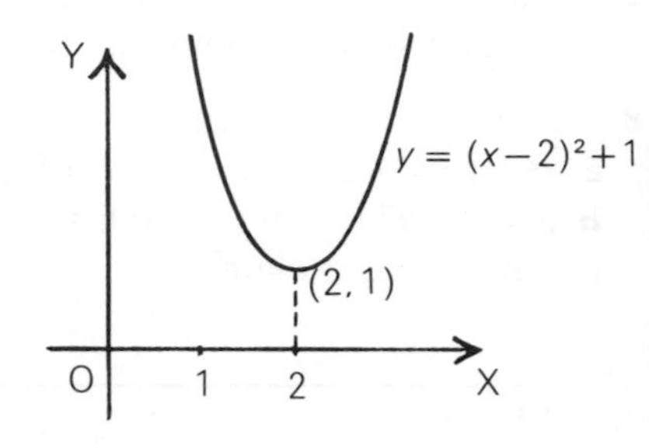

c State the minimum value of x^2-4x+5 and the corresponding value of x.

3 **a** Show that $4x-2x^2-3=-2(x-1)^2-1$.
b Check the graphs of this family of parabolas:

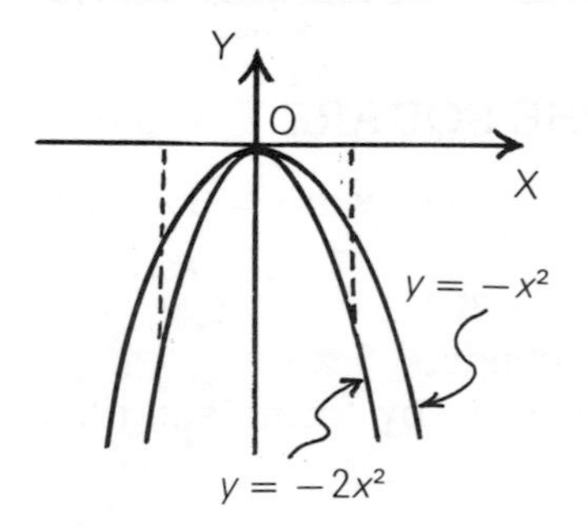

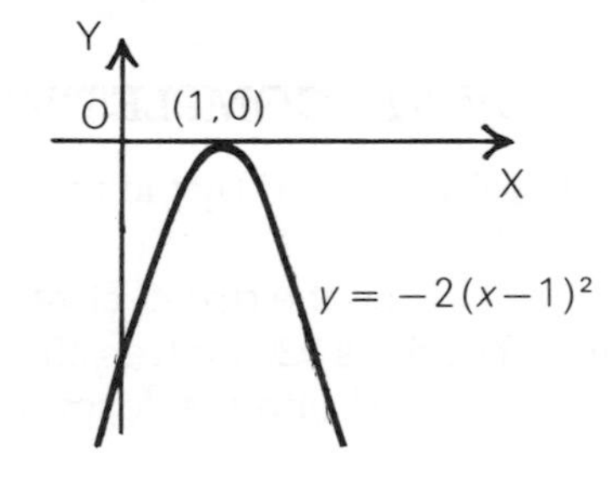

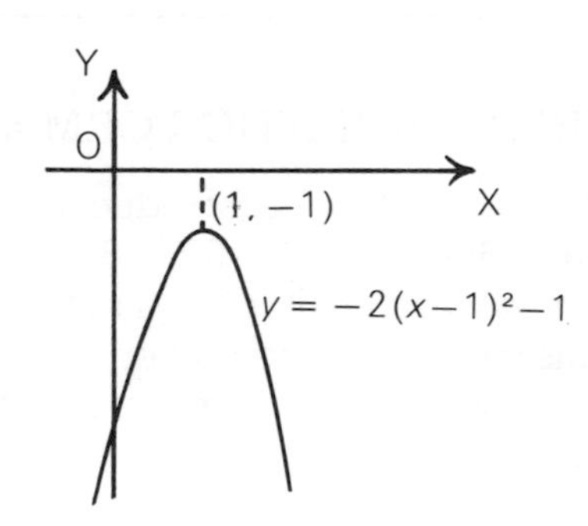

c State the maximum value of $4x-2x^2-3$ and the corresponding value of x.

4 Questions **2** and **3** suggest a rule: *The parabola $y=ax^2+bx+c$ is concave up or down according to whether $a>0$ or $a<0$.*
Check this rule for: **a** $y=x^2-4x$ **b** $y=2x-x^2$.

5 Sketch the graphs:
a $y=x^2$ **b** $y=x^2-4$ **c** $y=4-x^2$ **d** $y=(x+1)^2$ **e** $y=(x+1)^2+2$.

6 **a** Express these in the form $a(x-p)^2+b$.
b Write down the maximum or minimum value of each, and the corresponding value of x:
(i) $x^2+4x+10$ (ii) x^2-6x+1 (iii) $8-2x-x^2$ (iv) $12+8x-x^2$.

Monkey business!

A zoo wants to fence in an exercise area for monkeys.

a If 40 metres of fencing are available what is the largest rectangular area that can be formed? (Take x m for the length, and show that the area

$A = 20x - x^2 = 100 - (x-10)^2$.)

b If the area is bordered on one side by a moat with a straight edge, what is the largest rectangular area that can now be formed, assuming no fencing is needed along the edge of the moat?

c If the moat has a right-angled bend, what is the largest area that can now be enclosed using the moat for two sides?

d A **brainstormer!** Suppose now that the same two edges of the moat are used, but the 40 metres of fencing are fixed in a rigid straight line. What is the maximum area that can now be enclosed?

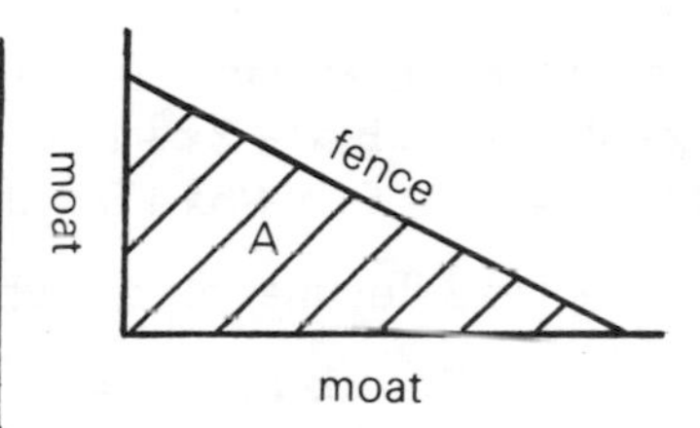

SOLVING QUADRATIC EQUATIONS

1 *By completing the square*

$x^2 - 2x - 4 = 0$

$(x^2 - 2x + 1^2 - 1^2) - 4 = 0$

$(x-1)^2 = 5$

$x - 1 = \pm\sqrt{5}$

$x = 1 \pm \sqrt{5}$

$= 1 + 2{\cdot}236$ or $1 - 2{\cdot}236$

$= 3{\cdot}24$ or $-1{\cdot}24$

(correct to 2 decimal places.)

2 *By formula*

$ax^2 + bx + c = 0$

$x^2 + \frac{b}{a}x + \frac{c}{a} = 0$

$x^2 + \frac{b}{a}x + \left(\frac{b}{2a}\right)^2 - \left(\frac{b}{2a}\right)^2 + \frac{c}{a} = 0$

$\left(x + \frac{b}{2a}\right)^2 = \frac{b^2}{4a^2} - \frac{c}{a} = \frac{b^2 - 4ac}{4a^2}$

$x + \frac{b}{2a} = \frac{\pm\sqrt{(b^2 - 4ac)}}{2a}$

$$x = \frac{-b \pm \sqrt{(b^2 - 4ac)}}{2a}$$

Example Solve $3x^2+4x-5=0$, using the quadratic formula

$$\left.\begin{aligned} a &= 3 \\ b &= 4 \\ c &= -5 \end{aligned}\right\} b^2-4ac = 16+60 = 76$$

$$\begin{aligned} x &= \frac{-b\pm\sqrt{(b^2-4ac)}}{2a} \\ &= \frac{-4\pm\sqrt{76}}{6} \\ &= \frac{-4\pm 8{\cdot}718}{6} \\ &= \frac{-4+8{\cdot}718}{6} \text{ or } \frac{-4-8{\cdot}718}{6} \\ &= 0{\cdot}79 \text{ or } -2{\cdot}12 \end{aligned}$$

(correct to 2 decimal places.)

Exercise 4

1 Solve these quadratic equations. Remember to check for factors first. Where necessary, give the solutions correct to 2 decimal places.

a $x^2-3x+2=0$ **b** $x^2+2x+1=0$ **c** $y^2-y-6=0$ **d** $2y^2+y-1=0$
e $x^2+2x-4=0$ **f** $x^2+5x+2=0$ **g** $2x^2+3x-4=0$ **h** $3x^2-6x+1=0$

2 Solve, correct to 2 decimal places where necessary:

a $y(3y+5)+2=0$ **b** $t(t+3)=1$ **c** $(2x-1)^2-1=0$ **d** $\frac{x}{2}+\frac{2}{x}=3$.

3 In question **3** of the Introduction, the guitar firm's costs (in £) in making x guitars are $C=10x+500$, and the income (in £) from their sale is $I=15x+x^2$. To break even, $C=I$. Make a quadratic equation, and solve it to find the break-even number of guitars that they must sell.

4 In the painting, the horizon is at B where $\frac{AB}{BC}=\frac{AC}{AB}$, i.e. at the 'golden section' ratio.

a Taking AC = 1 and AB = x, show that $x^2+x-1=0$.

b Calculate the value of x (exactly).

5 In still water, Rose can paddle her canoe at 12 mph. She travels upstream for 2 miles but after a while loses her paddle. Two hours after she set off she finds herself drifting back to where she started. What is the speed of the current?

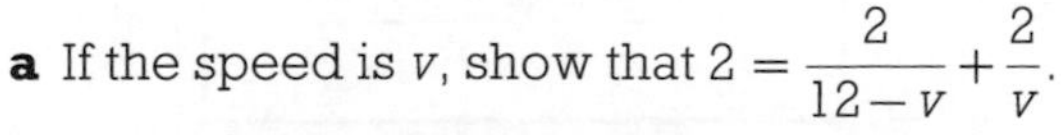

a If the speed is v, show that $2=\frac{2}{12-v}+\frac{2}{v}$.

b Solve the equation (correct to 1 decimal place) and explain the answer.

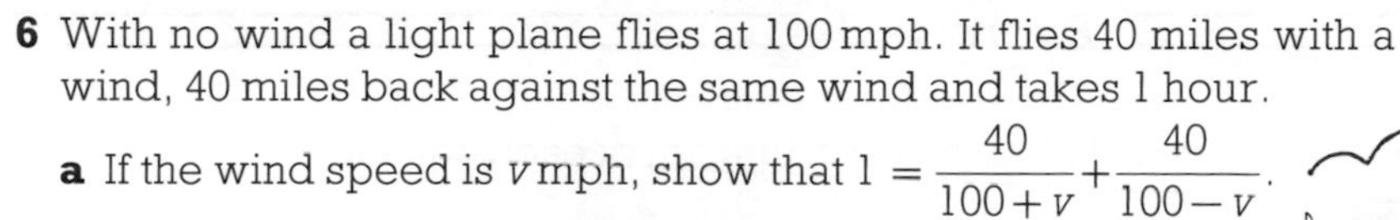

6 With no wind a light plane flies at 100 mph. It flies 40 miles with a wind, 40 miles back against the same wind and takes 1 hour.

a If the wind speed is v mph, show that $1 = \dfrac{40}{100+v} + \dfrac{40}{100-v}$.

b Calculate v.

7 Tim and Neil are on a racing track. Tim is 3 km/h faster than Neil. They race between two checkpoints 30 km apart. Tim arrives 30 minutes before Neil. What were their speeds?

> Solve the equation $x^4 - 2x^2 - 1 = 0$, giving the roots correct to 2 decimal places.

> Investigate the relationships between the sum and the product of the roots of the equation $ax^2 + bx + c = 0$ and the coefficients a, b and c.

THE DISCRIMINANT

Check this table:

Quadratic equation	b^2-4ac	Roots by formula	Nature of roots
$4x^2-2x-1=0$	20	$x=\frac{1}{8}(2\pm\sqrt{20})$	2 real roots (distinct)
$4x^2+4x+1=0$	0	$x=\frac{1}{8}(-4\pm\sqrt{0})=-\frac{1}{2}$	1 real root (repeated)
$4x^2-2x+1=0$	-12	$x=\frac{1}{8}(2\pm\sqrt{(-12)})$	no real roots

Did you notice that if b^2-4ac is
- (i) positive, there are 2 real roots
- (ii) zero, there is 1 real root (repeated)
- (iii) negative, there are no real roots.

In fact, b^2-4ac discriminates between real and non-real roots, and for this reason is called the *discriminant* of $ax^2+bx+c=0$.

$$ax^2+bx+c=0 \text{ has } \begin{cases} \text{2 real roots if } b^2-4ac>0 \\ \text{1 real root (repeated) if } b^2-4ac=0 \\ \text{no real roots if } b^2-4ac<0. \end{cases}$$

Exercise 5

1 Decide the nature of the roots by calculating the discriminant of each equation:
a $x^2-4x+1=0$ **b** $t^2-3t+4=0$ **c** $2y^2+5=0$
d $9x^2+6x+1=0$ **e** $3u^2+u-5=0$ **f** $3t^2+6t=-3$
g $2p(p+1)=3$ **h** $(t+1)^2+2t^2=8t$.

2 Find the real roots (if any) of each equation, correct to 2 decimal places where necessary:
a $2y^2-3y-2=0$ **b** $2p^2+p+1=0$ **c** $x^2-6x-2=0$.

3 In each set of three equations, one equation has 2 real roots, one has equal roots and one has no real roots. Identify the example of each type in the two sets.
a (i) $t^2-4t=1$ **b** (i) $3y^2-5y+4=0$
(ii) $x^2-8x=-16$ (ii) $3-8t+2t^2=0$
(iii) $p^2-2p=-3$ (iii) $4u(u-3)=-9$.

4 For what values of p does the equation $x^2-2x+p=0$ have:
a 2 real roots **b** equal roots **c** no real roots?

5 a Show that the line $y=mx$ meets the parabola $y=x^2+1$ at points given by $x^2-mx+1=0$.
b For the line to be a tangent, this quadratic equation has one (repeated) root, so '$b^2-4ac=0$'. Find possible values of m, and write down the equations of the tangents.

6 a Show that the line $y=x+c$ meets the parabola $y=x^2-3x$ where $x^2-4x-c=0$.
b Use the discriminant to find the value of c for which the line is a tangent to the parabola.
c Find the coordinates of the point of contact of the tangent.

7 a Show that the line $y=5x+c$ meets the parabola $y=2x^2+x-5$ where $2x^2-4x+(-5-c)=0$.
b Use the discriminant to find the value of c for which the line is a tangent to the parabola.
c Find the coordinates of the point of contact of this tangent.

QUADRATIC INEQUATIONS

A manufacturer's profit function, in making x machines, is $P(x)=100x-x^2$. How many machines does he have to produce to make: **a** a profit **b** the maximum profit?

For a profit, $P(x)>0$, so $100x-x^2>0$.
The graph $y=100x-x^2$ will help him.
When $x=0$, $y=0$.
When $y=0$, $100x-x^2=0$
$x(100-x)=0$
$x=0$ or 100.

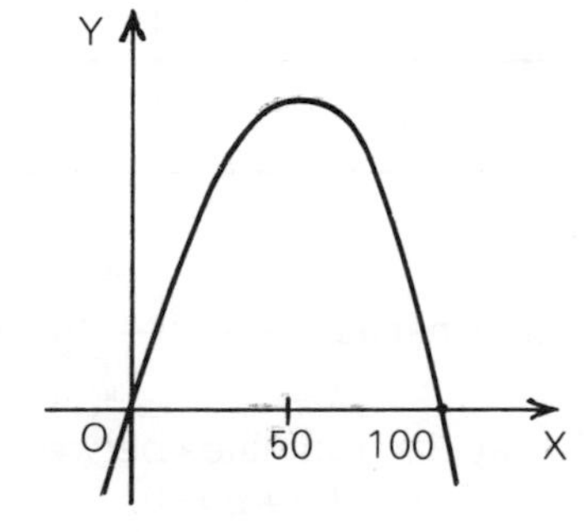

From symmetry, the maximum turning point is given by $x=50$, i.e. (50, 2500).
a For a profit, $100x-x^2>0$, when $0<x<100$.
If $x>100$, a loss is made.
b For maximum profit, $x=50$; i.e. 50 machines should be made.

Exercise 6

1 Find the values of x for which $x^2-2x-3>0$.

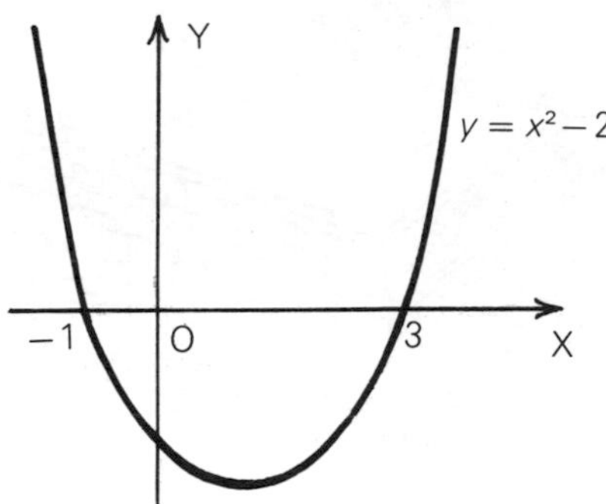

2 Find the values of x for which $4+5x+x^2 \leqslant 0$.

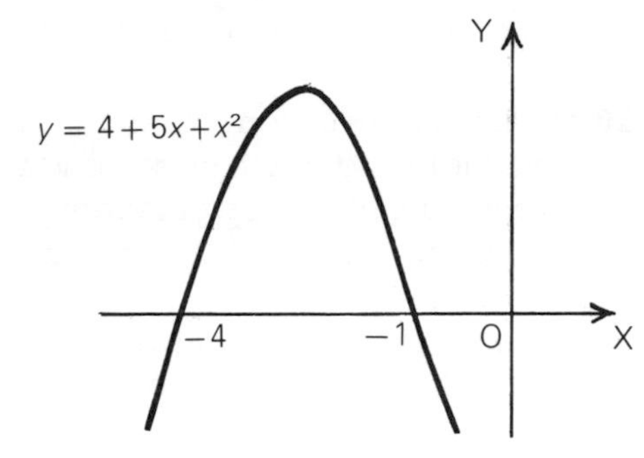

3 Write down the values of x for which:

a $x^2-9=0$
b $x^2-9>0$
c $x^2-9 \leqslant 0$.

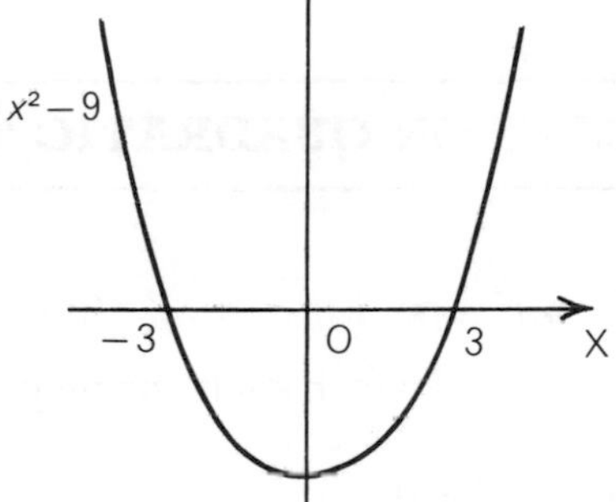

4 In each of these questions the sketch of the graph should be used. Find the values of x for which:

a $x(x-5) \geqslant 0$ **b** $(x-1)(x-4)<0$
c $(2x-3)(x+1)>0$ **d** $(2x-5)(2x-7) \leqslant 0$.

5 A function f is defined by $f(x)=4-3x-x^2 (x \in R)$.

a Find the interval over which the values of f are positive.
b Decide: (i) the values of t for which $t^2+3t \geqslant 4$
(ii) the values of p for which $p-2p^2 \leqslant -1$.

6 Each of these graphs is of the form $y=ax^2+bx+c$.

(i)
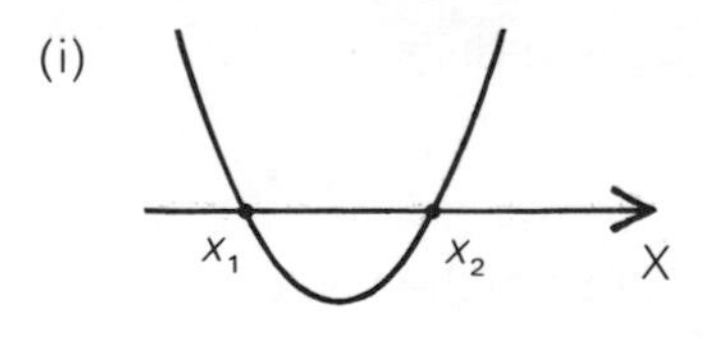

(ii)
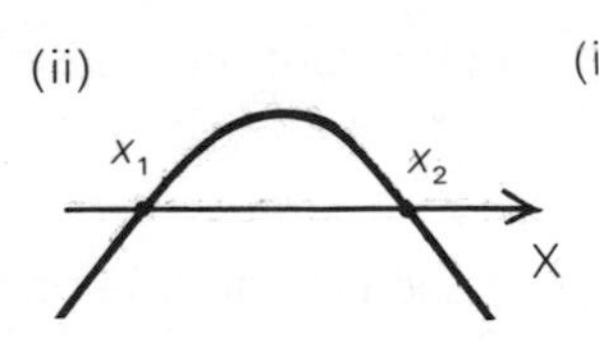

(iii)
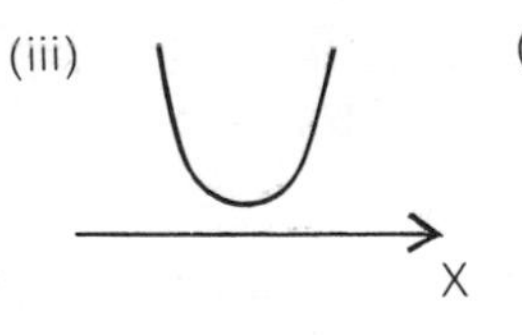

(iv)
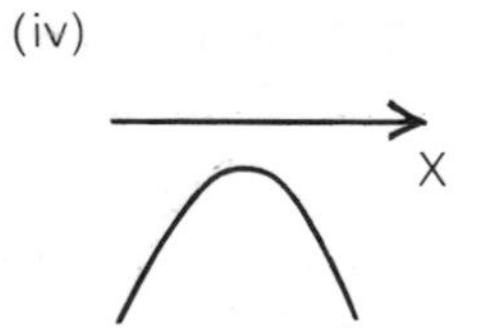

Match each of them with exactly one of:
$ax^2+bx+c \geqslant 0$ for: **a** $x_1 \leqslant x \leqslant x_2$ **b** $x \leqslant x_1$ or $x \geqslant x_2$
c for all $x \in R$ **d** for no $x \in R$.

7 Solve the inequations:

a $x-x^2 \leqslant 0$ **b** $x^2-2x+1>0$ **c** $x(x+4) \geqslant 5$.

8 Two numbers differ by 1 and their product is less than 6.

a Construct an inequation.
b Decide on the possible interval for the numbers.

9 For each of these pairs of functions f, g solve the inequations
(i) $f(x) \geqslant g(x)$ (ii) $f(x) < g(x)$.
Hint Deal with $f(x)-g(x)$.
a $f(x) = 5-x^2$, $g(x) = x^2-3$
b $f(x) = 2x^2+x-3$, $g(x) = 4x+2$.

10 In the construction of an oil rig, the designers laid down these conditions for a rectangular helicopter landing pad:
(i) length to be 10 metres more than breadth
(ii) area of pad to lie between 375 m^2 and 600 m^2.
Calculate the limits for the breadth of the pad.

CHECK-UP ON **QUADRATIC THEORY**

1 Factorise: **a** x^2+2x-3 **b** p^2-p-12 **c** $3t^2-2t-5$.

2 Find the coordinates of the points in which each line meets the curve:
a $y = 2x-5$, circle $x^2+y^2 = 10$
b $y = -5x-13$, parabola $y = 2x^2+3x-5$
c $y = 4(x-1)$, hyperbola $xy = 8$.
What is the relationship between line and curve in **b**?

3 Express each of these in the form $a(x-p)^2+b$, by completing the square:
a $x^2-4x+11$ **b** $3t^2+6t-5$ **c** $5-3x-2x^2$

4 Express each in the form $a(x-p)^2+b$, and then write down its maximum or minimum value, and the corresponding value of x.
a x^2-2x-6 **b** $6+4x-2x^2$ **c** x^2+3x+1

5 Sketch the parabolas:
a $y = x^2$ **b** $y = (x+1)^2$ **c** $y = (x+1)^2-2$.

6 Solve these quadratic equations, giving the roots correct to 2 decimal places:
a $x^2-2x-4 = 0$ **b** $t^2-8t+5 = 0$ **c** $3+2u-2u^2 = 0$.

7 For each equation:
(i) calculate the discriminant D (ii) use D to decide on the nature of the roots.
a $2x^2+3x+1 = 0$ **b** $3x^2+2x+1 = 0$ **c** $-9+12t-4t^2 = 0$
d $(t-1)(t-4) = 3$ **e** $u(3-2u) = 1$ **f** $(u-1)^2+3(u-1)-3 = 0$

8 A toy manufacturer has profit function (in £) $P(x) = -100+20x-0{\cdot}1x^2$. What is his maximum profit and the corresponding number of toys (x)?

9 a Prove that the line $y = 3x+t$ meets the parabola $y = x^2+4$ where $x^2-3x+(4-t) = 0$.
b Find the value of t for which the line is a tangent, and also the point of contact.

10 Solve these inequations for $x \in R$:
a $4x^2+5x+1 \leqslant 0$ **b** $x(x+1) > -2$.

REMINDERS

1 Definitions

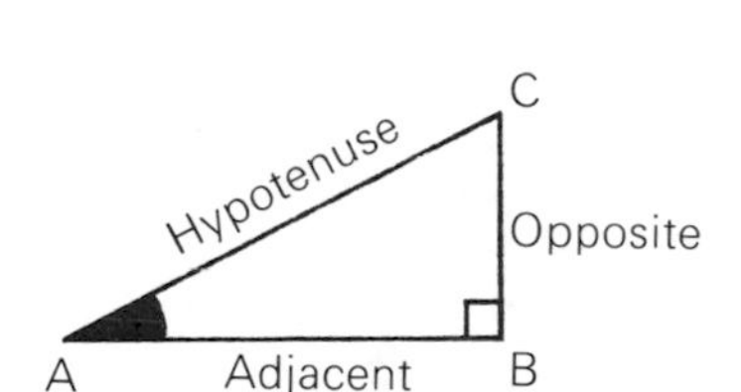

$$\textbf{S}\text{in A} = \frac{\textbf{O}\text{pposite}}{\textbf{H}\text{ypotenuse}}$$

$$\textbf{C}\text{os A} = \frac{\textbf{A}\text{djacent}}{\textbf{H}\text{ypotenuse}}$$

$$\textbf{T}\text{an A} = \frac{\textbf{O}\text{pposite}}{\textbf{A}\text{djacent}}$$

2 Triangle formulae

Sine rule $\frac{a}{\sin A} = \frac{b}{\sin B} = \frac{c}{\sin C}$

Cosine rule $a^2 = b^2 + c^2 - 2bc \cos A$

$$\cos A = \frac{b^2 + c^2 - a^2}{2bc}$$

Area of $\triangle$ABC $\triangle = \frac{1}{2}ab \sin C$

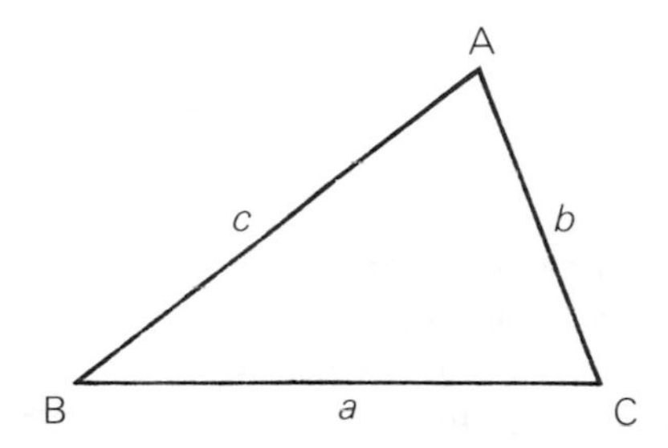

Exercise 1

1 Calculate x, correct to 1 decimal place:

a

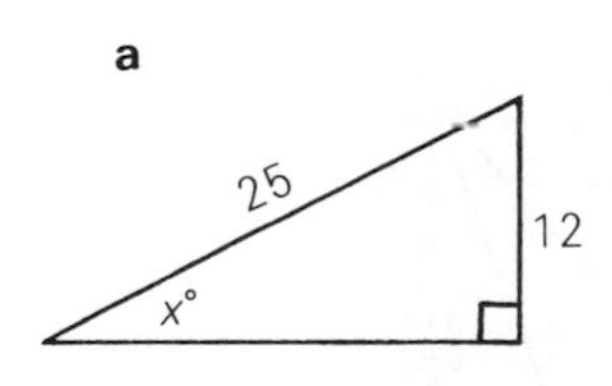

b

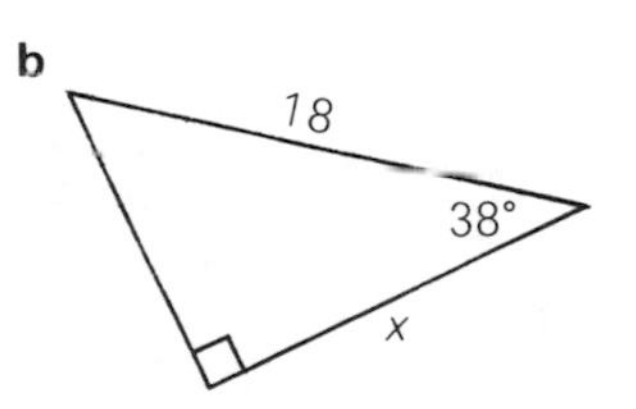

c

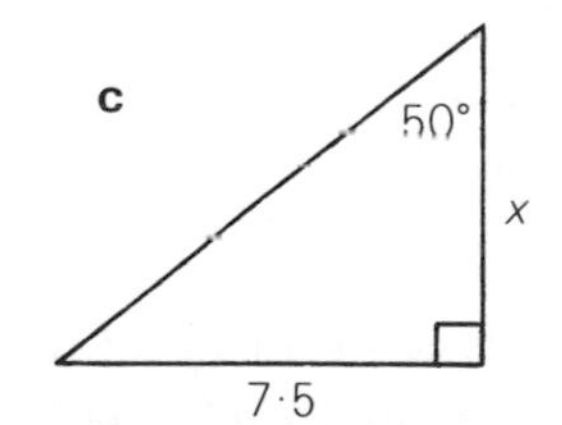

2 This is a trigometer. It is pointed at an object, in this case a tree.

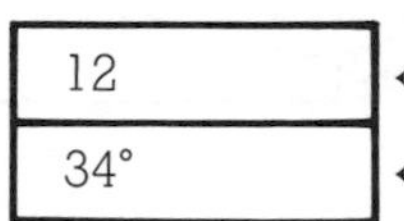

12	← distance of object in metres
34°	← angle from horizontal in degrees.

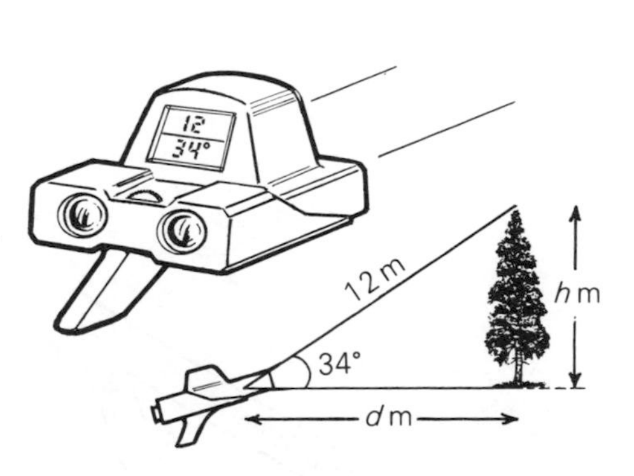

Calculate to the nearest metre:

a the height of the tree

b the distance from the trigometer to the tree along the ground.

3 The coastguard uses his trigometer.

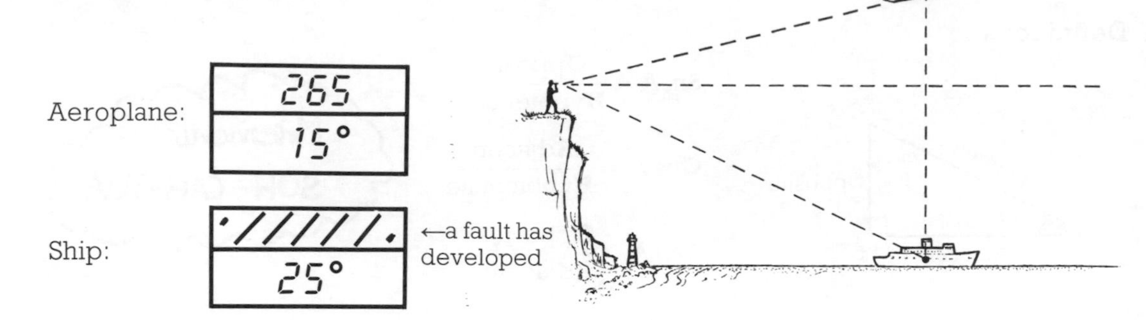

The ship is directly below the plane. Calculate, to the nearest metre:

a the horizontal distance from the coastguard to the point below the plane

b the faulty distance reading.

4 A video camera has a field of view covering a 35° angle. The operator is filming the side view of a house, and estimates that if he stands 1·3 m from the house he will get full coverage of the door and window.

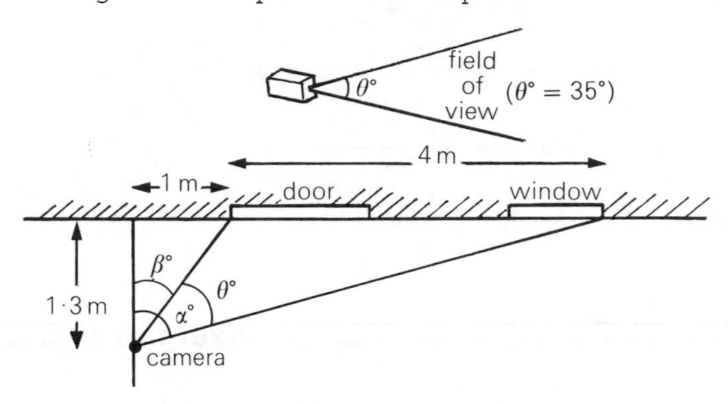

Calculate angles $\alpha°$ and $\beta°$, correct to 1 decimal place, to find out whether he is right.

5 In $\triangle ABC$, $\angle B = 35°$, $\angle C = 66°$ and $a = 8{\cdot}5$ cm. Calculate b, correct to 1 decimal place.

6 Measurements were made of the top of a spire. Calculate, to the nearest metre:

a QC

b AC.

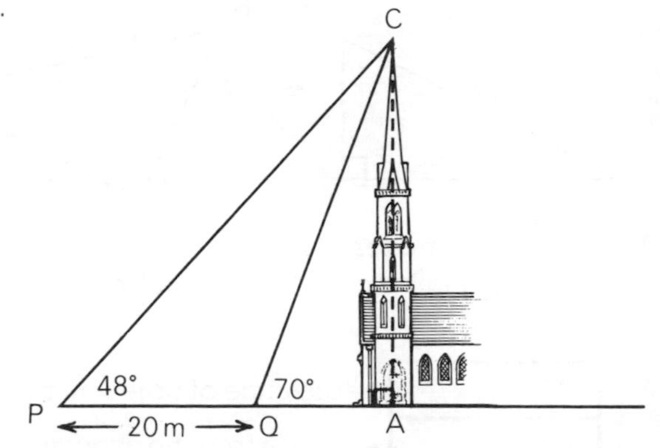

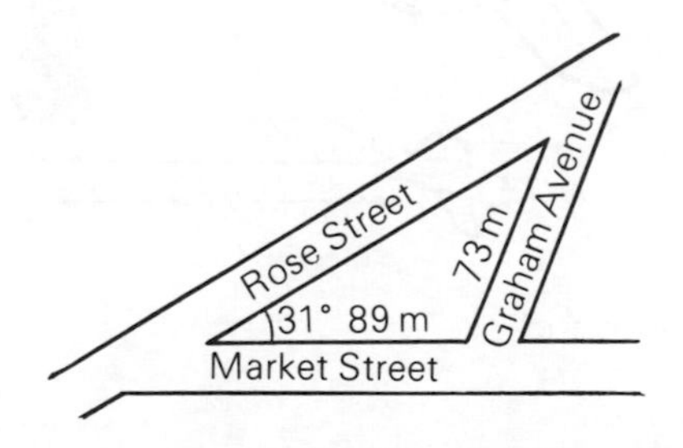

7 This is part of the Newtown road system. Calculate:

a the acute angle between Rose Street and Graham Avenue, correct to 0·1°

b the area of the large traffic island surrounded by the three roads, to the nearest m^2.

8 Calculate the size of the largest angle in the triangle with sides 7·5 cm, 9·5 cm and 15·5 cm long, correct to 0·1°.

9 Two ships sail from port O. The first follows a course 050° for 80 km to A. The second steers 315° for 40 km to B. Calculate:
a the distance AB, to the nearest km
b the bearing of A from B (to the nearest degree).

3 DIMENSIONS

ABCDEFGH is a cuboid.

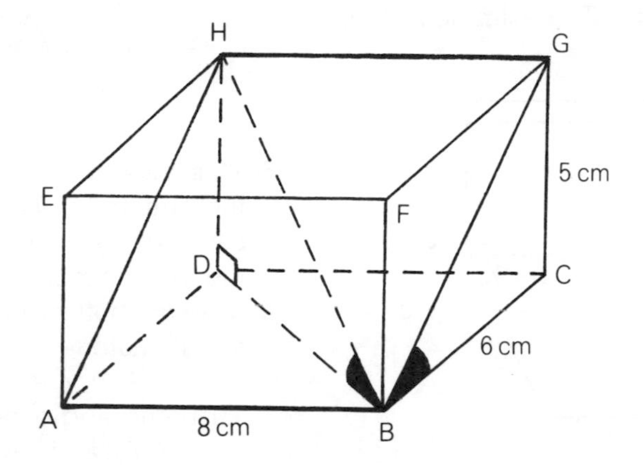

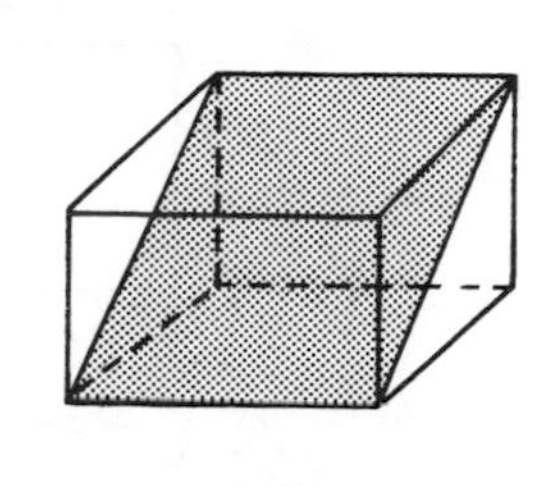

(i) *Angle between line and plane*
To find angle between HB and ABCD, find perpendicular HD from H to ABCD. *∠HBD is the required angle.*

(ii) *Angle between two planes*
To find angle between planes ABGH and ABCD, find their line of intersection AB, then two lines BC and BG perpendicular to AB. *∠CBG is the required angle.*

Exercise 2

1 In the cuboid above, AB = 8 cm, BC = 6 cm and CG = 5 cm. Calculate:
a BD **b** ∠HBD **c** ∠CBG, giving **b** and **c** correct to 0·1°.

2 PQRSTUVW is a cube.
Name the angle between:

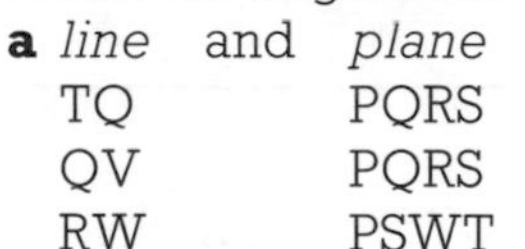

a *line* and *plane*
TQ PQRS
QV PQRS
RW PSWT

b *plane* and *plane*
PQVW PQRS
TWRQ PQRS
PQVW VWSR

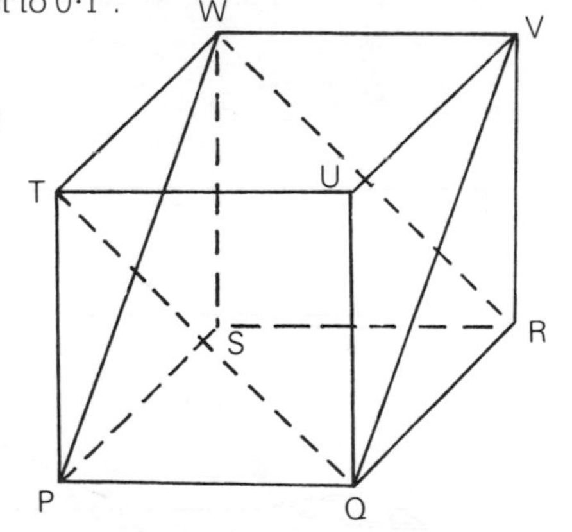

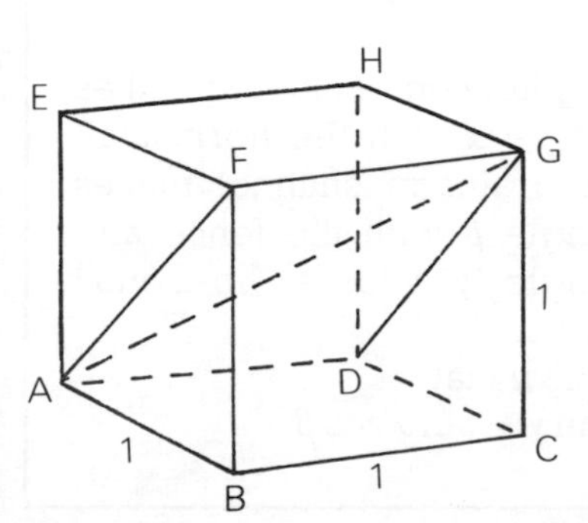

3 The cube has side 1 unit long. Calculate, to 0·1 degree, the size of the angle between:
a planes AFGD and ABCD
b face diagonal AF and space diagonal AG.

4 ABCD is a regular tetrahedron of side 2 m. AO is the perpendicular from A to base BCD, and DO meets BC at M.

a Why is M the midpoint of BC?

b Calculate the lengths of DM and MO, to 1 decimal place (MO = $\frac{1}{3}$MD).

c Calculate the size of the angle between the faces ABC and BCD (i.e. angle AMD), to 0·1°.

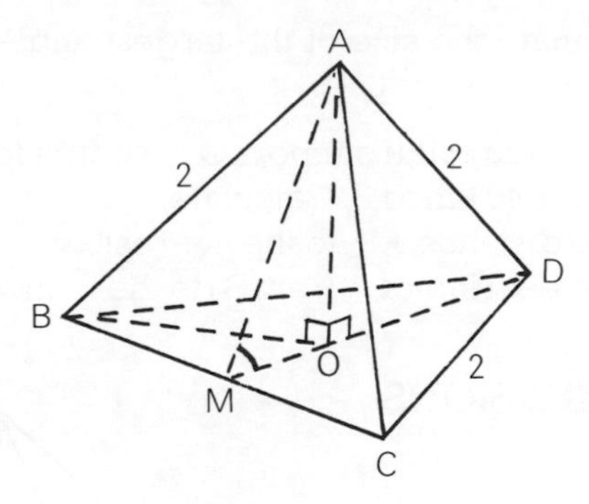

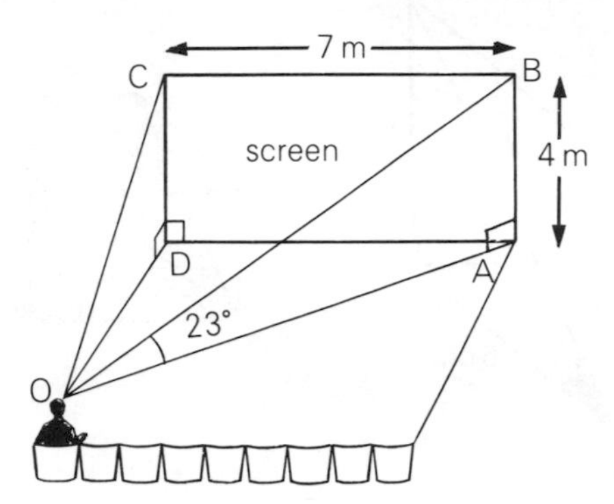

5 This viewer is sitting directly in front of the bottom left-hand corner of the screen. Satisfactory viewing in a cinema requires the eyes to rise through no more than 30° from the bottom to the top of the screen.

a Calculate:

(i) OA, correct to 1 decimal place

(ii) the angle between the two planes OAB and ABCD.

b Does the viewer have 'satisfactory viewing' straight ahead?

6 Calculate, correct to 1 decimal place:

a the length of the chalk line

b the angle between the chalk line and the top of the board

c the height of the top of the blackboard above the ground

d the angle between the front and back blackboards

e the angle the chalk line makes with the horizontal.

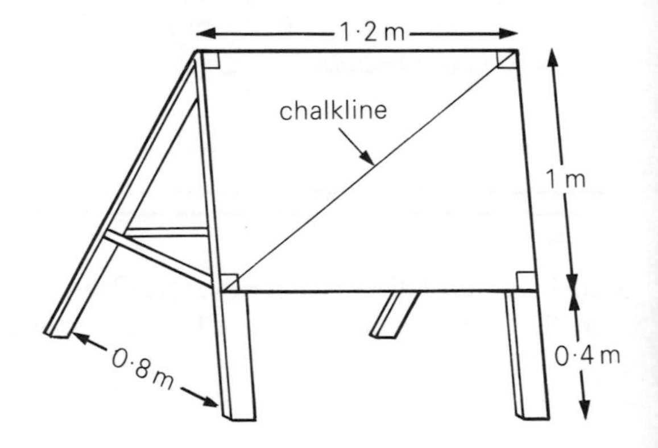

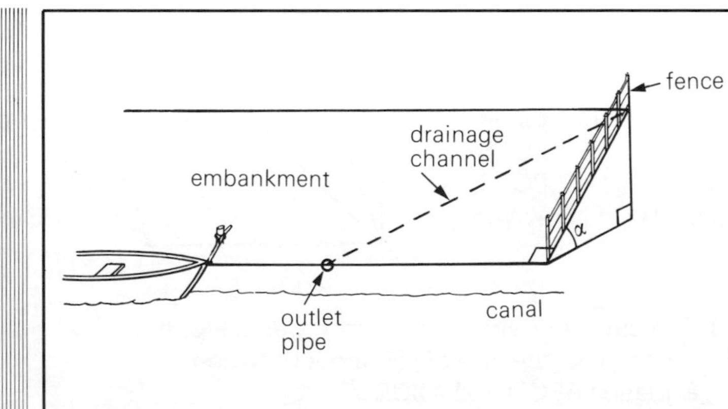

A plane embankment makes angle α with the horizontal. A drainage channel makes angle β with the fence and angle γ with a horizontal plane.
Show that
$\sin\gamma = \sin\alpha\cos\beta$.

RADIAN MEASURE

So far you have measured angles in *degrees*. Their definition of 360 in a complete turn is possibly connected with the number of days the earth takes to circle the sun. Another useful measure for angles is the *radian*, which is closely connected with the circle, as follows.

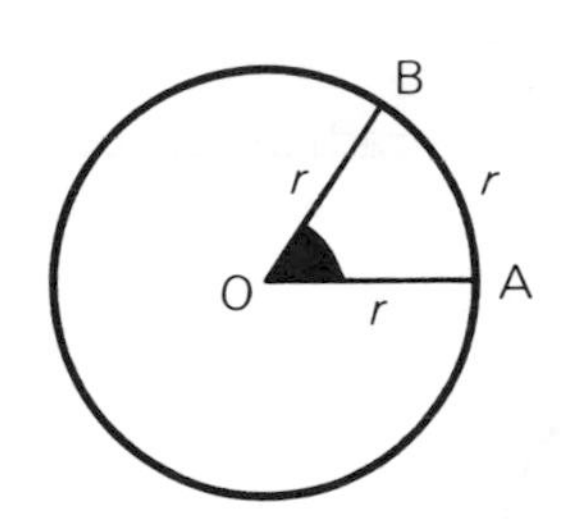

The circle has centre O, and radius r.

> If $\angle$AOB subtends an arc equal in length to the radius, then $\angle$AOB = 1 radian.

(i) $$\frac{\text{Arc AB}}{\text{Circumference}} = \frac{\angle \text{AOB}}{360°}$$

$$\frac{r}{2\pi r} = \frac{1 \text{ radian}}{360°}$$

Cross-multiplying,

> π radians = 180°

(ii) π radians = 180°

So 1 radian $= \frac{180°}{\pi} \doteqdot 57°$

(iii) 180° = π radians

So 90° $= \frac{\pi}{2}$ radians

60° $= \frac{\pi}{3}$ radians, etc.

Exercise 3

1 Change to radians, giving your answers in terms of π:
a 30° **b** 60° **c** 90° **d** 120° **e** 180° **f** 360°.

2 Change to degrees:
a $\frac{\pi}{4}$ radians **b** $\frac{\pi}{2}$ radians **c** $\frac{2\pi}{3}$ radians **d** $\frac{3\pi}{2}$ radians.

3 a Copy the triangles, and mark the lengths of all the sides.
b Then copy and complete the table of exact values.

a

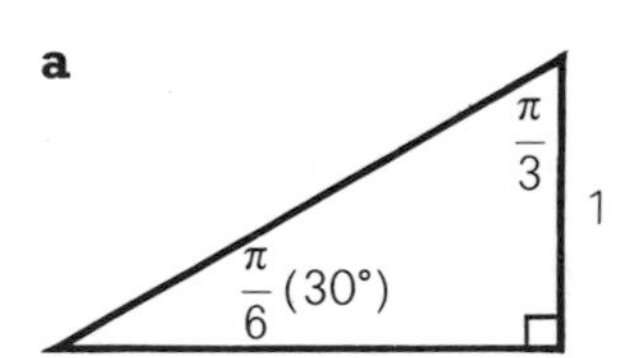

b

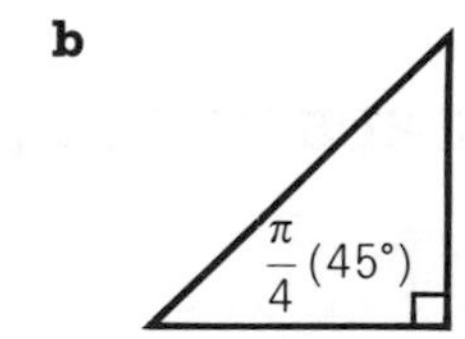

θ	$\frac{\pi}{6}$	$\frac{\pi}{4}$	$\frac{\pi}{3}$
$\sin\theta$			$\frac{\sqrt{3}}{2}$
$\cos\theta$			
$\tan\theta$			

4 a Using your calculator, write down the values of sin 30°, cos 45° and tan 60°.

b (i) Find out how to use radians on your calculator, and write down the values of $\sin\frac{\pi}{6}$, $\cos\frac{\pi}{4}$ and $\tan\frac{\pi}{3}$.

(ii) Compare your answers with those for part **a**.

5 A wheel rotates at 120 revolutions per minute. Copy and complete:
Speed of rotation = 120 revs/minute = ... revs/second = ... radians/second, correct to 3 significant figures.

6 a Copy and complete:

$$\frac{\theta}{2\pi} = \frac{\text{arc AB}\ (l)}{\dots} = \frac{\text{area of sector AOB}\ (A)}{\dots}$$

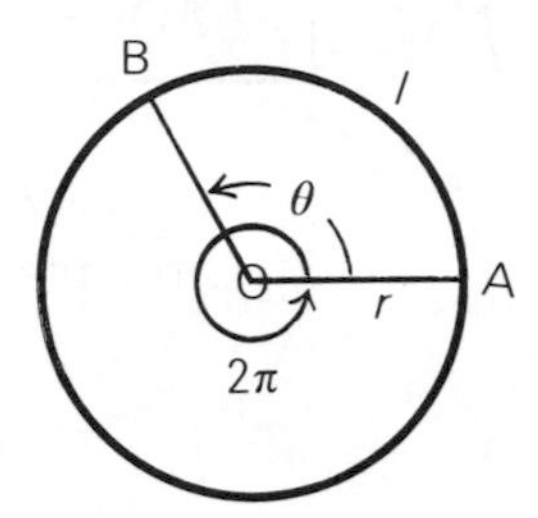

b Prove the formulae:
(i) length of arc AB, $l = r\theta$ (ii) area of sector AOB, $A = \frac{1}{2}r^2\theta$.

7 A circle has radius 35 mm and centre O. In sector AOB, ∠AOB = 2·4 radians. Calculate:
a the length of arc AB **b** the area of the sector.

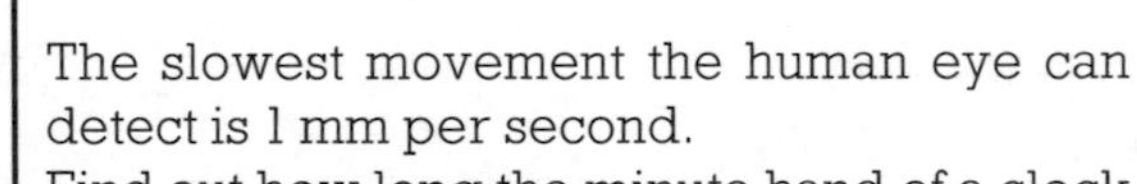

The slowest movement the human eye can detect is 1 mm per second.
Find out how long the minute hand of a clock has to be for the movement of its tip to be detected. (Use radian measure.)

ANGLES OF ALL SIZES—COORDINATE DEFINITIONS

In Book **4B** the definitions of the sine, cosine and tangent of an angle were extended from acute and obtuse angles to all sizes of angle by means of these definitions:

$$\sin\theta = \frac{y}{r},\ \cos\theta = \frac{x}{r},\ \tan\theta = \frac{y}{x},\ x \neq 0$$

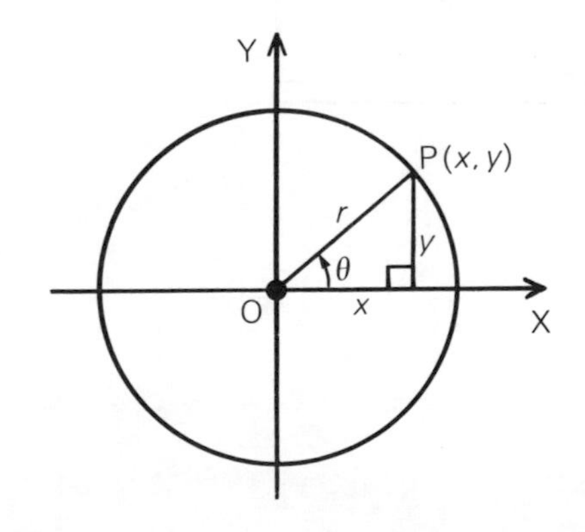

Exercise 4

1 Write down, as ratios, the values of $\sin\theta$, $\cos\theta$ and $\tan\theta$ for each diagram.

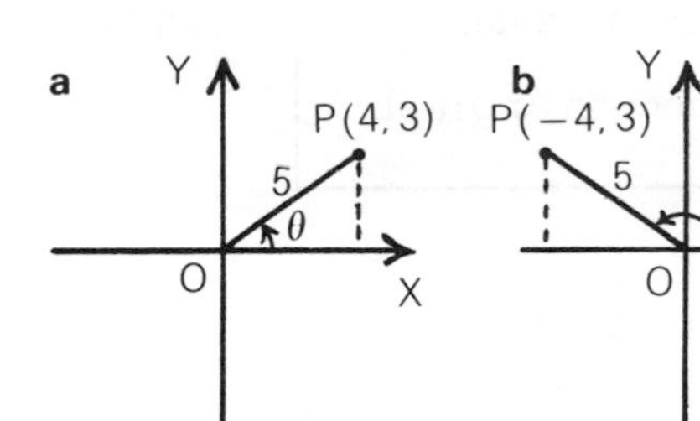

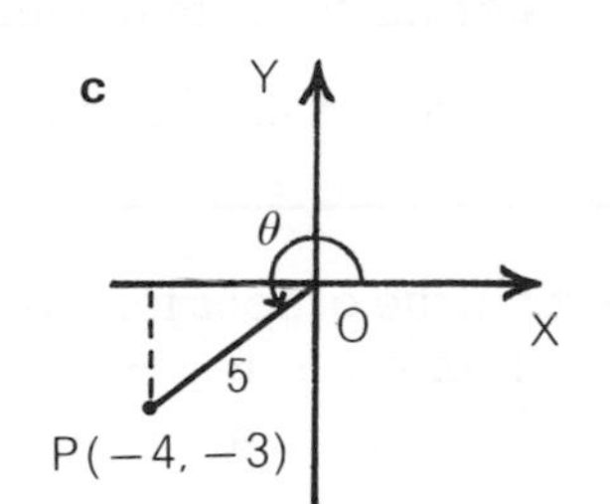

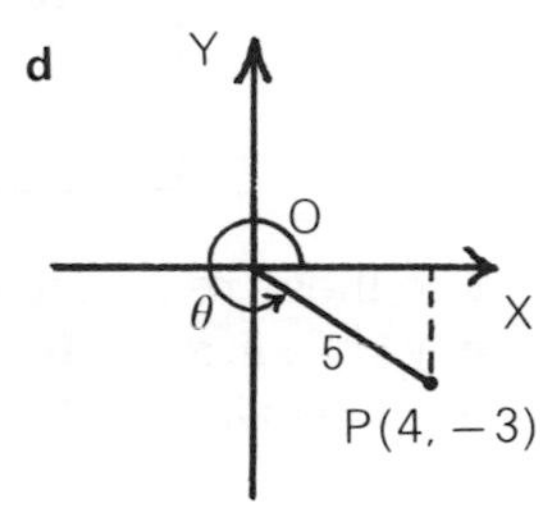

2 Check that the values you got in question **1** agree with this quadrant rule of signs:

$\frac{\pi}{2}$

Sin positive	**All** positive
Tan positive	**Cos** positive

π ... 0

$\frac{3\pi}{2}$

3 Say whether each of these is positive or negative (without using your calculator!):

a $\sin 340°$ **b** $\cos 125°$ **c** $\tan 200°$

d $\sin\frac{\pi}{3}$ **e** $\cos\frac{2\pi}{3}$ **f** $\tan\frac{5\pi}{4}$

4 Using the definitions, copy and complete:

θ	0	$\frac{\pi}{2}$	π	$\frac{3\pi}{2}$	2π
$\sin\theta$	0	1			
$\cos\theta$	1				
$\tan\theta$	0	—		—	

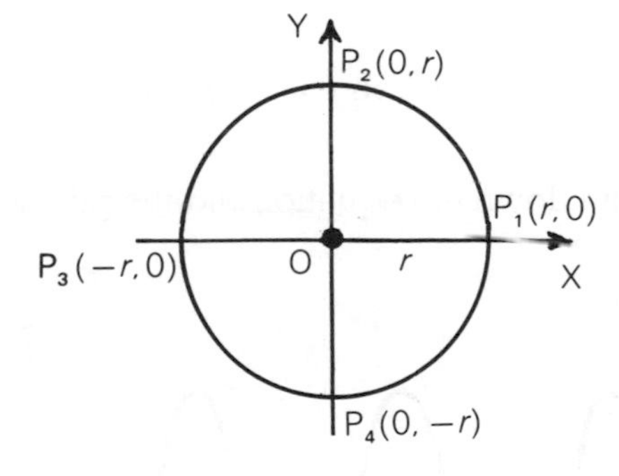

5 The rotating arm OP returns to its starting position after rotations about O of 2π, 4π, 6π . . . radians, also -2π, -4π, . . . radians. So $\sin\theta$ and $\cos\theta$ have period 2π. Using the information in question **4** also, we can sketch the graphs $y = \sin x$ and $y = \cos x$.

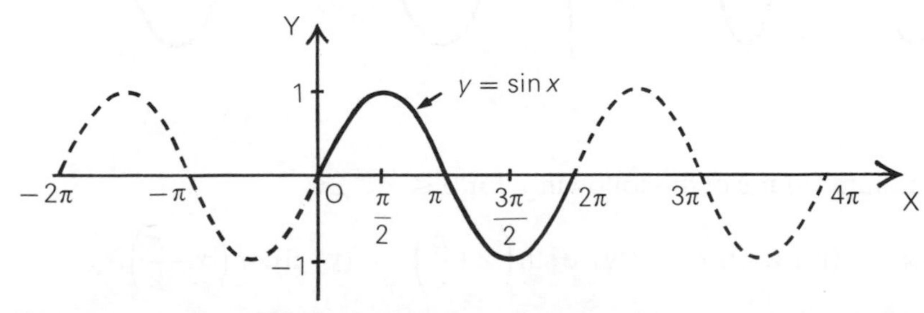

Sketch the graph of $y = \cos x$ from -2π to 4π radians in the same way.

6 In Book **4B**, we found that:

> In the equations: $y = a\sin nx$ and $y = a\cos nx$,
> (i) a determines the maximum and minimum values
> (ii) n determines the period, $\frac{2\pi}{n}$, of the function or graph.

Use these results to sketch the graphs for $0 \leqslant x \leqslant 2\pi$:

a $y = 2\sin x$ **b** $y = 4\sin x$ **c** $y = 2\cos x$ **d** $y = \frac{1}{2}\cos x$
e $y = \sin 2x$ **f** $y = 2\sin 2x$ **g** $y = \cos 3x$ **h** $y = \cos\frac{1}{2}x$.

7 a Use the definitions to prove that:
$\tan(\pi+\theta) = \tan\theta$.

b What is the period of the graph $y = \tan\theta$?

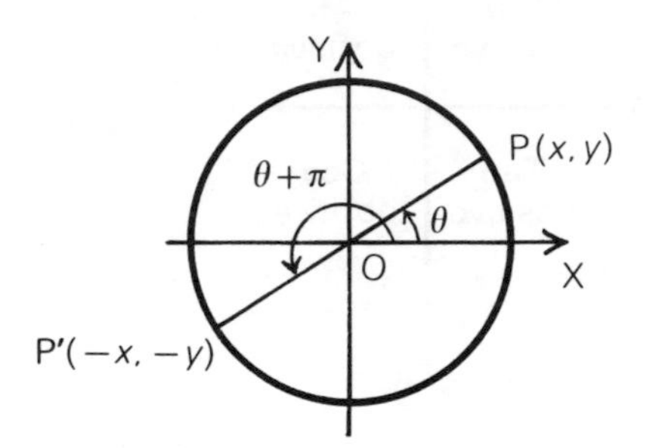

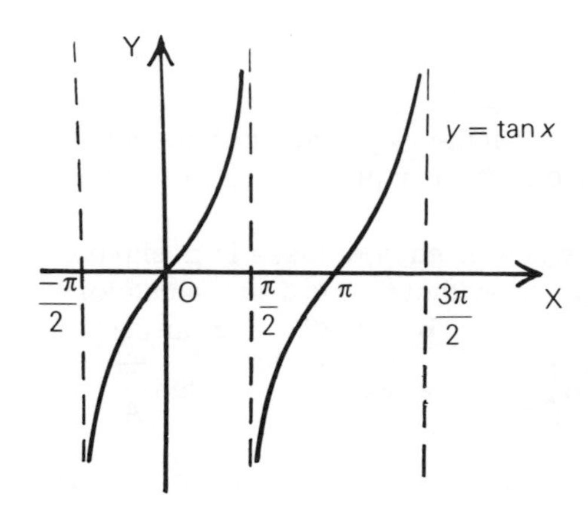

c Sketch the graph $y = \tan x$ for $-\frac{3\pi}{2} < x < \frac{3\pi}{2}\left(x \neq \pm\frac{\pi}{2}\right)$.

8 Write down the equation and the period of each graph. Each is a sine or cosine graph.

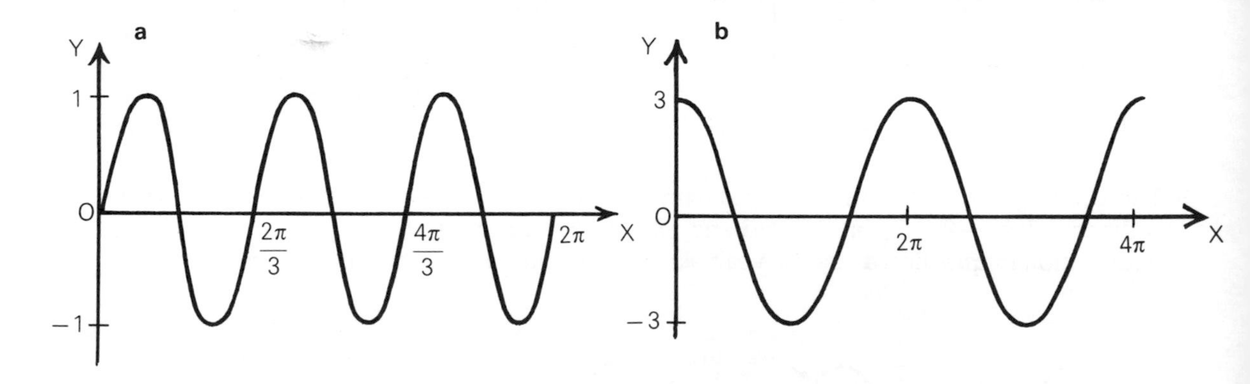

9 a Write down the maximum values of the following, for $0 \leqslant x \leqslant 2\pi$:

(i) $\sin x$ (ii) $\cos x$ (iii) $2\sin x$ (iv) $3\sin\left(x+\frac{\pi}{2}\right)$ (v) $6\cos\left(x-\frac{\pi}{2}\right)$.

b For what values of x, $0 \leqslant x \leqslant \pi$, does each have its maximum value?

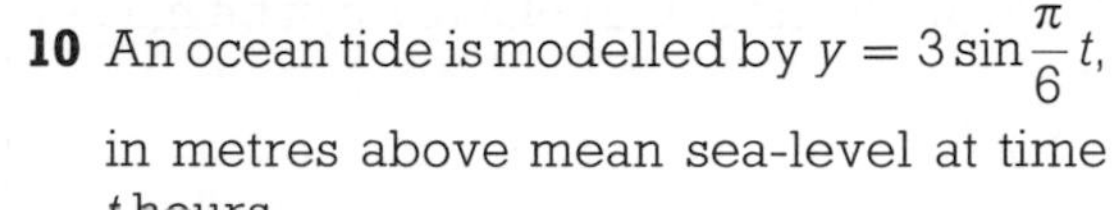

10 An ocean tide is modelled by $y = 3\sin\frac{\pi}{6}t$, in metres above mean sea-level at time t hours.

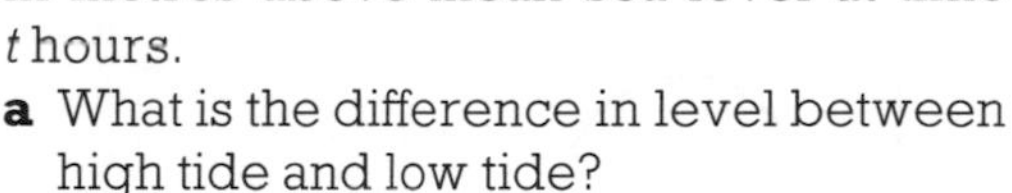

a What is the difference in level between high tide and low tide?
b Sketch the graph for $0 \leqslant t \leqslant 12$.

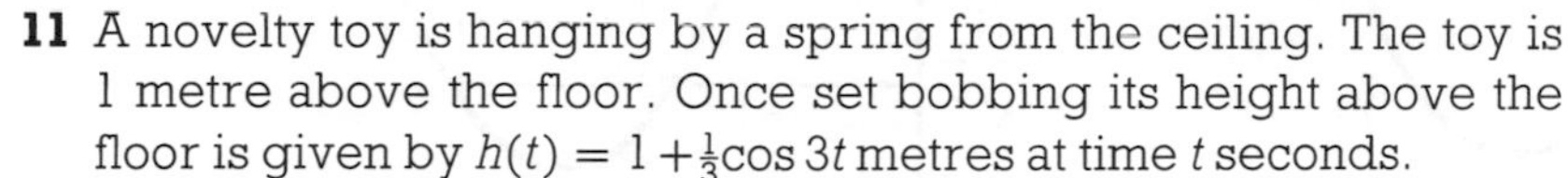

11 A novelty toy is hanging by a spring from the ceiling. The toy is 1 metre above the floor. Once set bobbing its height above the floor is given by $h(t) = 1 + \frac{1}{3}\cos 3t$ metres at time t seconds.
a Calculate the difference between its maximum and minimum heights.
b What is the period of its oscillation?
c Sketch the graph for $0 \leqslant t \leqslant \frac{2}{3}\pi$.

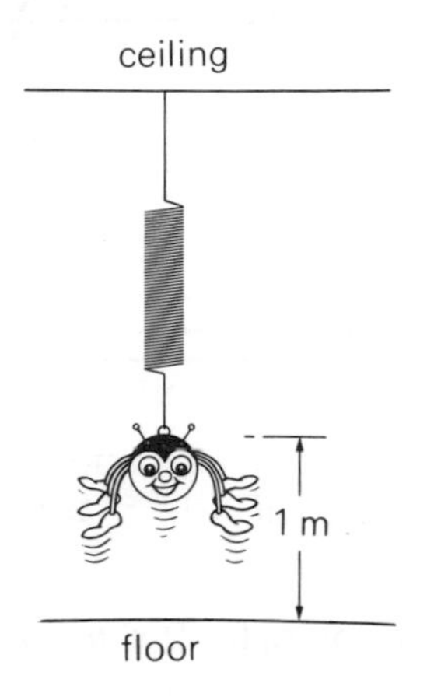

TRIGONOMETRIC EQUATIONS

Example 1 Solve $\sin x^\circ = \frac{1}{2}$, $0 \leqslant x \leqslant 360$, $\sin x^\circ$ is positive, so from the quadrant diagram x is in the first or second quadrants (shown by ✓s).

So $x = 30$ or $180 - 30$
$= 30$ or 150.

(Check both solutions using the $\boxed{\sin^{-1}}$ key.)

✓ Sin + | All + ✓
$(180-x)^\circ$ | x°
$(180+x)^\circ$ | $(360-x)^\circ$
Tan + | Cos +

Example 2 Solve $\cos\theta = -\frac{1}{\sqrt{2}}$, $0 \leqslant \theta \leqslant 2\pi$,

From the quadrant diagram, θ is in the second or third quadrants.

So $\theta = \pi - \frac{\pi}{4}$ or $\pi + \frac{\pi}{4}$
$= \frac{3\pi}{4}$ or $\frac{5\pi}{4}$.

(Check both solutions using the $\boxed{\cos^{-1}}$ key.)

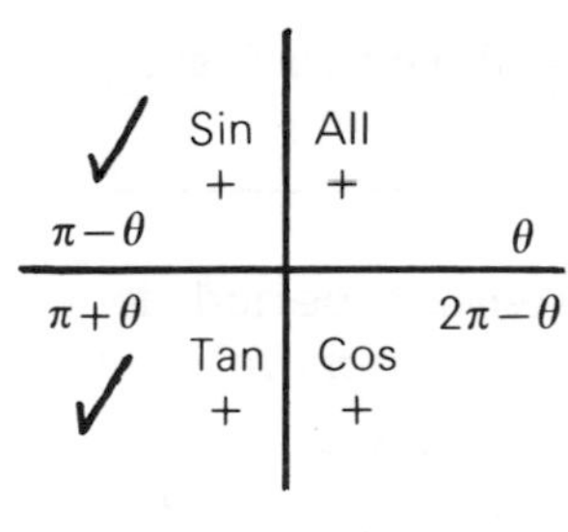

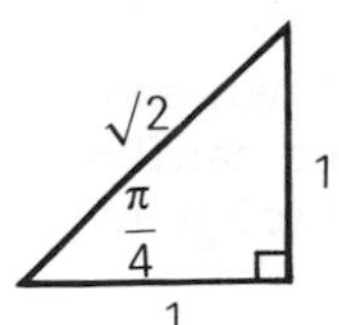

A rough check of the solutions can be made from the sine and cosine graphs:

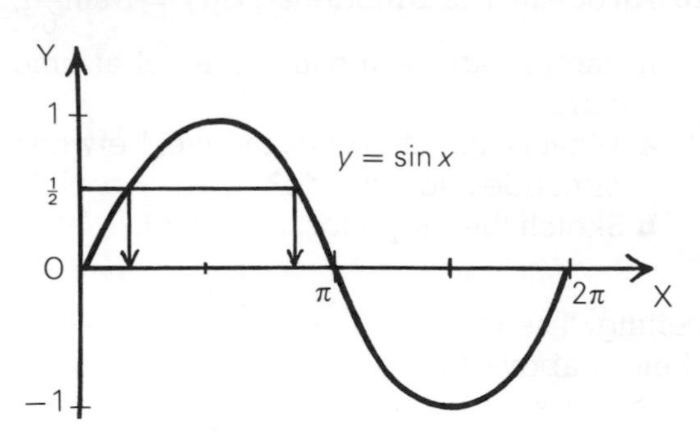

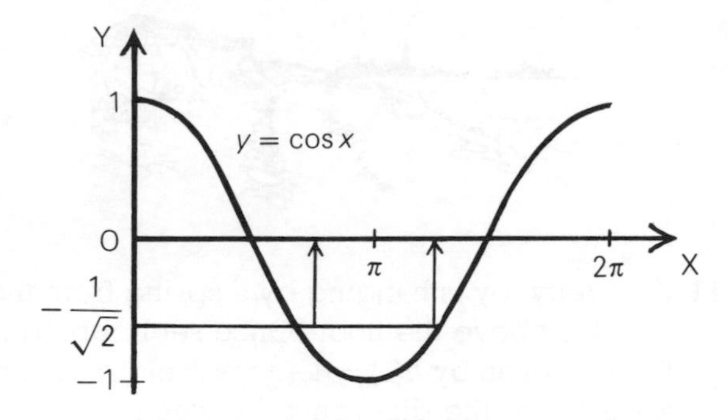

Exercise 5

1 Use the sine and cosine graphs to write down the solutions, for $0 \leqslant \theta \leqslant 2\pi$, of the equations:
a $\sin\theta = 1$ **b** $\cos\theta = 1$ **c** $\sin\theta = 0$ **d** $\cos\theta = -1$.

2 Solve, for $0 \leqslant \theta \leqslant 2\pi$:
a $\sin\theta = \frac{1}{\sqrt{2}}$ **b** $\sin\theta = -\frac{1}{\sqrt{2}}$ **c** $\cos\theta = -\frac{1}{2}$ **d** $\cos\theta = \frac{\sqrt{3}}{2}$.

3 Solve, for $0 \leqslant x \leqslant 360$:
a $2\sin x° + 1 = 0$ **b** $2\cos x° - 1 = 0$ **c** $2\sin x° + \sqrt{3} = 0$ **d** $\sqrt{2}\cos x° - 1 = 0$.

4 Solve, for $0 \leqslant \theta \leqslant 2\pi$:
a $\tan\theta = 1$ **b** $\tan\theta = -1$ **c** $\tan\theta = \sqrt{3}$ **d** $\sqrt{3}\tan\theta + 1 = 0$.

5 Find when the toy is first $1\frac{1}{6}$ metres from the floor in question **11** of Exercise 4.

6 Find when the tide is first $1\frac{1}{2}$ metres above mean sea-level in question **10** of Exercise 4.

From the periodicity of the rotating radius OP, the 'general solution' of $\sin\theta = \sin\alpha$ is

$$\theta = \begin{cases} \alpha + 2n\pi \\ (\pi - \alpha) + 2n\pi \end{cases}, \; n \in Z.$$

Write down similar statements for $\cos\theta = \cos\alpha$ and $\tan\theta = \tan\alpha$. Apply them to the general solutions of:

a $\sin\theta = \frac{1}{2}$ **b** $\cos\theta = \frac{1}{\sqrt{2}}$ **c** $\tan\theta = -1$.

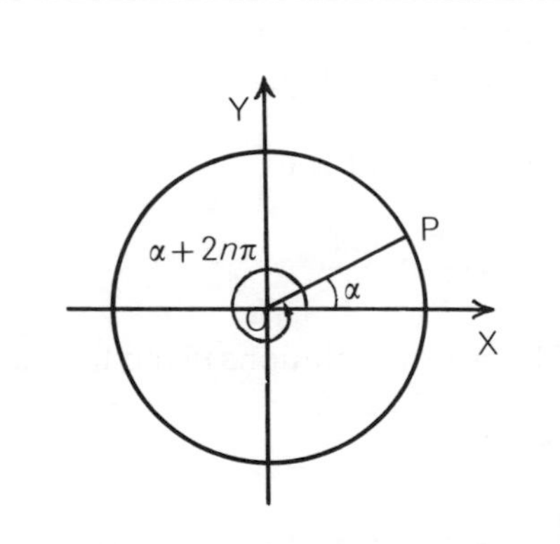

CHECK-UP ON **TRIGONOMETRY—REMINDERS AND RADIANS**

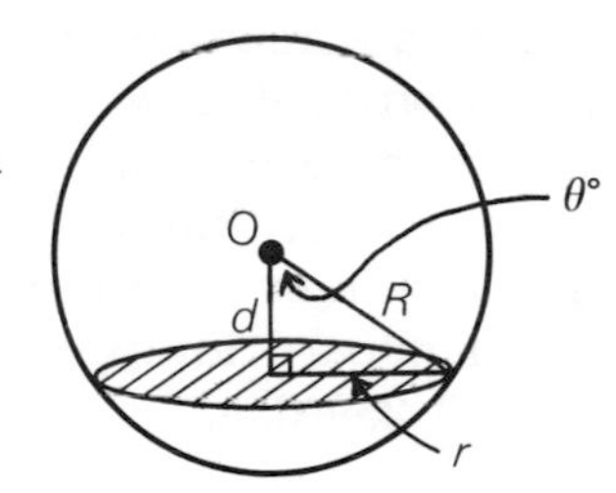

1 O is the centre of the sphere, radius R. The shaded circular cross-section has radius r, and is distance d from O. Express:
a r in terms of R and d
b $\sin\theta°$, $\cos\theta°$ and $\tan\theta°$ in terms of R and d.

2

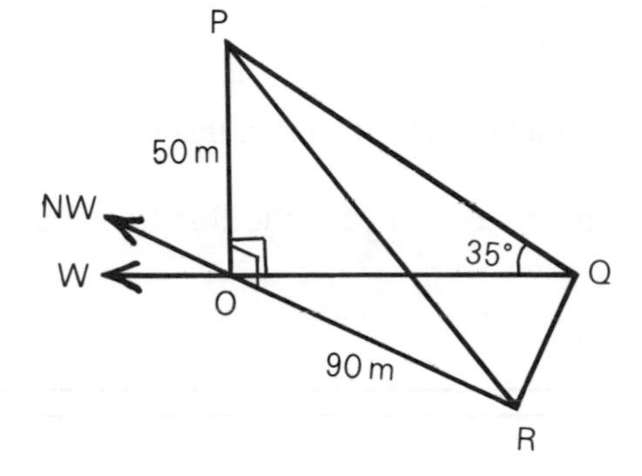

A tower OP stands on horizontal ground. Calculate (to the nearest whole unit):
a the length of its shadow OQ, when the sun is due west
b the angle of elevation of the sun ($\angle$ORP) when it shines from the north west
c the distance QR.

3 Express in radian measure:
a 60° **b** 120° **c** 150° **d** 270° **e** 315°.

4 A scanner rotates at 15 revolutions per minute. Find its speed of rotation in:
a degrees per second **b** radians per second.

5 Sketch each of these pairs of graphs on the same diagram, for $0 \leqslant x \leqslant 2\pi$:
a $y = \sin x$ and $y = 2\sin x$ **b** $y = 4\cos x$ and $y = 4\cos 2x$.

6 The distance moved by the piston from one end of the cylinder is modelled by the equation $x(t) = \frac{1}{4}\sin 3t$, in metres, with t in seconds.

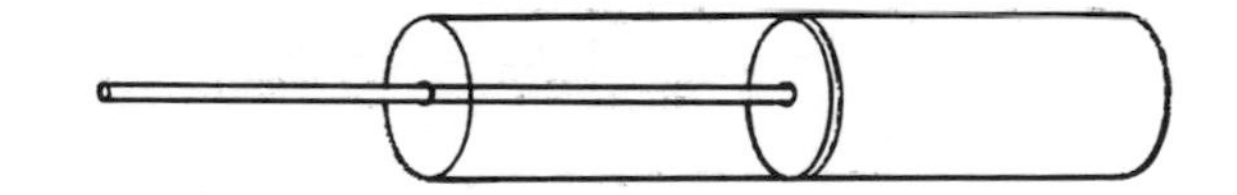

a Calculate the distance over which the piston moves in the cylinder.
b Find the times at which $x(t) = \frac{1}{8}$.

7 Solve these equations, for $0 \leqslant \theta \leqslant 2\pi$:
a $\sin\theta = -1$ **b** $\cos\theta = 0$ **c** $\sin\theta = -\frac{1}{2}$ **d** $2\cos\theta + \sqrt{3} = 0$.

SEQUENCES

INTRODUCTION

Everyone in the class should think of a number from 0 to 9, and write it down. Each person in turn should call out his or her number, and have it written in order on the blackboard; for example, 1, 5, 7, 0, 2, 9, 3, ...

You'll now have a *sequence* of numbers, 'at random'.
The first *term* is 1, second is 5, third is 7, and so on, but there is no pattern that can give you the nth term. Sequences like this are very important in statistical work, but in this chapter you will deal only with sequences that do have patterns.
u_1 will be used for the first term of a sequence, u_2 for the second, u_3 for the third, . . . , u_r for the rth, . . . , u_n for the nth.

Exercise 1

1 The first three terms of a sequence are 1, 2, 4.
What is the next term? You need some clue to the pattern.
Write down the fourth term of the above sequence, using these rules:
a The sequence is 1 followed by all the even numbers.
b The sequence is the powers of 2 i.e. $2^0, 2^1, 2^2, \ldots$.
c The sequence is the factors of 20 (in increasing order).
d The sequence is the factors of 44 (in increasing order).

2 Two of the sequences in **1** are finite (with a definite number of terms) and two are infinite.
a Identify these.
b Write out the finite sequences completely.

3 Write down u_1, u_2, u_3, u_4, u_5 for the sequences given by these formulae:
a $u_n = 2n$ **b** $u_n = 3n+1$ **c** $u_n = n^2+1$
d $u_n = \dfrac{1}{3^n}$ **e** $u_n = 1-n^3$ **f** $u_n = 2-(0{\cdot}5)^n$.

4 A sequence is defined by the formula $u_n = \sin\dfrac{n\pi}{6}$, $n \geqslant 1$. Find the first six terms in simplest form.

5 A sequence is given by the formula $u_n = an+b$ (a, b constants).
a If $u_1 = 5$ and $u_2 = 8$, find a, b.
b Write down the values of u_3, u_4, u_5.

6 A sequence is given by $u_n = \dfrac{n-1}{n}$, $n \geqslant 1$.
a Write down u_1, u_2, u_3, u_4, u_5.
b Write down $u_{10}, u_{100}, u_{1000}, u_{10\,000}, u_{100\,000}, u_{1\,000\,000}$.
c Write down the value which u_n approaches as n gets larger, i.e. as $n \to \infty$.

d $u_n = 1 - \frac{1}{n}$. Why does this make it clearer that $u_n \to 1$ as $n \to \infty$ ('u_n tends to 1 as n tends to "infinity"')?

7 Old Ben is panning for gold. After panning n times he finds he has extracted $0{\cdot}65(1-(0{\cdot}25)^n)$ grams of gold.

a How much gold did he extract on his first panning (to 2 decimal places)?

b How much on his second panning (to 2 decimal places)?

c What is the maximum amount he could aim to extract?

8 Each of the sequences defined by these formulae is *convergent*, i.e. has a limit value as $n \to \infty$:

a $u_n = \left(\frac{3}{4}\right)^n$ **b** $u_n = 2 - \frac{1}{n^2}$ **c** $u_n = 3 + \frac{1}{n^3}$.

State the limit value in each case.

9 $OP_0 = 1$ unit. A sequence of points $P_0, P_1, P_2, P_3, \ldots$ is constructed as shown $(0 < x < 90)$.

a Find the lengths of OP_1, OP_2, OP_3.

b State the length of OP_n.

c What happens to OP_n as $n \to \infty$.

d Find the lengths of P_0P_1, P_1P_2, P_2P_3.

e State the length of $P_{n-1}P_n$ and its limit as $n \to \infty$.

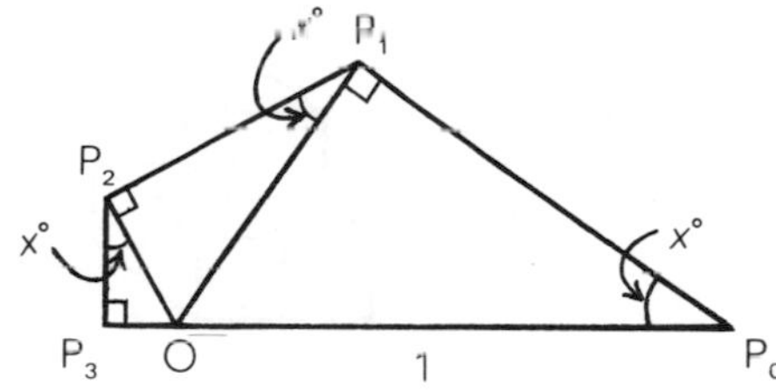

Sequences with constant differences

1 Look at this table:

n	1	2	3	4	5	6	...
$u_n = an+b$	$a+b$	$2a+b$	$3a+b$	$4a+b$	$5a+b$	$6a+b$	...
1st differences (u_2-u_1, u_3-u_2, etc)	a	a	a	a	a	a	

A sequence with formula $u_n = an+b$ (a, b constants) has constant first differences (a), and $b = u_1 - a$.
Check that each sequence has constant first differences, and then find a formula for u_n:

a $1, 3, 5, 7, 9, \ldots$ **b** $-2, 3, 8, 13, 18, \ldots$ **c** $4, -2, -8, -14, -20, \ldots$.

2 a What about this sequence: 1, 3, 6, 10, 15, 21, . . .?

n	1		2		3		4		5		6	. . .
u_n	1		3		6		10		15		21	. . .
1st differences		2		3		4		5		6		. . .
2nd differences			1		1		1		1			. . .

In this case the second differences have the constant value 1. There is only one sequence for which this is true, with first six terms 1, 3, 6, 10, 15, 21.
What kind of formula must this u_n have? Any idea? If not, read on.

b Show that a sequence with formula of the form $u_n = an^2 + bn + c$ (a, b, c constants) has constant second differences.
Find the formula for the sequence in **a** and for these sequences which have constant second differences:
(i) $-3, 1, 7, 15, 25, \ldots$ (ii) $1, \frac{11}{2}, 13, \frac{47}{2}, 37, \ldots.$

3 a Investigate sequences with constant third differences.

b Investigate the number of identical cubes in each pile below and the sequence generated by extending these piles.

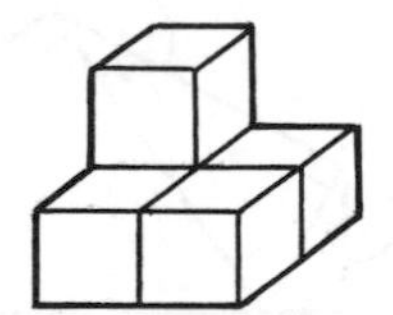

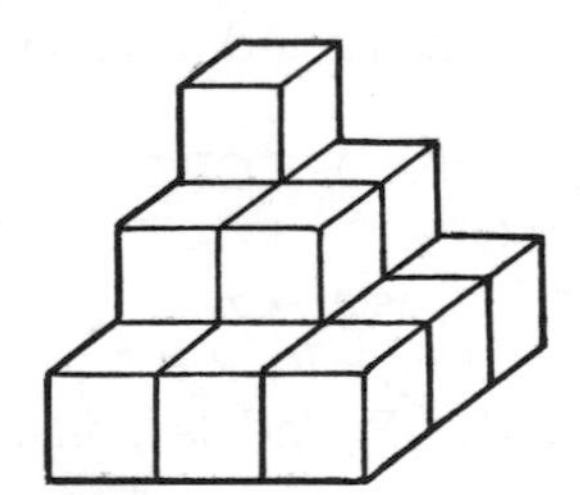

RECURRENCE RELATIONS

A pool of stagnant water has 100 chilodonella in it on Day 1. The number of chilodonella grows at the rate of 25% per day.

a How long will it be until there are at least 200 chilodonella in the water?
b How many will there be on Day 8 (to nearest whole number), Day n?
This table shows the growth:

Day n	1	2	3	4	. . .
Population P_n	100	$100(1+\frac{1}{4})$	$100(1+\frac{1}{4})^2$	$100(1+\frac{1}{4})^3$	. . .

If P_n is the population at the end of Day n, then P_{n+1} (the population at the end of Day $n+1$) is such that $P_{n+1} = P_n + \frac{25}{100}P_n = (1+\frac{1}{4})P_n = \frac{5}{4}P_n, n \geqslant 1$.
The sequence $P_1, P_2, P_3, P_4, P_5, \ldots$ is given by the *recurrence relation* $P_{n+1} = \frac{5}{4}P_n (n \geqslant 1)$. Once P_1 is given, then $P_2, P_3, P_4, \ldots$ can be calculated in turn by a **recurring** use of the relation.
Since $P_1 = 100$, $P_n = 100(\frac{5}{4})^{n-1}, n \geqslant 1$.
If $n = 1$, $P_2 = \frac{5}{4}P_1 = 100(\frac{5}{4})$.
If $n = 2$, $P_3 = \frac{5}{4}P_2 = 100(\frac{5}{4})^2$.
If $n = 3$, $P_4 = \frac{5}{4}P_3 = 100(\frac{5}{4})^3$.
. . . .
In particular, $P_8 = 100(\frac{5}{4})^7 \doteqdot 477$.
Also, $P_3 \doteqdot 195$ and $P_4 \doteqdot 244$, so the population will reach 200 on Day 4.

Exercise 2

1 Using each recurrence relation, write down: (i) u_1, u_2, u_3, u_4 (ii) a formula for u_n.
a $u_{n+1} = 2u_n$, and $u_1 = 1$ **b** $u_{n+1} = u_n + 5$, and $u_1 = 3$
c $u_{n+1} = \frac{1}{2}u_n$, and $u_1 = 6$ **d** $u_{n+1} = u_n - 4$, and $u_1 = 5$.

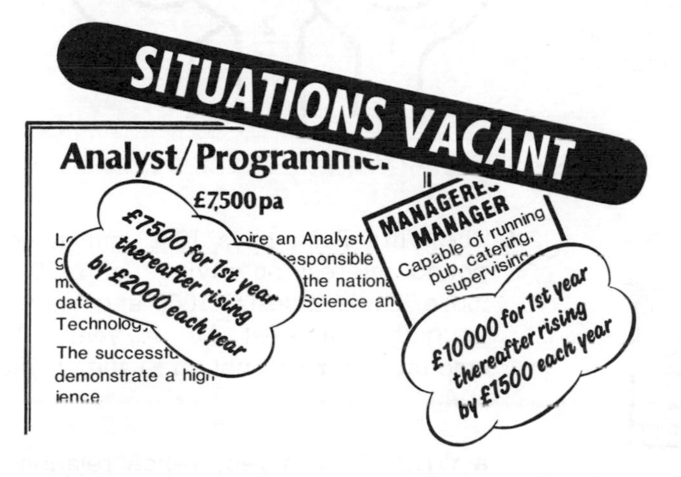

2 Wayne gets the Analyst's job and Linda the Manageress' job on the same day. W_n (in £) is Wayne's salary in year n and L_n (in £) is Linda's in the same year.

a (i) Explain why $W_{n+1} = W_n + 2000$, $n \geqslant 1$, is the recurrence relation for the sequence $\{W_n\} = \{W_1, W_2, W_3, \ldots\}$.
(ii) Write down the recurrence relation for $\{L_n\}$.
b Use the recurrence relation to write down W_2, W_3, W_4, and explain why $W_n = 7500 + (n-1)2000, n \geqslant 1$.
c In the same way, write down L_2, L_3, L_4, and find a formula for L_n.
d After how many years will their salaries be the same?

3 a Find l_2, l_3, l_4 in terms of θ.
b Write a recurrence relation for the sequence of lengths $\{l_n\}$.
c Find a formula for l_n.
d What happens to l_n as $n \to \infty$?

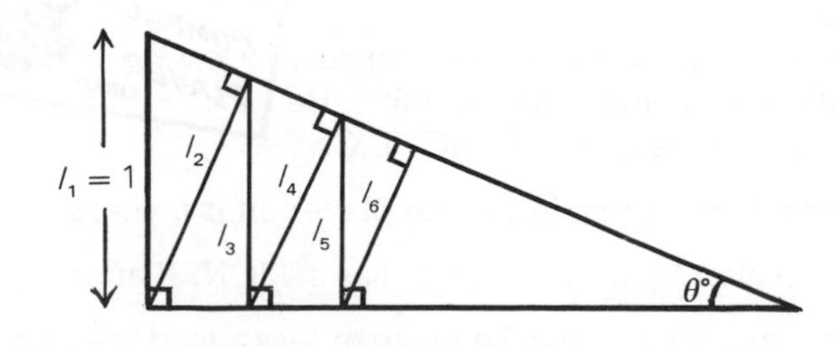

4 A child's toy consists of equilateral and circular shapes that just fit inside each other. The radius of the smallest circular shape is 1 cm.

a Show that the sequence $\{r_n\}$ of radii of the circles has the recurrence relation $r_{n+1} = 2r_n, n \geqslant 1$.
b Find a formula for r_n.
c Show that, if x_n is the length of a side of the nth triangle, then $x_n = 2r_n\sqrt{3}$, and find a formula for x_n.

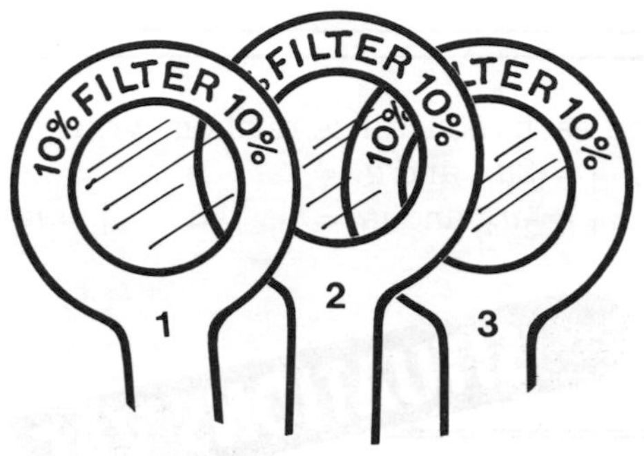

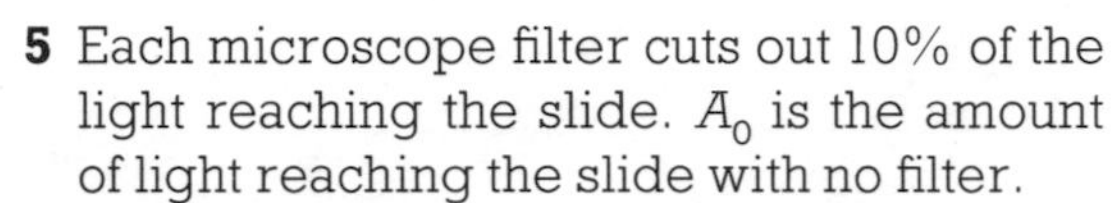

5 Each microscope filter cuts out 10% of the light reaching the slide. A_0 is the amount of light reaching the slide with no filter.
a Obtain a recurrence relation connecting A_n, A_{n+1}, where A_n is the amount of light getting through n filters.
b Find a formula for A_n in terms of A_0.
c How many filters would be needed to cut out at least 50% of the light?

6 The Ultra Bank pays 10% annual compound interest on savings of £1000 or more. Sadia invests £1000 and leaves it to gather interest for n years. The amount in her account (in £) after n years is A_n.

a Write down a recurrence relation involving A_n and A_{n+1}.

b Find a formula for A_n.

c After how many years would her account contain more than £2000 for the first time?

7 James has only £190 at the moment. He would like the CD player but he won't borrow money.
The price of the player goes up annually in line with the inflation rate of 4·5%. He decides to invest his money in MONEX.

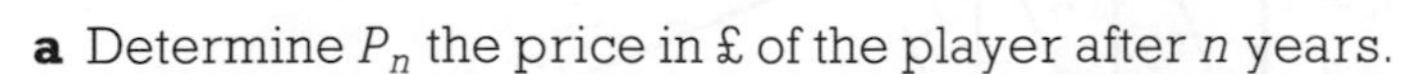

a Determine P_n the price in £ of the player after n years.

b Determine also M_n the amount he has in MONEX after n years.

c After how many years can he more or less afford the player?

8 Separate populations of 200 caterpillars are sprayed weekly with these pesticides.

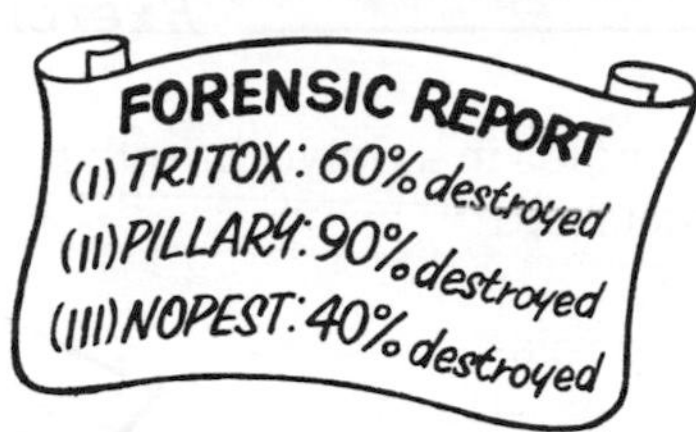

a Find formulae for the numbers left after n weeks.

b Calculate in each case the number of weeks needed before the caterpillar population is reduced to 10 or less.

9

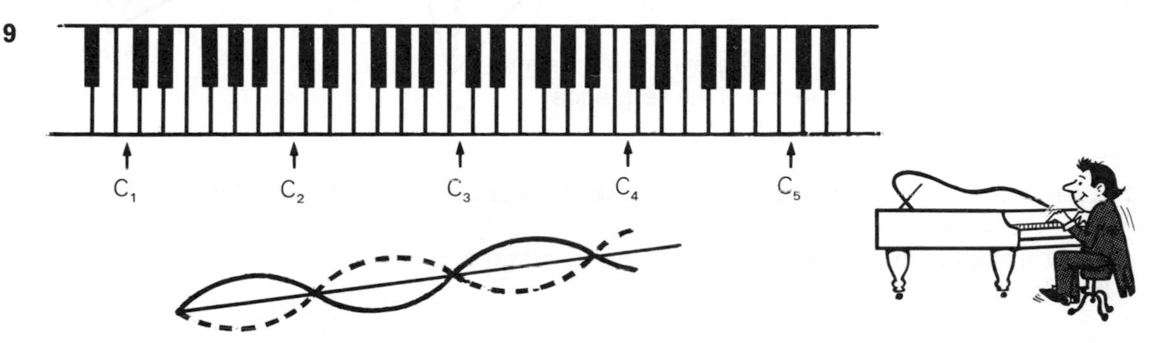

Every time Rosinzki strikes a key the piano strings vibrate at certain frequencies. For example, the frequency of C_2 is double that of C_1, and the frequency of C_3 is double that of C_2, and so on.

a Write down a recurrence relation for C_n.

b Low C (C_1) has frequency 32·7 cycles per second. Find the frequency of the other 7 C-notes on the piano.

c How can you find the frequency of C_8 directly from that of C_1?

Look back at question **8**, of Exercise 2, on the caterpillars. The problem is really more complicated than this, because each crop is invaded regularly by 100 new caterpillars each week.
Now find formulae for the numbers left after n weeks of application of the pesticides. (Remember to build up the recurrence relation first.)

LINEAR RECURRENCE RELATION $u_{n+1} = mu_n + c$ (m, c constants)

You have already met this recurrence relation (in different notation) in various places in the chapter:

Exercise 3

1 S_n is the sum of the angles in a polygon with n vertices.

a How can S_5 be calculated from S_4? S_6 from S_5?

b Write a recurrence relation, stating the initial value (for S_3).

c Find a formula for S_n in terms of n.

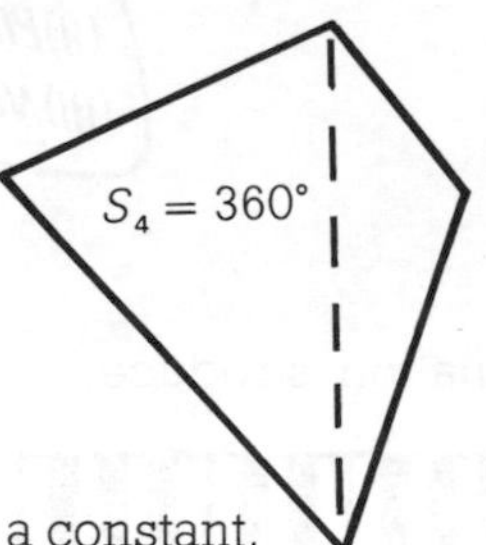

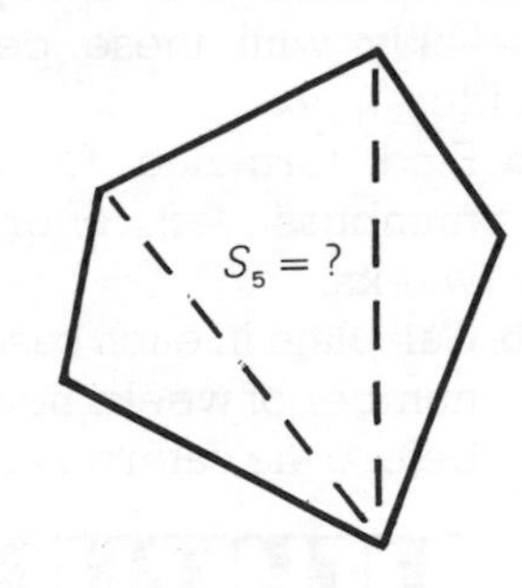

2 $u_{r+1} = ku_r (r \geqslant 0)$ and $u_0 = 5$, where k is a constant.

a Write down the first 5 terms of the sequence.

b Construct the formula for u_n.

3 A sequence $\{u_n\}$, $n \geqslant 0$, satisfies a recurrence relation $u_{n+1} = mu_n + c$ (m, c constants), and $u_0 = 1$, $u_1 = 3$, $u_2 = 7$.

a Form two equations, and solve them to find m and c.

b Calculate u_5.

4

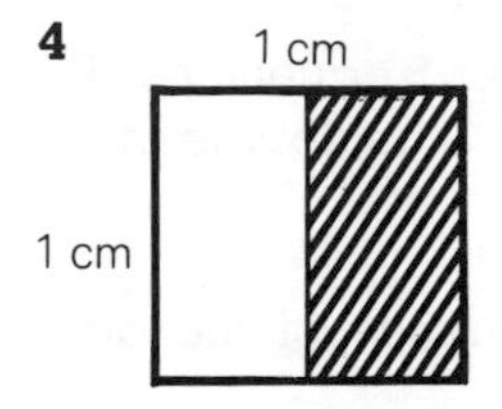

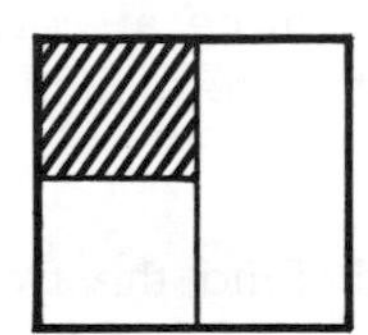

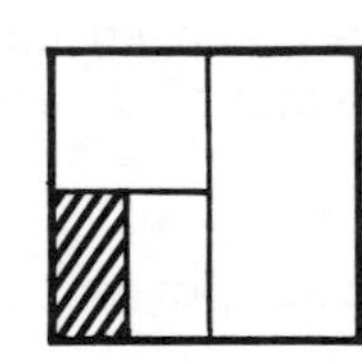

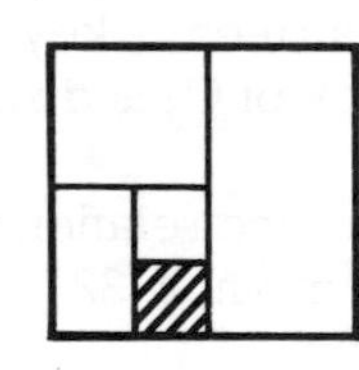

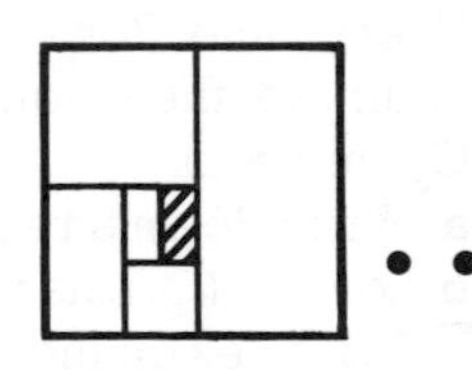

a_n is the area in cm² that is shaded in the nth diagram.

a Write down the relation between: (i) a_2 and a_1 (ii) a_3 and a_2 (iii) a_4 and a_3.

b Write down a recurrence relation for $\{a_n\}$, $n \geqslant 0$, stating the value of a_1.

c Find a formula for a_n in terms of n.

d What happens to a_n as $n \to \infty$?

5 In question **4**, t_n is the area in cm² that is *unshaded* in the nth diagram.

a From **4**, $a_n = \frac{1}{2^n}$. What is the value of t_n?

b What happens to t_n as $n \to \infty$?

c Show that $\{t_n\}$ has the recurrence relation $t_{n+1} = \frac{1}{2}t_n + \frac{1}{2}$, $n \geqslant 0$.

6 *The case $m = 1$—**arithmetic sequences***

$\{u_n\}$ has a recurrence relation $u_{n+1} = u_n + d$, $n \geqslant 1$ (d constant). Often u_1 is written as a, so the sequence is $a, a+d, a+2d, a+3d, \ldots$.

Since $d = u_{n+1} - u_n$, $n \geqslant 1$, d is called the *common difference*.

a Show that $u_n = a + (n-1)d$, $n \geqslant 1$.

b Find the first term and the common difference for these arithmetic sequences:
(i) 3rd term is 5, 7th term is 17 (ii) 2nd term is 6, 6th term is -10.

c S_n is the sum of the first n terms of the arithmetic sequence $\{a + (n-1)d\}$, $n \geqslant 1$. Find a recurrence relation for S_n (in terms of a, n, d).

7 *The case $c = 0$—***geometric sequences**
Here $\{u_n\}$ has a recurrence relation $u_{n+1} = ru_n$, $n \geqslant 1$ (r constant). Again u_1 is written as a, so the sequence is $a, ar, ar^2, ar^3, \ldots$.

Since $r = \dfrac{u_{n+1}}{u_n}$, $n \geqslant 1$, r is called the *common ratio*.

a Show that $u_n = ar^{n-1}$, $n \geqslant 1$.
b Find the first term and the common ratio for these geometric sequences:
(i) 3rd term is 12, 6th term is 96 (ii) 2nd term is -3, 5th term is 81.
c S_n is the sum of the first n terms of the geometric sequence $\{ar^{n-1}\}$, $n \geqslant 1$.

$$S_n = a + ar + ar^2 + \ldots + ar^{n-2} + ar^{n-1}$$
$$rS_n = \quad ar + ar^2 + \ldots + ar^{n-2} + ar^{n-1} + ar^n.$$

By subtraction, show that $S_n = \dfrac{a(1-r^n)}{1-r}$, $r \neq 1$.

What is the value of S_n if $r = 1$?
d If $-1 < r < 1$, what happens to S_n as $n \rightarrow \infty$?

A mushroom bed starts with A mushrooms.
Each day 70% of the mushrooms are picked.
Each day 300 new mushrooms become ready for picking.
a Show that the number N_n left at the end of Day n is

$$N_n = A\left(\frac{3}{10}\right)^n + 300\left[\left(\frac{3}{10}\right)^{n-1} + \left(\frac{3}{10}\right)^{n-2} + \ldots + \left(\frac{3}{10}\right) + 1\right].$$

b What happens to N_n as $n \rightarrow \infty$?

Hint Question **7** of Exercise 3 should be useful.

RECURRENCE RELATIONS WITH 3 OR MORE TERMS

A male wasp has this ancestral tree:

	Parents	Grandparents	Great Grandparents	Great Great Grandparents	Great Great Great Grandparents
Numbers:	1	2	3	5	8

Including the male wasp itself this gives the sequence $1, 1, 2, 3, 5, 8, \ldots$.
Can you see the pattern?

This is one of the most famous sequences of all—the *Fibonacci* sequence. Did you notice that from 2 onwards, each term is the sum of the two preceding terms. If the sequence is $g_0, g_1, g_2, g_3, \ldots$, then $g_2 = g_1 + g_0$, $g_3 = g_2 + g_1, \ldots, g_{n+1} = g_n + g_{n-1}$, $n \geqslant 1$.
When you know g_0 and g_1, the sequence can start to grow.

Exercise 4

1 Write down the first 10 terms of the sequences of Fibonacci type given by:
a $g_0 = 1, g_1 = 3$ **b** $g_0 = -4, g_1 = 5$.

2 $\{a_n\}, n \geqslant 1$, is the sequence $1, 3, 7, 17, 41, 99, \ldots$, and
$\{b_n\}, n \geqslant 1$, the sequence $1, 2, 5, 12, 29, 70, \ldots$.
There are various recurrence relations connecting the sequences. Try to spot at least one of them before you read on for help.
a Check that $a_{n+2} = 2a_{n+1} + a_n, n \geqslant 1$, and use this to extend the sequence for another 6 terms.
b Find a similar relation for $\{b_n\}$ and extend it by 6 terms.
c Check that the relations $a_{n+1} = a_n + 2b_n, n \geqslant 1$ and $b_{n+1} = a_n + b_n, n \geqslant 1$, also hold. There are others!

3 **a** Express each of $1+\sqrt{2}$, $(1+\sqrt{2})^2$, $(1+\sqrt{2})^3$, $(1+\sqrt{2})^4$ in the form $a + b\sqrt{2}$ with a, b integers.
b Writing $(1+\sqrt{2})^n = a_n + b_n\sqrt{2}, n \geqslant 1$, show that

$$a_{n+1} + b_{n+1}\sqrt{2} = (1+\sqrt{2})(a_n + b_n\sqrt{2}),\ n \geqslant 1.$$

Hint $(1+\sqrt{2})^{n+1} = (1+\sqrt{2})(1+\sqrt{2})^n$.

c Show that $a_{n+1} = a_n + 2b_n$, $b_{n+1} = a_n + b_n, n \geqslant 1$.
d Check that the sequences $\{a_n\}, \{b_n\}$ are those of question **2**.

4 **a** By considering $(1+\sqrt{3})^n = c_n + d_n\sqrt{3}, n \geqslant 1$, create sequences $\{c_n\}, \{d_n\}$, giving the first 6 terms of each.
b State the recurrence relations of the type given in question **3c** connecting them.

CHECK-UP ON **SEQUENCES**

1 Write down u_1, u_2, u_3, u_4 for the sequences given by:
a $u_n = 3 - 5n$ **b** $u_n = 7 - 2n^2$ **c** $u_n = 1 + (0{\cdot}25)^n$.
In **c**, what happens to u_n as $n \to \infty$?

2 The sequence 0, 9, 22, 39, 60, 85, . . . has constant second differences. Find a formula for the sequence.

3 West Bank pays 12% compound interest per year for their 'golden' accounts. Theo's account starts with £2000.
a If the amount in his account after n years is £A_n, write down a recurrence relation involving A_n and A_{n+1}.
b Obtain a formula for A_n.
c After how many years will his account for the first time exceed £3000?

4 The Blue Loch is badly polluted; in fact, it contains 21 tonnes of pollutant. A process of reducing this to 1 tonne is set in motion. It reduces the pollutant 40% per week.

a If P_n(tonnes) is the amount present at the end of n weeks, write down a recurrence relation for $\{P_n\}$.

b Obtain a formula for P_n.

c How many weeks are needed to get the pollution below 1 tonne?

5 A sequence $\{u_n\}$, $n \geqslant 0$, satisfies a recurrence relation $u_{n+1} = mu_n + c$ (m, c constants), and $u_0 = -1$, $u_1 = 5$, $u_2 = -7$.

a Form two equations, and solve them for m and c.

b Calculate u_4.

6 For an arithmetic sequence, $u_n = a + (n-1)d$, $n \geqslant 1$.

a Find the first term a and the common difference d, given that the third term is 13 and the seventh term is 33.

b Write down the first seven terms of the sequence.

7 For a geometric sequence, $u_n = ar^{n-1}$, $n \geqslant 1$.

a Find the two possible values of a and of r, given that the second term is 6 and the fourth term is 54.

b Write down the first four terms of each sequence.

INTRODUCTION

Throughout your study of mathematics you have been using *functions*, which were defined by formulae, graphs or tables.

a The *formula* for the volume of a box (V cm^3) with square base of side x cm and depth $2x$ cm is $V(x) = 2x^3$.

b The temperature at Sunnycove from 2 pm to 10 pm on 25th May, 1988, is given by this *graph*.

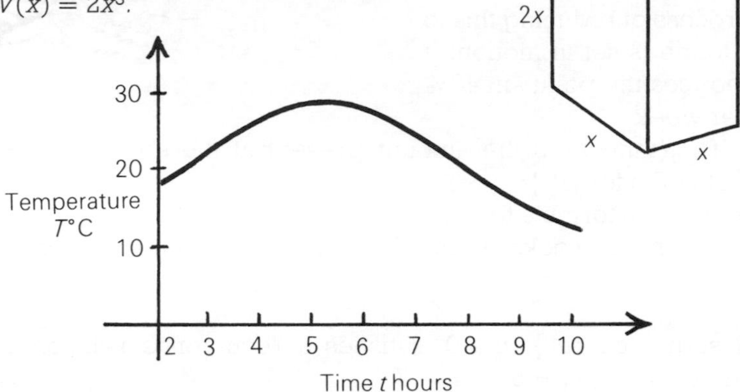

c The stopping distances in feet for a car travelling at u mph are given by the table:

Speed u (mph)	0	20	40	60	80	100
Stopping distance D (ft)	0	40	120	240	400	600

These are all different ways of describing a function.

Exercise 1

1 a Check that the formula for the surface area A(cm^2) of the box in **a** above is $A(x) = 10x^2$.

b Calculate: (i) $A(1)$ (ii) $A(3)$ (iii) $A(6)$ (iv) $A(\frac{1}{2})$.

c Comment on $A(-2)$.

2 Think of a number between -10 and 10. Square it. Multiply the square by 2. Add 5.

a Calculate the answer if you think of: (i) 1 (ii) -2 (iii) $\frac{1}{4}$.

b What is the answer if your number is x?

c Which is the correct formula?

(i) $N(x) = (2x)^2+5$ (ii) $N(x) = 2x^2+5$ (iii) $N(x) = 2(x+5)^2$.

3 A rectangle has constant area and breadth x cm, where $10 \leqslant x \leqslant 40$.
The graph of its perimeter P(cm) against breadth is shown.

a Estimate: (i) $P(10)$ (ii) $P(20)$ (iii) $P(40)$.

b What breadth gives the minimum value of P, and what is this perimeter?

c Using **b**, find the constant area of the rectangle.

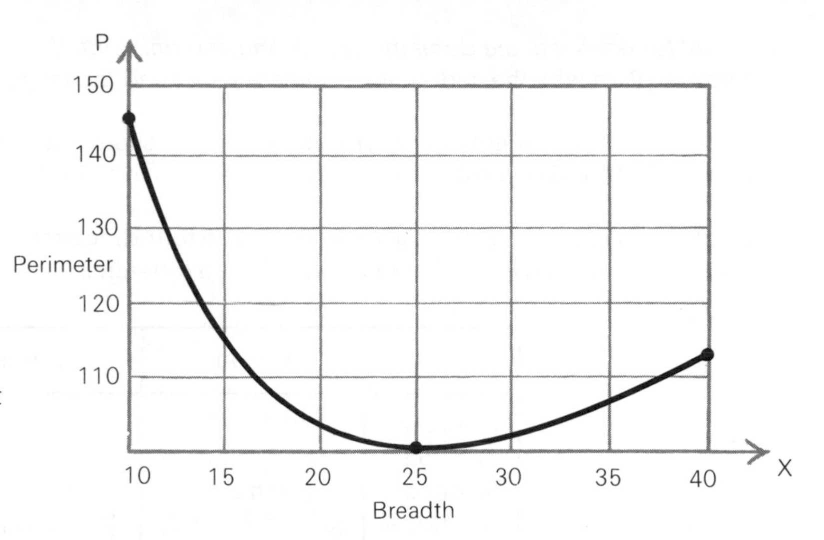

4 This table lists the side lengths of squares with given area.

Area A (cm²)	0	1	4	9	16	25
Side s (cm)	0	1	2	3	4	5

a Graph s against A.

b Why doesn't the table contain negative values of A?

5 The formula for a function is $f(x) = x^2 + x + c$.

a Calculate $f(2)$.

b Given that $f(2) = 10$, calculate c.

LOOKING AT A FUNCTION IN 'DEPTH'

A square has area x cm². The length of its side is given by the formula $f(x) = x^{1/2} = \sqrt{x}$. What are the 'nuts and bolts' of f? What must you decide before you have a full picture of f?

First decision: What is the set of replacements for x (the *domain*)? (The set of whole numbers (W), or integers (Z), or real numbers (R), or what?)

Second decision: What is the set to which the values of f (i.e. $f(x)$) belong (the *codomain*)? (W, or Z, or R, or what?)

Last decision: As x 'runs through' the whole domain, what set is formed by all the values $f(x)$ (the *range*)?

Does a picture help?

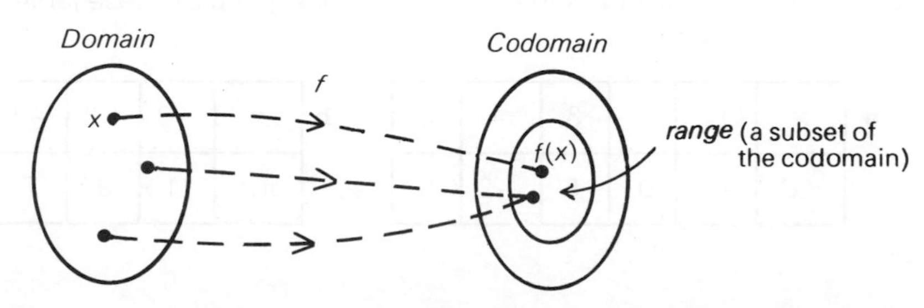

Note that for each x in the domain there is one and only one f(x).
Can you explain why the codomain is sometimes called the target set?

Important note From now on your functions will usually have R, the real numbers, as their codomain. This makes life a little simpler—you can concentrate on domain and range only.

Example 1 Back to the square above—here are four different functions with the same formula $f(x) = \sqrt{x}$. Check the ranges listed for the chosen domains:

	domain	range
function 1	$\{0, 1, 4, 9\}$	$\{0, 1, 2, 3\}$
function 2	$\{1, 25, 100\}$	$\{1, 5, 10\}$
function 3	$\{n^2 : n \in W\}$	$\{n : n \in W\}$
function 4	$\{a^2 : a \in R, a \geqslant 0\}$	$\{a : a \in R, a \geqslant 0\}$

For any function given by $f(x) = \sqrt{x}$ no negative number can be in the domain, e.g. -1, since $\sqrt{(-1)}$ is not in the codomain R.

Example 2 What is the largest possible domain for function g with formula $g(x) = \sqrt{(x-2)}$? For $g(x)$ to be in R you need $x - 2 \geqslant 0$, so the domain is $\{x \in R : x \geqslant 2\}$.

Exercise 2

1 Find the ranges of the functions with these formulae and domains:

a $f(x) = x + 5$, domain $= \{0, 1, 2\}$

b $f(x) = 2x - 1$, domain $= \{-1, 0, 1\}$

c $f(x) = \dfrac{1}{x}$, domain $= \{1, 2, 3\}$

d $f(x) = (x + 5)^2$, domain $= \{-5, 0, 5\}$

e $f(x) = x^2$, domain $= \{0, 1, 2\}$

f $f(x) = x^2$, domain $= \{-2, -1, 0, 1, 2\}$

g $f(x) = x^2$, domain $= R$

h $f(x) = \sin x$, domain $= \left\{0, \dfrac{\pi}{4}, \dfrac{\pi}{2}\right\}$

i $f(x) = \sin x$, domain $= \left\{-\dfrac{\pi}{2}, -\dfrac{\pi}{6}, 0, \dfrac{\pi}{6}, \dfrac{\pi}{2}\right\}$

j $f(x) = 2 \sin x$, domain $= R$.

2 *List* the domains and ranges of the functions given by these tables:

a

x	1	2	3	4	5
$f(x)$	-2	0	5	-2	6

b

x	-3	-2	-1	0	1	2	3
$g(x)$	11	9	7	5	-5	1	7

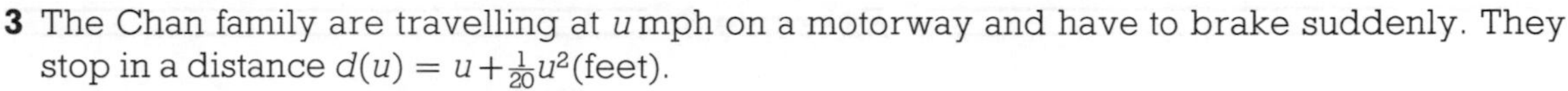

3 The Chan family are travelling at u mph on a motorway and have to brake suddenly. They stop in a distance $d(u) = u + \frac{1}{20}u^2$ (feet).

a What is the domain of the function d? (According to the law!)

b Find the corresponding range.

4 These formulae define functions. Decide in each case on the largest possible domain:

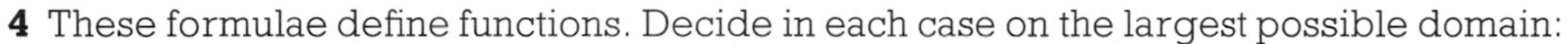

a $f(x) = \sqrt{x}$ **b** $g(x) = \sqrt{(x-1)}$ **c** $h(x) = \sqrt{(3-x)}$

d $i(x) = \sqrt{(x^2-1)}$ **e** $j(x) = \dfrac{1}{x}$ **f** $k(x) = \dfrac{2}{x^2-1}$.

NOTATIONS AND GRAPHS

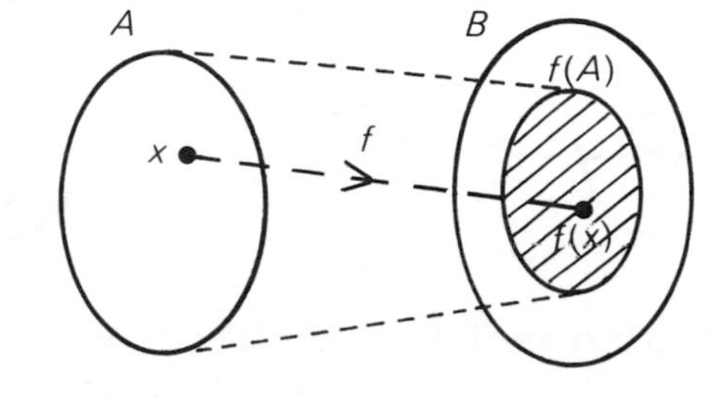

The arrow diagram tells you that for the function f:

a the domain is set A

b the codomain is set B

c the range $f(A)$ (the shaded part) is a subset of B.

It is useful to write f: $A \rightarrow B$ and say 'f *maps* A to B' or 'f is a *function* (or *mapping*) from A to B'.

Points to note

(i) x runs through *all* the members of the domain A.

(ii) For each x in A, f has one and only one (i.e. a unique) corresponding value $f(x)$ in the codomain B.

When A is a subset of R and $B = R$ the *graph* $y = f(x)$ is a good way of presenting properties of a function f.

Example f is defined for $x \geqslant 0$ by $f(x) = 2x^2 + 3$.
Sketch the graph and state the range of f.

The graph is the 'half parabola' $y = 2x^2 + 3 (x \geqslant 0)$.

The domain is on the x-axis and the codomain is the y-axis. The range is the 'projection' of the graph onto the y-axis.

The range is $\{y \in R : y \geqslant 3\}$.

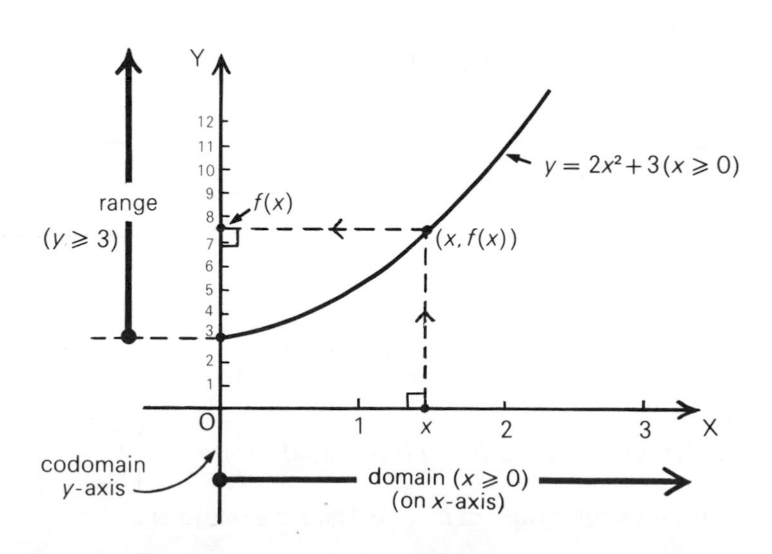

Exercise 3

1 Decide which of these are diagrams of functions. Remember that for *every* x in the domain there is one and only one $f(x)$.

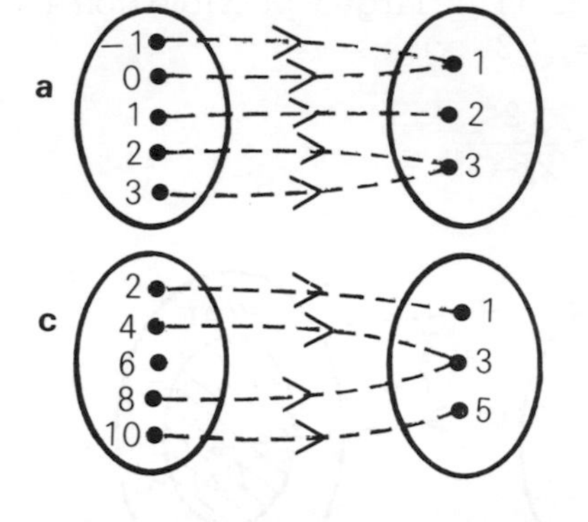

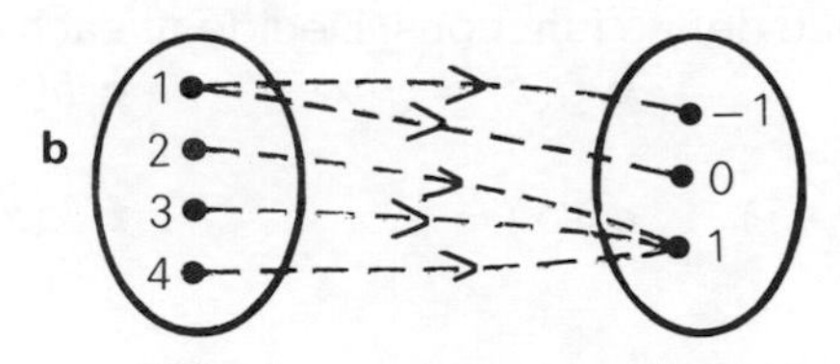

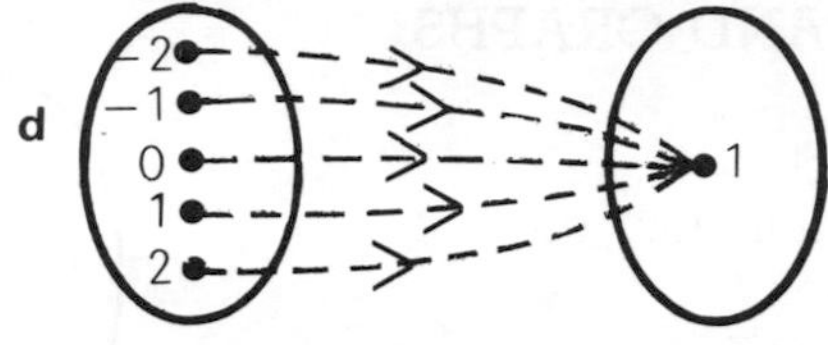

2 **a** A function g is given by the formula $g(x) = x^2 + 1$ and has domain $\{x \in R : x \geqslant 0\}$.
(i) Draw the graph of g. (ii) State its range.
b Change the domain of g to R. How will you have to adjust:
(i) your graph (ii) the range of g?

3 Each function is defined on R by the given formula.
(i) Sketch its graph. (ii) State the range.
a $f(x) = 4x$ **b** $g(x) = 2x + 3$ **c** $h(x) = x^2$ **d** $i(x) = 4 - x^2$.

4 The function h is defined on $\{x \in R : x \geqslant 3\}$ by the formula $h(x) = 3x + 1$.
a Draw the graph of h. **b** State the range of h.

5 The domain of a function g is R. Its formula is

$$g(x) = \begin{cases} 1 & \text{when } x \geqslant 1 \\ -1 & \text{when } x < 1. \end{cases}$$

a Copy its graph.
b Why does one branch begin with a full dot and the other with an open dot?
c State the range of g.

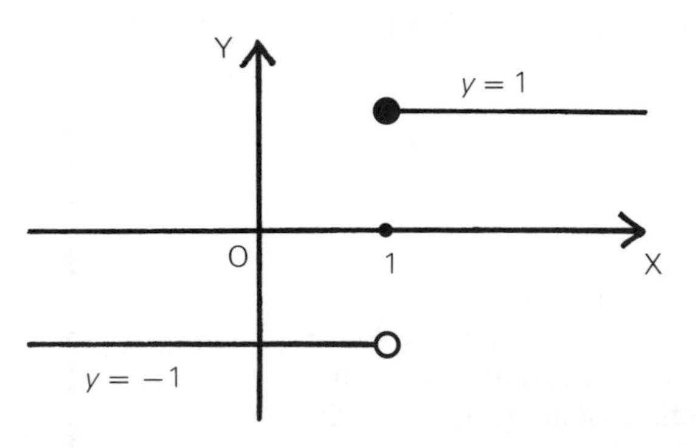

6 The domain of i is R and i is defined by $i(x) = \begin{cases} 2 \ (x > 0) \\ 1 \ (x = 0) \\ 0 \ (x < 0). \end{cases}$

a Draw the graph of i. **b** State the range of i.

7 Some of these graphs can represent functions and some cannot. Say which is which and give reasons. (See the reminder in question **1**.)

a

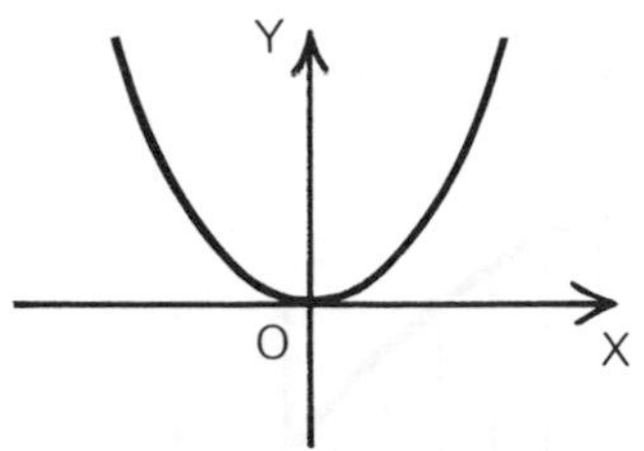

b

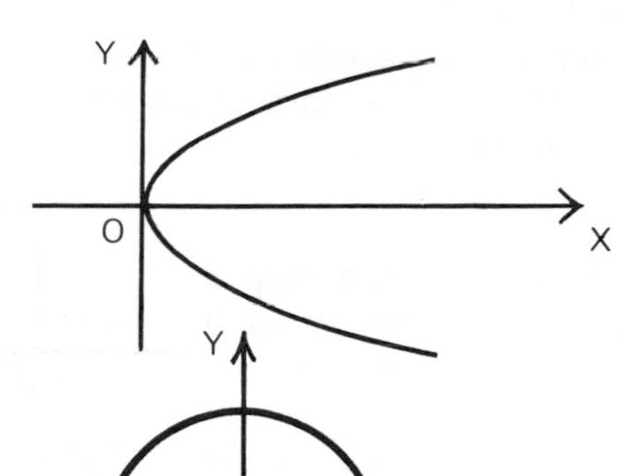

c

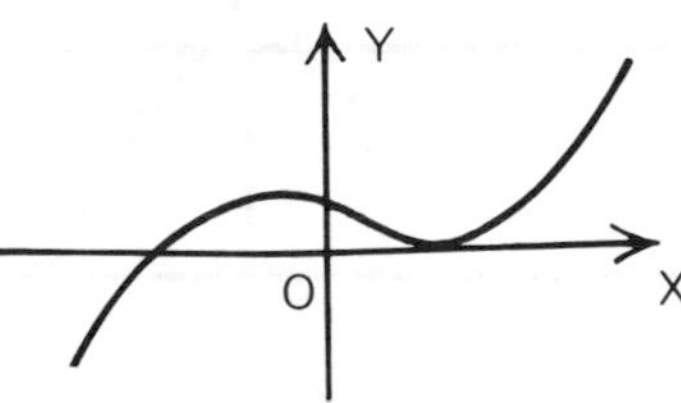

d

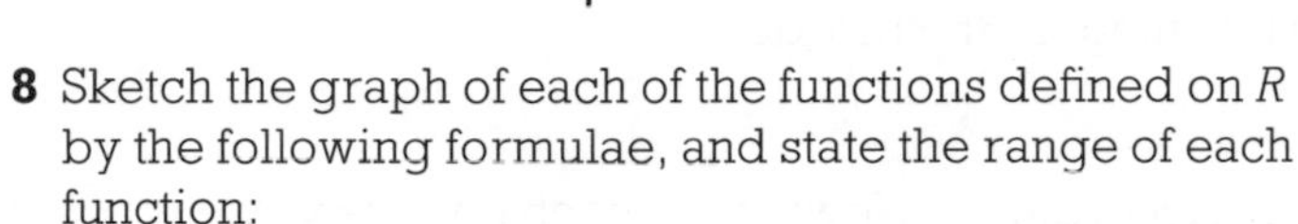

8 Sketch the graph of each of the functions defined on R by the following formulae, and state the range of each function:

a $f(x) = |x| = \begin{cases} x \; (x \geqslant 0) \\ -x \; (x < 0) \end{cases}$ $\left(\begin{array}{l} f \text{ is the absolute value function;} \\ \text{e.g. } f(-3) = |-3| = 3 = |3| = f(3). \end{array} \right)$

b $j(x) = \begin{cases} 3-2x \; (x \geqslant 0) \\ x+1 \; (x < 0) \end{cases}$ **c** $k(x) = \begin{cases} x^2+3 \; (x \geqslant 1) \\ x+3 \; (x < 1). \end{cases}$

MODELLING WITH FUNCTIONS

Exercise 4

1 Electricity charges are: 7p per unit for the first 500 units and 3p per unit thereafter. If x is the number of units used and $f(x)$ the charge in pence, then

$$f(x) = \begin{cases} 7x & (0 \leqslant x \ldots) \\ \ldots & (x > \ldots). \end{cases}$$

a Copy and complete the formula for the function f.
b Draw the graph of f and state its range.

2 The cost of advertising an item for sale in the *Daily Star* is: nil for items $<$£10; £1 for items $\geqslant$£10 and $<$£50; £5 for items $\geqslant$£50.

a If £$C(x)$ is the cost of an item selling at £x, then

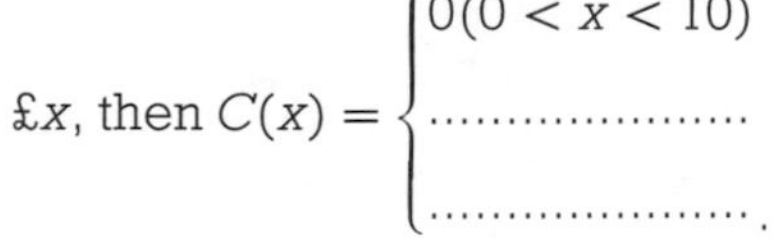

$$C(x) = \begin{cases} 0 \, (0 < x < 10) \\ \ldots\ldots\ldots\ldots \\ \ldots\ldots\ldots\ldots . \end{cases}$$

Complete the formula for $C(x)$.

b Draw the graph of C, and state the range of the function.

3 The Walkers pulled into a B&B in Wales. Mrs Jones, the owner, told them the cost per person was £10 per day, with a discount of 10% for a stay of 5 days or more. £$B(x)$ stands for the cost for x days.

a Write down a value for: (i) $B(2)$ (ii) $B(4)$ (iii) $B(6)$.

b Write down the formula for $B(x)$.

c Sketch the graph of B.

4 The profile of the support of a suspension bridge is modelled by the formula $f(x) = 8x - \frac{1}{2}x^2$ $(-3 \leqslant x \leqslant 19)$, where the x-axis is the road.

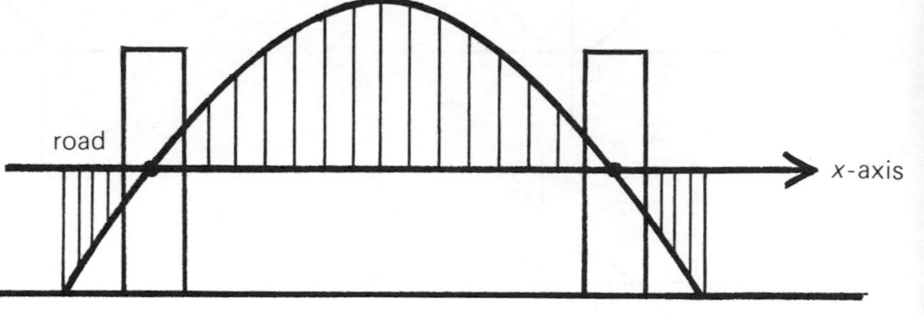

a With the help of calculus draw the graph of f.

b Find the range of f, and explain its meaning for the bridge.

5 A rocket is being launched into space. According to Newtonian mechanics its velocity V m/s after t seconds into flight is $V(t) = 10t^2$ $(t \geqslant 0)$.

a Draw the graph of V.

b What does the model say about V as $t \to \infty$?

Einstein said that nothing can travel faster than the constant speed c of light (3×10^{10} cm/s). What does this suggest about Newton's model?

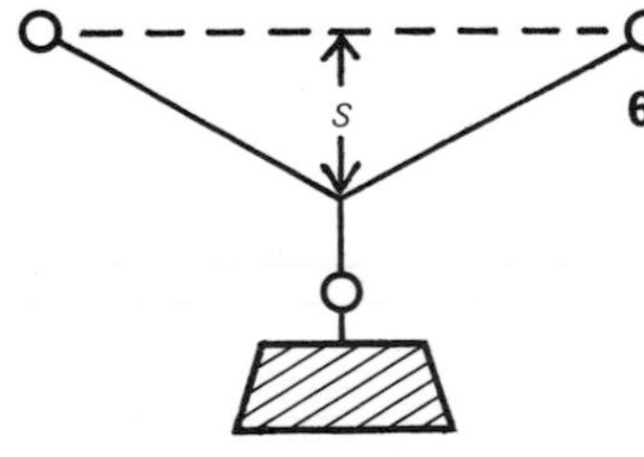

6 Attach a weight to the middle of a piece of string. The force F needed to stretch the string is inversely proportional to the sag, s, in the string.

a Write down a suitable formula for F in terms of s.

b As s gets smaller and smaller, what happens to the values of F?

c With the weight attached, the string will never be horizontal. Why not?

7 The cost, £C million, of laying one kilometre of pipe for a water main is given by the formula $C = \frac{200}{9a} + 2a$, where a is the cross-sectional area of the pipe in square metres.

a The cross-section should be neither too small nor too large. Why not?

b State a simple approximate formula for C for: (i) a large (ii) a small.

c Sketch the graph of C (using calculus to locate a stationary point).

d Find the minimum cost of laying one kilometre of pipe.

COMPOSITE FUNCTIONS

Think of a number x; square it; *then* add 3.
Your answer should be $x^2 + 3$.
By the two steps you are combining *two* functions: a 'squaring' function f where $f(x) = x^2$ and an 'add on 3' function g where $g(x) = x + 3$.

Take as domain of f, $A = \{1, 2, 3, 4\}$. Then the diagram combining the functions is:

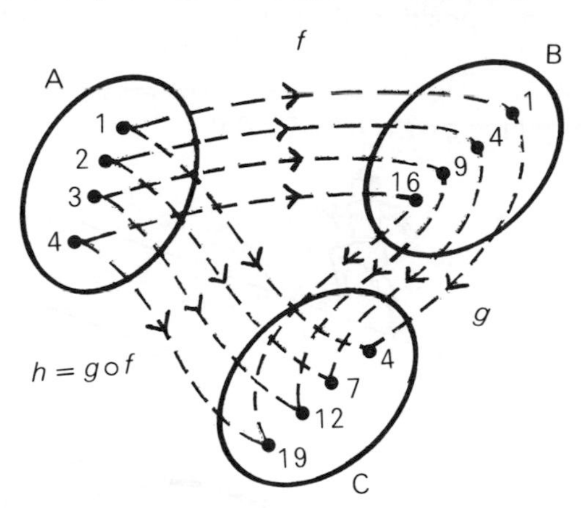

By operating with f first and then bringing in g you have produced another function h which maps A to C where $h(x) = x^2+3$. It is usual to write this as $h(x) = g(f(x))$ and to call h a *composite function* (obtained by *composition*).

Check: $h(1) = g(f(1)) = g(1) = 4$
$h(x) = g(f(x)) = g(x^2) = x^2+3$.

h is often written $g \circ f$ ('g circle f'). You may find it more helpful to read it as 'g after f' (f first, g second).

Example If $f(x) = x-2$ and $g(x) = x^3$, find formulae for $g(f(x))$ and $f(g(x))$.
$g(f(x)) = g(x-2) = (x-2)^3$
$f(g(x)) = f(x^3) = x^3-2$.

Exercise 5

1 $f(x) = x+3$, $g(x) = 2x$ and the domain of f is $\{0, 1, 2\}$.
Calculate: **a** $g(f(0))$ **b** $g(f(1))$ **c** $g(f(2))$ **d** $g(f(x))$.

2 A composite function $h = f \circ g$, where $f(x) = x^2$ and $g(x) = 2x$.
The domain of g is $\{-1, 0, 1\}$.
Calculate: **a** the range of h **b** the formula for $h(x)$.

3 $f(x) = 2x$ and $g(x) = x+1$.
a Construct $h(x) = g(f(x))$ and $k(x) = f(g(x))$.
b Any comments on your answers?

4 For each pair of functions below obtain formulae for (i) $g(f(x))$ (ii) $f(g(x))$:

a $f(x) = x+3$, $g(x) = x+2$ **b** $f(x) = 2x+3$, $g(x) = 3x+2$

c $f(x) = x^2$, $g(x) = x^3$ **d** $f(x) = x-2$, $g(x) = x^2$

e $f(x) = x^3$, $g(x) = 3x$ **f** $f(x) = x^2+3$, $g(x) = 2x+5$

g $f(x) = \sin x$, $g(x) = 2x$ **h** $f(x) = \cos x$, $g(x) = 1-x^2$.

5 For how many pairs in question **4** was $g(f(x)) = f(g(x))$? Note that, in general, $g \circ f \neq f \circ g$ (when they both exist), i.e. composition of functions is *not commutative*.
Let $f(x) = 3x+1$ and $g(x) = 5x+a$, where a is a constant.
a Form $g(f(x))$ and $f(g(x))$.
b There is a value of a for which $g \circ f = f \circ g$. Find it.

6 Sheelagh, during her evening training, runs up a gentle 1 in x slope, covering the s metres in 10 seconds at a steady speed of v m/s.
a Express v as a function of s.
b Express s as a function of x.
c Use composition of functions to express v in terms of x.

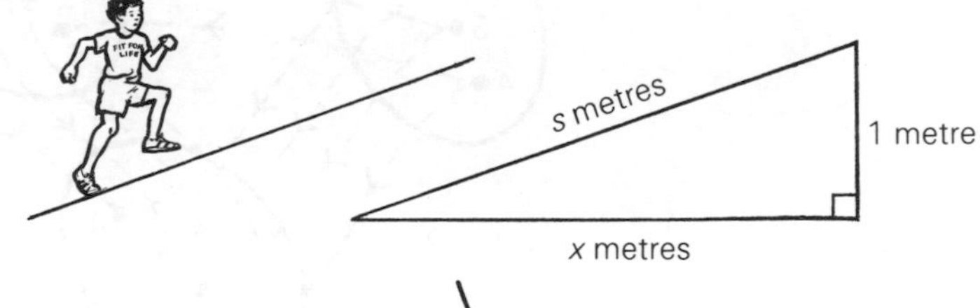

7 A fairground's Big Wheel has radius 10 metres, and its centre C is 12 metres above the ground. One of the Big Wheel's chairs is at A. The sun's rays strike the chair at 20° to the vertical.
s (metres) is the distance from D, directly below A, to the edge of the chair's shadow. Express:
a d as a function of x **b** h as a function of d
c s as a function of h **d** s as a function of: (i) d (ii) x (using composition).

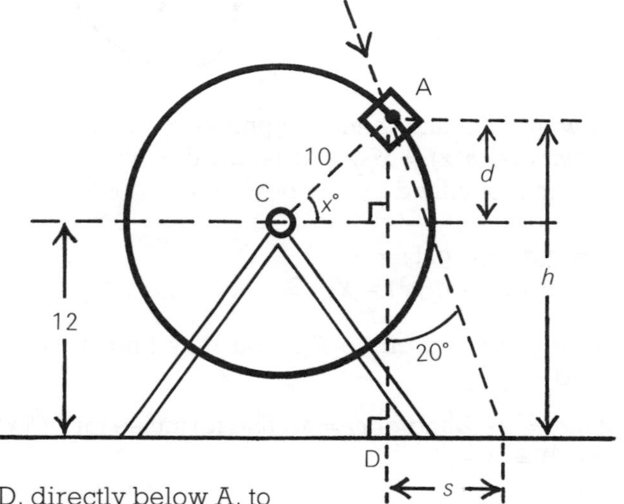

1 Working with real numbers, you know that:
$7+(3+2) = (7+3)+2$, and $7 \times (3 \times 2) = (7 \times 3) \times 2$, but $7-(3-2) \neq (7-3)-2$.
So what about composition of functions? Is $h \circ (g \circ f) = (h \circ g) \circ f$?
Investigate this, taking $f(x) = 2x$, $g(x) = x+1$ and $h(x) = x^2$.
Try it again for $f(x) = 2x+3$, $g(x) = 3x+2$ and $h(x) = x+4$.
Experiment with some functions of your own choice.

2 a $a + \square = \square + a = a$ where $a \in R$. Which number goes in each box?
b $a \times \square = \square \times a = a$ where $a \in R$. Which number goes in each box?
0 is the identity under addition.
What do you think 1 is called?
c Can you discover something similar for functions?
For example $f: R \rightarrow R$ where $f(x) = 2x+3$. Is there an identity function i such that:

$$f(i(x)) = f(x) = i(f(x))?$$

Remember that you will have to get a formula for $i(x)$ $(x \in R)$.
d If you have the correct i, then $f \circ i = i \circ f$ for every function $f: R \rightarrow R$.
e What is the graph of $i: R \rightarrow R$?

FUNCTIONS WITH INVERSES

Look at the diagram.

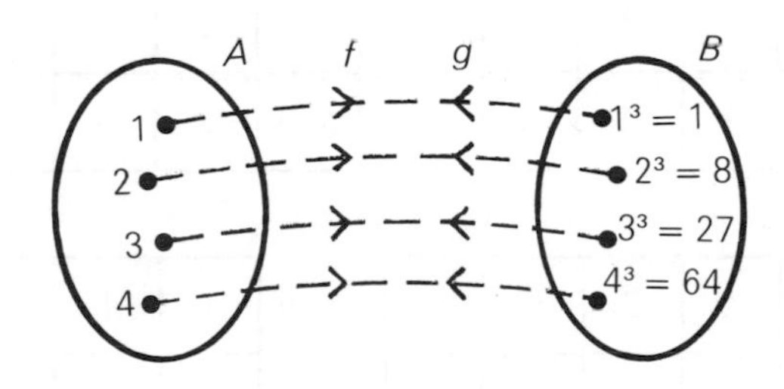

What is the function f that maps A to B? Can you see that it has the formula $f(x) = x^3$?
If you want to go back to where you started, is there a function g from B to A that undoes f?
You need:

$$g(1^3) = 1,\ g(2^3) = 2,\ g(3^3) = 3,\ g(4^3) = 4.$$

Can you think of a formula for g?
In fact, $g(x) = x^{1/3}$. Check that:

$$g(f(1)) = 1,\ g(f(2)) = 2,\ g(f(3)) = 3,\ g(f(4)) = 4;\ g(f(x)) = x, \text{ i.e. } (x^3)^{1/3} = x.$$

Also $f(g(x)) = (x^{1/3})^3 = x$.
g is the *inverse* of f. (It 'inverts' it or 'turns it back'.) The symbol used for it is f^{-1} ('f inverse').

$$f(x) = x^3,\ f^{-1}(x) = x^{1/3};\ f^{-1}(f(x)) = x \text{ and } f(f^{-1}(x)) = x.$$

Exercise 6

1 Decide whether each of these functions has or has not an inverse $f^{-1}: B \to A$.
Remember that for *every* element in the domain there is one and only one value of the function.

a
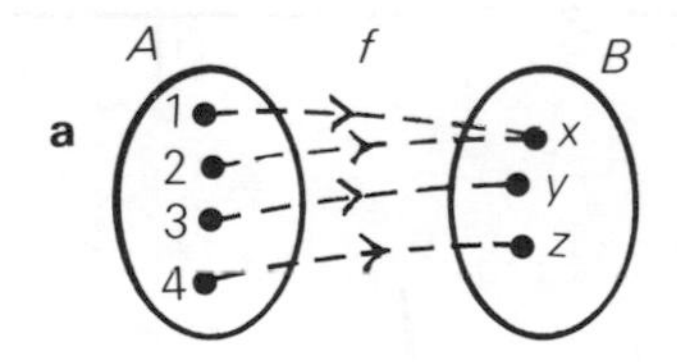

b
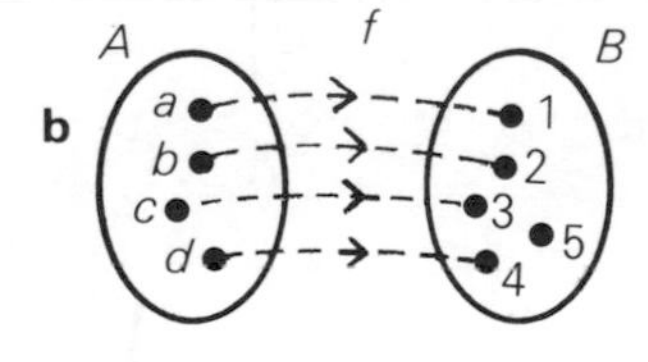

c
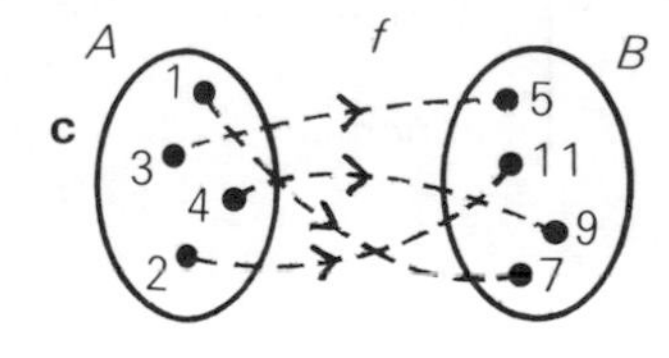

2 Here are the answers to question **1**; check yours.

a f does not have an inverse because from x in B there are two 'strings' back to elements in A ($x = f(1) = f(2)$).

b This f also has no inverse since 5 in B is not in the range of f, so there is no string from 5 back to A.

c f has an inverse f^{-1} because: (i) B is the range of f
(ii) there is one and only one 'string' joining each pair of members in A and B (and no points remain unpaired).

3 $A = \{-2, 0, 2\}$ and $B = \{0, 4\}$. $f: A \to B$ has formula $f(x) = x^2$.
Does f have an inverse f^{-1}? An arrow diagram could be useful.

4 Each table defines a function. Decide which have inverses.

a

x	-2	-1	0	1	2	3
$f(x)$	2	1	0	1	2	3

b

x	1	2	3	4	5	6	7
$f(x)$	2	3	4	5	6	7	6

c

x	1	2	3	4	5	6	7
$f(x)$	2	3	4	5	6	7	8

d

x	1	2	3	4	5	6
$f(x)$	10	10^2	10^3	10^4	10^5	10^6

5 In question **4**, take each function f that has an inverse f^{-1} and construct a table. For example, in **c**:

y	2	3
$f^{-1}(y)$	1	2

6 Each graph represents a function f: $R \to R$. None of the three has an inverse. Discuss why.

a

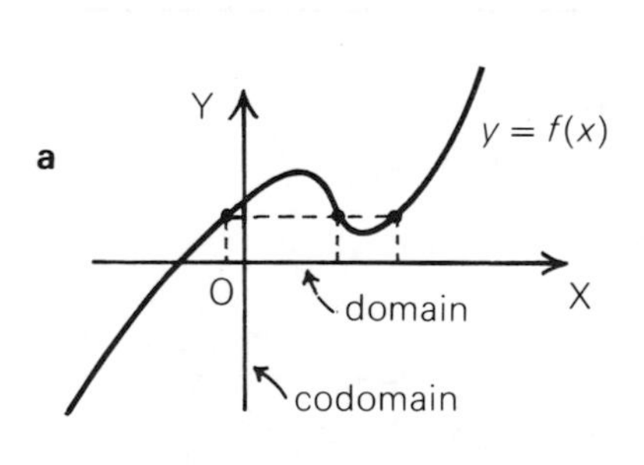

b

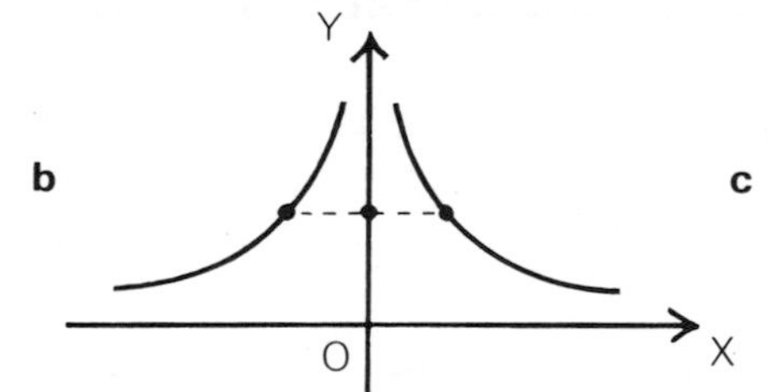

c

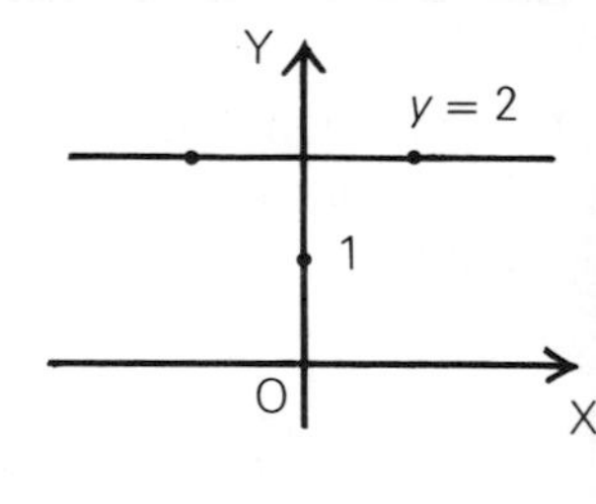

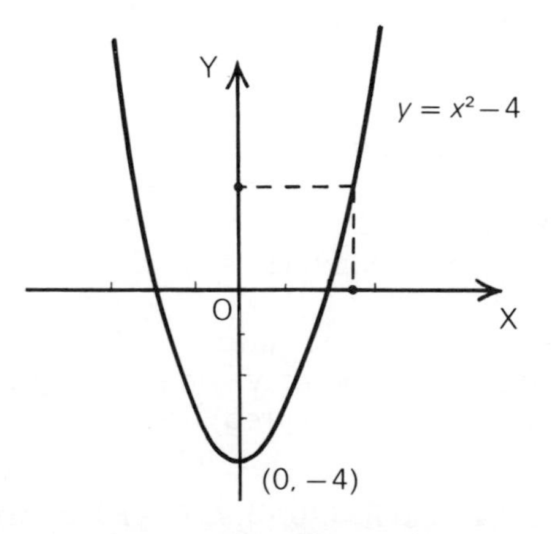

7 a This is the graph of the function $f(x) = x^2 - 4$. Why does f: $R \to R$ not have an inverse?

b State the range B of f.

c $A = \{x \in R : x \geqslant 0\}$. Describe the graph of f: $A \to B$. Explain why this function has an inverse. (It is a '*restriction*' of f: $R \to R$.)

d Can you think of another subset A_1 of R for which f: $A_1 \to B$ also has an inverse?

FORMULAE FOR INVERSES

You should now have some idea of when a function $f: A \to B$ has an inverse.

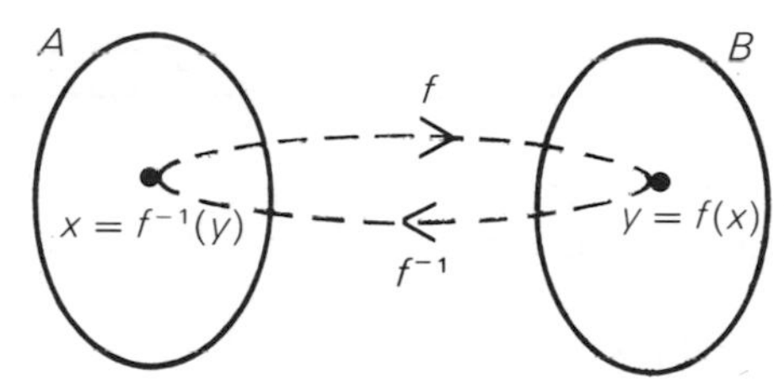

(i) B has to be the range of f.
(ii) f has to set up a '*one-to-one correspondence*' between A and B (For each $x \in A$ there is only one $y \in B$ and for each $y \in B$ there is only one $x \in A$.)

If f takes x to y, then f^{-1} takes y back to x.

$$y = f(x) \Leftrightarrow x = f^{-1}(y)$$

Example $f: R \to R$ has formula $f(x) = 2x - 4$.
Its graph is the line $y = 2x - 4$.

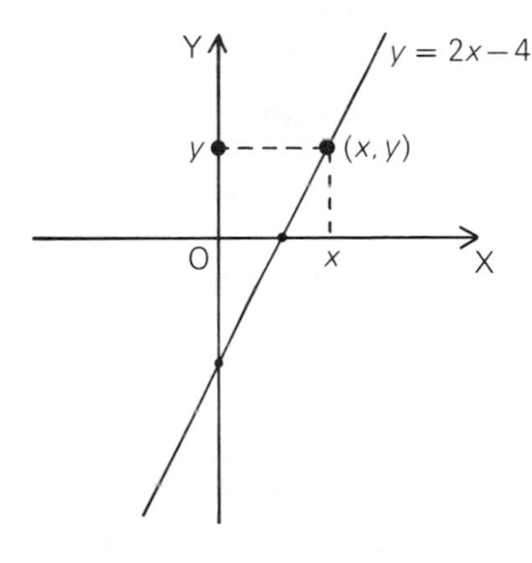

a Explain why f^{-1} exists.
(i) The range of f is the y-axis.
(ii) Each line parallel to the x-axis (through points of the range) meets the graph in exactly one point.

So f is a one-to-one correspondence and f^{-1} exists.

b Find a formula for $f^{-1}: R \to R$.
We have simply to solve $y = 2x - 4$ for x in terms of y.
$y = 2x - 4 \Leftrightarrow 2x = y + 4 \Leftrightarrow x = \frac{1}{2}(y+4)$.
So $f^{-1}(y) = \frac{1}{2}(y+4)$.
In x-notation, $f^{-1}(x) = \frac{1}{2}(x+4)$.

Exercise 7

1 Each of these functions has an inverse. Find a formula for each inverse.

a $f(x) = 2x$ **b** $g(x) = 4x - 2$ **c** $h(x) = 3x + 7$ **d** $i(x) = 6x + 2$
e $j(x) = \frac{1}{2}(1 - 5x)$ **f** $k(x) = \frac{1}{2}x + 3$ **g** $m(x) = x^3$ **h** $n(x) = x^5 + 1$
i $p(x) = \dfrac{1}{x+1}$ **j** $q(x) = \dfrac{2}{x-3}$ **k** $r(x) = \dfrac{x}{x+2}$ **l** $s(x) = \dfrac{2}{3-2x}$.

2 $f(x) = x^2 - 1$ has graph $y = x^2 - 1$, where $A = \{x \in R : x \geqslant 0\}$ and $B = \{y \in R : y \geqslant -1\}$.

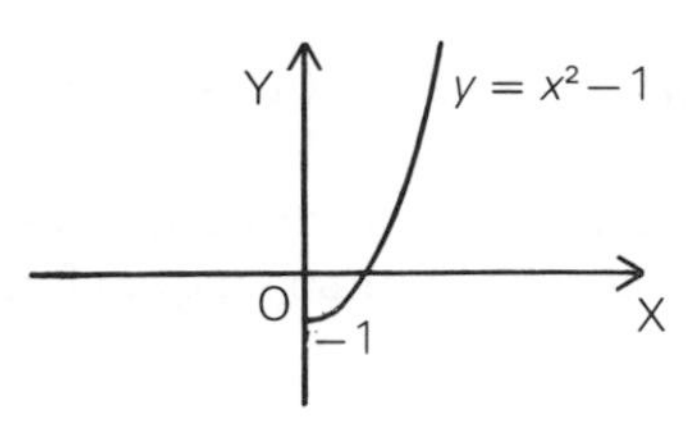

a Explain why $f: A \to B$ has an inverse f^{-1}.

b To find a formula for $f^{-1}(y)$ you have to solve $y = x^2 - 1$ for x in terms of y.
$$y = x^2 - 1 \Leftrightarrow x^2 = y + 1 \Leftrightarrow x = \pm\sqrt{(y+1)}.$$
Use this to find a formula for $f^{-1}: B \to A$.

Investigate the connection between the graphs of a function, its inverse and the line $y = x$, as follows.

a If $f(x) = 2x-4$, draw the graphs of $y = f(x)$ and $y = f^{-1}(x)$, and the line $y = x$, all in the same diagram.

b If $f(x) = x^3$, draw the graphs $y = f(x)$, $y = f^{-1}(x)$ and $y = x$ in the same diagram.

c Have you discovered the geometrical connection between the graphs $y = f(x)$, $y = f^{-1}(x)$ and $y = x$?

d If you have, try to show that this connection always holds.

CHECK-UP ON **FUNCTIONS**

1 Find the range of each of these functions:

a $f(x) = 3x+4$, domain $\{0, 1, 2, 3\}$ **b** $f(x) = x^{1/3}$, domain $\{0, 1, 8, 27\}$

c $f(x) = x^2+2$, domain $\{-3, 0, 3\}$ **d** $f(x) = \sin 2x$, domain R.

2 These formulae define functions. Decide in each case on the largest possible domain:

a $f(x) = \sqrt{(2-x)}$ **b** $g(x) = \dfrac{3}{x+1}$ **c** $h(x) = \sqrt{(x^2-4)}$ **d** $k(x) = \dfrac{1}{x(x-2)}$.

3 Sketch the graphs of these functions, and state the range of each:

a $f(x) = x+2\ (x \in R)$ **b** $g(x) = 4-x^2\ (x \in R)$ **c** $h(x) = \begin{cases} 2x+1 & (x \geqslant 0) \\ 0 & (x < 0) \end{cases}$.

4 For each of these pairs of functions obtain formulae for:

(i) $g(f(x))$ (ii) $f(g(x))$ (iii) $f(f(x))$:

a $f(x) = 3x+5$, $g(x) = x^2$ **b** $f(x) = \sin x$, $g(x) = \cos x$.

5 a Sketch the graph of each of these functions: $R \rightarrow R$, and explain why each has an inverse:

(i) $f(x) = 3x$ (ii) $g(x) = 2x+7$ (iii) $h(x) = -x^3$ (iv) $k(x) = x^5$.

b Obtain a formula for each inverse function.

6 Explain why each of the following 'restrictions' of the sine, cosine and tangent functions has an inverse (draw the graph in each case):

a $\sin x$, where $x \in A_1 = \left\{x: -\dfrac{\pi}{2} \leqslant x \leqslant \dfrac{\pi}{2}\right\}$

b $\cos x$, where $x \in A_2 = \{x: 0 \leqslant x \leqslant \pi\}$

c $\tan x$, where $x \in A_3 = \left\{x: -\dfrac{\pi}{2} < x < \dfrac{\pi}{2}\right\}$.

Note These inverse functions are denoted by $\sin^{-1}$, $\cos^{-1}$, $\tan^{-1}$, and called the inverse sine, inverse cosine and inverse tangent functions. You'll see them on your calculator.

7 COMPOUND ANGLE FORMULAE

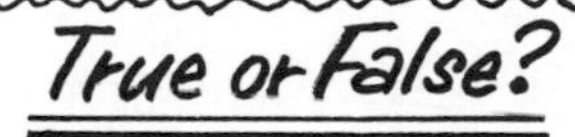

$-40 > -30 \quad 40 > 30$

$\frac{1}{40} > \frac{1}{30} \quad \cos 40^\circ > \cos 30^\circ$

$\sin(40^\circ + 30^\circ) = \sin 40^\circ + \sin 30^\circ$

All of these statements *look* reasonable. But only one of them is true. This chapter contains some formulae that may surprise you.

REMINDERS, FOR REFERENCE

1 Related angles

(i) $180° - A$ (ii) $90° - A$ (iii) $-A$

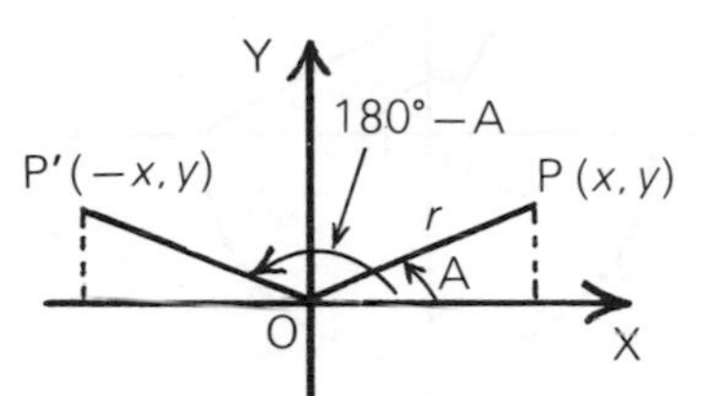

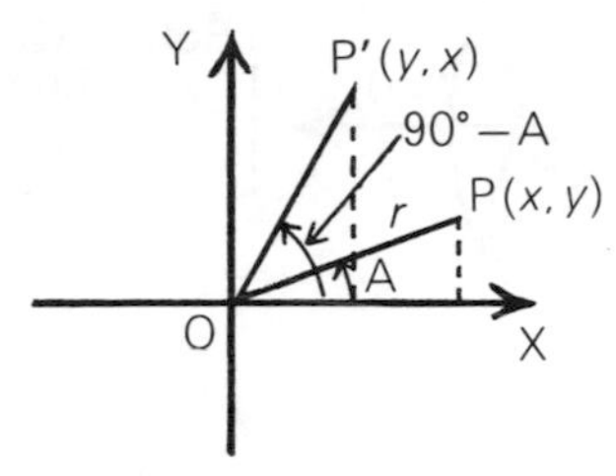

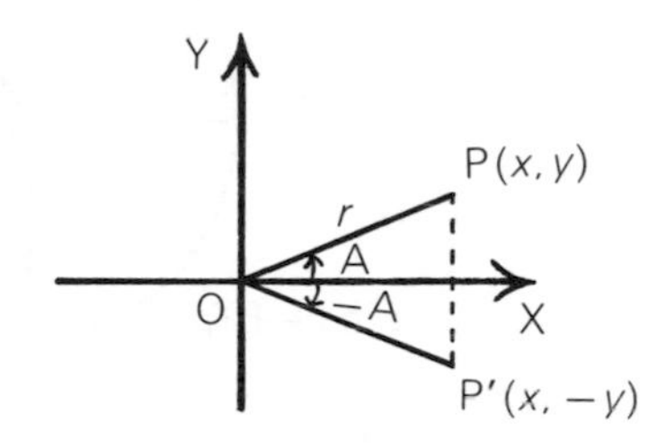

$\sin(180° - A) = \frac{y}{r} = \sin A$ $\sin(90° - A) = \frac{x}{r} = \cos A$ $\sin(-A) = \frac{-y}{r} = -\sin A$

$\cos(180° - A) = \frac{-x}{r} = -\cos A$ $\cos(90° - A) = \frac{y}{r} = \sin A$ $\cos(-A) = \frac{x}{r} = \cos A$

2 Sin – cos – tan formulae

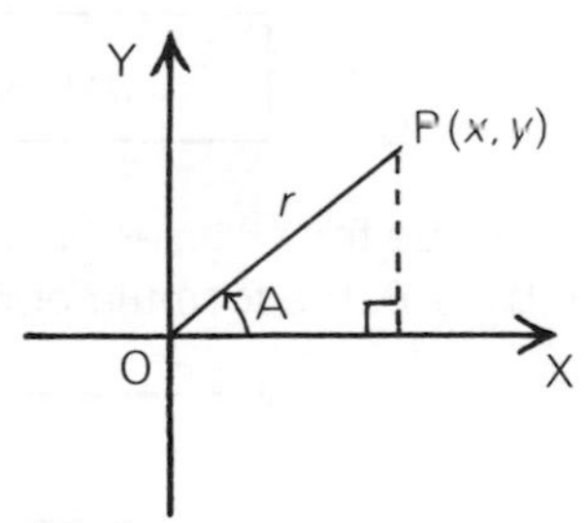

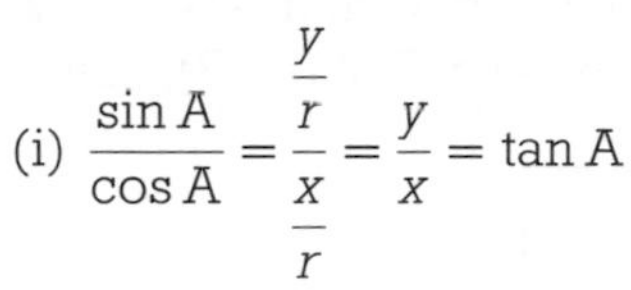

(i) $\frac{\sin A}{\cos A} = \frac{\frac{y}{r}}{\frac{x}{r}} = \frac{y}{x} = \tan A$

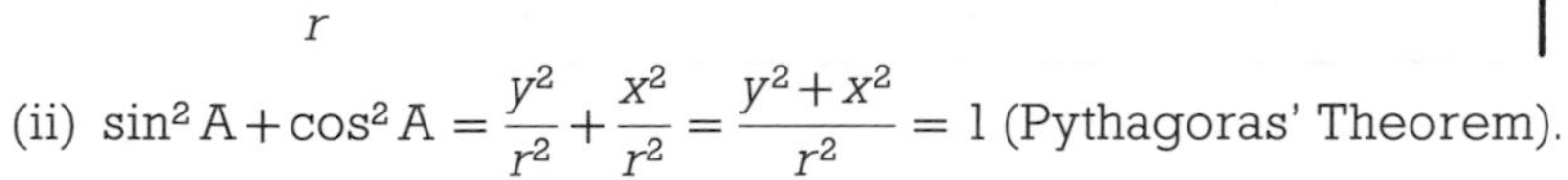

(ii) $\sin^2 A + \cos^2 A = \frac{y^2}{r^2} + \frac{x^2}{r^2} = \frac{y^2 + x^2}{r^2} = 1$ (Pythagoras' Theorem).

Exercise 1

1 Simplify:

a $\sin(180 - x)°$ **b** $\cos(90 - y)°$ **c** $\sin(-t)°$ **d** $\sin(90 - n)°$.

2 Write in simpler form:

a $\frac{\sin 2u°}{\cos 2u°}$ **b** $\sin^2 3v + \cos^2 3v$ **c** $1 - \sin^2 4w$.

3 Express these as sines or cosines of acute angles:
a $\sin 100°$ **b** $\cos 160°$ **c** $\sin 95°$ **d** $\cos 111°$ **e** $\sin 177°$.

4 Use your calculator to check that:

a $\dfrac{\sin 40°}{\cos 40°} = \tan 40°$ **b** $\sin^2 40° + \cos^2 40° = 1$.

5 Reminders **1** (i) and (ii) concern supplementary and complementary angles. State the results in words, using these terms.

FORMULAE GALORE

cos (A + B) and cos (A − B)

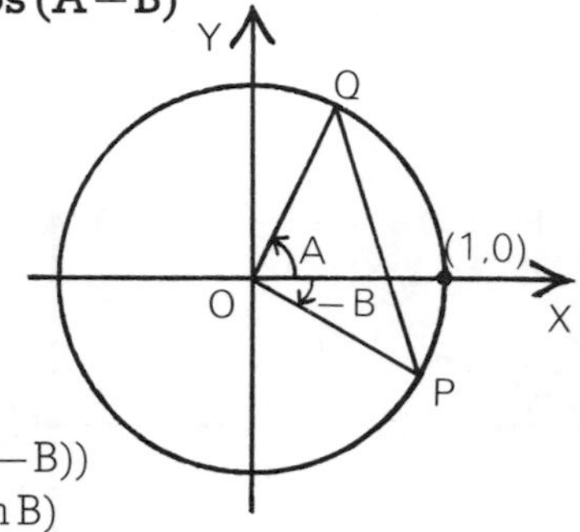

Rotate ΔOPQ through angle B

Y
Q′
A+B
P′(1, 0)
O
X

P is $(\cos(-B), \sin(-B))$
i.e. P is $(\cos B, -\sin B)$
Q is $(\cos A, \sin A)$

$$\begin{aligned}PQ^2 &= (\cos A - \cos B)^2 + (\sin A + \sin B)^2\\ &= \cos^2 A + \sin^2 A + \cos^2 B + \sin^2 B\\ &\quad - 2\cos A\cos B + 2\sin A\sin B\\ &= 2 - 2(\cos A\cos B - \sin A\sin B)\end{aligned}$$

P′ is $(1, 0)$
Q′ is $(\cos(A+B), \sin(A+B))$

$$\begin{aligned}(P'Q')^2 &= (1 - \cos(A+B))^2 + (\sin(A+B))^2\\ &= 1 + \cos^2(A+B) + \sin^2(A+B)\\ &\quad - 2\cos(A+B)\\ &= 2 - 2\cos(A+B).\end{aligned}$$

But $PQ^2 = (P'Q')^2$.

It follows that

$$\boxed{\cos(A+B) = \cos A\cos B - \sin A\sin B}$$

The formula is true for all sizes of angles A and B, whether in degrees or in radians.
Replacing B by −B, the formula becomes $\cos(A-B) = \cos A\cos(-B) - \sin A\sin(-B)$, that is:

$$\boxed{\cos(A-B) = \cos A\cos B + \sin A\sin B}$$

sin (A + B) and sin (A − B)

Reminder **1** (ii) tells you how to change a sine to a cosine.

$$\begin{aligned}\text{So, } \sin(A+B) = \cos[90° - (A+B)] &= \cos[(90° - A) - B]\\ &= \cos(90° - A)\cos B + \sin(90° - A)\sin B\\ &= \sin A\cos B + \cos A\sin B\end{aligned}$$

Find a formula for $\sin(A-B)$, by replacing B by −B.

$$\boxed{\begin{aligned}\sin(A+B) &= \sin A\cos B + \cos A\sin B\\ \sin(A-B) &= \sin A\cos B - \cos A\sin B\end{aligned}}$$

Exercise 2

1 Write down formulae for:
a $\sin(X+Y)$ **b** $\cos(X+Y)$ **c** $\cos(C-D)$ **d** $\sin(P-Q)$.

2 Make sketches of:
a the '30°, 60°, 90°' and '45°, 45°, 90°' triangles, using both degrees and radians, and marking in the 1, 2, $\sqrt{3}$ and 1, 1, $\sqrt{2}$ side lengths;
b the sine and cosine graphs from 0°–360° (0–2π radians).

Use these in the following questions where necessary. **Don't use a calculator.**

3 Verify the formulae for $\cos(A+B)$ and $\cos(A-B)$, using:

a $A = B = 90°$ **b** $A = 30°, B = 60°$ **c** $A = B = \frac{\pi}{4}$ **d** $A = \frac{\pi}{3}, B = 0$.

4 Write in shorter form, and simplify where possible:
a $\sin 70° \cos 20° + \cos 70° \sin 20°$ **b** $\cos 170° \cos 50° + \sin 170° \sin 50°$
c $\cos 2B \cos B + \sin 2B \sin B$ **d** $\sin 5y \cos y - \cos 5y \sin y$.

5 Prove that:
a $\sin\left(x+\frac{\pi}{6}\right) = \frac{1}{2}(\cos x + \sqrt{3}\sin x)$ **b** $\sin(x+60)° - \cos(x+30)° = \sin x°$
c $\cos(A+B) + \cos(A-B) = 2\cos A \cos B$ **d** $\sin(A+B) - \sin(A-B) = 2\cos A \sin B$.

6 Using the right-angled triangles, show that

$$\sin(C+D) = \frac{63}{65}.$$

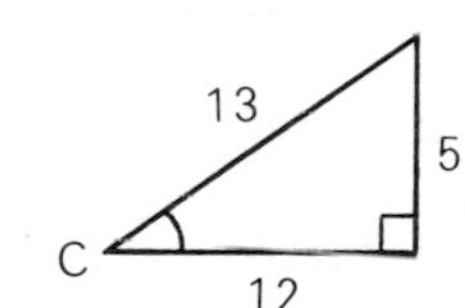

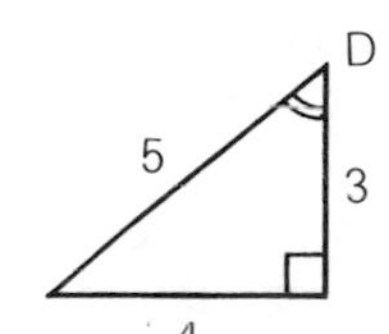

7 Given $\sin X = \frac{3}{5}$ and $\sin Y = \frac{7}{25}$, where $0 \leqslant X \leqslant \frac{\pi}{2}$ and $0 \leqslant Y \leqslant \frac{\pi}{2}$, show that $\cos(X+Y) = \frac{3}{5}$.

8 Which of these are true, and which false?
a $\cos(x-30)° - \cos(x+30)° = \sqrt{3}\sin x°$
b $\cos(x+45)° + \cos(x-45)° = \sqrt{2}\cos x°$
c $\sin(x+30)° - \sin(x-30)° = \sqrt{3}\sin x°$.

9 Using $15° = 45° - 30°$, prove that these are exact values:

a $\cos 15° = \frac{\sqrt{6}+\sqrt{2}}{4}$ **b** $\sin 15° = \frac{\sqrt{6}-\sqrt{2}}{4}$.

10 Using $75° = 45° + 30°$, show that:

a $\cos 75° = \frac{\sqrt{6}-\sqrt{2}}{4}$ **b** $\sin 75° = \frac{\sqrt{6}+\sqrt{2}}{4}$.

11 Prove that $(\cos A + \cos B)^2 + (\sin A - \sin B)^2 = 2[1 + \cos(A+B)]$.

12 Prove that: **a** $\tan 3A + \tan A = \dfrac{\sin 4A}{\cos 3A \cos A}$

b $\tan A - \tan B = \dfrac{\sin(A-B)}{\cos A \cos B}$.

13 In many branches of engineering the problem of making bends in pipes often arises. One view of this pipe is obtained by projecting vertically down onto the drawing board, to give AB and BC.

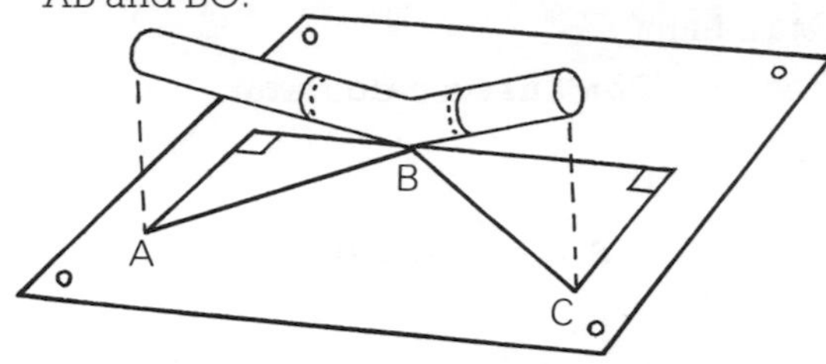

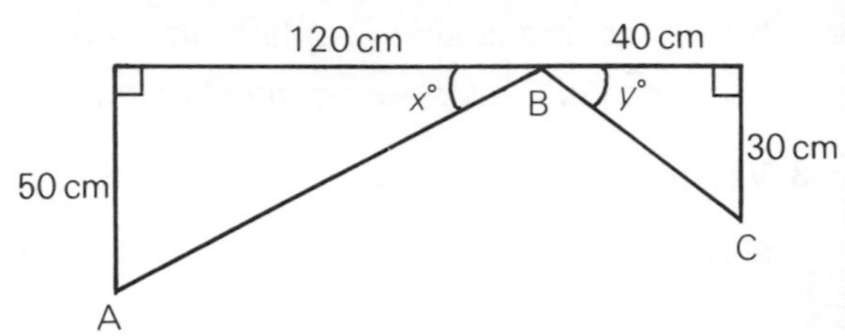

a Find the lengths AB and BC.
b State the relation between x, y and $\angle$ABC.
c Find the exact value of cos ABC.
d Calculate the length AC.

14

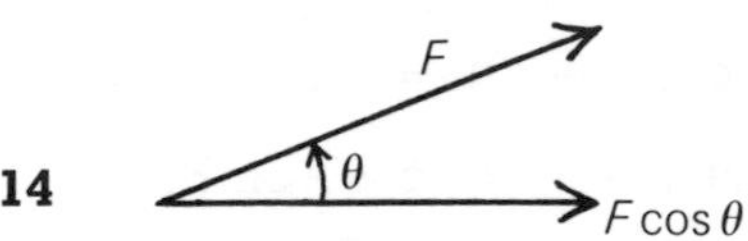

The horizontal effect of the force F is $F\cos\theta$.

Two equal forces (F) are pulling a crate up a slope. Calculate:
a the horizontal effect of each force
b the total horizontal effect of the two forces, in its simplest form.

15 For this snooker shot,
a prove that:
(i) $\sin y° = \sin 2x°$
(ii) $\cos y° = -\cos 2x°$
b calculate the exact values of:
(i) $\sin y°$ (ii) $\cos y°$.

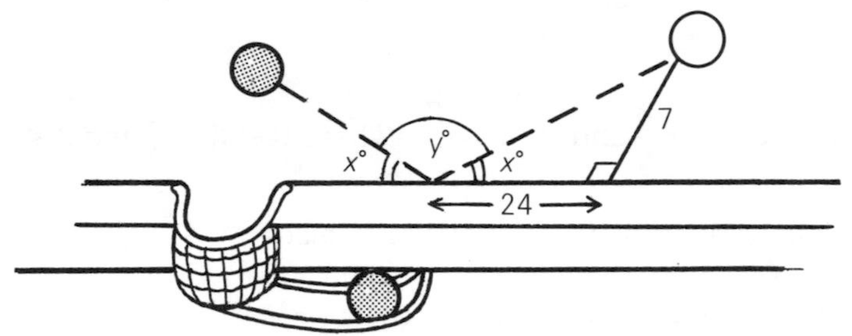

The gradients of these lines are m_1 and m_2 and the angle between the lines is $\theta°$.

a Prove that $\tan\theta° = \dfrac{m_2 - m_1}{1 + m_1 m_2}$.

Hint Try to find a formula for tan (B − A) in terms of tan A and tan B.

b Apply the $\tan\theta°$ formula to the lines $y = 2x + 1$ and $y = -3x + 2$.

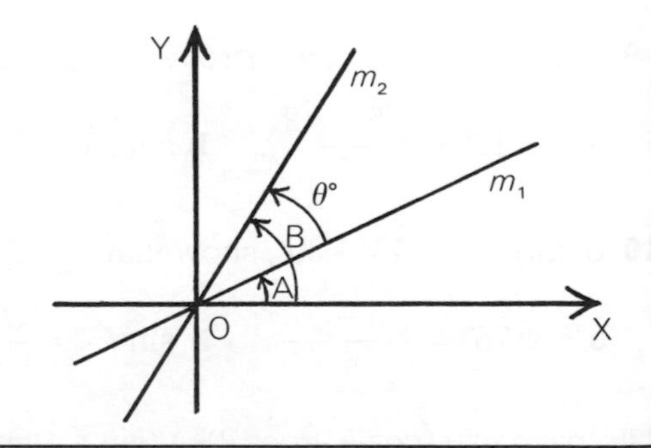

sin 2A and cos 2A

Replacing B by A in $\sin(A+B) = \sin A\cos B + \cos A\sin B$,

$$\sin(A+A) = \sin A\cos A + \cos A\sin A$$

So $\boxed{\sin 2A = 2\sin A\cos A}$

Replacing B by A in $\cos(A+B) = \cos A\cos B - \sin A\sin B$,

$$\cos(A+A) = \cos A\cos A - \sin A\sin A$$

So $\cos 2A = \cos^2 A - \sin^2 A$.

But $\cos^2 A = 1 - \sin^2 A$, so $\cos 2A = 1 - \sin^2 A - \sin^2 A = 1 - 2\sin^2 A$

or $\sin^2 A = 1 - \cos^2 A$, so $\cos 2A = \cos^2 A - (1 - \cos^2 A) = 2\cos^2 A - 1$

$$\boxed{\cos 2A = \cos^2 A - \sin^2 A = 1 - 2\sin^2 A = 2\cos^2 A - 1}$$

From $\cos 2A = 1 - 2\sin^2 A$,

$$\boxed{\sin^2 A = \tfrac{1}{2}(1 - \cos 2A)}$$

From $\cos 2A = 2\cos^2 A - 1$,

$$\boxed{\cos^2 A = \tfrac{1}{2}(1 + \cos 2A)}$$

Exercise 3

1 Write down all the formulae (in terms of angle X or angle Y) for:

a $\sin 2X$ **b** $\cos 2X$ **c** $\sin 2Y$ **d** $\cos 2Y$.

2 Write down formulae for:

a $\sin A$ in terms of $\frac{A}{2}$ **b** $\cos A$ in terms of $\frac{A}{2}$ (3 formulae).

3 Check all the formulae for sin 2A and cos 2A for A = (i) 30° (ii) 45° (iii) 60°.

4 Use the '2A' formulae to simplify and evaluate (without using a calculator):

a $2\sin 15°\cos 15°$ **b** $2\cos^2\frac{\pi}{6} - 1$ **c** $2\sin 45°\cos 45°$.

5 If $\sin A = \frac{1}{\sqrt{5}}$, $0 < A < \frac{\pi}{2}$, find the exact values of: **a** $\cos A$ **b** $\sin 2A$ **c** $\cos 2A$.

6 Prove that:

a $(\sin A + \cos A)^2 = 1 + \sin 2A$ **b** $\sin^3 x\cos x + \sin x\cos^3 x = \frac{1}{2}\sin 2x$.

7 Given $\cos 2A = \frac{1}{4}$, use the formulae for $\cos^2 A$ and $\sin^2 A$ in terms of $\cos 2A$ to show that $\cos A = \frac{\sqrt{5}}{2\sqrt{2}}$ and $\sin A = \frac{\sqrt{3}}{2\sqrt{2}}$. Check that, for these values, $\cos^2 A + \sin^2 A = 1$.

8 Using $3A = 2A + A$, prove that:
a $\sin 3A = 3\sin A - 4\sin^3 A$ **b** $\cos 3A = 4\cos^3 A - 3\cos A$.

9 Prove that:

a $\dfrac{2\tan\theta}{1+\tan^2\theta} = \sin 2\theta$ **b** $\dfrac{1-\tan^2\theta}{1+\tan^2\theta} = \cos 2\theta$ **c** $\dfrac{2\tan\theta}{1-\tan^2\theta} = \tan 2\theta$.

10 The equal sides of an isosceles triangle are each a cm long, and the angles opposite them are $\alpha°$. Prove that its area is $\frac{1}{2}a^2 \sin 2\alpha°$.

> $\cos^2 x = \frac{1}{2}(1+\cos 2x)$. Express $\cos^4 x$ in the form $a + b\cos 2x + c\cos 4x$.
> Try to find a corresponding expression for $\sin^4 x$.

SOLVING EQUATIONS

Example Solve the equation $\cos 2x° - \cos x° + 1 = 0$, for $0 \leqslant x \leqslant 360$.

$$\cos 2x° - \cos x° + 1 = 0$$
$$(2\cos^2 x° - 1) - \cos x° + 1 = 0$$
$$2\cos^2 x° - \cos x° = 0$$
$$\cos x°(2\cos x° - 1) = 0$$

$\cos x° = 0$ or $2\cos x° - 1 = 0$

$\cos x° = 0$ $\qquad$ $\cos x° = \frac{1}{2}$

$x = 90, 270$ (from cosine graph) $\qquad$ $x = 60$ (from calculator, or 30°, 60°, 90°△) or $360 - 60 = 300$

Sin +	All + ✓ $x°$
Tan +	$(360-x)°$ Cos + ✓

Solutions, for $0 \leqslant x \leqslant 360$, are 60, 90, 270, 300.

Exercise 4

Solve these equations for $0 \leqslant x \leqslant 360$, giving answers correct to 1 decimal place where relevant:

1 a $\sin 2x° - \cos x° = 0$ **b** $\sin 2x° - 3\sin x° = 0$ **c** $\cos 2x° + \sin x° = 0$.

2 a $\cos 2x° + \cos x° + 1 = 0$ **b** $\cos 2x° - 3\sin x° + 1 = 0$ **c** $\cos 2x° + 3\cos x° + 2 = 0$.

Solve these equations for $0 \leqslant \theta \leqslant 2\pi$:

3 a $\cos 2\theta + \sin\theta = 0$ **b** $\cos 2\theta - \sin\theta = 0$ **c** $\sin 2\theta - \sin\theta = 0$.

4 a $\cos 2\theta - 4\cos\theta - 5 = 0$ **b** $\sin 2\theta + \cos\theta = 0$ **c** $\cos 2\theta + \cos\theta = 0$.

TRIGONOMETRY IN ACTION

Example In $\triangle ABC$, prove that $a = 4b \cos x \cos 2x$.

Using the sine rule,

$$\frac{b}{\sin x} = \frac{a}{\sin(\pi - 4x)}$$

so $\quad a = b\dfrac{\sin 4x}{\sin x} = b\dfrac{2\sin 2x \cos 2x}{\sin x} = b\dfrac{2(2\sin x \cos x)\cos 2x}{\sin x}$

i.e. $a = 4b\cos x \cos 2x$.

Exercise 5

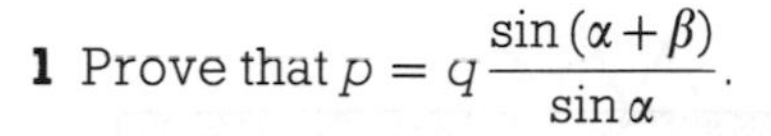

1 Prove that $p = q\dfrac{\sin(\alpha + \beta)}{\sin \alpha}$.

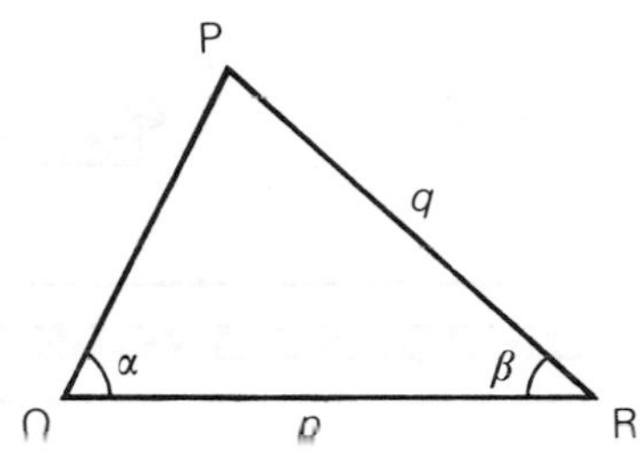

2 a Assuming a formula for $\cos 2\theta$, prove that $\cos^2\theta = \frac{1}{2}(1 + \cos 2\theta)$.

b A circle, centre O, has radius 2 units. BD is perpendicular to the radius OC. Prove that $AD^2 = 2(5 + 3\cos 2\theta)$.

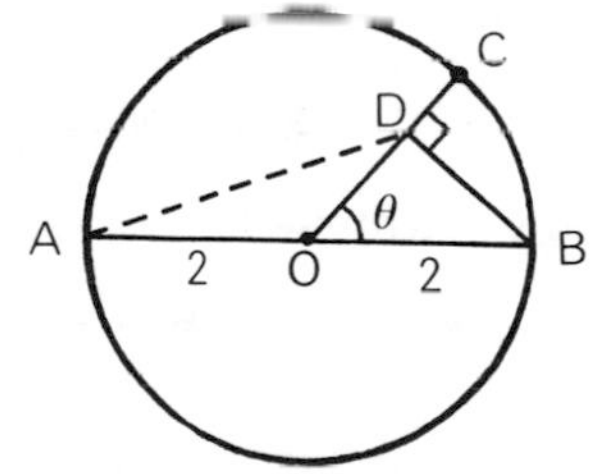

3 Prove that:

a $QC = \dfrac{d \sin x}{\sin(y - x)}$

b $AC = \dfrac{d \sin x \sin y}{\sin(y - x)}$.

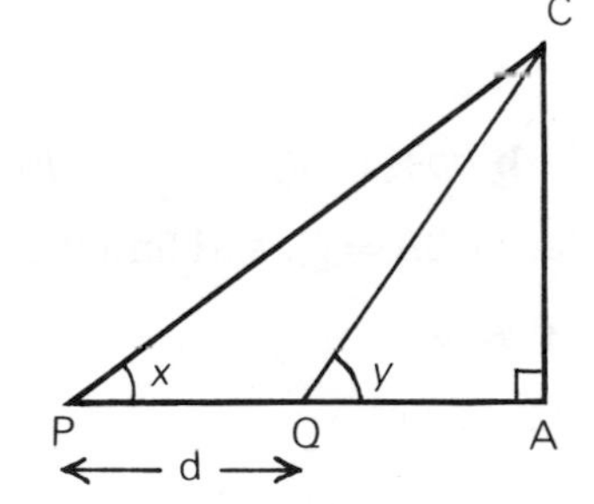

4 In $\triangle ABC$, BX is perpendicular to AB. Prove that $XC = a(\cos\theta° - \sin\theta°)$.

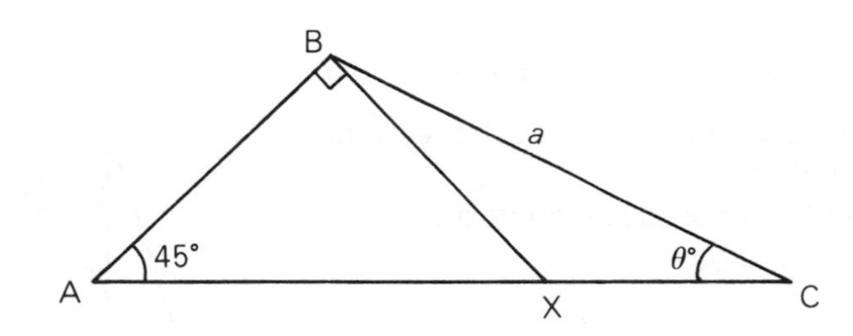

5 Light meets a plate glass surface at i° to the normal, and leaves at r° to the normal. Physicists will tell you that $\dfrac{\sin i^\circ}{\sin r^\circ}$ is a constant, called the refractive index.

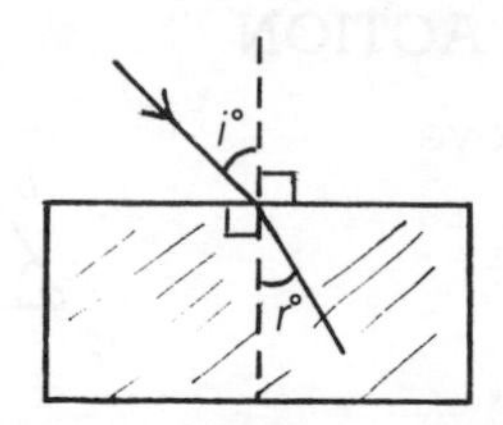

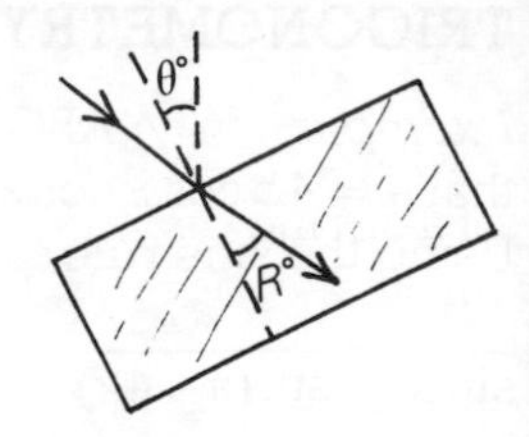

The glass surface is rotated through angle θ°, so that the light meets the glass at $(i-\theta)^\circ$ to the normal, and leaves at R° to the normal.

a Write down an equation using the refractive index.

b Prove that $\sin R^\circ = \left(\cos\theta^\circ - \dfrac{\sin\theta^\circ}{\tan i^\circ}\right)\sin r^\circ$.

6 Two mirrors are at x° to each other. A ray of light strikes one mirror at x° and is reflected at B onto the second mirror at C.

Prove that $\sin \text{BCD} = -\sin 4x^\circ$.

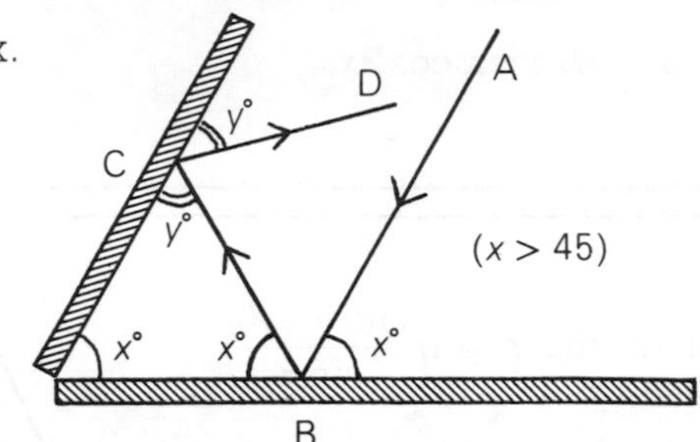

CHECK-UP ON **COMPOUND ANGLE FORMULAE**

1 Simplify:
a $\sin(180-u)^\circ$ **b** $\sin(90-v)^\circ$ **c** $\sin(-v)^\circ$ **d** $\cos 100^\circ$ (as the cosine of an acute angle).

2 Verify the formulae for $\cos(\text{A}\pm\text{B})$ and $\sin(\text{A}\pm\text{B})$ for:

a $\text{A} = 45^\circ, \text{B} = 90^\circ$ **b** $\text{A} = \dfrac{\pi}{4}, \text{B} = \pi$.

3 A and B are acute angles, $\sin \text{A} = \frac{4}{5}$ and $\sin \text{B} = \frac{5}{13}$. Show that $\sin(\text{A}+\text{B}) = \frac{63}{65}$, and find the exact value of $\cos(\text{A}+\text{B})$.

4 Prove that:
a $\sin(\alpha+\beta)-\sin(\alpha-\beta) = 2\cos\alpha\sin\beta$ **b** $\cos(\alpha-\beta)-\cos(\alpha+\beta) = 2\sin\alpha\sin\beta$.

5 Angle A is acute, and $\sin \text{A} = \frac{3}{5}$. Show that $\sin 2\text{A} = \frac{24}{25}$, and find the exact value of $\cos 2\text{A}$.

6 Prove that $(\cos\theta+\sin\theta)(\cos\theta-\sin\theta) = \cos 2\theta$.

7 Change the subject of $\cos 2\text{A} = 1-2\sin^2 \text{A}$ to $\sin^2 \text{A}$.

8 Using $\tan \text{A} = \dfrac{\sin \text{A}}{\cos \text{A}}$, prove that $\dfrac{1}{\tan 3x}+\dfrac{1}{\tan x} = \dfrac{\sin 4x}{\sin 3x \sin x}$.

9 Solve, for $0 \leqslant x \leqslant 360$:
a $\sin 2x^\circ - 3\cos x^\circ = 0$ **b** $\cos 2x^\circ - \sin x^\circ - 1 = 0$ **c** $\cos 2x^\circ - 5\cos x^\circ - 2 = 0$.

10 Use the formula $\triangle = \frac{1}{2}ab\sin \text{C}$ three times to prove that:
$\sin(x+y) = \sin x\cos y + \cos x \sin y$.

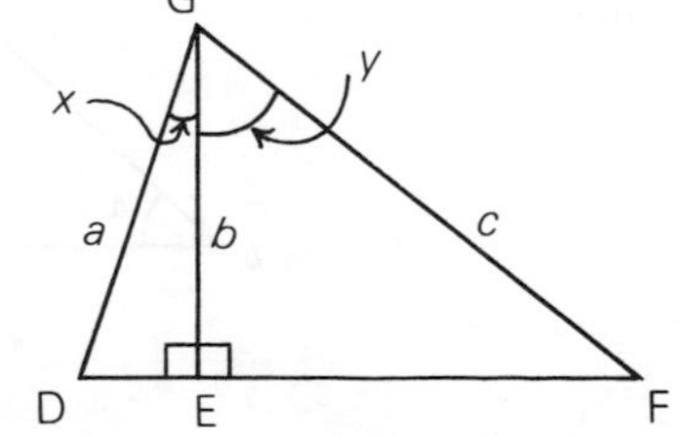

A practical problem

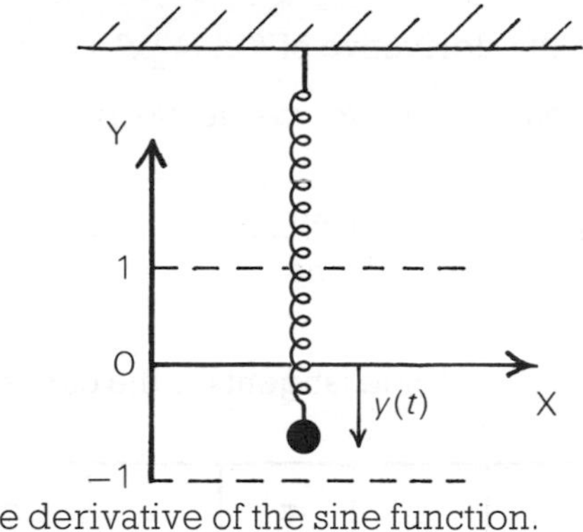

A weight is bobbing up and down at the end of a spring. Its displacement y cm from OX at time t seconds is $y(t) = \sin t$.

(i) What is its velocity at time t?

(ii) When is it stationary?

The velocity is $\frac{dy}{dt} = \frac{d}{dt}(\sin t)$, so we have to find the derivative of the sine function.

THE DERIVATIVES OF $f(x) = \sin x$ AND $f(x) = \cos x$

(i) *A practical approach*—***using the graph of the derived function***

Start with a function whose derivative *is* known.

If $f(x) = x^2$, $f'(x) = 2x$.

Is it possible to find this formula using the graphs $y = f(x)$ and $y = f'(x)$?

Remember that *the gradient of the tangent at $(x, f(x))$ on the graph $y = f(x)$ is $f'(x)$.*

By measuring the gradients of several tangents to $y = x^3$, this table is constructed:

x	−2	−1	0	1	2
$f'(x)$	−4	−2	0	2	4

Using the table, the graph of f' (the derived function) can be drawn.

Do you agree that it is the straight line with equation $y = 2x$?

This verifies that the derived function is $f'(x) = 2x$.

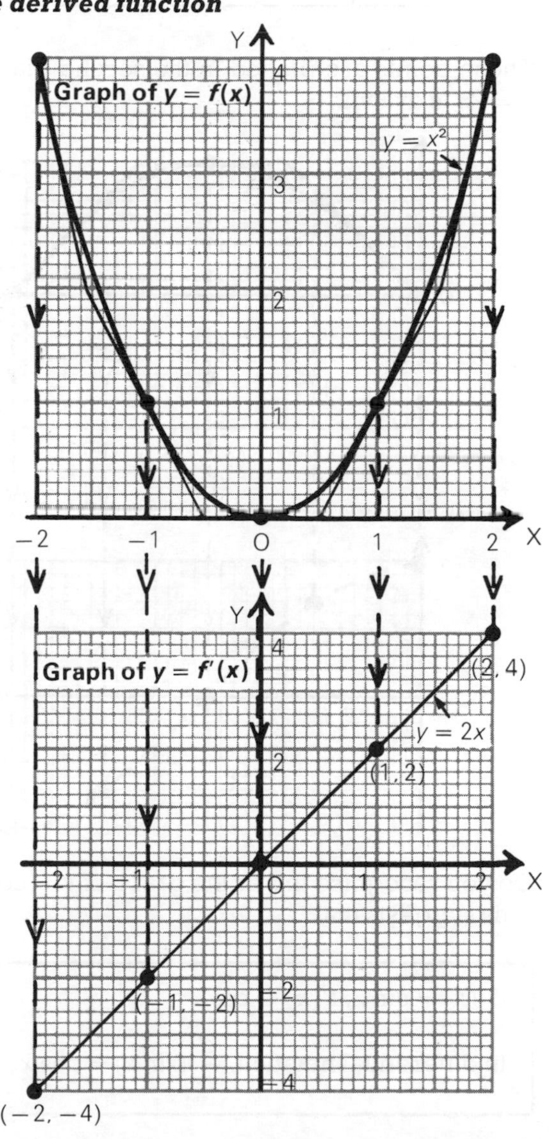

Try to find the derivative of $f(x) = \frac{1}{2}x^2$ in the same way.

Will this method work for the derivative of $y = f(x) = \sin x$?

Warning: In calculus, $\sin x$ and $\cos x$ $(x \in R)$ mean the sine and cosine of an angle of *x radians.*

The gradients of some tangents to the curve $y = \sin x$ are measured, giving:

x	0	$\frac{\pi}{4}$	$\frac{\pi}{2}$	$\frac{3\pi}{4}$	π	$\frac{5\pi}{4}$	$\frac{3\pi}{2}$	$\frac{7\pi}{4}$	2π
$f'(x)$	1	0·7	0	−0·7	−1	−0·7	0	0·7	1

These points are plotted in the second diagram, and the part of the graph of f' from $x = 0$ to $x = 2\pi$ is drawn.

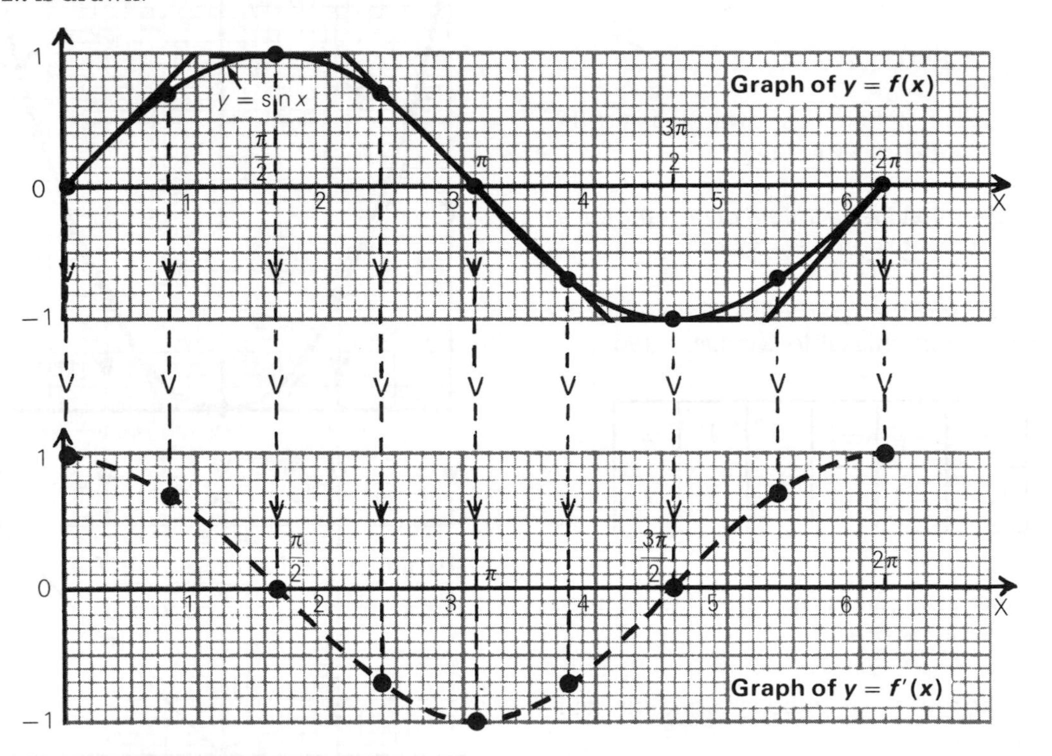

Do you recognise this as the graph of $y = \cos x$, for $0 \leqslant x \leqslant 2\pi$?
This suggests that:

> If $f(x) = \sin x$, $f'(x) = \cos x$.
>
> In differential notation, $\frac{d}{dx}(\sin x) = \cos x$.

Draw the graph of $f(x) = \cos x$ on squared paper, for $0 \leqslant x \leqslant 2\pi$.
Use the same scales as for the sine graph.

Measure the gradients of the tangents at $x = 0, \frac{\pi}{2}, \pi, \frac{3\pi}{2}$ and 2π.

Use them to draw the graph of the derived function f'.
You should obtain the graph of $y = -\sin x$.

> If $f(x) = \cos x$, $f'(x) = -\sin x$.
>
> In differential notation, $\frac{d}{dx}(\cos x) = -\sin x$.

(ii) ***Using the definition of a derivative***

Take $f(x) = \sin x$.

$$f'(x) = \lim_{h \to 0} \frac{f(x+h)-f(x)}{h} = \lim_{h \to 0} \frac{\sin(x+h)-\sin x}{h}$$

$$= \lim_{h \to 0} \frac{\sin x \cos h + \cos x \sin h - \sin x}{h}$$

$$= \lim_{h \to 0} \frac{\sin x(\cos h - 1) + \cos x \sin h}{h}$$

$$= \sin x \lim_{h \to 0} \left(\frac{\cos h - 1}{h}\right) + \cos x \lim_{h \to 0} \left(\frac{\sin h}{h}\right) \qquad (\mathbf{A})$$

What happens to $\frac{\cos h - 1}{h}$ and $\frac{\sin h}{h}$ as $h \to 0$?

Calculate their values for $h = 0{\cdot}1, 0{\cdot}01, 0{\cdot}001, 0{\cdot}0001$, using your calculator in radian mode.

What can you say about the values of $\lim_{h \to 0} \frac{\cos h - 1}{h}$ and $\lim_{h \to 0} \frac{\sin h}{h}$?

Do you agree that line (**A**), above, becomes $\cos x$? If so, $f'(x) = \cos x$.

> Follow through the process above for $f(x) = \cos x$.
>
> Check that $f'(x) = \cos x \lim_{h \to 0} \frac{\cos h - 1}{h} - \sin x \lim_{h \to 0} \frac{\sin h}{h}$, and deduce that $f'(x) = -\sin x$.
>
> $$\frac{d}{dx}(\sin x) = \cos x, \quad \frac{d}{dx}(\cos x) = -\sin x.$$

Example 1 Differentiate $5\sin x - 3\cos x$.

$$\frac{d}{dx}(5\sin x - 3\cos x) = 5\frac{d}{dx}(\sin x) - 3\frac{d}{dx}(\cos x) = 5\cos x + 3\sin x.$$

Example 2 $f(x) = 4\sin^2 \tfrac{1}{2}x$. Calculate $f'(x)$.

$f(x) = 4\sin^2 \tfrac{1}{2}x = 4 \times \tfrac{1}{2}(1 - \cos x) = 2 - 2\cos x.$

$$f'(x) = \frac{d}{dx}(2) - \frac{d}{dx}(2\cos x) = 0 - (-2\sin x) = 2\sin x.$$

Exercise 1

1 Differentiate with respect to the relevant variable:

a $3\sin x$ **b** $4\cos x$ **c** $\sin x + \cos x$ **d** $u - \cos u$
e $\cos t - \sin t$ **f** $5\sin t + 3\cos t$ **g** $x^3 - 3\cos x$ **h** $\tfrac{1}{2}u^2 + 3\sin u$.

2 Calculate $f'(x)$ in each case:

a $f(x) = 5x^2 - 2\sin x$ **b** $f(x) = \dfrac{1}{2x} - 3\cos x$ **c** $f(x) = \sqrt{x} + \dfrac{1}{2}\cos x$

d $f(x) = \dfrac{1}{2}\sin x + \dfrac{1}{3}\cos x$ **e** $f(x) = 10\cos x - 11\sin x$ **f** $f(x) = \dfrac{1}{\sqrt{x}} - \cos x$

g $f(x) = \dfrac{x}{2} + \dfrac{2}{x} - 4\sin x$ **h** $f(x) = \dfrac{1 + x\sin x}{x}$ **i** $f(x) = \dfrac{3 + 2x^2\cos x}{x^2}$.

3 Calculate:

a $\dfrac{d}{dx}(\cos^2 \tfrac{1}{2}x)$ **b** $\dfrac{d}{dt}(3\sin^2 \tfrac{1}{2}t)$ **c** $\dfrac{d}{du}(2\sin \tfrac{1}{2}u \cos \tfrac{1}{2}u)$.

[Use a cos 2A or a sin 2A formula.]

4 Reminder: With the help of the diagram, write down the values of:

a $\sin\dfrac{\pi}{2}$ **b** $\sin \pi$ **c** $\cos 0$ **d** $\cos \pi$ **e** $\cos\dfrac{\pi}{3}$

f $\sin\dfrac{\pi}{4}$ **g** $\sin\dfrac{\pi}{6}$ **h** $\cos\dfrac{3\pi}{2}$ **i** $\sin\dfrac{2\pi}{3}$ **j** $\sin\dfrac{4\pi}{3}$.

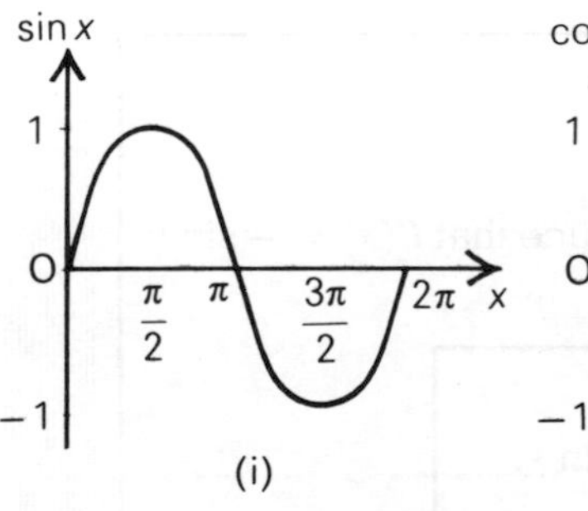

(i)

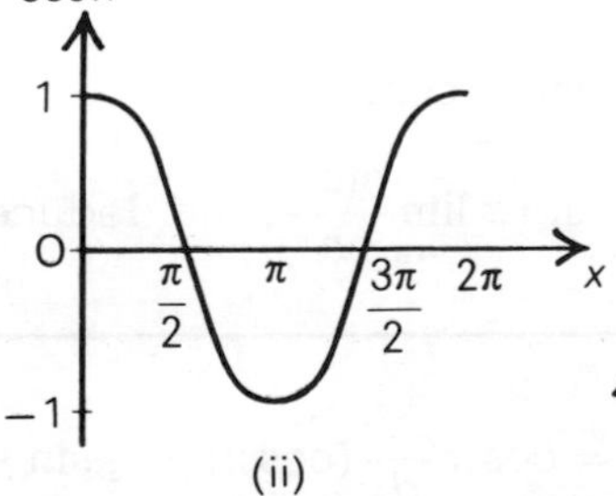

(ii)

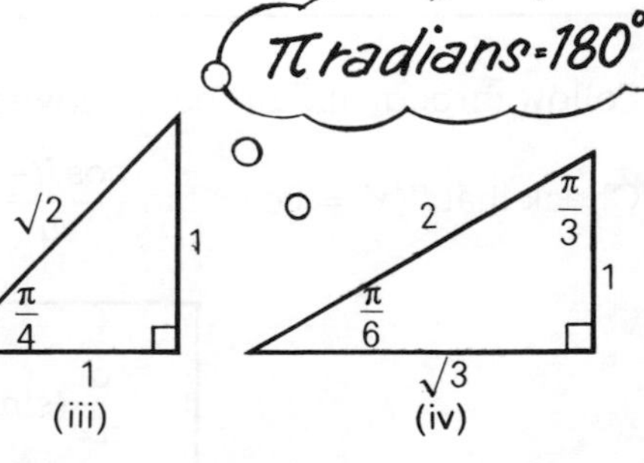

(iii) (iv)

5 a Calculate the gradient of the tangent at $x = 0$ on the curve $y = 3 - \cos x$.
b Find the equation of the tangent.

6 Find the equation of the tangent at the given point on each curve:

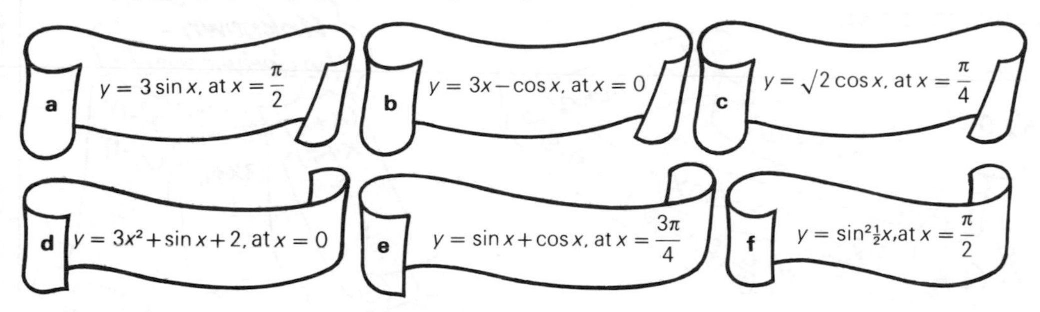

a $y = 3\sin x$, at $x = \frac{\pi}{2}$
b $y = 3x - \cos x$, at $x = 0$
c $y = \sqrt{2}\cos x$, at $x = \frac{\pi}{4}$
d $y = 3x^2 + \sin x + 2$, at $x = 0$
e $y = \sin x + \cos x$, at $x = \frac{3\pi}{4}$
f $y = \sin^2 \frac{1}{2}x$, at $x = \frac{\pi}{2}$

7 Look back at the weight on the spring at the beginning of the chapter. Its displacement is $y(t) = \sin t$, and its velocity is $\frac{dy}{dt}$.

a Calculate, correct to 2 decimal places, the velocity of the weight after:
(i) 1 second (ii) 2 seconds.

b After what time is the velocity first zero?

8 The height of the ocean above or below mean sea-level is given by $y(t) = 2\sin t$, measured in metres at t hours.

a What is the difference in metres between high tide and low tide?

b Find the rate of rise or fall at time t.

c At what level is this rate at maximum (numerical) value?

9 A door of width l feet is being opened smoothly through θ radians from OA to OB.

a Write down the area of $\triangle$OAB.

b Find:
(i) the rate at which the area of triangle AOB is increasing, relative to θ
(ii) the position at which this rate is a maximum.

10 A regular polygon with n sides is inscribed in a circle of radius 1 unit (case $n = 6$ is shown).

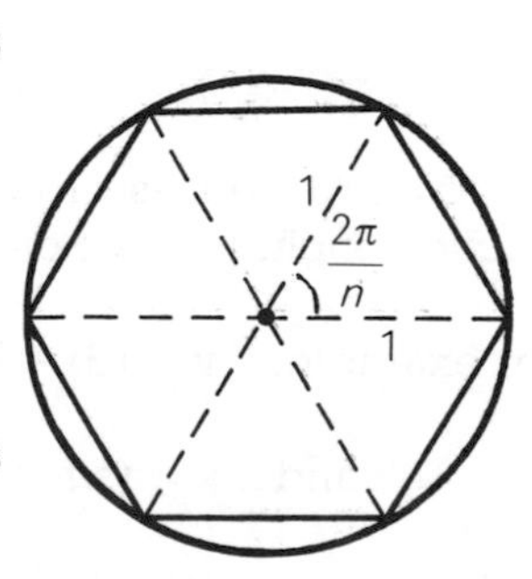

a Show that the area A_n of the polygon is

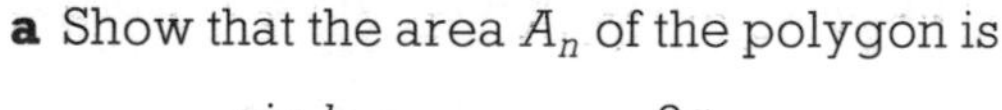

$$A_n = \pi\frac{\sin h}{h}, \text{ where } h = \frac{2\pi}{n}.$$

b Assuming that $A_n \to$ the area of the circle as $n \to \infty$, and so as $h \to 0$, deduce that $\lim_{h\to 0} \frac{\sin h}{h} = 1$.

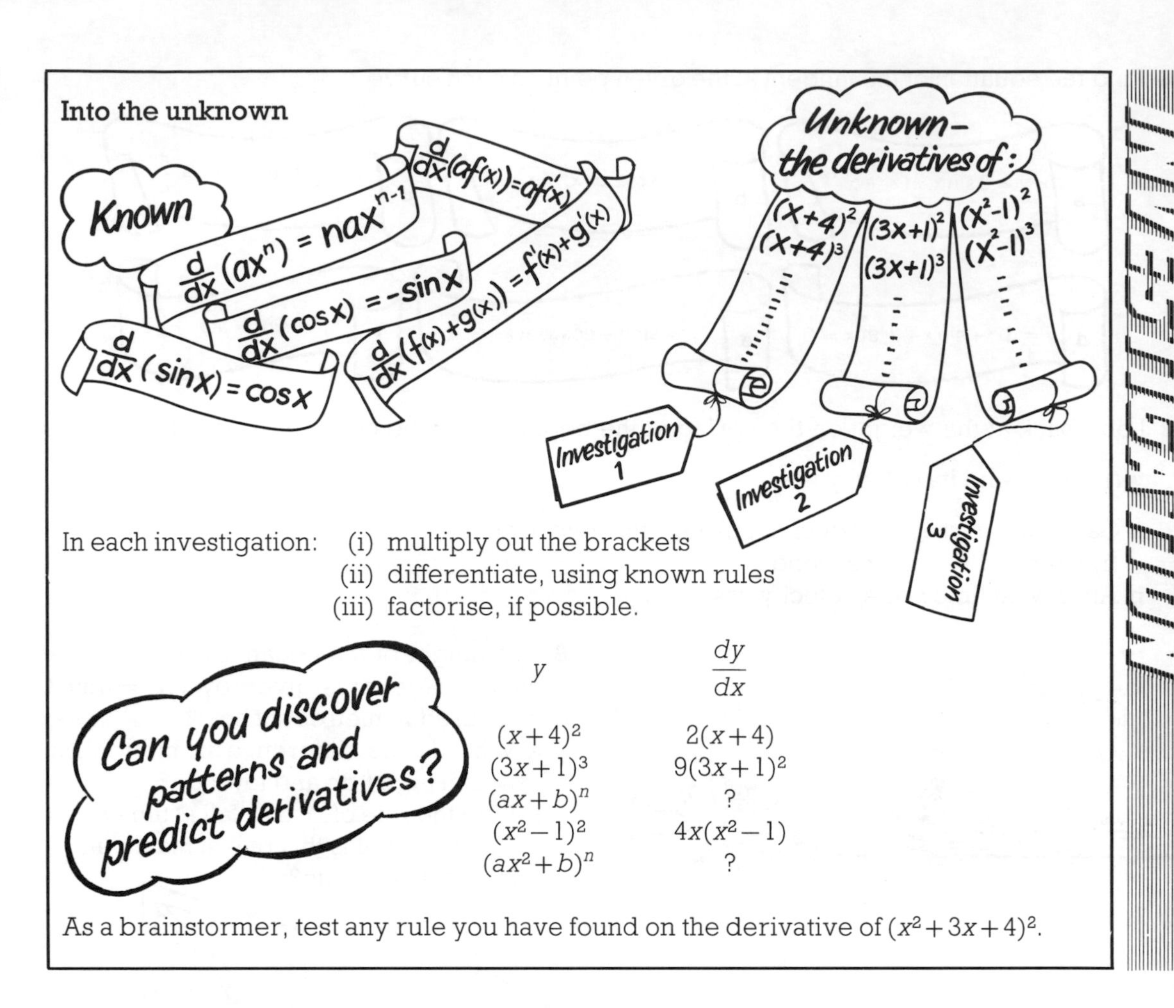

In each investigation: (i) multiply out the brackets
(ii) differentiate, using known rules
(iii) factorise, if possible.

y	$\frac{dy}{dx}$
$(x+4)^2$	$2(x+4)$
$(3x+1)^3$	$9(3x+1)^2$
$(ax+b)^n$	?
$(x^2-1)^2$	$4x(x^2-1)$
$(ax^2+b)^n$	?

As a brainstormer, test any rule you have found on the derivative of $(x^2+3x+4)^2$.

THE CHAIN RULE FOR DIFFERENTIATION

In the Investigation above you may have found that:

if $y = (3x+1)^2$, $\dfrac{dy}{dx} = 6(3x+1) = 2 \times 3(3x+1)$,

if $y = (3x+1)^3$, $\dfrac{dy}{dx} = 9(3x+1)^2 = 3 \times 3(3x+1)^2$, and deduced that

if $y = (ax+b)^n$, $\dfrac{dy}{dx} = na(ax+b)^{n-1}$.

The *chain rule* described below will enable you to differentiate many types of *composite function*, including the ones above.

For example, if $y = (3x+1)^2$, then $y = u^2$, where $u = 3x+1$,

and if $y = (\sin x)^3$, then $y = u^3$, where $u = \sin x$.

Each is a composite function; y is a function of u, where u is a function of x, so y is a function of x.

In each you can find $\frac{dy}{du}$ and $\frac{du}{dx}$, but how can these be combined to give you $\frac{dy}{dx}$?

Use the 'delta' idea used in chapter **2**, Differentiation—1, and follow the chain reaction.

A change Δx in x produces a change Δu in u, which in turn produces a change Δy in y. And, of course, as $\Delta x \to 0$, $\Delta u \to 0$.

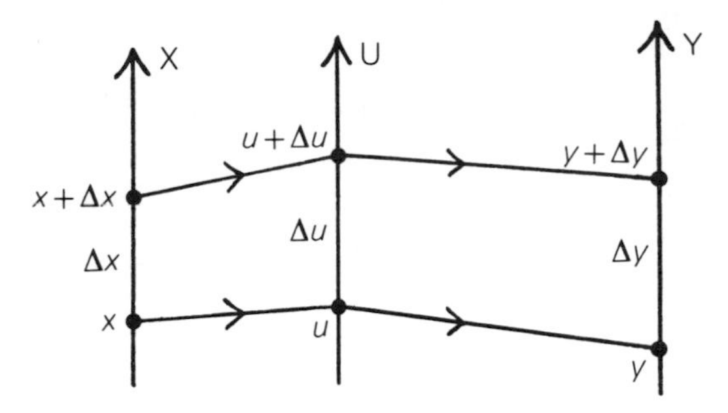

$$\frac{\Delta y}{\Delta x} = \frac{\Delta y}{\Delta u} \times \frac{\Delta u}{\Delta x}, \text{ so } \lim_{\Delta x \to 0} \frac{\Delta y}{\Delta x} = \lim_{\Delta u \to 0} \frac{\Delta y}{\Delta u} \times \lim_{\Delta x \to 0} \frac{\Delta u}{\Delta x}. \ (\Delta u \neq 0).$$

That is, $\boxed{\frac{dy}{dx} = \frac{dy}{du}\frac{du}{dx}.}$

This is called the *chain rule* for differentiation.

Example 1 Differentiate $(3x+1)^2$.

Let $y = (3x+1)^2 = u^2$, where $u = 3x+1$.

$$\frac{dy}{dx} = \frac{dy}{du}\frac{du}{dx} = 2u \times 3 = 6(3x+1).$$

Example 2 Differentiate $\sin^3 x$.

Let $y = (\sin x)^3 = u^3$, where $u = \sin x$.

$$\frac{dy}{dx} = \frac{dy}{du}\frac{du}{dx} = 3u^2 \times \cos x = 3\sin^2 x \cos x.$$

The chain rule in function notation

If $y = f(g(x))$, a composite function, then $y = f(u)$, where $u = g(x)$.

$\frac{dy}{dx} = \frac{dy}{du}\frac{du}{dx}$ becomes:

$$\frac{d}{dx}f(g(x)) = f'(u)\frac{d}{dx}(g(x)) = f'(g(x))\frac{d}{dx}(g(x)),$$

That is

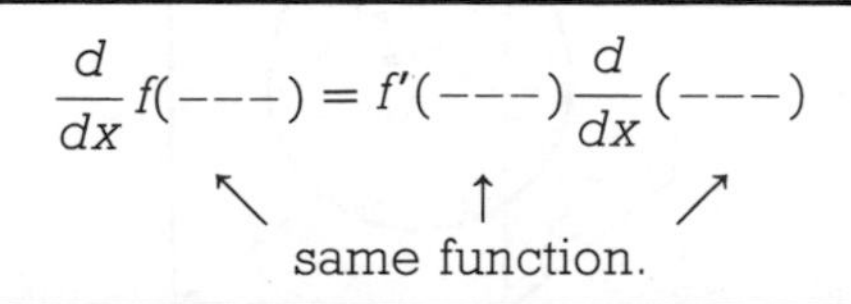

In particular, $\frac{d}{dx}(---)^n = n(---)^{n-1}\frac{d}{dx}(---)$

$$\frac{d}{dx}\sin(---) = \cos(---)\frac{d}{dx}(---)$$

$$\frac{d}{dx}\cos(---) = -\sin(---)\frac{d}{dx}(---).$$

Examples **1** $\frac{d}{dx}(3x+1)^2 = 2(3x+1)\frac{d}{dx}(3x+1) = 6(3x+1)$

2 $\frac{d}{dx}(2-x^2)^{1/3} = \frac{1}{3}(2-x^2)^{-2/3}\frac{d}{dx}(2-x^2) = \frac{-2x}{3}(2-x^2)^{-2/3}$

3 $\frac{d}{dx}\left(\frac{1}{2x^3+1}\right) = \frac{d}{dx}(2x^3+1)^{-1} = (-1)(2x^3+1)^{-2}\times 6x^2 = \frac{-6x^2}{(2x^3+1)^2}.$

Exercise 2

In questions **1–5** find the derivatives with respect to the relevant variables.

1 a $(x+1)^8$ **b** $(2+u)^4$ **c** $(2t-1)^3$ **d** $(3v+1)^5$ **e** $(1-2x)^{10}$

2 a $(x^2+2)^4$ **b** $(1+u^2)^3$ **c** $(4-t^3)^5$ **d** $(2v^4-1)^3$ **e** $(x^2+3x+4)^2$

3 a $(3x-1)^{-2}$ **b** $(x^2+2)^{-1}$ **c** $(2u^2-5)^{-3}$ **d** $\frac{1}{2x+3}$ **e** $\frac{2}{(3t+4)^2}$

4 a $(2x+1)^{1/2}$ **b** $(3u-2)^{1/3}$ **c** $(x^2+x)^{1/2}$ **d** $(t^3+t)^{4/3}$ **e** $(2v^2+4v)^{1/4}$

5 a $(x^3-2x^2)^4$ **b** $\sqrt{(4u+3)}$ **c** $\frac{1}{\sqrt[3]{(6t+1)}}$ **d** $\frac{3}{(2x^2-1)^2}$ **e** $\left(u+\frac{1}{u}-1\right)^2$

6 A curve has equation $y=(x^2+9)^{1/2}$. Find the equation of the tangent at the point where $x=4$.

7 Find the stationary value of $f(x) = 9x-(2x-1)^{3/2}$, and determine its nature.

8 Calculate:

a $\frac{d}{dx}(x^3+3-2\sin x)$ **b** $\frac{d}{dt}\frac{1}{\sqrt{(\cos t)}}$ **c** $\frac{d}{du}\left(\frac{1}{3u+1}+3\cos u\right)$.

9 In a model of a lung, air is drawn from a box, and as a result a balloon inside the box expands.

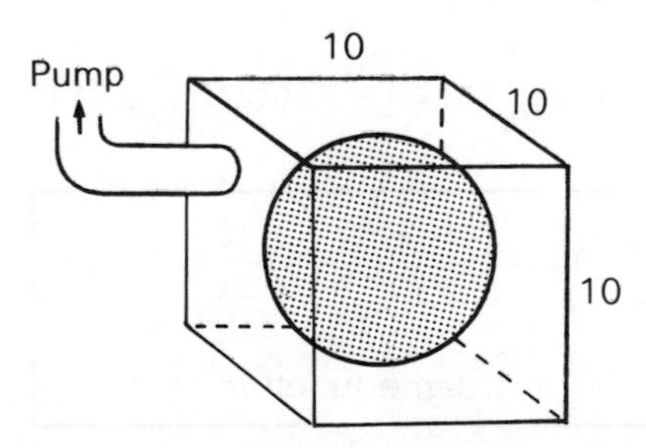

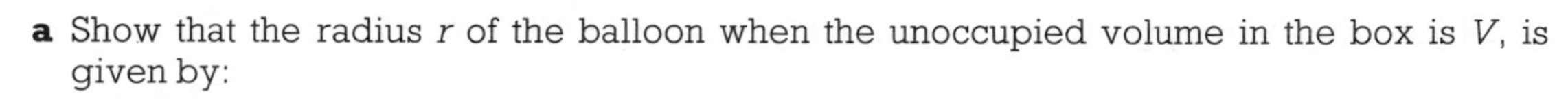

a Show that the radius r of the balloon when the unoccupied volume in the box is V, is given by:

$$r = \left[\frac{3}{4\pi}(1000 - V)\right]^{1/3}.$$

$$\left(\text{For a sphere, } V = \frac{4\pi}{3}r^3.\right)$$

b Find $\dfrac{dr}{dV}$, and explain its sign.

10 A ladder of length 10 feet has its base on the ground and its top against a wall. The base is being pulled along the ground at a steady speed of 2 ft/s.

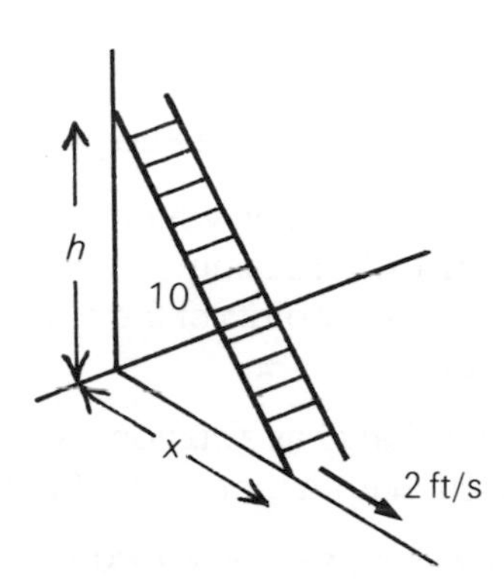

a Write down the value of $\dfrac{dx}{dt}$.

b Show that $h = \sqrt{(100 - x^2)}$.

c Prove that $\dfrac{dh}{dt} = \dfrac{-2x}{\sqrt{(100 - x^2)}}$.

The chain rule can be used in the same way for composite trigonometric functions.

Examples **1** $\dfrac{d}{dx}\sin^3 x = 3\sin^2 x \dfrac{d}{dx}(\sin x) = 3\sin^2 x \cos x$

2 $\dfrac{d}{dx}\cos(2x+3) = -\sin(2x+3) \times 2 = -2\sin(2x+3)$

Exercise 3

In questions **1–3**, find the derivatives with respect to the relevant variables.

1 a $\sin 2x$ **b** $\cos 2x$ **c** $\sin(3u+1)$ **d** $\cos(4v-1)$ **e** $\sin(ax+b)$

2 a $\sin^5 x$ **b** $\cos^2 u$ **c** $\sin^4 x$ **d** $\cos^4 x$ **e** $\sqrt{\sin x}$

3 a $(1+\cos x)^3$ **b** $(\sin x - \cos x)^4$ **c** $\dfrac{1}{\sin x}$ **d** $\dfrac{2}{\cos t}$ **e** $\dfrac{3}{\sin^2 x}$

4 a Explain why, without differentiating, you can say that $\dfrac{d}{dx}(\cos^2(ax) + \sin^2(ax)) = 0$ for each constant a.

b Check by differentiating that the derivative is zero.

5 Find $\dfrac{d}{dx}(\sin x°)$ and $\dfrac{d}{dx}(\cos x°)$.

Hint $\sin x° = \sin\left(\dfrac{\pi}{180} x \text{ radians}\right) = \sin\left(\dfrac{\pi}{180} x\right)$.

6 a Find the stationary points on the graph of $y = \sin x$ for $0 \leqslant x \leqslant 2\pi$.

b Show that the one at $x = \dfrac{\pi}{2}$ is a maximum turning point.

7 a Show that the graphs of $y = \sin x$ and $y = \cos x$ intersect at $x = \dfrac{\pi}{4}$.

b Find the gradients of the tangents to the curves at this point.

c Calculate the acute angle between the tangents.

8 The equation of a curve is $y = \sin^2 x$.

a Show that $\dfrac{dy}{dx} = \sin 2x$.

b Find the equation of the tangent at the point given by $x = \dfrac{\pi}{4}$.

c If $0 \leqslant x \leqslant \pi$, find the interval on which y is:
(i) increasing (ii) decreasing.

d Investigate the stationary points of the curve for $0 \leqslant x \leqslant \pi$.

e Sketch the curve for $0 \leqslant x \leqslant \pi$.

9 If $f(x) = x + \sin x$, show that f has a stationary point at $x = \pi$, and that this is a point of inflexion. Explain why the function is an increasing function on R.

10 Sketch the graph of $y = \sin x + \cos x$ for $0 \leqslant x \leqslant 2\pi$.

MORE MAXIMUM AND MINIMUM VALUES

In Differentiation—1 you developed techniques for finding stationary values. You can now take this a little further by considering values of functions over given intervals.
Have a careful look at the following two graphs on the given intervals.

1 $y = x^2$ on the closed interval $-1 \leqslant x \leqslant 3$, written $[-1, 3]$.
On $[-1, 3]$ the maximum value of $f(x) = x^2$ is $f(3) = 9$ and the minimum value is $f(0) = 0$. $f(3)$ is a value at an end point and $f(0)$ is a minimum *turning* value.

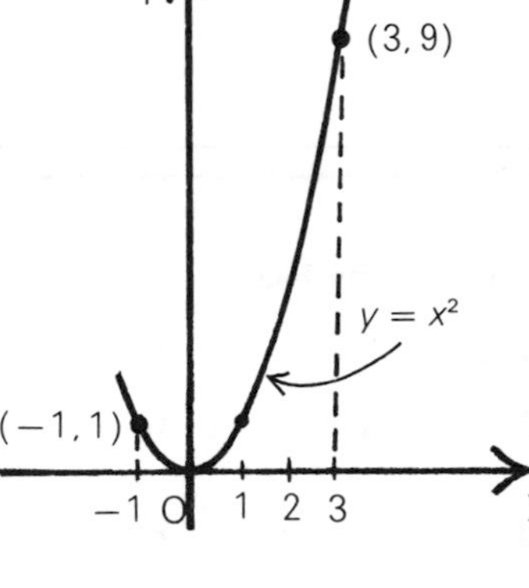

2 The graph $y = f(x)$ of a function f is drawn below on the interval $[-4, 3]$, i.e. for $-4 \leqslant x \leqslant 3$.

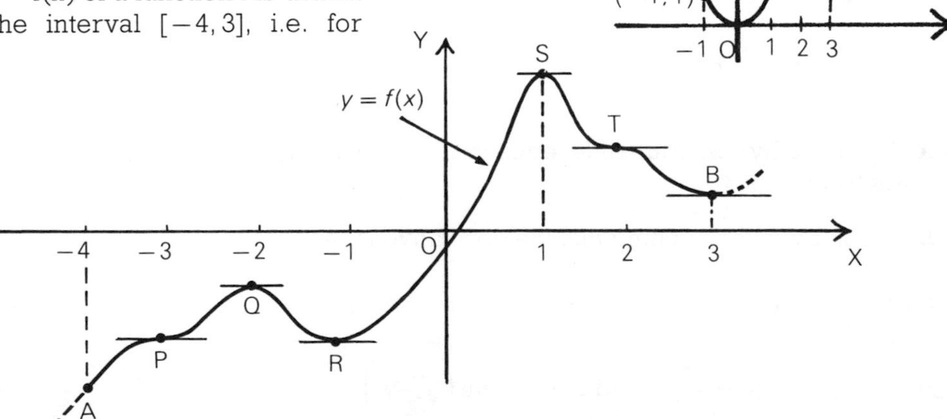

Check these facts:

(i) The end points are A, B.
(ii) The stationary points are P, Q, R, S, T, B.
(iii) The maximum *turning* points are Q, S, the minimum *turning* points R, B and the horizontal points of inflexion P, T.
(iv) On $[-4, 3]$, f has maximum value $f(1)$ (given by the maximum turning point S) and minimum value $f(-4)$ (given by the end point A).

> Notice, from these two graphs, that the maximum and minimum values of f on a *closed* interval can occur at end points or at stationary points in the interval.

Example Find the maximum value and the minimum value of $f(x) = 2x^3 - 3x^2 - 12x + 5$ on $[-2, 1]$, i.e. for $-2 \leqslant x \leqslant 1$.

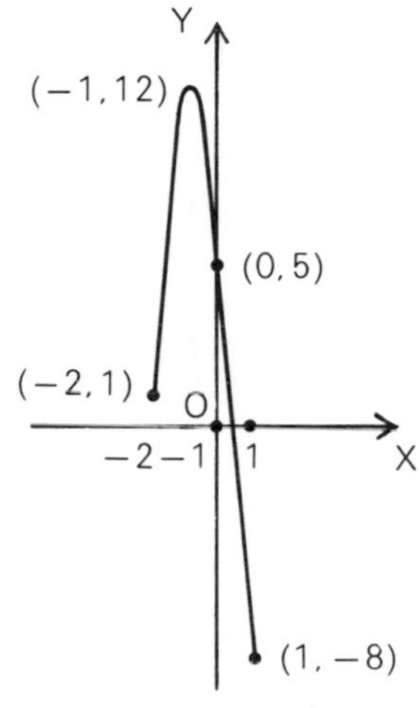

$f'(x) = 6x^2 - 6x - 12 = 6(x^2 - x - 2) = 6(x+1)(x-2)$.
For stationary values, $f'(x) = 0$, so $x = -1$ or $x = 2$.
$x = 2$ does not lie in the interval, so is discarded.
$f(-1) = 2(-1)^3 - 3(-1)^2 - 12(-1) + 5 = 12$.
The values at the end points are:
$f(-2) = 2(-2)^3 - 3(-2)^2 - 12(-2) + 5 = -16 - 12 + 24 + 5 = 1$,
and $f(1) = -8$.
The maximum value of f on $[-2, 1]$ is 12, and the minimum value is -8.

Exercise 4

1 The graph $y = f(x)$ of function f is shown for $-4 \leqslant x \leqslant 2$.

a In $[-4, -1]$, the maximum value is $f(-4)$. Write down the minimum value.

b Write down the maximum and minimum values of f on the interval $[0, 2]$.

c Write down the maximum and minimum values on $[-2, 1]$.

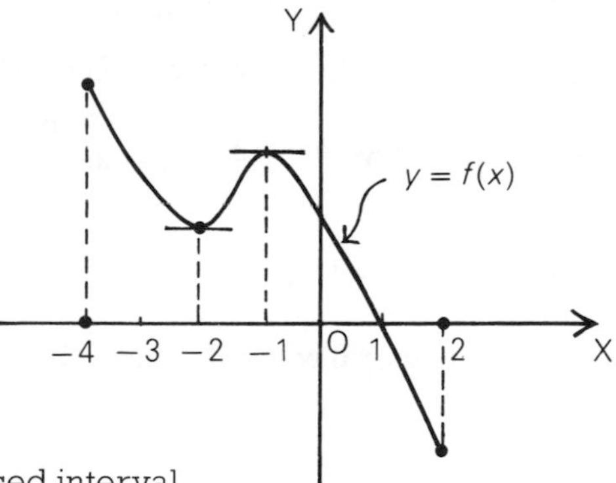

2 Calculate for each function:
(i) the values of $f(x)$ at the end points of the closed interval
(ii) the stationary value(s) lying within the interval.
Then (iii), write down the maximum and minimum values of f in the interval.

a $f(x) = 5 - 2x^2$ on $[-1, 2]$
b $f(x) = x^3 - 3x$ on $[-2, 3]$
c $f(x) = 3x^5 - 20x^3$ on $[-1, 3]$
d $f(x) = \sin x + \cos x$ on $[0, \pi]$
e $f(x) = x + \sin 2x$ on $[0, \pi]$
f $f(x) = 3\cos 2x$ on $\left[-\frac{\pi}{2}, \frac{\pi}{2}\right]$.

MAXIMUM AND MINIMUM VALUES ON NON-CLOSED INTERVALS

Example Discuss the maximum and minimum values of:
$f(x) = 7+2x-x^2$ on R.

a There are no end points this time. Why?

b For large x, positive or negative, $f(x) \doteqdot -x^2$, which is itself large and negative.
$f(x) \to -\infty$ as $x \to \infty$ and as $x \to -\infty$.

c *Stationary values* $f'(x) = 2-2x = 2(1-x)$.
For SVs, $f'(x) = 0$, so $x = 1$. Then $f(1) = 8$.

Nature of SV

x	$\longrightarrow$	1	$\longrightarrow$
$f'(x)$	+	0	−

Shape of graph of f:

$f(x)$ has maximum value 8, when $x = 1$.
$f(x)$ has no minimum value on R.

Exercise 5

Investigate maximum and minimum values of:

1 $f(x) = 7+2x+x^2$ on R

2 $f(x) = 2(x+1)^2$ on R

3 $f(x) = x^4+4x^3+3$ on R

4 $f(x) = x+\dfrac{9}{x}$ on $x > 0$

5 $f(x) = x^2+\dfrac{1}{x^2}$ on $x > 0$

6 $f(x) = 4+\sin^2 x$ on R.

(Can you also answer **6** without using calculus?)

OPTIMIZATION

This is the word used for maximum–minimum problems, especially those of a practical nature. In many cases differential calculus can be used to obtain solutions. The approach is to *model the problem* by bringing in suitable variables and functions.

Example

A ship at 1305 hours is 4 km due South of Arran lighthouse and is steering a straight course in a N.E. direction. What is its shortest distance from the lighthouse?

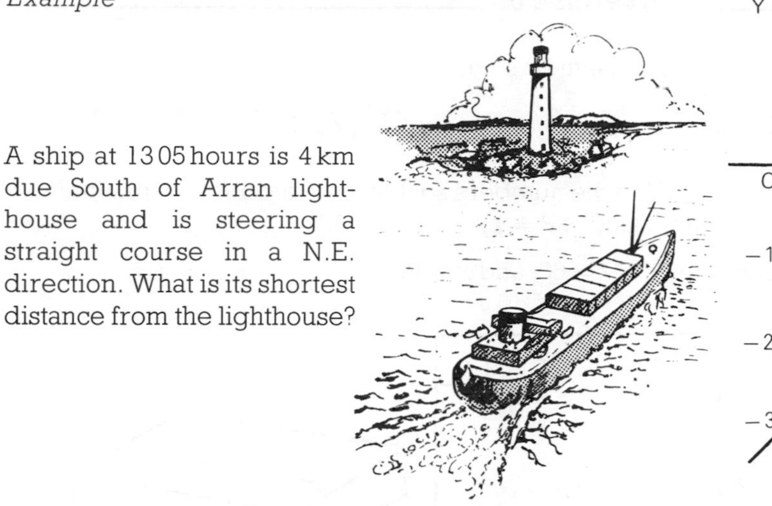

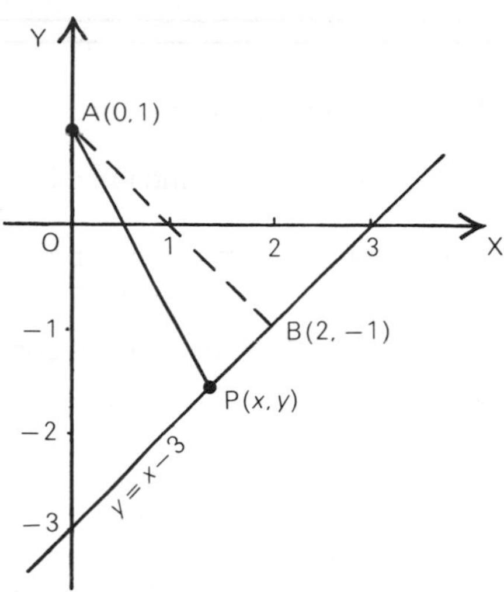

The problem can be modelled as shown, so that you have to find the shortest distance from the point A(0, 1) to the line $y = x - 3$. Why?
As P(x, y) varies on the line, you have to find the minimum value of AP. We use AP² to avoid square roots (noting that AP > 0).

$$AP^2 = (x-0)^2 + (y-1)^2 = x^2 + (x-3-1)^2 = x^2 + (x-4)^2$$
$$= x^2 + x^2 - 8x + 16 = 2x^2 - 8x + 16.$$

The problem now is that of finding the minimum value of $D = 2x^2 - 8x + 16$ for $x \in R$.

$\dfrac{dD}{dx} = 4x - 8 = 4(x-2)$. If $\dfrac{dD}{dx} = 0$, $x = 2$, then $D(2) = 8$.

x	⟶	2	⟶
$\dfrac{dD}{dx}$	−	0	+

Shape of graph of D:

It follows that $D(2) = 8$ is the least possible value of D, so $\sqrt{8} = 2\sqrt{2}$ km is the minimum distance from the lighthouse.

(Check that, in the model, the ship is then at the point B(2, −1), and AB is perpendicular to the line $y = x - 3$, as geometry would suggest.)

Exercise 6

1 The product of two numbers is 36. Minimize their sum.
$\left(\textit{Model: } xy = 36\text{; find the minimum value of } S = x+y = x+\dfrac{36}{x}\ (x > 0).\right)$

2 The difference of two numbers is 12. Find the numbers so that their product is a minimum. (*Model:* $x-y = 12$; find the minimum value of $P = xy = y(y+12)$.)

3 Maximize x^2y subject to the condition $x+y = 60$, and $x > 0$.

4 Four congruent squares are cut out at the corners of a square of cardboard of side 12 cm, and then the cardboard is folded to make a small open tray. What should the length of side, x cm, of a cut out square be in order to maximize the volume of the tray?
(*Model:* Use a formula for the volume $V(x)$ of the box.)

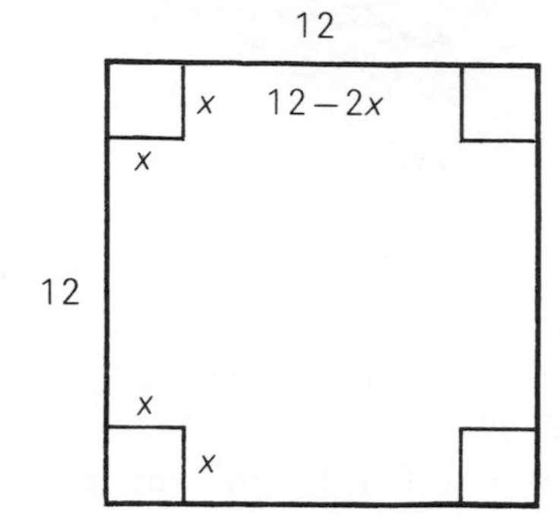

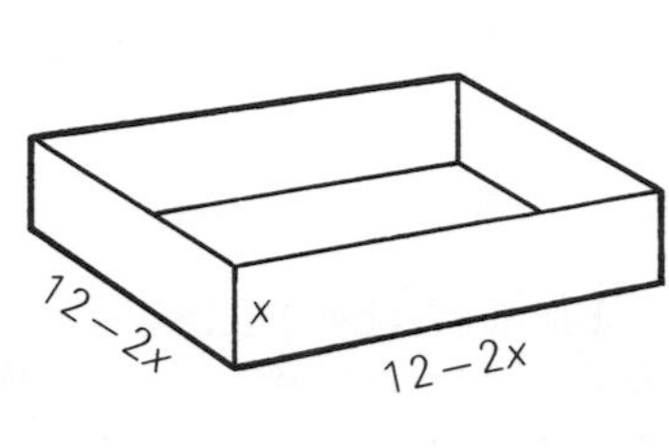

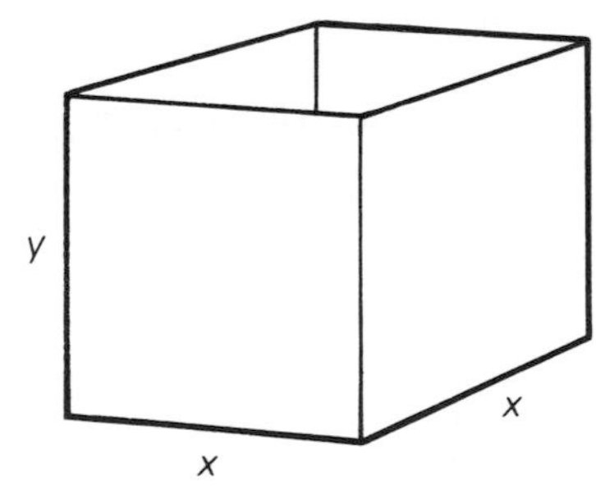

5 A box, cuboid in shape, has a square base and no lid. Its volume is 13·5 cm³. Choose the dimensions which will minimize the external surface area of such a box.

6 The mathematical model for the height h metres of a rocket t seconds into flight is:

$h(t) = 5+195t+96t^2-t^3\ (t \geqslant 0)$.

When does the rocket reach its maximum height?

7 Market researchers have determined that the fraction of market share of a new product x months after its introduction is $F(x) = \dfrac{2x-4}{x^2}\ (x \geqslant 2)$.

After how many months does the product have a maximum market share?

8 The manufacturer of SUNSHINE fruit juice is reconsidering the size of the container to reduce costs. The container must be cylindrical and hold 128π ($\doteqdot 402$) cm³ of juice. The cost is proportional to the surface area of the container. What are the best dimensions?
(*Model:* $\pi r^2h = 128\pi$; minimize $S = 2\pi r^2+2\pi rh$.)

9 The manager of a fast-food restaurant wishes to estimate the number of employees he should hire for the busy lunch hour. Labour cost is £5 per hour per employee. He estimates that the lost profit per hour by having only x employees is £300/x. What staffing level will minimize the total cost per hour?

$\left(\textit{Model:}\ \text{Minimize } 5x + \dfrac{300}{x}\ (x > 0).\right)$

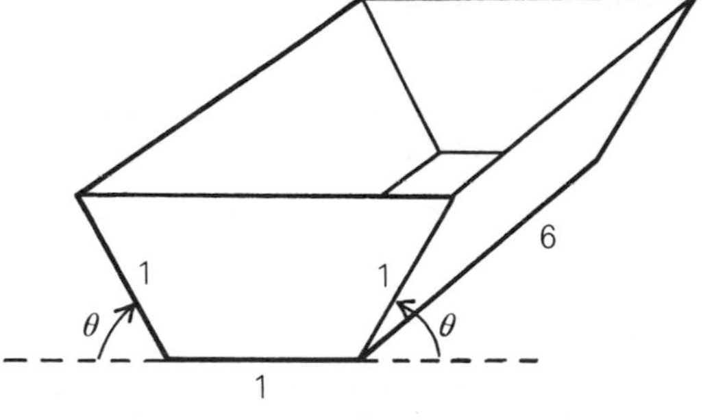

10 A trough is constructed in the shape shown with a trapezoidal cross-section and length 6 feet, three sides of the trapezium having length 1 foot. Find the value of θ which maximizes capacity $\left(0 < \theta < \dfrac{\pi}{2}\right)$.

(*Model:* Show that the capacity of the trough is $V(\theta) = 6(1 + \cos\theta)\sin\theta$.)

11 A triangular field is bounded by two roads meeting at right angles and by a hedge. A rectangular play area has to be constructed with two sides along the roads. The layout is modelled as shown, using coordinates.

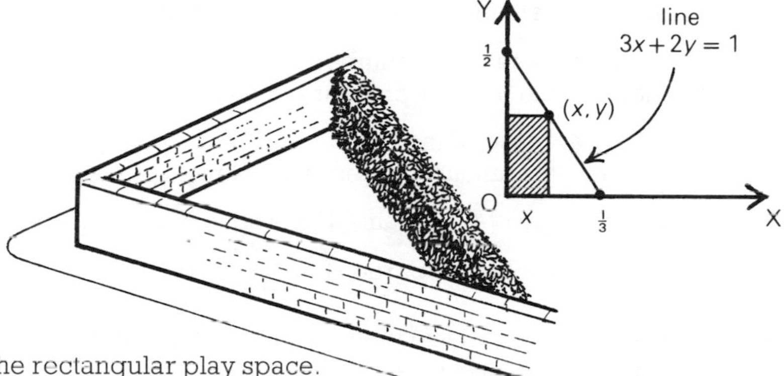

Find the maximum area of the rectangular play space.
(*Model:* The area is $A = xy$, where $3x + 2y = 1$.)

BRAINSTORMER

You met this tricky problem in the Introduction to Differentiation—1.

Sunnyside Cottage and Canal View Villa are at A, B on opposite sides of a straight canal. Where should a bridge CD be placed to minimize the walking distance AC+CD+DB between the two houses (distances in kilometres)?

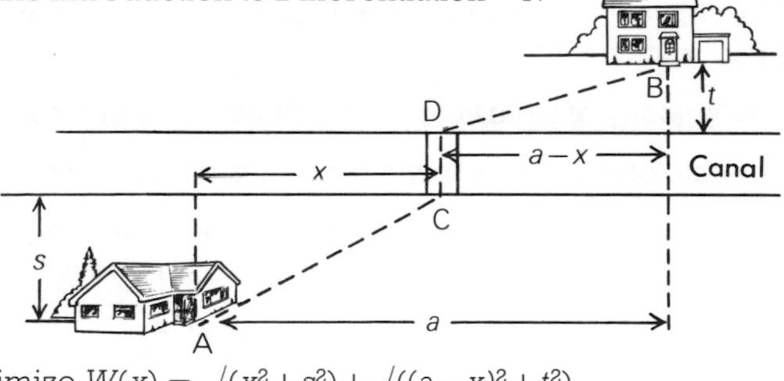

a Explain why you have to minimize $W(x) = \sqrt{(x^2 + s^2)} + \sqrt{((a-x)^2 + t^2)}$.

b Check that $\dfrac{dW}{dx} = \dfrac{x}{\sqrt{(x^2+s^2)}} - \dfrac{(a-x)}{\sqrt{((a-x)^2+t^2)}}$.

c Show that, at a stationary point, $x^2((a-x)^2 + t^2) = (a-x)^2(x^2 + s^2)$.

d Deduce that the distance is minimized when $\dfrac{x}{a-x} = \dfrac{s}{t}$, and that AC, DB are then parallel.

e Where should the bridge be placed if A is on the bank of the canal?

MATHEMATICAL MODELS INVOLVING DERIVATIVES

You should now be thoroughly used to the idea of modelling in mathematics. You always have to translate a problem statement into some mathematical form—an equation, a graph, a formula, etc. Many models involve rates of change which can be written as derivatives.

Example

Statement: The rate of decrease in temperature (T) of a body is proportional to the difference between the temperature of the body and the temperature of the surroundings (T_o).

Model: $\frac{dT}{dt} = -k(T - T_o)$ (t time, k a positive constant since T is decreasing).

Exercise 7

Write down a mathematical model for each statement:

1 The rate of increase of the number (N) of bacteria present in a culture at time t seconds is proportional to the number present at time t.

2 The rate of decrease of the quantity (Q) of radium present in a substance at any time t is proportional to the amount present at that time.

3 The rate of change of the volume V of a sphere with respect to its radius r is proportional to r^2. Find the constant of proportion $\left(V = \frac{4\pi}{3}r^3\right)$.

4 The rate of change of the volume V of a cube with respect to its edge length l is proportional to l^2. Find the constant of proportion.

5 The rate of change of the volume V of a cube with respect to its surface area A is proportional to $\sqrt{A}$. Find the constant of proportion.

6 The rate of change of $\cos x$ with respect to x is proportional to $\sin x$. What is the constant of proportion?

7 The velocity of a particle at time t moving on the x-axis is proportional to t.

8 The rate of change of y with respect to x is inversely proportional to x^3.

9 In blowing up a balloon (assumed to be spherical) the radius r changes at the rate 1 cm/s. Find the rate at which the volume V is changing when $r = 10$.

CHECK-UP ON **DIFFERENTIATION—2**

1 Differentiate:

a $3\sin x - 4\cos x$ **b** $x + 2\sin x$ **c** $5\cos x + \frac{1}{2x}$ **d** $\sin 2x + \cos 3x$.

2 Differentiate:

a $(x^2 + 3x + 1)^{1/3}$ **b** $\sqrt{(x - x^3)}$ **c** $\sin(2x + 3)$ **d** $\sin^4 x$ **e** $\sqrt{(\cos 3x)}$.

3 A curve has equation $y = 1+\cos^2 x$.

a Show that $\dfrac{dy}{dx} = -\sin 2x$.

b Find the equation of the tangent at $x = \dfrac{\pi}{3}$.

c Find, for $0 \leqslant x \leqslant \pi$, where y is increasing and where it is decreasing.
d Discuss stationary points for $0 \leqslant x \leqslant \pi$.
e Sketch the curve for $0 \leqslant x \leqslant \pi$.

4 Find the maximum value and the minimum value of:

a $f(x) = x^2+4x-1$ on $[-3, 1]$ **b** $f(x) = x+\cos 2x$ on $\left[0, \dfrac{\pi}{2}\right]$, to 2 decimal places.

5 Discuss maximum values and minimum values of:

a $f(x) = 2x^2+12x+5$ on R **b** $f(x) = x^2+\dfrac{16}{x}$ for $x > 0$

c $f(x) = \sin^3 x$ on R. (Do you need calculus to answer **c**?)

6 Ian Thomson decides to fence in a rectangular section of his garden. One edge is against the side of the house, so no fence is needed there. Material for the fence costs £2/metre for the two ends and £5/metre for the edge opposite the house. He can afford to spend £1000 but naturally he wants as large an area fenced off as possible. Help him to find the dimensions of his largest fenced area.

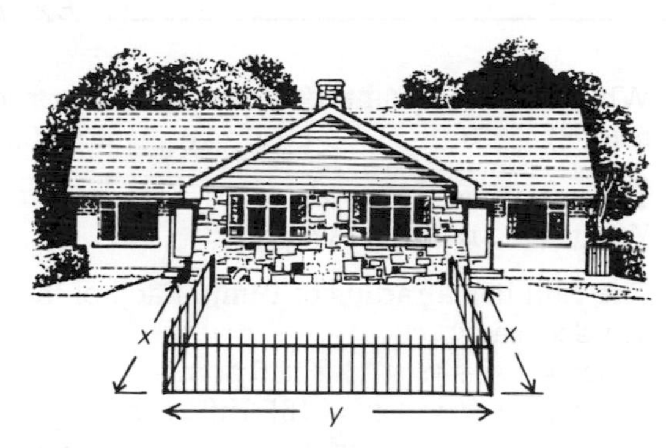

(*Model:* $4x+5y = 1000$; find the maximum value of $A = xy = \frac{1}{5}x(1000-4x)$).

7 Write down a mathematical model for each statement:

a In a large city of population A the number of people who hear a piece of news at time t is $N(t)$. The rate at which $N(t)$ is changing is proportional to the number who have not yet heard the news.

b The rate of change of the area of a circle with respect to the circumference of the circle is proportional to that circumference. Find the constant of proportion.

CIRCULAR REGIONS OF THE PLANE

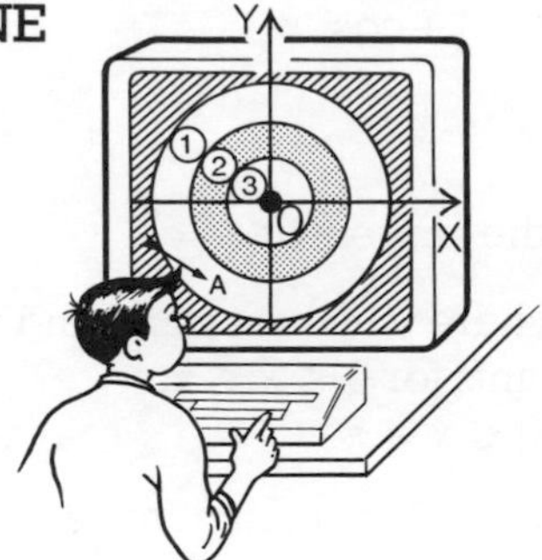

In a computer simulation of an archery target the arrow lands randomly on the screen. A bull scores 5, the inner ring scores 3, the middle 2 and the outer 1.
The radius r of the target's outer circle is 10 units.

The computer records A$(-5, -8)$ as a *hit*, having checked that $OA^2 = (-5-0)^2+(-8-0)^2 = 89 \leqslant 100\ (= r^2)$.
It would record (a, b) as a hit after checking that the point (a, b) belongs to the set $\{(x, y): x^2+y^2 \leqslant 100,\ x, y \in R\}$.

Exercise 1

1 Which of these points does the computer record as hits?
a $(4, 8)$ **b** $(8, 7)$ **c** $(9, -3)$ **d** $(-6, -6)$ **e** $(-7, 7)$.

2 The computer records (a, b) as a *miss* after checking that $(a, b) \in \{(x, y): x^2+y^2 \ldots 100\}$. What is the missing symbol?

3 The bull has a radius of 1 unit. Each of the three rings is 3 units wide. Copy and complete, as necessary:
a $\{(x, y): x^2+y^2 \leqslant 1\}$. Hit, 5 points
b $\{(x, y): 1 < x^2+y^2 \leqslant 16\}$. Hit, ... points
c $\{(x, y): \ldots < x^2+y^2 \leqslant \ldots\}$. Hit, 2 points
d $\{(x, y): \ldots < x^2+y^2 \leqslant \ldots\}$. Hit, 1 point
e $\{(x, y): x^2+y^2 > \ldots\}$. Miss, 0 points.

4 a Use your 'rules' in question **3** to write down the scores for arrows landing at:
(i) $(3, 4)$ (ii) $(-1, 2)$ (iii) $(6, 8)$ (iv) $(1, 12)$ (v) $(-6{\cdot}2, 8)$.
b Three arrows land at $\{(1, 1), (6, 9), (-2, -8)\}$. Calculate the total score.

5 Make separate sketches of these circular regions:
a $\{(x, y): x^2+y^2 \leqslant 1\}$ **b** $\{(x, y): 1 \leqslant x^2+y^2 \leqslant 16\}$
c $\{(x, y): 16 \leqslant x^2+y^2 \leqslant 25\}$ **d** $\{(x, y): x^2+y^2 = 1\}$
e $\{(x, y): x^2+y^2 = 25\}$ **f** $\{(x, y): x^2+y^2 = 100\}$.

CIRCLE, CENTRE O(0, 0) AND RADIUS r

In questions **5d**, **e**, **f** of Exercise **1**, the sets define the rim of a region.
In **d**, $x^2+y^2 = 1$ is called 'the equation of a circle, centre $(0, 0)$ and radius 1 unit.
In **e** the radius is 5, and in **f** the radius is 10.

C is the set of points at distance r from O.

$C = \{P(x, y) : OP = r\}$
$= \{P(x, y) : OP^2 = r^2\}$
$= \{P(x, y) : x^2 + y^2 = r^2\}$.

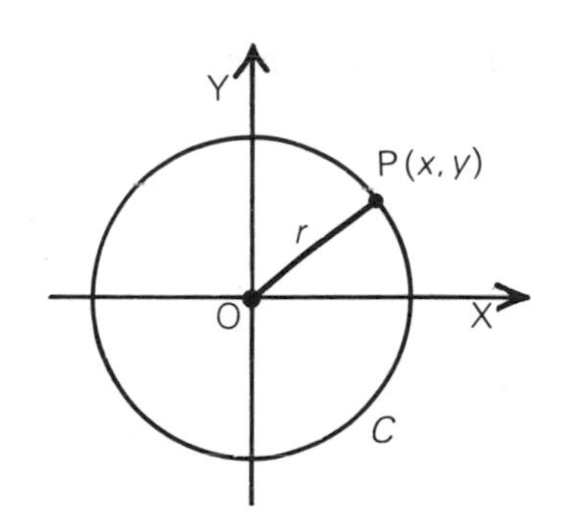

C has equation $x^2 + y^2 = r^2$.

Exercise 2

1 Write down the equation of the circle, centre the origin and radius:
a 2 **b** 7 **c** 12 **d** 25

2 Find the radius and the equation of the circle, centre the origin, passing through:
a (3, 4) **b** (−5, 12) **c** (−3, −3) **d** (0, 7).

3 State the centre and radius of each of these circles:
a $x^2 + y^2 = 64$ **b** $x^2 + y^2 = 169$ **c** $y^2 = 36 - x^2$ **d** $3x^2 + 3y^2 = 12$
(In **c** and **d** you have to arrange the equation in the 'standard' form $x^2 + y^2 = r^2$.)

4 A stone is dropped into water. A circular ripple travels out at a rate of 2 cm per second.
Using the point where the stone entered the water as the origin, write down an equation for the ripple after:
a 1 second **b** 5 seconds **c** 7 seconds **d** t seconds.

5 **a** Check that A(3, 4), B(−4, 3) and C(5, 0) are all the same distance from the origin.
b Write down the equation of the circle, centre O, passing through A, B and C.
c Sketch the circle, and on your sketch shade the region defined by $\{(x, y) : x^2 + y^2 \leqslant 25\}$.
d Find the equation of the circle whose radius is half the radius of the circle in your sketch.

6 A circular paddock has an area of 64π m^2.
A farmer uses a 10π metres length of fencing to form a circular enclosure in the middle of the paddock.
Taking the centre of the paddock as origin, find the equation of:
a the perimeter of the paddock
b the perimeter of the enclosure.

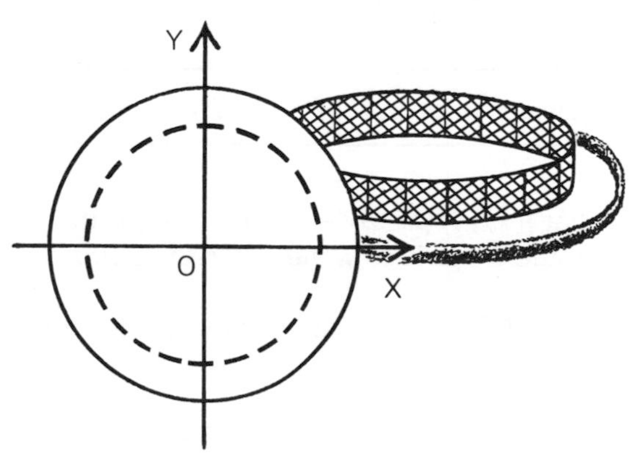

7 Dr Shepherd, an astronomer, focuses her telescope on the crescent moon, setting the crosswires of the telescope at the moon's centre (already computed). The photograph is faulty, with white flecks on it. The radius of the moon's photograph is 3·6 cm.

a Write down the equation of the moon's outer rim using the crosswires as axes. (The 'unseen' part is shown dotted.)

b White flecks are found at these points on the plate. Which of them cannot be stars?
(i) (3, 1) (ii) (2, −3) (iii) (1·5, 2) (iv) (3·5, −1·5) (v) (2·5, −2·5).

A car jack is shaped like a rhombus. To change a wheel, one side of the car has to be lifted, which means that the vertical diagonal of the rhombus increases as the horizontal one decreases. A is the centre of a strut.

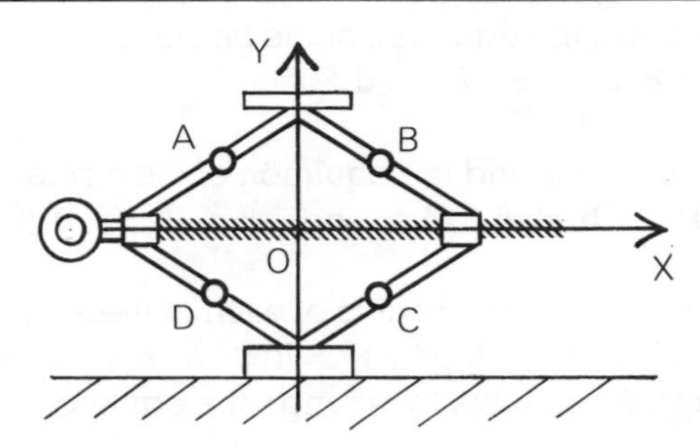

a The jack is placed in position and then wound up. Describe the locus of A from the *x*-axis to the *y*-axis.

b B, C and D are the centres of the other struts.
Describe the union of the loci of A, B, C and D, and find its equation, using the axes shown.

CIRCLE S, CENTRE $C(a, b)$ AND RADIUS r

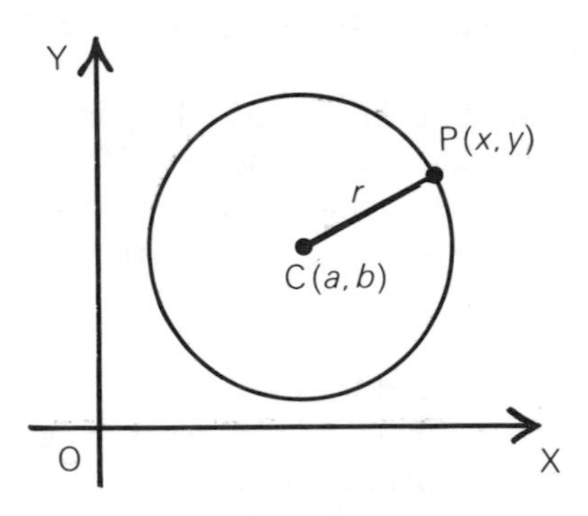

S is the set of all points at distance r from C.

$S = \{P(x, y) : CP = r\}$
$= \{P(x, y) : CP^2 = r^2\}$
$= \{P(x, y) : (x-a)^2+(y-b)^2 = r^2\}$.

> S has equation $(x-a)^2+(y-b)^2 = r^2$.

Note that $\{(x, y) : (x-a)^2+(y-b)^2 < r^2\}$ is the *interior*, the region inside the circle. You should be able to write down the *exterior* in set notation.

Example Find the equation of the circle, centre C(1, −2), passing through A(4, 7).

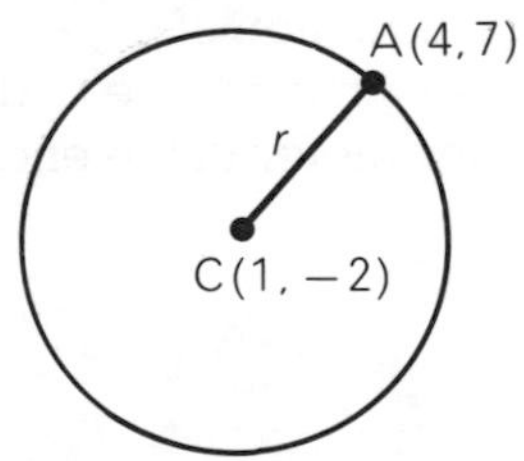

$$r^2 = CA^2 = (4-1)^2+(7-(-2))^2 = 9+81 = 90.$$

The circle has equation $(x-1)^2+(y+2)^2 = 90$.

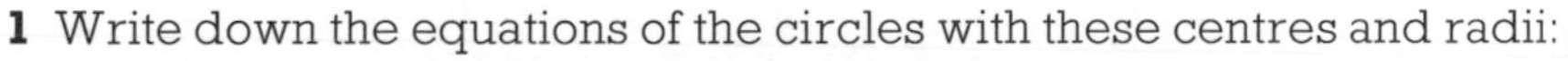

Exercise 3

1 Write down the equations of the circles with these centres and radii:
a $(2, 7)$, $r = 4$ **b** $(3, -1)$, $r = 7$ **c** $(-5, -7)$, $r = 2{\cdot}5$.

2 State the centre and radius of each of these circles:
a $(x-3)^2+(y-1)^2 = 25$ **b** $(x-4)^2+(y-4)^2 = 1$
c $(x+5)^2+(y-2)^2 = 4$ **d** $(x+2)^2+(y+1)^2 = 1{\cdot}44$
e $(x-1)^2+y^2 = 16$ **f** $x^2+(y+3)^2 = 36$.

3 For each pair of points, A is the centre of a circle passing through B.
(i) Calculate the radius of the circle. (ii) Write down its equation.
a A(2, 2), B(5, 4) **b** A(1, −5), B(3, 1) **c** A(−2, −3), B(6, 1)
d A(−3, −2), B(3, −2) **e** A(−3, 0), B(3, 0) **f** $A(a_1, a_2)$, $B(b_1, b_2)$.

4 P is (2, −2) and Q is (8, 6). A circle is drawn on PQ as diameter.
Find: **a** the centre **b** the radius **c** the equation, of the circle.

5 David took a photograph of the Big Wheel at the Fairground. Using the axes shown, he estimated A to be (1, 6) and B(11, 10).
Find the equation of the circular part of the Big Wheel through A and B.

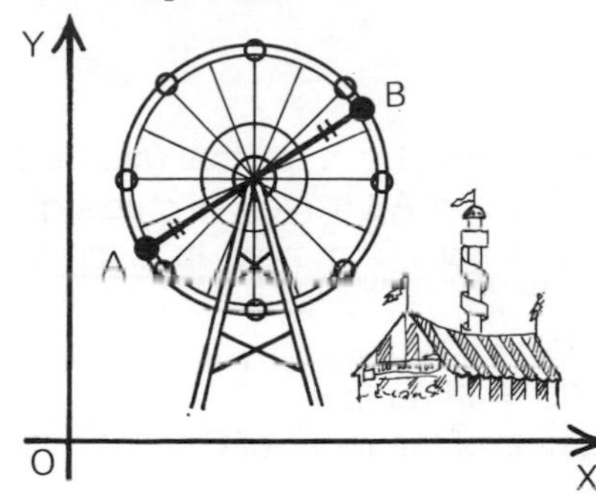

6

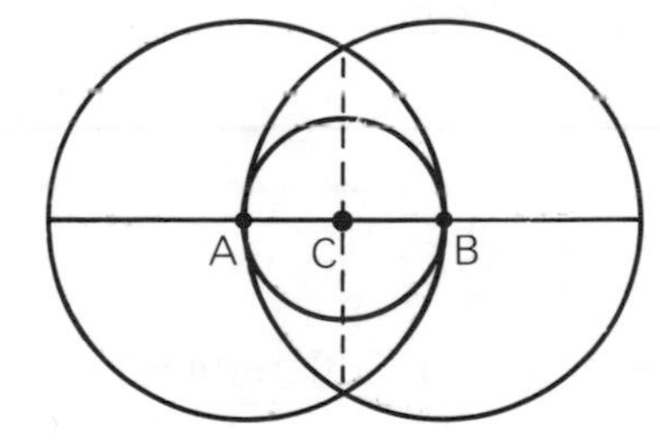

The small circle, centre C, has equation
$(x+2)^2+(y+1)^2 = 25$.
The large circles, centres A and B, touch the small circle, and AB is parallel to the x-axis. Find:

a the centre and radius of each circle
b the equations of the large circles.

7 In a computer game a tractor is drawn. The front wheel's rim has equation
$$(x-5)^2+(y-3)^2 = 4.$$
a The ground is a line parallel to the x-axis.
Write down the equation of this line.

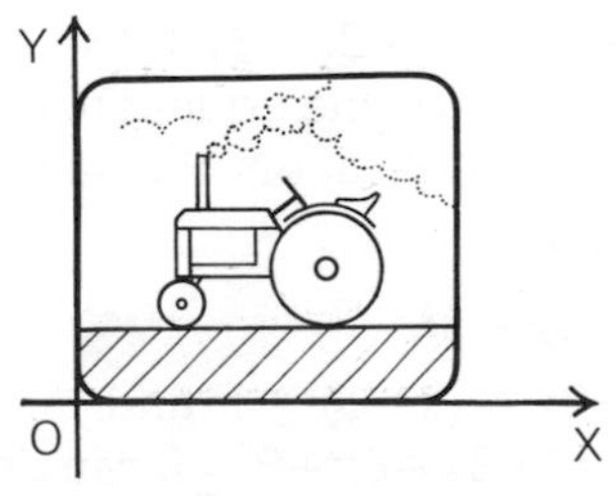

b The rear wheel's radius is 3 times the size of the front one's. If the points of contact of the wheels with the ground are 10 units apart, find the equation of the rear wheel's rim.

c The tractor moves 2 units to the left. Find the new equation of the rim of:
(i) the front wheel (ii) the rear wheel.

8

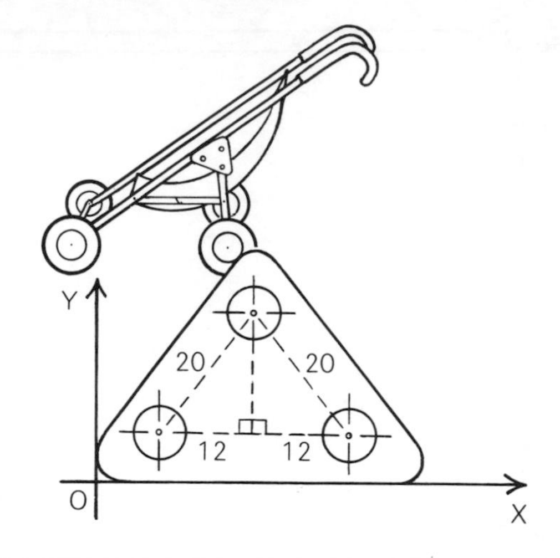

A metal plate is used to strengthen the struts of a push-chair. The holes have radius 4 units and their centres form an isosceles triangle with sides 20, 20 and 24 units long. If the equation of the edge of the top hole is $(x-20)^2+(y-18)^2=16$, find the equations of the edges of the other two holes.

A linkage is driven by a piston (shaded) 3 units long. The rod AB travels in a straight line. D is fixed and C moves in a circle centre D.
In *Figure 1*, AD = 20 units.
In *Figure 2*, AD = 14 units.
Find the equation of the locus of C.

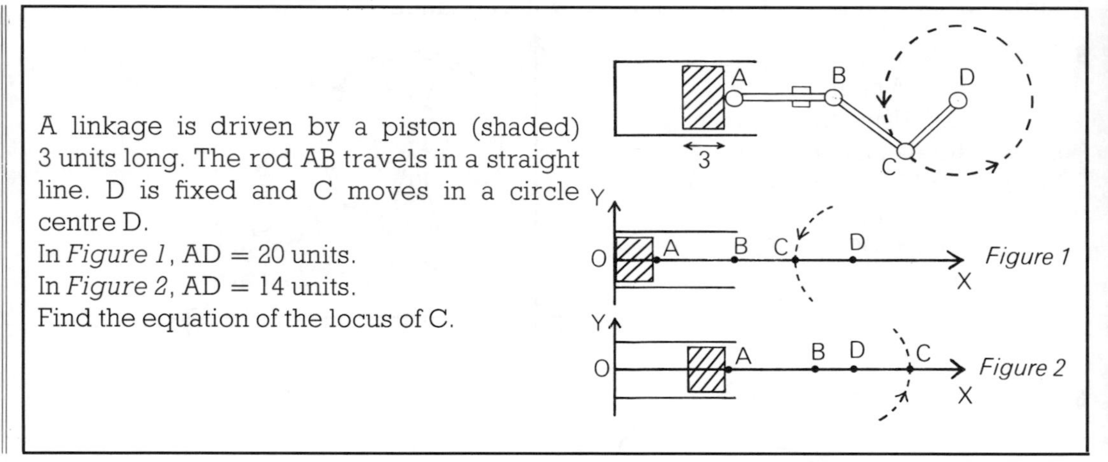

THE CIRCLE DISGUISED

The equation $(x-4)^2+(y+3)^2=4$ represents the circle, centre $(4, -3)$, radius 2.
Multiplying out, $(x^2-8x+16)+(y^2+6y+9)=4$,

so, $x^2+y^2-8x+6y+21=0$ represents the same circle.

But what about the other way round?
For example, does $x^2+y^2+2x-8y+8=0$ represent a circle?
You will have to rearrange the equation in 'standard' form $(x-a)^2+(y-b)^2=r^2$.
In Chapter **3** on Quadratic Theory you met the useful trick of completing the square:

$x^2+2ax+a^2=(x+a)^2$,

so, $x^2+2ax=(x+a)^2-a^2$.

For example, $x^2+2x=(x+1)^2-1^2$ and $y^2-8y=(y-4)^2-4^2$.
The equation $x^2+y^2+2x-8y+8=0$ can be rearranged:

$$(x^2+2x)+(y^2-8y)=-8$$
$$(x+1)^2-1^2+(y-4)^2-4^2=-8$$
$$(x+1)^2+(y-4)^2=9.$$

The equation represents the circle, centre $(-1, 4)$, radius 3.

From the particular to the general

Start with $x^2+y^2+2gx+2fy+c=0$ and rearrange as above:

$$(x^2+2gx)+(y^2+2fy) = -c$$
$$(x+g)^2-g^2+(y+f)^2-f^2 = -c$$
$$(x+g)^2+(y+f)^2 = g^2+f^2-c$$
$$(x-(-g))^2+(y-(-f))^2 = g^2+f^2-c.$$

> As long as $g^2+f^2-c>0$, the equation $x^2+y^2+2gx+2fy+c=0$ represents a circle, with centre $(-g, -f)$ and radius $\sqrt{(g^2+f^2-c)}$.

A point to note: Coefficients of x^2 and y^2 both $= 1$.

Example Show that this equation represents a circle, and write down the centre and radius: $x^2+y^2-4x+8y-1=0$.
$g=-2$, $f=4$, $c=-1$; $g^2+f^2-c=4+16+1=21>0$, so the equation represents a circle.
The centre is $(2, -4)$ and the radius is $\sqrt{(g^2+f^2-c)}=\sqrt{(21)}$.

Exercise 4

1 Check which of these equations represent circles. When the equation represents a circle, write down the coordinates of its centre and the length of its radius.

a $x^2+y^2-4x-8y+17=0$ **b** $x^2+y^2-6x-10y+35=0$
c $x^2+y^2-8x-2y+13=0$ **d** $x^2+y^2-2x+2y-5=0$
e $x^2+y^2+6x-2y+10=0$ **f** $x^2+y^2+4x+6y+4=0$
g $x^2+y^2+2x+4y+13=0$ **h** $2x^2+2y^2+6x+2y-13=0$.

2 Match up each circle with its equation:
a $x^2+y^2+6x-8y+24=0$
b $x^2+y^2+2x-2y-2=0$
c $x^2+y^2+2x+2y-14=0$
d $x^2+y^2-6x-2y+1=0$
e $x^2+y^2-10x+8y+32=0$
f $x^2+y^2+8x-8y-49=0$.

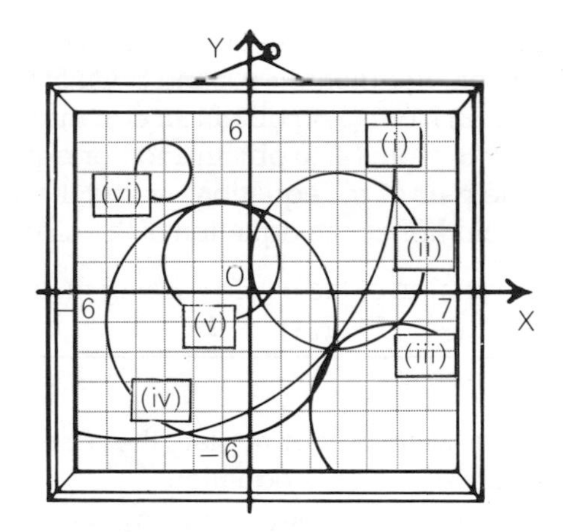

3 A circle has equation $x^2+y^2-6x-4y-12=0$.
a Check that $A(-1, 5)$ lies on the circle.
b Find the point on the circle diametrically opposite A.

4 a Write down the centre and radius of the circle $x^2+y^2-2x-8y-8=0$.
b Which of the points $A(1, 1)$, $B(4, 7)$, $C(-1, -1)$, $D(3, 2)$, $E(-3, 0)$, $F(5, 1)$ lie:
(i) inside the circle (ii) on the circle (iii) outside the circle?

5 A tangent is drawn from A(9, −4) to the circle $x^2+y^2+8x+2y-8=0$.

a Find the centre and radius of the circle.

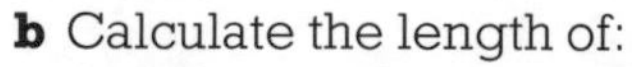

b Calculate the length of:
(i) AC (ii) the tangent AB.

c Repeat the above steps to develop a formula for the length of the tangent from $A(x_1, y_1)$ to the circle $x^2+y^2+2gx+2fy+c=0$.

6 Use the formula you developed in question **5c** to calculate the length of the tangent from:

a (5, −1) to $x^2+y^2-2x-4y-20=0$ (Explain!)
b (−4, −1) to $x^2+y^2-10x-7=0$
c (8, −9) to $x^2+y^2+6x+4y-21=0$.

7 a Point (8, *a*) lies on the circle $x^2+y^2-8x-6y+9=0$. Find the possible values of *a*.
b Point (*b*, 1) lies on the circle $x^2+y^2+2x+2y-11=0$. Calculate the possible values of *b*.
c By sketching the circles, illustrate why there is one possibility in **a** and two in **b**.

8 a Check that A(−3, 4) lies on the circle $x^2+y^2=25$, centre O.
b Calculate the gradient of: (i) radius OA (ii) the tangent at A.
c Find the equation of the tangent at A.

9 Find the equation of the tangent at A to each circle:

a A(−2, 3), $x^2+y^2=13$
b A(−1, 3), $x^2+y^2=10$
c A(2, 3), $(x-2)^2+(y-6)^2=9$
d A(3, −2), $(x-2)^2+(y+1)^2=2$
e A(3, −1), $x^2+y^2+8x-6y-40=0$
f A(−1, 2), $x^2+y^2-2x-3y-1=0$.

10 Points (2, 5) and (−2, 1) lie on the circle $x^2+y^2+2gx+2fy+7=0$.

a Make two equations in *g* and *f*.
b Solve this pair of equations to find the values of *g* and *f*.
c Hence find the equation of the circle, and give its centre and radius.

11 The points (−1, 1), (3, 3) and (6, 2) lie on the circle $x^2+y^2+2gx+2fy+c=0$.

a Form three equations in *g*, *f* and *c*. Call them (1), (2), (3).
b Subtract (2) from (1) to obtain an equation in *g* and *f*.
c Subtract (3) from (2) to obtain another equation in *g* and *f*.
d Solve the pair of *g*, *f* equations to find the values of *g* and *f*.
e Find *c*, and hence the equation of the circle through the three points.

12 Find the equation of the circle passing through the points:

a (−5, −2), (−2, 7), (7, 4) **b** (−1, −1), (1, 3), (0, 6).

On the watchmaker's plan the rim of the medium cog, B, has equation

$$4x^2+4y^2-28x-28y+97=0.$$

The radii of the rims of the cogs are in the ratio 7 : 4 : 2. Find the equations of the rims of cogs A and C.

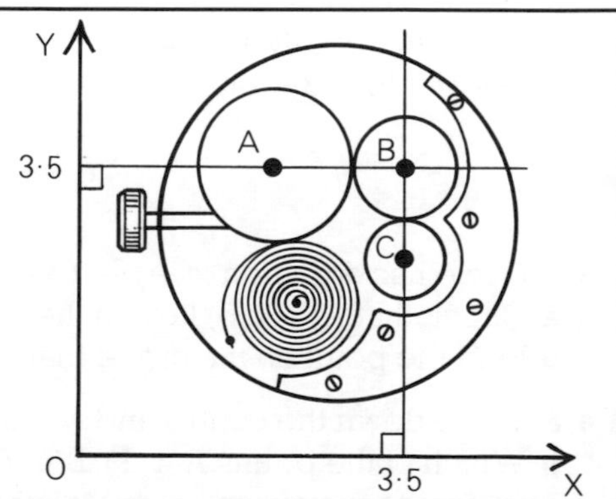

INTERSECTIONS OF LINES AND CIRCLES

If you sketch a line and a circle you will find that your diagram will look like one of these:

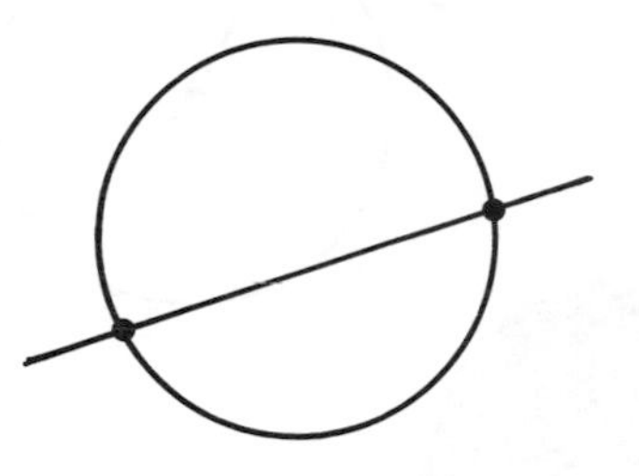

2 points of intersection

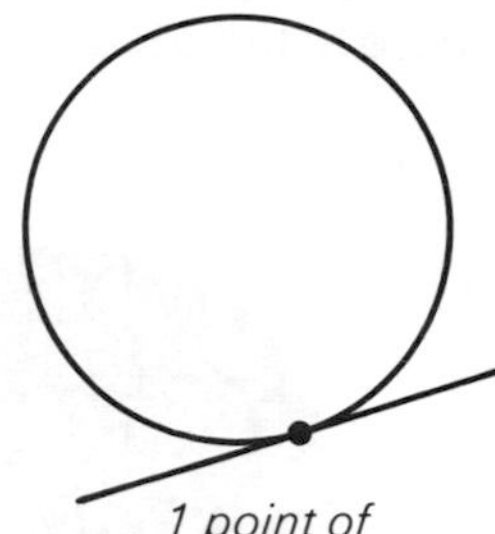

1 point of intersection—tangency

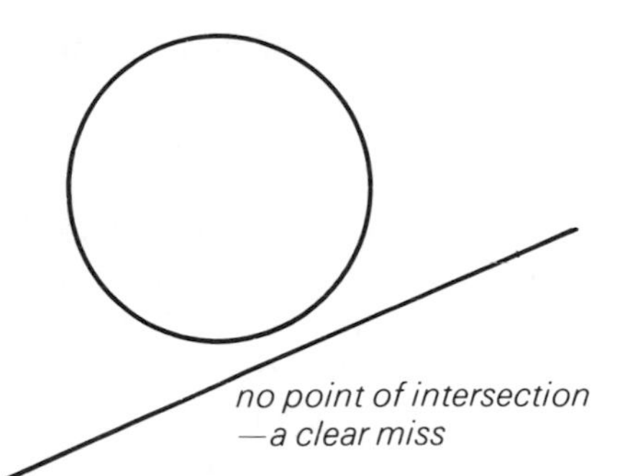

no point of intersection —a clear miss

Take the line $5y-x+7=0$ and the circle $x^2+y^2+2x-2y-11=0$.

They meet where $\begin{cases} 5y-x+7=0, \text{ i.e. } x=5y+7 & (1) \\ x^2+y^2+2x-2y-11=0 & (2) \end{cases}$

At the points of intersection, substituting (1) into (2),

$$(5y+7)^2+y^2+2(5y+7)-2y-11=0$$
$$25y^2+70y+49+y^2+10y+14-2y-11=0$$
$$26y^2+78y+52=0$$
$$y^2+3y+2=0.$$

So $(y+2)(y+1)=0$ and $y=-2$ or $y=-1$.
From (1), $x=-3$ when $y=-2$, and $x=2$ when $y=-1$.
The line meets the circle at the points $(-3, -2)$ and $(2, -1)$.

Exercise 5

1 Do the lines and circles intersect? If they do, find the coordinates of the intersections.
a $y=1$, $x^2+y^2=10$ **b** $y=2$, $x^2+y^2=4$ **c** $y=4$, $x^2+y^2=9$
d $y=x$, $x^2+y^2=8$ **e** $y=x+1$, $x^2+y^2=5$ **f** $2x+y=5$, $x^2+y^2=5$
g $y-x=1$, $x^2+y^2+2x+2y+1=0$ **h** $y=9-2x$, $x^2+y^2-2x-4y-5=0$
i $y-3x+10=0$, $x^2+y^2+2x+6y-10=0$ **j** $2y-x+8=0$, $x^2+y^2+2x+4y=0$.

2 For each circle, calculate the length of the chord cut off by:
(i) the x-axis (ii) the y-axis.
a $x^2+y^2-6x-8y=0$ **b** $x^2+y^2-24x+10y=0$
c $x^2+y^2-7x+8y+12=0$ **d** $x^2+y^2-6x-10y+9=0$.
What is special about **d**?

3 In each case prove that the line is a tangent to the given circle and find the point of contact:
a $x=3$, $x^2+y^2=9$ **b** $y=3x-10$, $x^2+y^2=10$
c $4x-y+7=0$, $x^2+y^2-6x-4y-4=0$
d $y-2x+4=0$, $x^2+y^2-2x-6y+5=0$
e $3x+2y-6=0$, $x^2+y^2+2x+4y-8=0$.

4 a Prove that the line $x-3y+20=0$ is a common tangent to the circles $x^2+y^2=40$ and $x^2+y^2-22x-14y+160=0$.
b Find the points of contact and the length of the tangent between these points.

5 a Write down the coordinates of the centre of the circle $x^2+y^2-8x-2y-3=0$.
b Find the equation of the tangent to the circle at A(8, 3).

A (8, 3)
C

6 Find the equation of the tangent to each circle at the given point:
a $x^2+y^2-4x-2y+3=0$, (3, 2)
b $x^2+y^2+6x+2y+6=0$, $(-1, -1)$.

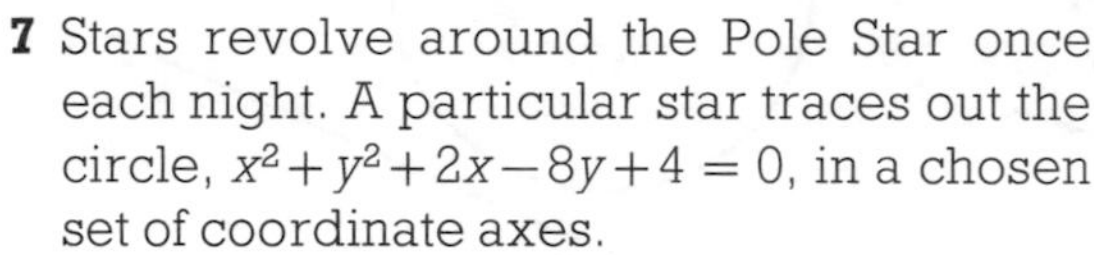

7 Stars revolve around the Pole Star once each night. A particular star traces out the circle, $x^2+y^2+2x-8y+4=0$, in a chosen set of coordinate axes.
The horizon has equation $y = 1$.
a State the coordinates of the Pole Star.
b Calculate the coordinates of the points of rising and setting of the moving star.

8 The circle $x^2+y^2=10$ meets a circle with centre $(-2, 4)$ at A and B, where AB has equation $x-2y+5=0$.
a Find the coordinates of A and B.
b Find the equation of the second circle.

9 The position of a quayside derrick is monitored by a computer, which calculates angle α. The locus of the end of the arm is a circle with equation

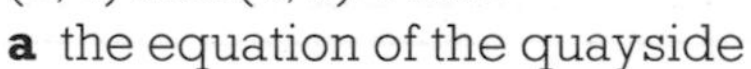

$$x^2+y^2-10x-8y+31=0$$

for the axes chosen. The line representing the quayside passes through the points (0, 4) and (4, 6). Find:
a the equation of the quayside
b the coordinates of A and B
c the range of values of α, to the nearest degree, for which the end of the derrick's arm projects beyond the edge of the quay.

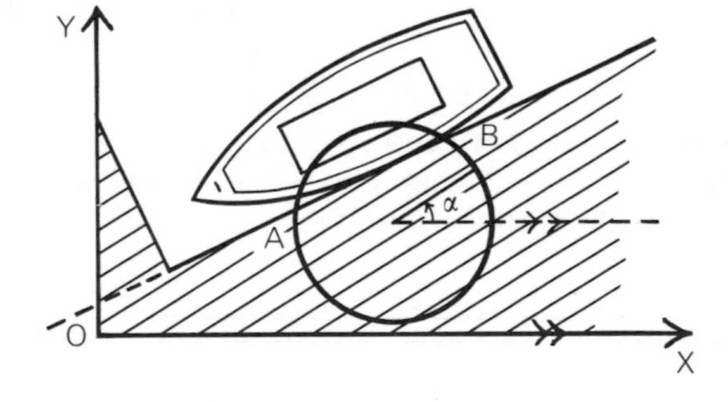

1 If you ask the professionals they'll tell you that the snooker ball heads towards a specific point provided the cue ball strikes it at a point diametrically opposite the target point.
For example, the cue ball C_1 is just able to strike the snooker ball S at P, sending it to target T. You can't do this with ball C_2. Investigate the locus of the centres of cue balls able to strike the point P. Assume that each ball has a radius of 3 cm.

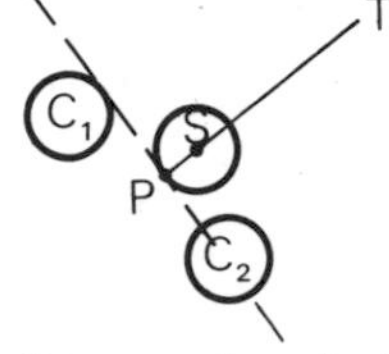

2 Investigate the condition that the circle $x^2+y^2+2gx+2fy+c=0$:
a passes through the origin
b has its centre on: (i) the x-axis (ii) the y-axis (iii) the line $y = x$
c touches: (i) the y-axis (ii) the x-axis (iii) both axes
d has radius 1, etc.

3 Systems of circles

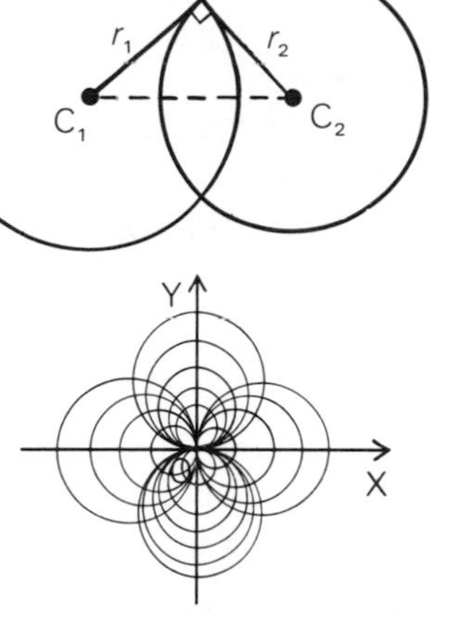

a Two circles are *orthogonal* if they intersect at right angles. Investigate the relationship between g_1, g_2, f_1, f_2, c_1, c_2 if the circles $x^2+y^2+2g_1x+2f_1y+c_1=0$ and $x^2+y^2+2g_2x+2f_2y+c_2=0$ are orthogonal.

b Construct a system of circles passing through the origin, and touching:
(i) the x-axis (ii) the y-axis.
Why are these orthogonal systems?

c There are other orthogonal systems of circles. Find out what you can about them and their applications in physics.

CHECK-UP ON **THE CIRCLE**

1 Shade the region of the plane defined by:
a $\{(x, y): x^2+y^2 \leqslant 9\}$ **b** $\{(x, y): 4 \leqslant x^2+y^2 \leqslant 16\}$.

2 a State the equation of the circle, centre the origin and radius 5 units.
b State the centre and radius of the circle $4x^2+4y^2=9$.

3 a Write down the centre and radius of the circle $(x-4)^2+(y+3)^2=16$.
b Write down the equation of the circle, centre $(-5, 7)$ and radius 2 units.

4 On a street plan two roads are represented by straight lines with equations $y=-x+4$ and $y=x$.
The intersection of these lines marks the centre of a roundabout whose diameter is 2 units on the plan. Find the equation of the circle representing the roundabout.

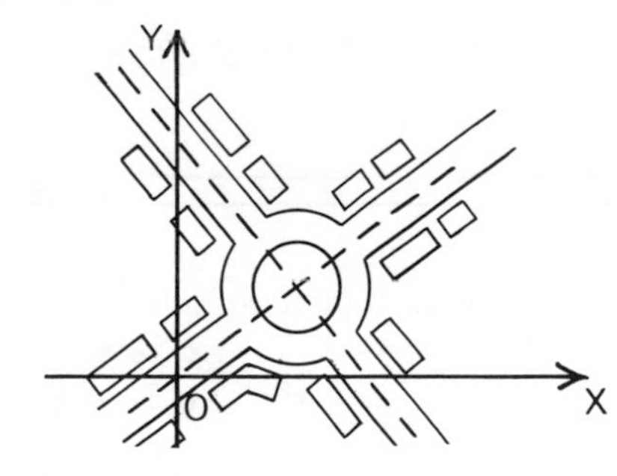

5 a Write down the centre and radius of the circle $x^2+y^2+6x+2y-15=0$.
b Which of the following points lie:
(1) inside, (2) on, (3) outside, the circle $x^2+y^2-2x-2y-11=0$?
(i) $(-2, -1)$ (ii) $(-1, 4)$ (iii) $(2, 4)$ (iv) $(5, 1)$ (v) $(3, -2)$.

6 Find the points in which the line $y=x+1$ intersects the circle $x^2+y^2-4x-8y+7=0$.

7 a Prove that the line $x-7y-50=0$ is a tangent to the circle $x^2+y^2=50$, and find the point of contact.
b Find the equation of the tangent to the circle $x^2+y^2-6x-8y-9=0$ at the point $(0, -1)$.

8 Find the equation of the circle passing through the points $(1, 1)$, $(6, -2)$, $(3, -7)$.

POLYNOMIALS

What do these expressions in x have in common?
They are all sums of constants and constants times x^n ($n \geqslant 0$).
They are examples of *polynomials*.

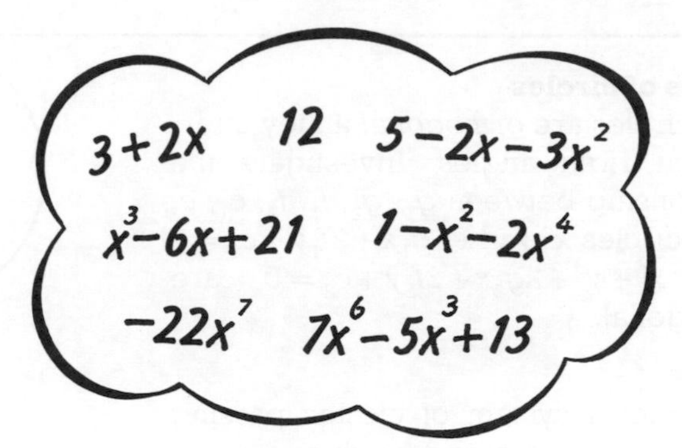

12 is a constant polynomial.
$3+2x$ is a polynomial of *degree* 1. (x^1 is the highest power of x in it.)
$5-2x-3x^2$ is a polynomial of degree 2. (x^2 is its highest power of x.)
$x^3-6x+21$ has degree 3, and $1-x^2-2x^4$ has degree 4.

The 'general polynomials' form a tree structure:

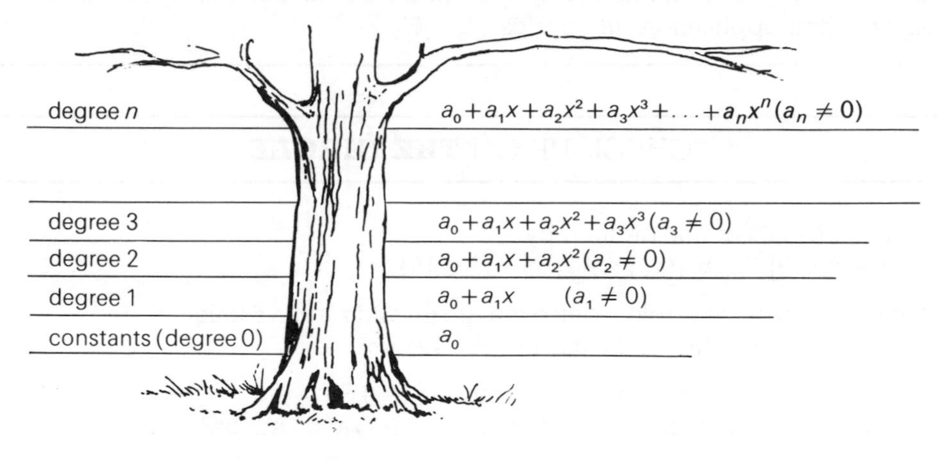

$a_0, a_1, a_2, \ldots$ are the *coefficients* of the polynomials.

Exercise 1

1 State the degree, and the coefficient of x^2, for each of these polynomials:
a x^2-1 **b** $4-2x^2+3x^3$ **c** $5x^4+x^3+7x^2$ **d** x^5-3x^4+2x+9
e 17 **f** $-1-3x-x^2$ **g** $7x-x^{11}$ **h** $3x^2-2x^{10}+1$.

2 Polynomials can be disguised. For example,
$x^2(2x+1) = 2x^3+x^2$; $(x^2+1)^2 = x^4+2x^2+1$.
Rewrite these as polynomials in 'standard' form with powers of x in *decreasing* order:
a $x(x-1)$ **b** $(2x+1)(3-x)$ **c** $(2x-1)^2$ **d** $(2x-1)^3$
e $(3x+1)(5x^3-1)$ **f** $(4-x)(x^4-3x^2+x-1)$ **g** $(x-1)(x-2)(x-3)$.

3 Each term of a polynomial is of the form a_nx^n, where $n \geqslant 0$. Simplify the following if necessary, and where you have a polynomial, state its degree:

a $(4x)^4$ **b** $x^2-x-(x^2+x)$ **c** $\dfrac{1}{x}$ **d** $\dfrac{x^4-x^3}{x^2}$

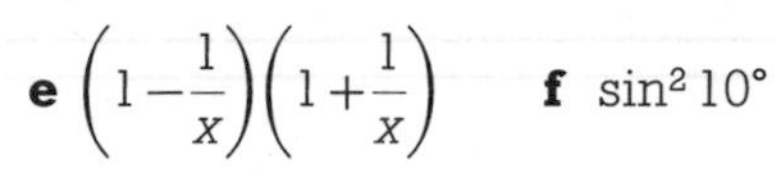

e $\left(1-\frac{1}{x}\right)\left(1+\frac{1}{x}\right)$ **f** $\sin^2 10°$ **g** $4(2x-1)$ **h** $(x-1)^6$

i $x\left(1-\frac{1}{x}\right)$ **j** $\frac{x-1}{x-2}$ **k** $\pi^2-\pi$ **l** $x^2\left(2-\frac{1}{x}\right)^2$.

4 You can add, subtract and multiply polynomials.
For these three polynomials, find in standard (decreasing) form:
a the sum **b** the difference (two possibilities in each case)
c the product for each of the three possible pairs:
$f(x) = 3x^3-2x+1$, $g(x) = x^2+2x$, $h(x) = x-1$.

VALUES OF POLYNOMIALS

If $f(x) = 4x^3-3x^2+2x-7$, its value for $x = h$ is $f(h) = 4h^3-3h^2+2h-7$.
To calculate $f(5)$ you can calculate 4×5^3, -3×5^2, 2×5, -7, in turn, adding each result to the memory of your calculator. How many key-stroke operations did you need?
The work can be done with fewer operations, using the *nested form* of $f(h)$:

$$\begin{aligned} f(h) &= 4h^3-3h^2+2h-7 \\ &= (4h^2-3h+2)h-7 \\ &= [(4h-3)h+2)]h-7. \end{aligned}$$

Calculate $f(5)$ using this nested form, and check the number of operations needed.

A nested calculation scheme

Again $f(x) = 4x^3-3x^2+2x-7$
For $f(h) = [(4h-3)h+2]h-7$, the calculator's order of operation is:

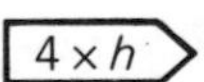

$4\times h$ → Subtract 3 → $\times h$ → Add 2 → $\times h$ → Subtract 7 → $f(h)$

This can be shown in a nested calculation scheme:

h	4		-3		2		-7
	0	$\times h$	$4h$	$\times h$	$(4h-3)h$	$\times h$	$[(4h-3)h+2]h$
	4		$4h-3$		$(4h-3)h+2$		$[(4h-3)h+2]h-7 = f(h)$

Example Calculation for $f(5)$.

5	4		-3		2		-7
	0	$\times 5$	20	$\times 5$	85	$\times 5$	435
	4		17		87		$428 = f(5)$

This is written out as:

5	4	-3	2	-7
	0	20	85	435
	4	17	87	$428 = f(5)$

Exercise 2

1 Use the nested calculation scheme to calculate the value of each polynomial:

a x^2+3x+1, when $x=4$ **b** $3x^2-2x+5$, when $x=2$
c $5x^2-3x-11$, when $x=-1$ **d** x^3+x^2+3x-6, when $x=3$
e x^3-2x^2-2x-3, when $x=4$ **f** u^3-3u^2-9u-3, when $u=-2$
g $5t^3-2t+3$, when $t=5$ (*Warning*: coefficients are 5 0 −2 3.)
h t^3+3t^2-5, when $t=-2$ **i** $x^4-3x^3+2x^2+x+1$, when $x=-3$.

2 Use the nested calculation scheme to check that each equation has the given number as a root:

a $2x^4-3x^3-17x^2+27x-9=0$, $x=1$ (Show that $f(1)=0$.)
b $2t^3+t^2-25t+12=0$, $t=3$
c $6u^3+13u^2-14u+3=0$, $u=-3$.

3 $f(x)=x^4+3ax^3-3a^2x^2-11a^3x-6a^4$. Show that $x=-a$, $x=2a$ and $x=-3a$ are all roots of the equation $f(x)=0$.

DIVISION OF POLYNOMIALS

If $f(x)=ax^2+bx+c$,

$$f(h)=ah^2+bh+c$$
$$f(x)-f(h)=a(x^2-h^2)+b(x-h)$$
$$f(x)=(x-h)[a(x+h)+b]+f(h)$$

i.e. $f(x)=(x-h)[ax+(ah+b)]+ah^2+bh+c.$ (1)

↑ divisor ↑ quotient ↑ remainder

Using the nested calculation scheme for $f(h)$:

h	a	b	c
	0	ah	$(ah+b)h$
	a	$ah+b$	ah^2+bh+c

↖ ↗ coefficients of quotient in (1) — ↑ remainder in (1)

So the nested calculation scheme provides the quotient and the remainder when $f(x)$ is divided by $x-h$. This process is called *synthetic division*.

Example Find the quotient and remainder when $x^3+6x^2+3x-15$ is divided by $x+3$.

The divisor is $x-(-3)$.

−3	1	6	3	−15
	0	−3	−9	18
	1	3	−6	3

The quotient is x^2+3x-6, and the remainder is 3.
So $x^3+6x^2+3x-15=(x+3)(x^2+3x-6)+3$.

Exercise 3

Use synthetic division to find the quotient and remainder when each polynomial is divided by the given divisor:

	Polynomial	*Divisor*		*Polynomial*	*Divisor*
1	x^2+2x-2	$x-1$	**2**	$2x^2-x-3$	$x+2$
3	x^3+x^2+x-1	$x+1$	**4**	x^3-x^2+3x-4	$x-2$
5	$2t^3+6t^2+33$	$t+4$	**6**	$3u^3-13u^2-50$	$u-5$
7	$2x^2+x-2$	$x-\frac{1}{2}$	**8**	$4x^2+2x+3$	$x+\frac{1}{2}$
9	x^3+8	$x+2$	**10**	t^6-1	$t-1$

Division by $ax+b$

Example Find the quotient and remainder when $4x^3-2x+5$ is divided by $2x+1$.

$2x+1 = 2(x+\frac{1}{2}) = 2(x-(-\frac{1}{2}))$

$-\frac{1}{2}$	4	0	-2	5
	0	-2	1	$\frac{1}{2}$
	4	-2	-1	$5\frac{1}{2}$

So $4x^3-2x+5 = (x+\frac{1}{2})(4x^2-2x-1)+5\frac{1}{2}$
$= (2x+1)(2x^2-2x-\frac{1}{2})+5\frac{1}{2}$.

The quotient is $2x^2-2x-\frac{1}{2}$ and the remainder is $5\frac{1}{2}$.

11 Find the quotients and remainders for:

	Polynomial	*Divisor*		*Polynomial*	*Divisor*
a	$2x^2-x-3$	$2x-1$	**b**	$3x^2-5x+7$	$3x+1$
c	$4t^3+6t^2-2t-1$	$2t+1$	**d**	$3x^3-2x+5$	$2x+1$

THE REMAINDER THEOREM AND FACTOR THEOREM

In the last section you saw that when a quadratic polynomial $f(x) = ax^2+bx+c$ is divided by $x-h$, the remainder is $f(h)$. It is easy to prove that this is true for all polynomials.

Proof If R is the remainder and $q(x)$ the quotient when polynomial $f(x)$ is divided by $x-h$, R is a constant (of degree 0) and

$f(x) = (x-h)q(x)+R$ — an *identity*, i.e. true for all values of x

$f(h) = (h-h)q(h)+R$ — replacing x by h
$= 0\times q(h)+R$
$f(h) = R.$

$$f(x) = (x-h)q(x)+f(h)$$

When a polynomial f(x) is divided by x−h the remainder is f(h). This is the **remainder theorem**.

If $f(h) = 0$, then $f(x) = (x-h)q(x)$; so $x-h$ is a factor of $f(x)$.
Conversely, if $f(x) = (x-h)q(x)$, then $f(h) = 0$.

$$x-h \text{ is a factor of polynomial } f(x) \Leftrightarrow f(h) = 0.$$

This is the **factor theorem**.

Example Find the rational factors of $f(x) = 2x^3-11x^2+17x-6$.

Try factors of 6, i.e. ± 1, ± 2, ± 3, ± 6.

1	2	−11	17	−6
	0	2	−9	8
	2	−9	8	2 ≠ 0

$x-1$ is not a factor

−1	2	−11	17	−6
	0	−2	13	−30
	2	−13	30	−36 ≠ 0

$x+1$ is not a factor

2	2	−11	17	−6
	0	4	−14	6
	2	−7	3	0

$x-2$ is a factor

So $f(x) = (x-2)(2x^2-7x+3)$
$\quad = (x-2)(2x-1)(x-3)$, by factorising the quadratic.

Note The roots of the equation $f(x) = 0$, i.e. $2x^3-11x^2+17x-6 = 0$, are $x = 2, \frac{1}{2}, 3$.

Only rational factors, like $x-3$ and $2x+1$, in which the coefficients are rational numbers, are included here.

Exercise 4

Factorise the polynomials in questions **1–9**:

1 x^3-x^2-x+1 **2** x^3-3x-2 **3** x^3-7x+6

4 x^3-2x^2+4x-8 **5** $3x^3-2x^2-19x-6$ **6** $x^3+3x^2-9x-27$

7 x^4-1 **8** x^3+3x^2+3x+1 **9** $6x^3+13x^2+x-2$.

10 $x+1$ is a factor of $f(x) = x^3+x^2-4x+a$. Find a, and factorise $f(x)$ fully.

11 $2x-1$ is a factor of $f(x) = 2x^3-17x^2+40x+b$. Find b, and factorise $f(x)$ fully.

Exercise 5

Solve the equations in questions **1–6**. (This Exercise deals with real roots.)

1 $x^3+x^2-x-1 = 0$ **2** $x^3-2x^2-2x+4 = 0$ **3** $x^3+2x^2-4x-8 = 0$

4 $x^3-3x^2-16x+48 = 0$ **5** $x^3+3x^2+6x+18 = 0$ **6** $2x^3-3x^2-18x+27 = 0$.

7 $x = 1$ is a root of the equation $x^3-4x^2+5x+k = 0$. Find k, and the other two roots of the equation.

8 Show that $x^3+x^2+2x+2 = 0$ has only one real root.

9 The tangent at A(1, 2) to the curve $y = x^3 + x^2$ meets the curve again at B.

a Use calculus to show that AB has equation $y = 5x - 3$.

b Show that the tangent and curve meet where $x^3 + x^2 - 5x + 3 = 0$.

c Solve the equation, and write down the coordinates of B.

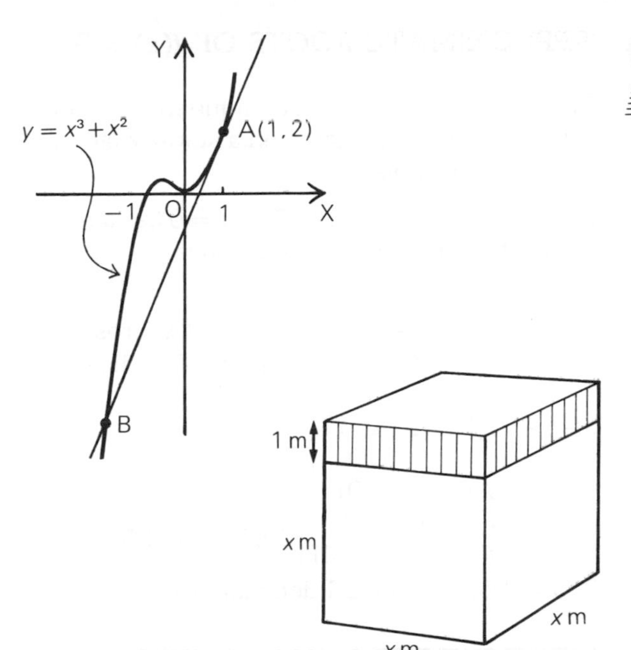

10 The tangent to the curve $y = x^3$ at A (1, 1) meets the curve again at B. Find:

a the equation of the tangent

b the coordinates of B.

11 A rectangular tank has a square base of side x m and height $(x + 1)$ m. Its volume is 36 m³.

a Form a polynomial equation in x.

b Solve the equation and find the length of a side of the base.

12 A new national daily was launched. It was found that the value, in £ million, of the daily print run in year y was modelled by the function, $V(y) = y^3 - 6y^2 + 11y + 4$.
Find the years in which the print run value was £10 million.

a Set up the polynomial equation.

b Solve the equation and complete the problem.

13 t seconds after the van hits the motorway crash barrier the sideways displacement, d cm, of the barrier is given by:
$d = 15t(t^3 - 6t - 9)$.
How long after impact does the barrier take to return to its original position?

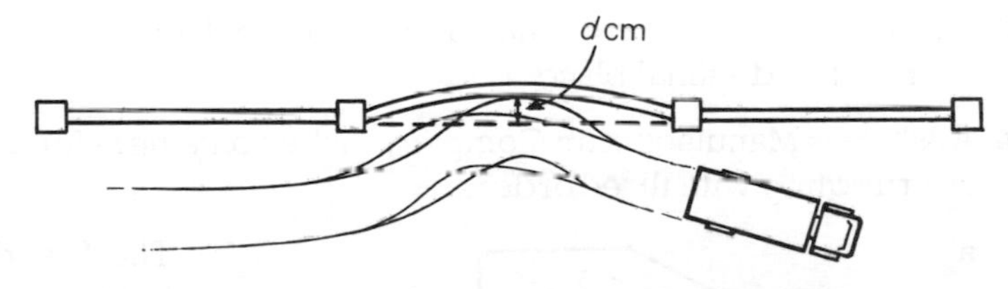

14 A planet's sun is becoming warmer. It is estimated that the planet's surface temperature t°C after y million years can be modelled by $t = 10y^3 - 100y^2 + 270y - 180$.

a How long will it be until the present ice age on the planet ends (i.e. the ice melts)?

b (i) When will the next ice age begin?
(ii) How long will it last?

15 An iceberg under tow from Antarctica to Africa has a volume V after n days, given by

$$V = \frac{500\pi}{3}(2000 - 100n + 20n^2 - n^3).$$

How long will it be until the iceberg melts completely?

APPROXIMATE ROOTS OF $f(x) = 0$

The last section gave a technique for solving the equation $f(x) = 0$, provided that the roots are rational. If they are not, then a step-by-step method can be used in order to find approximate values of the roots.

Example Show that $x^3 - 3x + 1 = 0$ has a real root between 1 and 2. Find an approximation for the root, correct to 1 decimal place.
Let $f(x) = x^3 - 3x + 1$.

Step 1 $\left.\begin{array}{l} f(1) = -1 \\ f(2) = 3 \end{array}\right\}$ so graph $y = f(x)$ crosses the x-axis between $x = 1$ and $x = 2$, indicating a root α there.

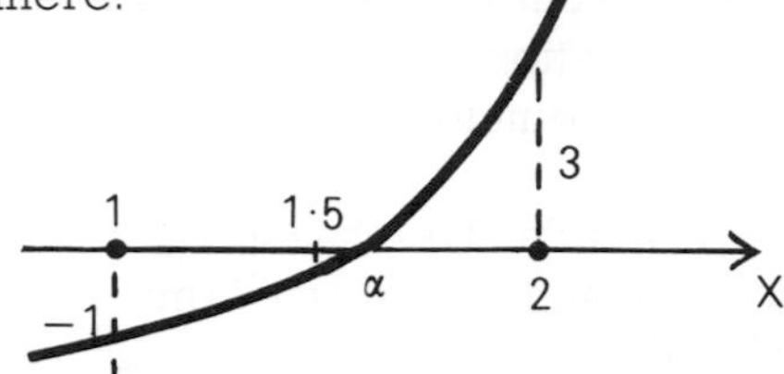

Step 2 $\left.\begin{array}{l} f(1{\cdot}5) \doteqdot -0{\cdot}13 \\ f(1{\cdot}6) \doteqdot 0{\cdot}30 \end{array}\right\} 1{\cdot}5 < \alpha < 1{\cdot}6.$

$f(1{\cdot}55) \doteqdot 0{\cdot}07$

$\left.\begin{array}{l} f(1{\cdot}54) \doteqdot 0{\cdot}03 \\ f(1{\cdot}53) \doteqdot -0{\cdot}01 \end{array}\right\} 1{\cdot}53 < \alpha < 1{\cdot}54.$

So $\alpha = 1{\cdot}5$, correct to 1 decimal place.

Exercise 6

1 Show that $x^3 - x - 1 = 0$ has a real root between 1 and 2. Find an approximation for the root, correct to 1 decimal place.

2 Show that $x^3 + x - 3 = 0$ has a real root between 1 and 1·5, and find an approximation, correct to 1 decimal place.

3 Show that $x^3 - 4x + 2 = 0$ has a real root between 0·5 and 1, and find an approximation, correct to 1 decimal place.

4 Allshapes Manufacturing Company make containers for fruit juice. Their design department is struggling with three orders:

a

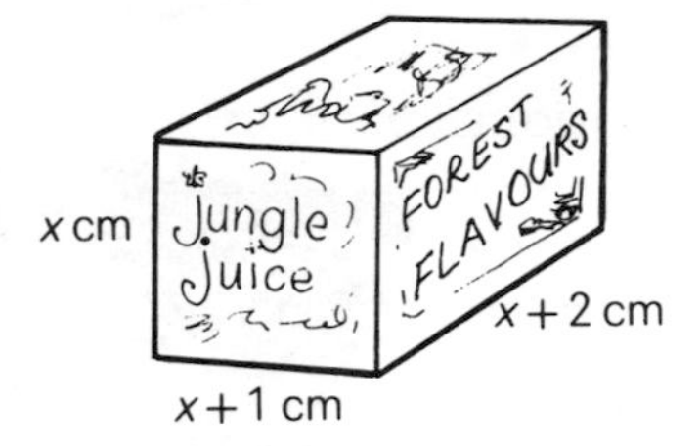

The first customer wants cuboid containers holding 200 ml of juice, with each edge differing in length by 1 cm as shown. Show that the design department has to solve the equation $x^3 + 3x^2 + 2x - 200 = 0$, and help them to find x, to the nearest millimetre.

b The second customer wants his product to be sold in cylindrical cans holding 1 litre, with height 10 cm more than the radius. Show that the equation this time is $r^3 + 10r^2 - 318 = 0$, and find r, to the nearest millimetre.

$\left(\text{Take 318 as an approximation for } \dfrac{1000}{\pi}.\right)$

ENER-JUICE
$r + 10$ cm
r cm

c

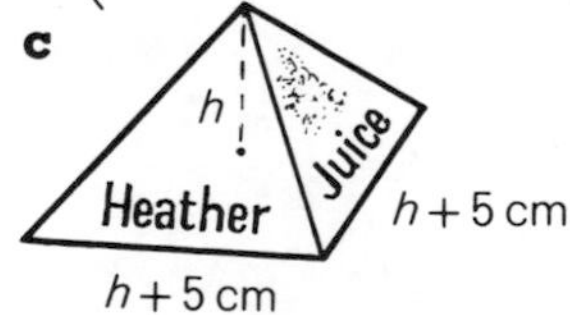

The third customer has asked for containers in the shape of square pyramids, holding 250 ml, with side of base 5 cm more than the height.
Find an equation, and hence an approximation for h, correct to the nearest millimetre.

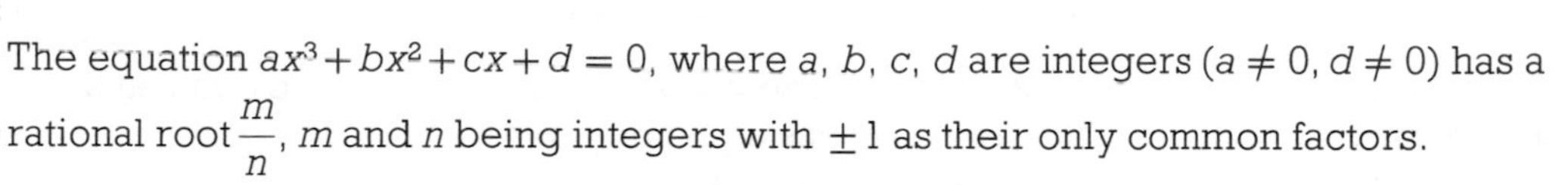

The equation $ax^3+bx^2+cx+d=0$, where a, b, c, d are integers ($a \neq 0$, $d \neq 0$) has a rational root $\frac{m}{n}$, m and n being integers with ± 1 as their only common factors.

a Show that $am^3+bm^2n+cmn^2+dn^3=0$.
b Prove that there is an integer k with $mk=dn^3$.
c Now explain why m is a factor of d (positive or negative).
d Show similarly that n is a factor of a.

Divisibility Tests

a 198 is a multiple of 9; 17397 is a multiple of 9.
$1+9+8$ is a multiple of 9; $1+7+3+9+7$ is a multiple of 9.

Rule: 'abcd' is a multiple of 9 if $a+b+c+d$ is a multiple of 9.

$abcd = a\times 10^3+b\times 10^2+c\times 10+d=f(10)$, where $f(x)=ax^3+bx^2+cx+d$.
$f(1)=a+b+c+d$, and $f(x)=(x-1)q(x)+f(1)$ (for some $q(x)$).
You now have all the tools needed to prove the rule.
See what you can do.
What modifications do you have to make for the 'general' rule?

b Investigate a similar rule for multiples of 3.
c Go through the same reasoning but with $f(-1)$ instead of $f(1)$. What divisibility test does this lead to?

CHECK-UP ON **POLYNOMIALS**

1 State the degree of the polynomial and the coefficient of x for each of these:
a $2x^2-3x$ **b** 11 **c** $2-5x$ **d** $7x^4+x^2-2x+1$.

2 Write each polynomial in 'standard' form with powers of x in increasing order:
a $x^2(2-x)$ **b** $(3x-1)^2$ **c** $(x^2+3)(4-x-2x^2)$.

3 Use the nested calculation scheme to calculate the value of:
a $2x^3-x+7$ when $x=5$ **b** $t^4-3t^3+5t^2+4$ when $t=-2$.

4 Use synthetic division to find the quotient and remainder in each case:
a $3x^3-x^2+2x-5$ with divisor $x-2$
b $2x^4+5x^3+x+7$ with divisor $x+3$
c $2x^4+x^3-3x^2+3x-4$ with divisor $2x-1$.

5 Find the rational factors of:
a x^3-x^2-5x-3 **b** $x^3-4x^2-11x+30$
c $2x^4-5x^2-12$ **d** $2x^3+x^2+x-1$.

POLYNOMIALS

6 Find the real roots of the equations:
a $x^3+3x^2+2x+6=0$ **b** $4x^4-9x^3-2x^2+9x-2=0$.

7 The tangent at the point A(1, 2) on the curve $y=x^3+x$ meets the curve again at B.
a Use calculus to show that the tangent has equation $y=4x-2$.
b Show that the tangent and curve meet where $x^3-3x+2=0$.
c Solve this equation, and find the coordinates of B.

8 Show that $x^3+2x^2-5=0$ has a root between 1 and 2, and find an approximation for the root, correct to 1 decimal place.

A FAMILY OF PARABOLAS

The gradient of the tangent to a parabola at the point (x, y) is given by $\frac{dy}{dx} = 8x$.

Check that the equation of the parabola could be $y = 4x^2$, or $y = 4x^2 + 1$, or $y = 4x^2 - 3$. In fact there is a family of parabolas, with equation $y = 4x^2 + C$ where C is a constant. To choose a particular member of the family, more information is needed. For example, if it passes through the point $(1, -2)$,

$y = 4x^2 + C \Rightarrow -2 = 4 + C \Rightarrow C = -6$.

This particular parabola is $y = 4x^2 - 6$.

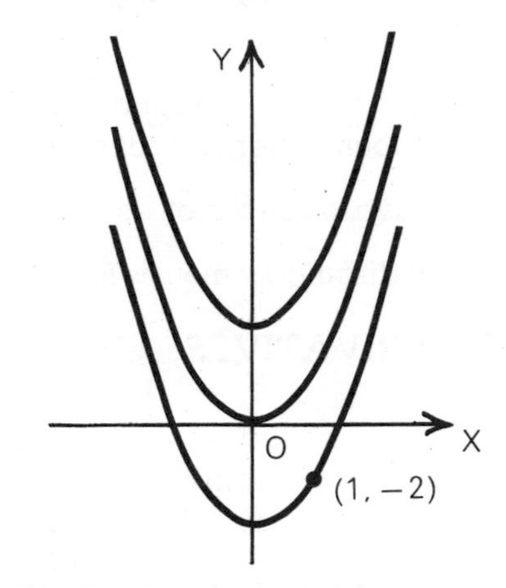

Notation

$\frac{dy}{dx} = 8x$ is a *differential equation.*
$y = 4x^2 + C$, where C is a constant, is the *general solution.*
$y = 4x^2 - 6$ is a *particular solution* of the differential equation.

Exercise 1

1 Find the general solutions of these differential equations:

a $\frac{dy}{dx} = 2x$ **b** $\frac{dy}{dx} = 4x$ **c** $\frac{dy}{dx} = 6x + 1$ **d** $\frac{dy}{dx} = 3 - 4x$.

Check each answer by differentiation.

2 For each of these differential equations find the particular solution satisfying the given condition.

For example $\frac{dy}{dx} = 3 - 8x$, given that $y = 4$ when $x = 1$.
The general solution is $y = 3x - 4x^2 + C$.
But $y = 4$ when $x = 1$. So $4 = 3 \times 1 - 4 \times 1^2 + C$, from which $C = 5$.
The particular solution is $y = 3x - 4x^2 + 5$.

a $\frac{dy}{dx} = 3 + 6x$, given that $y = 2$ when $x = -1$.

b $\frac{dy}{dx} = 12x + 1$, given that $y = 16$ when $x = 2$.

c $\frac{dy}{dx} = 5 - 8x$, given that $y = -2$ when $x = 1$.

3 The gradient of the tangent to a parabola at the point (x, y) is given by $\frac{dy}{dx} = 2x - 1$. If the parabola passes through the point $(2, 1)$, find its equation.

4 The gradient at (x, y) on a curve is $\frac{dy}{dx} = 8x$. If the curve passes through $(-2, 8)$, find its equation.

5 Kate and Mike make a simultaneous parachute jump.
Their velocity after x seconds is $v = 5 + 10x$ m/s.
If they have fallen y metres, then $v = \frac{dy}{dx}$, so $\frac{dy}{dx} = 5 + 10x$.
Find the distance y metres fallen in x seconds, given $y = 0$ when $x = 0$.

ANTI-DERIVATIVES

$5x + 5x^2 + C$ is called an *anti-derivative* of $5 + 10x$ since $\frac{d}{dx}(5x + 5x^2 + C) = 5 + 10x$. It comes from $5 + 10x$ by 'undoing' differentiation.
Leibniz invented a useful notation for anti-derivatives, namely $\int(5 + 10x)dx = 5x + 5x^2 + C$.

In general, $\int f(x)dx = F(x) + C$ means $F'(x) = f(x)$.

Notation

The process of calculating an anti-derivative is called *integration*.
The anti-derivative $F(x)$ is called the *integral*, $f(x)$ the *integrand* and C the *constant of integration*.
$F(x)$ is obtained from $f(x)$ by *integrating with respect to x*.

A USEFUL RULE

In Chapter **2**, Differentiation—1 you saw that $\frac{d}{dx}(x^n) = nx^{n-1} (n \in Q)$. What about $\int x^n \, dx$? In other words, what gives x^n when it is differentiated?

$$\frac{d}{dx}(x^{n+1}) = (n+1)x^n, \text{ so } \frac{d}{dx}\left(\frac{x^{n+1}}{n+1}\right) = x^n, \ n \neq -1.$$

If $n \neq -1$, $\int x^n dx = \frac{x^{n+1}}{n+1} + C$.

Also, $\int (f(x) + g(x))dx = \int f(x)dx + \int g(x)dx$,
and $\int kf(x)dx = k\int f(x)dx$.

Examples

1 a $\displaystyle\int x dx = \frac{x^2}{2} + C$ **b** $\displaystyle\int x^2 dx = \frac{x^3}{3} + C$ **c** $\displaystyle\int x^3 dx = \frac{x^4}{4} + C$.

2 If $n = 0$, $\displaystyle\int x^0 dx = \frac{x^1}{1} + C = x + C$, that is $\displaystyle\int 1 dx = x + C$.

3 $\displaystyle\int (\sqrt[3]{x} - 5)dx = \int (x^{1/3} - 5)dx = \frac{x^{4/3}}{\frac{4}{3}} - 5x + C = \frac{3}{4}x^{4/3} - 5x + C$.

Exercise 2

1 Write down the anti-derivatives of the following. (Remember C.)

a x^2 **b** x^4 **c** x^{-3} **d** x^{-6} **e** $x^{1/2}$ **f** $x^{4/3}$

g $2x$ **h** $8x^3$ **i** 5 **j** $\dfrac{1}{x^2}$ **k** $\dfrac{1}{\sqrt{x}}$ **l** $\dfrac{3}{x^{3/2}}$.

2 Find:

a $\int x^5 dx$ **b** $\int x^{-5} dx$ **c** $\int \sqrt[3]{x}\, dx$ **d** $\int \dfrac{1}{x^4} dx$ **e** $\int \dfrac{2}{x^{3/4}} dx$.

3 Find:

a $\int (1-2x) dx$ **b** $\int (x^2+x) dx$ **c** $\int (x^3-1) dx$ **d** $\int (6x^2-4x+2) dx$.

4 Find:

a $\int (2x-3x^2) dx$ **b** $\int (4x^3-1) dx$ **c** $\int (2x^5-3x+1) dx$

d $\int \left(\dfrac{2}{t^2}-t\right) dt$ **e** $\int (\sqrt{u}-3) du$ **f** $\int (v^{3/2}-2v^{-1/2}) dv$.

5 Find the general solutions of the differential equations:

a $\dfrac{dy}{dx} = 4x^3+2x$ **b** $\dfrac{dy}{dx} = 6x^5-3x^2$ **c** $\dfrac{dy}{dx} = 3x^2-1$

d $\dfrac{dy}{dt} = 2t+\dfrac{1}{t^2}$ **e** $\dfrac{dv}{dt} = 5t^4-\dfrac{1}{2\sqrt{t}}$ **f** $\dfrac{dp}{dz} = z+\dfrac{1}{4z^{3/4}}$.

6 For each of these differential equations find the particular solution satisfying the given condition:

a $\dfrac{dy}{dx} = x^2+\dfrac{1}{x^2}$ given that $y = \frac{4}{3}$ when $x = 1$

b $\dfrac{dy}{dt} = 4t-6t^2$ given that $y(1) = 0$, i.e. that $y = 0$ when $t = 1$

c $\dfrac{dp}{du} = \dfrac{1}{\sqrt{u}}+3\sqrt{u}$ given that $p = 12$ when $u = 4$.

7 The gradient at the point (x, y) on a curve is $\dfrac{dy}{dx} = 3x^2-6x+1$, for each $x \in R$. Given that the curve passes through the point (2, 4), find its equation.

8 The gradient at the point (x, y) on a curve is $\dfrac{dy}{dx} = 4x^3-6x^2+5$. If the curve passes through the point (2, 9), find its equation.

9 If $f'(x) = \frac{1}{2}\sqrt{x}-5x^{3/2}$ and $f(0) = 3$, find $f(x)$.
(*Note* $f(x) = \int f'(x)\, dx$.)

10 Integrate these with respect to the relevant variables:

a $x^2(3-4x)$ **b** $(2t+1)^2$ **c** $\dfrac{u^3+1}{u^3}$ **d** $v^2\left(2-\dfrac{1}{v}\right)$.

INTEGRATION

A useful anti-derivative

Remember that, if $F'(x) = f(x)$, then $\int f(x)dx = F(x)+C$.

$$\frac{d}{dx}(ax+b)^{n+1} = (n+1)(ax+b)^n a, \text{ so}$$

$$\boxed{\int (ax+b)^n dx = \frac{(ax+b)^{n+1}}{(n+1)a}+C}$$

where a, b, n are constants with $a \neq 0$, $n \neq -1$.

Example $\displaystyle\int (3+2x)^{1/2}dx = \frac{(3+2x)^{3/2}}{\frac{3}{2}\times 2}+C = \frac{1}{3}(3+2x)^{3/2}+C.$

Exercise 3

1 Integrate:
a $(x+1)^4$ **b** $(2x+1)^3$ **c** $(14+x)^2$ **d** $(1-x)^5$ **e** $(3x-4)^6$.

2 Find:

a $\displaystyle\int (x-3)^{-2}dx$ **b** $\displaystyle\int \sqrt{(x+4)}\,dx$ **c** $\displaystyle\int \frac{dx}{(2x+3)^4}$ **d** $\displaystyle\int \frac{dx}{\sqrt{(4x-1)}}$.

3 Integrate with respect to the relevant variables:

a $(2t-1)^4$ **b** $7(1-4x)^{3/4}$ **c** $(1-4u)^{1/2}$ **d** $\dfrac{2}{(3v-1)^{1/3}}$ **e** $1+\dfrac{1}{(x+1)^2}$.

4 Find the general solutions of:

a $\dfrac{dy}{dx} = (3-4x)^5$ **b** $\dfrac{du}{dt} = \sqrt{(2t+1)}$ **c** $\dfrac{dy}{dv} = \dfrac{1}{(1-2v)^3}$.

5 Find the particular solution of each differential equation:

a $\dfrac{dy}{dx} = \dfrac{1}{(x-1)^2}$, given that $y = 3$ when $x = 2$

b $\dfrac{dy}{dx} = \sqrt{(x+1)}$, given that $y = 20$ when $x = 8$.

6 Alison's answer to $\displaystyle\int (x^2+1)^2 dx$ is $\dfrac{1}{5}x^5+\dfrac{2}{3}x^3+x+C$.

Sammy's answer is $\dfrac{1}{3}\times\dfrac{1}{2x}\times(x^2+1)^3+C$. Who is correct?

More useful anti-derivatives

$\frac{d}{dx}(\sin x) = \cos x$, so

$$\int \cos x\,dx = \sin x + C.$$

$\frac{d}{dx}(\cos x) = -\sin x$, so

$$\int \sin x\,dx = -\cos x + C.$$

As extensions of these, you can easily check by differentiation, that

$$\int \cos(ax+b)\,dx = \frac{1}{a}\sin(ax+b) + C$$

$$\int \sin(ax+b)\,dx = -\frac{1}{a}\cos(ax+b) + C$$

where a, b are constants with $a \neq 0$. (Remember to use the Chain Rule!)

Examples

1 $\int(2\sin x + 5)\,dx = -2\cos x + 5x + C,$

2 $\int 3\cos\left(6x+\frac{\pi}{12}\right)dx = 3\times\frac{1}{6}\sin\left(6x+\frac{\pi}{12}\right)+C = \frac{1}{2}\sin\left(6x+\frac{\pi}{12}\right)+C.$

Exercise 4

1 Integrate:

a $\sin x$ **b** $\cos x$ **c** $3\cos x$ **d** $4\sin x$ **e** $\frac{1}{2}\sin x$
f $\cos 2x$ **g** $\sin 3x$ **h** $4\cos 4x$ **i** $6\sin 2x$ **j** $\cos\frac{1}{2}x$.

2 Find:

a $\int\cos(x+2)\,dx$ **b** $\int\sin(2x-1)\,dx$ **c** $\int\sin(3-4x)\,dx$ **d** $\int a\cos(ax+b)\,dx$.

3 Find:

a $\int 5\sin x\,dx$ **b** $\int(2\cos x + 3\sin x)\,dx$ **c** $\int(\cos 5\theta - 3\sin 3\theta)\,d\theta$
d $\int(\sin 2x + 3\cos 3x)\,dx$ **e** $\int(t^2 + \cos 2t)\,dt$ **f** $\int 2\sin(3-4u)\,du$.

4 $\cos 2x = 2\cos x^2 - 1 = 1 - 2\sin^2 x$.

a Use this identity to express $\cos^2 x$ in terms of $\cos 2x$ and find $\int\cos^2 x\,dx$.
b Similarly find $\int\sin^2 x\,dx$.
c From **a** and **b** find $\int(\sin^2 x + \cos^2 x)\,dx$. Explain your answer.

5 Find the general solutions of:

a $\frac{dy}{dx} = 2\sin 2x$ **b** $\frac{dy}{dx} = \cos\left(1+\frac{x}{3}\right)$ **c** $\frac{dy}{dx} = 4\sin\left(2x+\frac{\pi}{3}\right)$.

6 For each of these differential equations find the particular solution satisfying the given conditions:

a $\dfrac{dy}{dx} = 3\sin 3x$, given that $y = 2$ when $x = \dfrac{\pi}{3}$

b $\dfrac{dy}{dt} = \sin\left(t - \dfrac{\pi}{4}\right) + \cos\left(t - \dfrac{\pi}{4}\right)$, given that $y = 1$ when $t = \dfrac{\pi}{4}$.

c $\dfrac{dy}{dt} = 3\cos(2t + \alpha)$ (α a constant), given that $y(0) = \sin\alpha$.

MODELLING WITH DIFFERENTIAL EQUATIONS

Exercise 5

1 The rate of growth per month (t) of the population $P(t)$ of Carlos Town is given by the differential equation $\dfrac{dP}{dt} = 5 + 8t^{1/3}$.

a Find the general solution of this equation.
b Find the particular solution, given that at present ($t = 0$), $P = 5000$.
c What will the population be 8 months from now?

2 Mr Isaacs produces x jewel boxes per week. His weekly profit function $P(x)$, in £, is such that $\dfrac{dP}{dx} = 70 - 0{\cdot}4x$.

(Economists call $\dfrac{dP}{dx}$ the marginal profit.)

Given that $P(0) = -1200$ (i.e. the 'fixed costs' are £1200 per week), find the profit function $P(x)$.

3 A dam begins to release water into a stream at time $t = 0$. The total volume of water released by time t satisfies the differential equation $\dfrac{dV}{dt} = 1400 + 60t$.

Solve this equation to find V.

4 A flu epidemic hits the small town of Wilmore. On a certain day there are 42 cases. The medical officer estimates that x days from then there will be $7 + 6x$ new cases per day. The total number of cases at the end of x days is $N(x)$.

a Construct a model of this involving a differential equation.
b Solve the equation to find a formula for $N(x)$.

5 During the first 15 weeks from birth the weight $w(t)$ (in ounces) at week t of a new type of white mouse increases at a rate of $\frac{1}{30}(2 + t)$ ounces per week. If the weight at birth is $0{\cdot}3$ ounces, find a formula for the mouse's weight after t weeks. Again you'll need a differential equation.

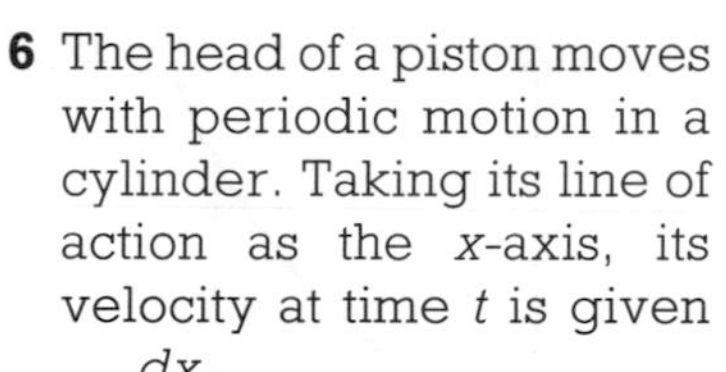

6 The head of a piston moves with periodic motion in a cylinder. Taking its line of action as the x-axis, its velocity at time t is given by $\dfrac{dx}{dt} = 4\sin 2t$.

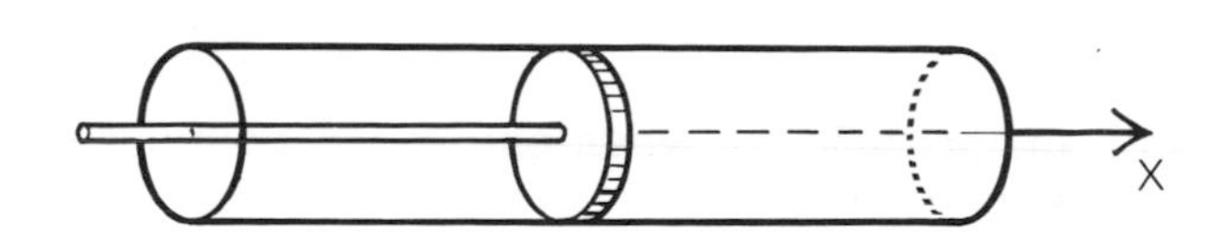

If it starts from the point given by $x = 1$, find its displacement $x(t)$ at time t.

7 The velocity of a car after t seconds has the formula $v(t) = 30 + 6t$ m/s. The distance travelled in t seconds is $s(t)$ metres.

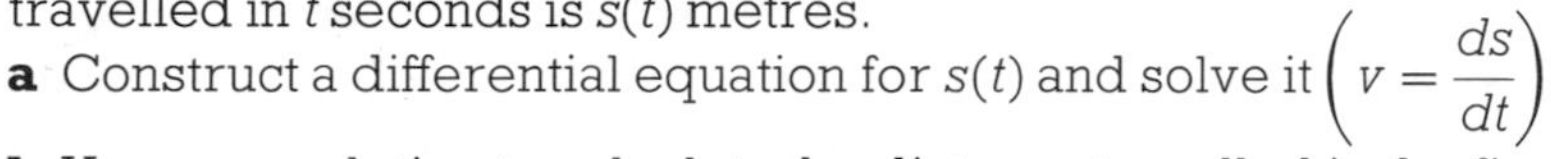

a Construct a differential equation for $s(t)$ and solve it $\left(v = \dfrac{ds}{dt}\right)$.

b Use your solution to calculate the distance travelled in the first 3 seconds.

8

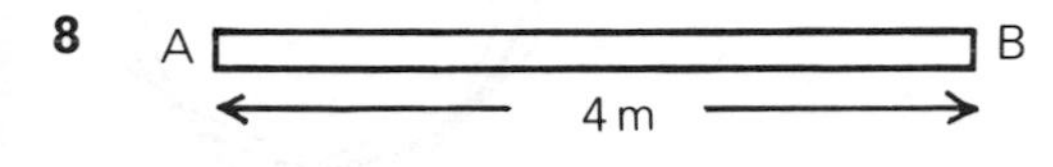

If you put a weight at the end, B, of a light metal beam of length 4 metres, the beam will sag. The sag at x metres along the beam is y cm. The model for this, when A is fixed, is $\dfrac{dy}{dx} = 2x\left(4 - \dfrac{1}{2}x\right)$.

Solve this equation and find the sag at B.

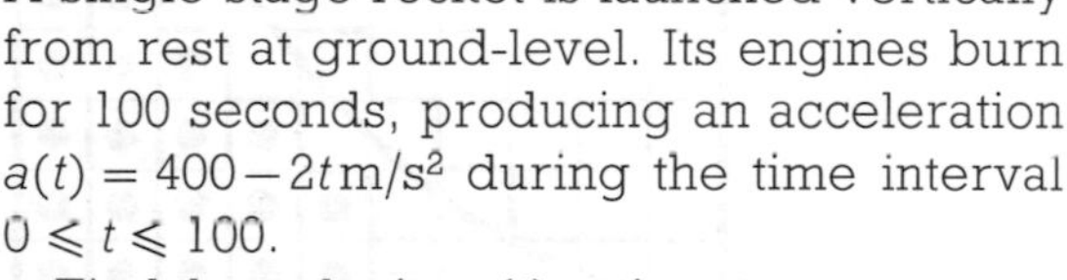

A single-stage rocket is launched vertically from rest at ground-level. Its engines burn for 100 seconds, producing an acceleration $a(t) = 400 - 2t$ m/s² during the time interval $0 \leqslant t \leqslant 100$.

a Find the velocity $v(t)$ at time t.

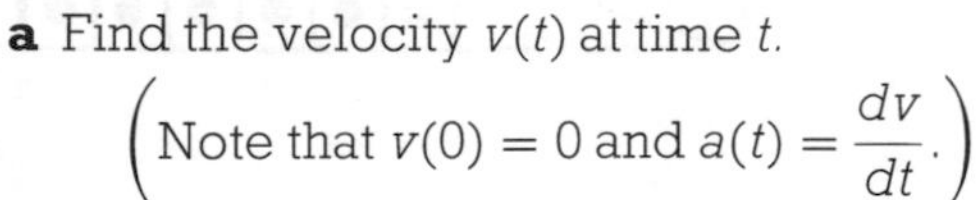

$\left(\text{Note that } v(0) = 0 \text{ and } a(t) = \dfrac{dv}{dt}.\right)$

b If $y(t)$ is the height at time t write down the differential equation for $y(t)$.

c Solve the equation in **b**, and find the height $y(100)$ reached at burnout.

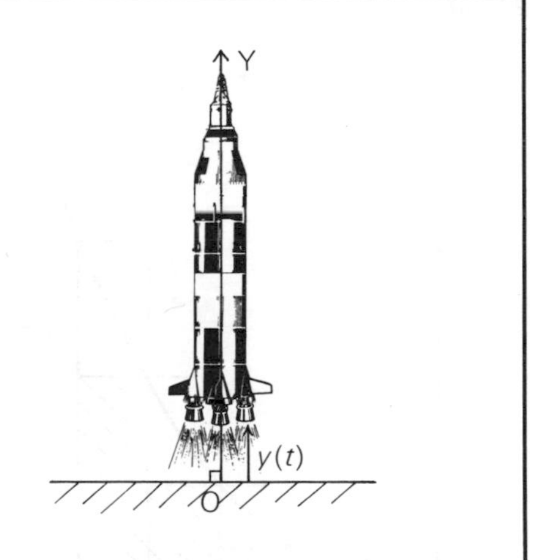

THE AREA PROBLEM—AND A SURPRISING LINK BETWEEN DIFFERENTIATION AND INTEGRATION

There is no difficulty in calculating the area of a rectangle, or triangle, or parallelogram, or any shape bounded by straight lines. For example, the area of this triangle is $A = \frac{1}{2}a^2$.

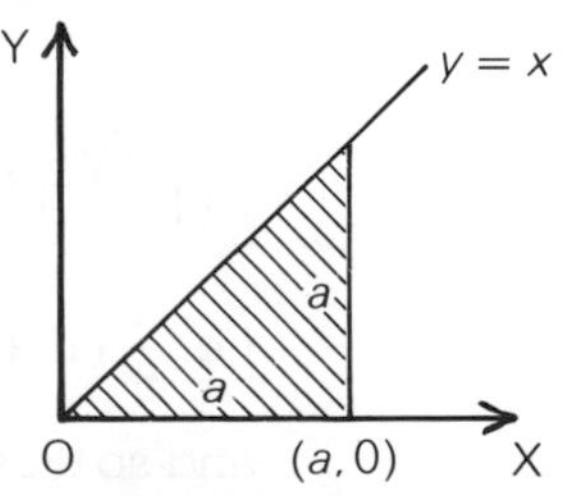

But there is a major difficulty when one of the boundary lines is curved, as in the diagram.

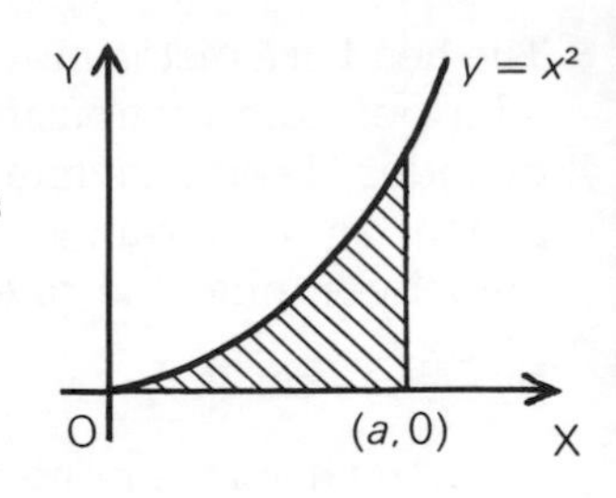

More than 2000 years ago the Greeks faced this problem in finding the area of a circle. They used the ingenious method of finding an upper bound and a lower bound of the area by calculating the areas of inscribed and escribed polygons. The area of the circle lay between these. We'll try a similar method to find the area bounded by the parabola $y = x^2$, the x-axis and the line $x = 1$.
The area $A < 0{\cdot}5$ square unit. Why?

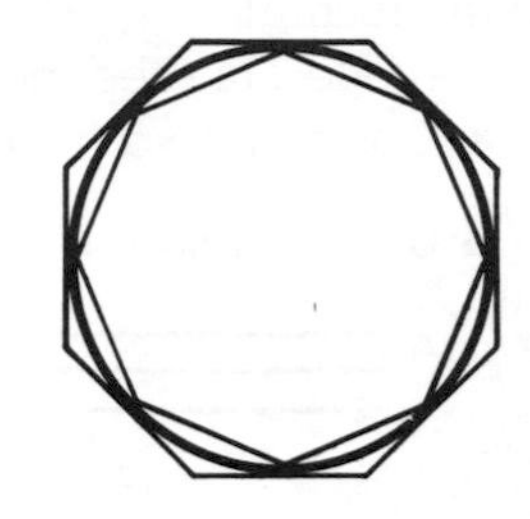

a Count the number of whole squares under the parabola.

b Count the number of whole squares including the 'staircase' steps.

The area of each square $= 0{\cdot}1 \times 0{\cdot}1 = 0{\cdot}01$ square unit.

Do you agree that $0{\cdot}24 < A < 0{\cdot}43$? That's a start anyway!

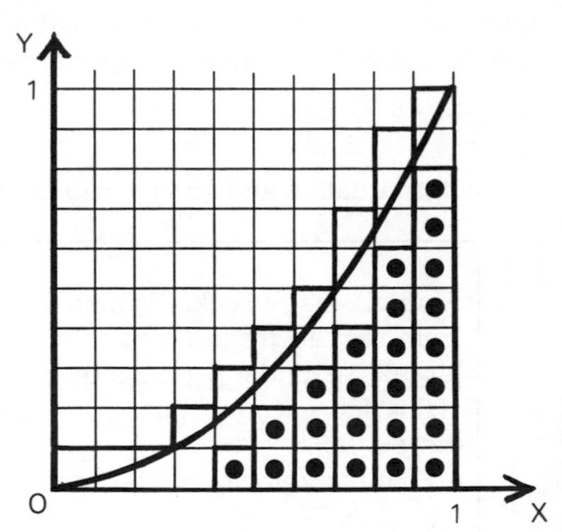

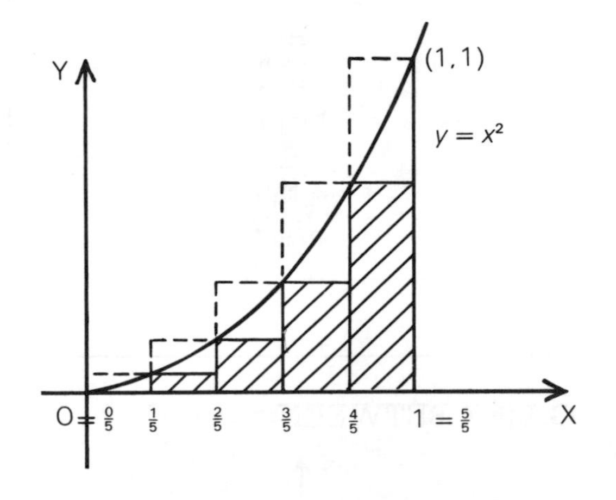

Ahmed prefers his own method. He breaks the area into 5 strips of equal width, and calculates the areas below and above the parabola, like this:

$$\frac{1}{5}\left(\left(\frac{0}{5}\right)^2+\left(\frac{1}{5}\right)^2+\left(\frac{2}{5}\right)^2+\left(\frac{3}{5}\right)^2+\left(\frac{4}{5}\right)^2\right) < A < \frac{1}{5}\left(\left(\frac{1}{5}\right)^2+\left(\frac{2}{5}\right)^2+\left(\frac{3}{5}\right)^2+\left(\frac{4}{5}\right)^2+\left(\frac{5}{5}\right)^2\right),$$

i.e. $\frac{1}{5^3}(1^2+2^2+3^2+4^2) < A < \frac{1}{5^3}(1^2+2^2+3^2+4^2+5^2)$,

and so $0{\cdot}24 < A < 0{\cdot}44$. Not much better than before!

Toni says: 'I'll find a more accurate answer with 10 strips.'

After working with her calculator for some time she claims that $0{\cdot}285 < A < 0{\cdot}385$. Do your calculations agree with this?

If you have time, organise a group to calculate lower and upper bounds for A using (i) 100 strips (ii) 1000 strips. Are you reaching a conclusion about the value of A?

Gordon has an inspiration. He uses n strips, and claims that:

$$\frac{1}{n^3}(1^2+2^2+\ldots+(n-1)^2) < A < \frac{1}{n^3}(1^2+2^2+\ldots+n^2).$$

Remembering the formulae:

$$1^2+2^2+\ldots+n^2 = \frac{1}{6}n(n+1)(2n+1), \text{ and } 1^2+2^2+\ldots+(n-1)^2 = \frac{1}{6}n(n-1)(2n-1)$$

he uses some algebra to prove that:

$$\frac{1}{3}\left(1-\frac{1}{n}\right)\left(1-\frac{1}{2n}\right) < A < \frac{1}{3}\left(1+\frac{1}{n}\right)\left(1+\frac{1}{2n}\right).$$

He says 'I can make n as large as I like, so clearly $A = \frac{1}{3}$'. And from the diagram, each area strip up to the line $x = a$, has a factor a^3. So $A = \frac{1}{3}a^3$, up to $x = a$.

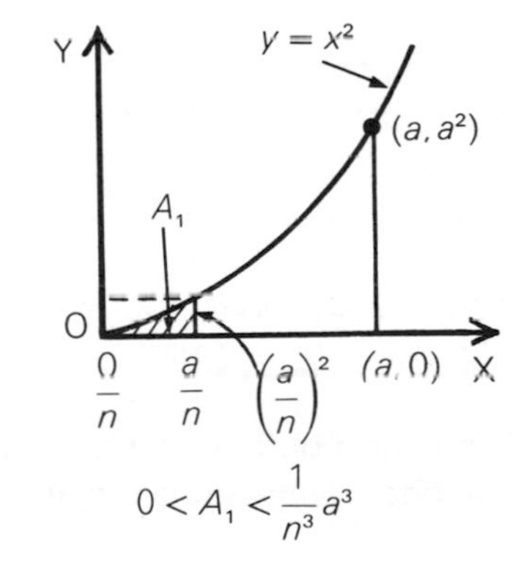

Working as a group, Ahmed, Toni and Gordon dealt with $y = x^3$ in the same way. They used the formula $1^3+2^3+\ldots+n^3 = \frac{1}{4}n^2(n+1)^2$. They found $A = \frac{1}{4}$ up to the line $x = 1$, and $A = \frac{1}{4}a^4$ up to the line $x = a$.

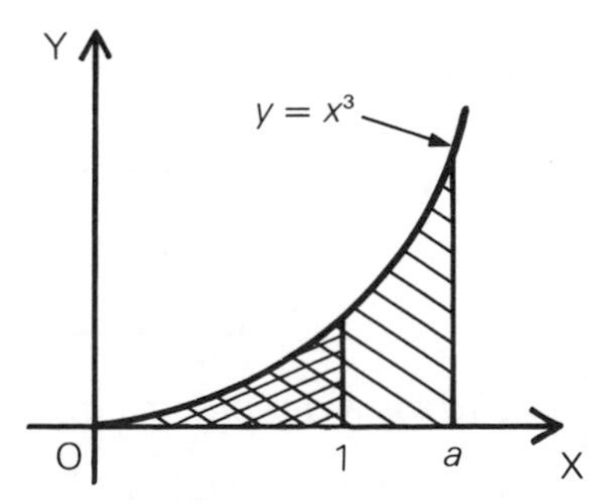

Summing up

The area of the region bounded by the x-axis, the line $x = a$ and:

(i) the line $y = x$ is $\frac{1}{2}a^2$; . . . and $\int x\,dx = \frac{1}{2}x^2 + C$

(ii) the parabola $y = x^2$ is $\frac{1}{3}a^3$; . . . and $\int x^2\,dx = \frac{1}{3}x^3 + C$

(iii) the curve $y = x^3$ is $\frac{1}{4}a^4$; . . . and $\int x^3\,dx = \frac{1}{4}x^4 + C$.

Areas between $x = a$ and $x = b$:

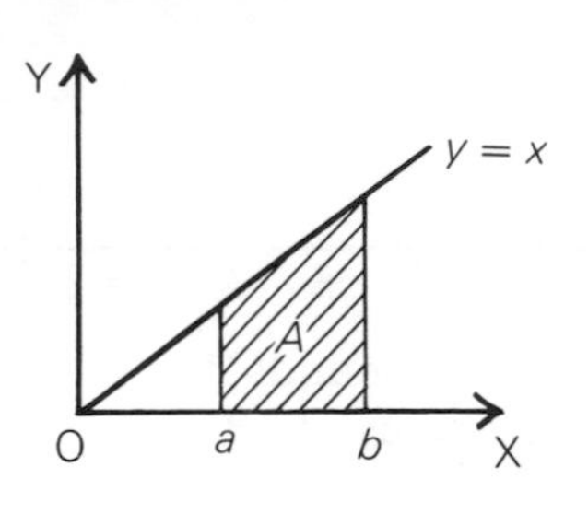

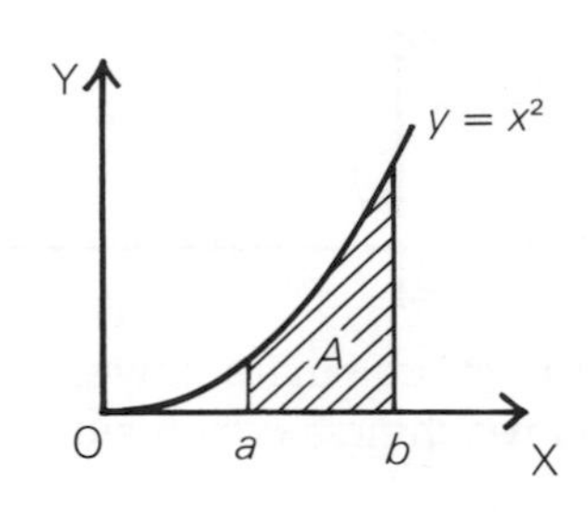

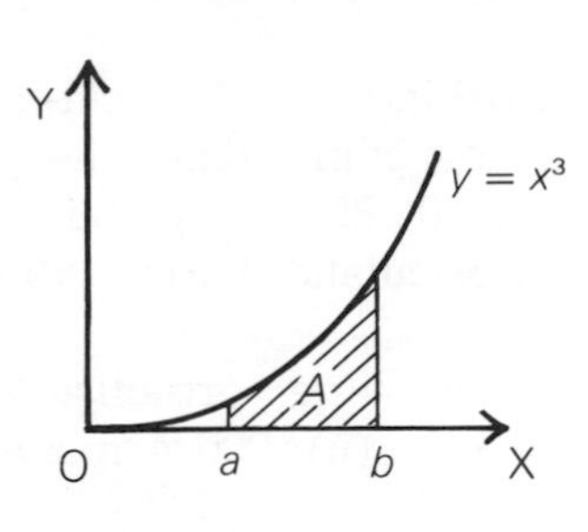

$A = \frac{1}{2}b^2 - \frac{1}{2}a^2$
$= F(b) - F(a)$,
where $F'(x) = x$

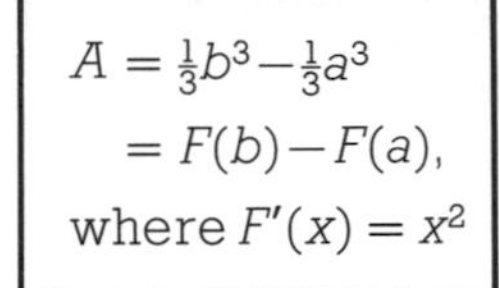

$A = \frac{1}{3}b^3 - \frac{1}{3}a^3$
$= F(b) - F(a)$,
where $F'(x) = x^2$

$A = \frac{1}{4}b^4 - \frac{1}{4}a^4$
$= F(b) - F(a)$,
where $F'(x) = x^3$

What would be the areas bounded by $x = a$, $x = b$ and the curves: (i) $y = x^4$ (ii) $y = x^5$?

Not only is there a pattern here, there is an important general result:

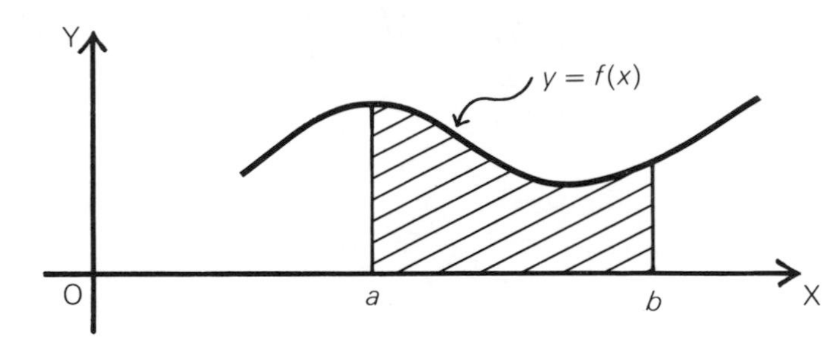

To calculate the area bounded by $y = f(x)$, the x-axis and the lines $x = a$, $x = b$ we:

(i) *find $F(x)$ such that $F'(x) = f(x)$, i.e. an anti-derivative of $f(x)$*
(ii) *calculate $F(b) - F(a)$.*

This is a rough outline of what is called the Fundamental Theorem of Calculus which links the two branches, Differentiation and Integration.

Notation

Noting that $F(x) = \int f(x)\,dx$, the difference $F(b) - F(a)$ is denoted by $\int_a^b f(x)\,dx$ or by $\Big[F(x)\Big]_a^b$, so

$$\int_a^b f(x)\,dx = \Big[F(x)\Big]_a^b = F(b) - F(a), \text{ where } F'(x) = f(x).$$

$\int_a^b f(x)\,dx$ is a *definite integral* with *lower limit* a and *upper limit* b. In contrast, $\int f(x)\,dx$ is often called an *indefinite integral.*

Although you first met the symbol $\int$ in anti-differentiation, it is a lengthened 's' and is suggested by the 'sum process' for areas and definite integrals.

Examples Evaluate the integrals:

1 $\int_1^2 (6x^2+2x)\,dx = \left[2x^3+x^2\right]_1^2 = (16+4)-(2+1) = 17$

2 $\int_0^4 \sqrt{(2x+1)}\,dx = \left[\frac{(2x+1)^{3/2}}{\frac{3}{2}\times 2}\right]_0^4 = \left[\frac{1}{3}(2x+1)^{3/2}\right]_0^4 = \frac{1}{3}(27-1) = 8\frac{2}{3}$

3 $\int_{\pi/6}^{\pi/4} (1+\sin 2x)\,dx = \left[x-\frac{1}{2}\cos 2x\right]_{\pi/6}^{\pi/4} = \left(\frac{\pi}{4}-\frac{1}{2}\cos\frac{\pi}{2}\right)-\left(\frac{\pi}{6}-\frac{1}{2}\cos\frac{\pi}{3}\right)$

$$= \left(\frac{\pi}{4}-0\right)-\left(\frac{\pi}{6}-\frac{1}{2}\cdot\frac{1}{2}\right) = \frac{\pi}{12}+\frac{1}{4}.$$

Exercise 6

1 Evaluate:

a $\int_1^3 x\,dx$ **b** $\int_{-1}^2 2x\,dx$ **c** $\int_0^2 3\,dx$ **d** $\int_1^2 \frac{dx}{2x^2}$

e $\int_1^4 \sqrt{t}\,dt$ **f** $\int_{-2}^0 u^2\,du$ **g** $\int_0^1 (5-3t^2)\,dt$ **h** $\int_{-1}^1 (3x^2+5x^4)\,dx.$

2 Evaluate:

a $\int_0^1 (2x+1)^4\,dx$ **b** $\int_0^{1/2} \sqrt{(1-2x)}\,dx$ **c** $\int_1^2 \frac{dx}{(1+x)^2}$

d $\int_{-1}^0 \frac{1}{\sqrt{(3x+4)}}\,dx$ **e** $\int_0^2 (1+4x)^{3/2}\,dx$ **f** $\int_{-1}^1 \frac{2}{(x-2)^2}\,dx.$

3 Evaluate:

a $\int_0^{\pi/2} \cos x\,dx$ **b** $\int_0^{\pi/2} \cos t\,dt$ **c** $\int_0^{\pi/2} \cos u\,du.$

Any comment?

4 Evaluate:

a $\int_0^{\pi/4} \sin 2x\,dx$ **b** $\int_0^{\pi/3} \cos 3x\,dx$ **c** $\int_0^{\pi} (\sin t+\cos t)\,dt$

d $\int_0^{\pi/4} (\sin 4x+\cos 4x)\,dx$ **e** $\int_{1/2}^1 (u^2+\sin \pi u)\,du$ **f** $\int_{\pi/6}^{\pi/4} \sin\left(2t-\frac{\pi}{3}\right)dt.$

5 Evaluate:

a $\int_1^3 \frac{x+1}{x^3}\,dx$ $\left(\textit{Hint}\ \frac{x+1}{x^3} = \frac{x}{x^3}+\frac{1}{x^3} = x^{-2}+x^{-3}\right)$ **b** $\int_0^1 (x^2+1)^2\,dx$

c $\int_1^9 \frac{1+2x}{\sqrt{x}}\,dx$ **d** $\int_1^2 \frac{t^2+2}{t^4}\,dt$ **e** $\int_0^{\pi/4} \sin^2 x\,dx$ (See Exercise **4**, question **4**).

6 Express the areas of these shaded regions as definite integrals. Do not evaluate them.

a

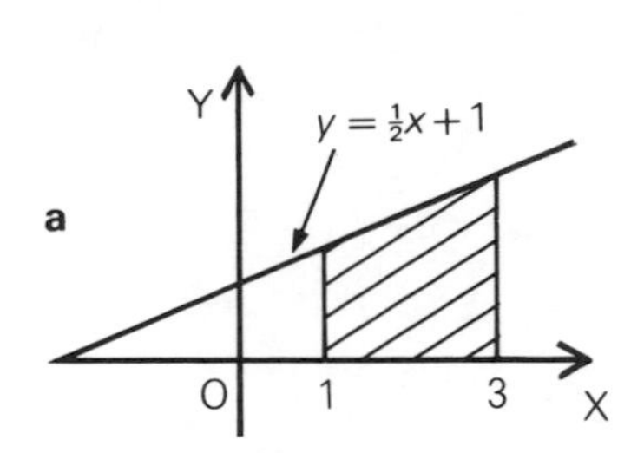

b

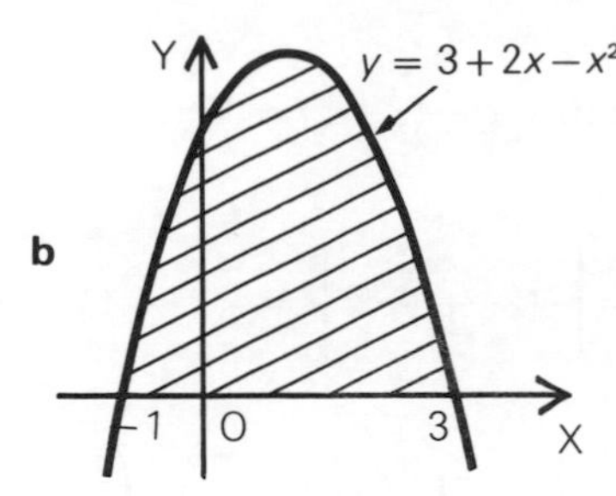

c

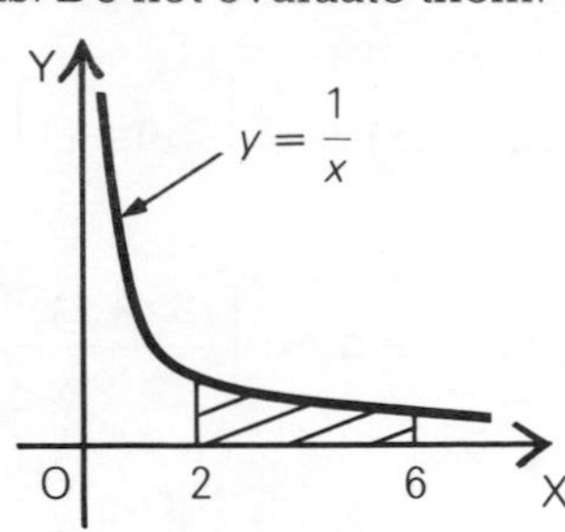

USE OF SKETCHES

Example Calculate the area of the region:

a bounded by the parabola $y = x^2 - 1$, the x-axis and the line $x = 2$ (A)

b enclosed by the parabola and the x-axis (B).

a Area A is above the x-axis, and has area

$$\int_1^2 (x^2 - 1)\,dx = \left[\tfrac{1}{3}x^3 - x\right]_1^2 = (\tfrac{8}{3} - 2) - (\tfrac{1}{3} - 1) = 1\tfrac{1}{3}.$$

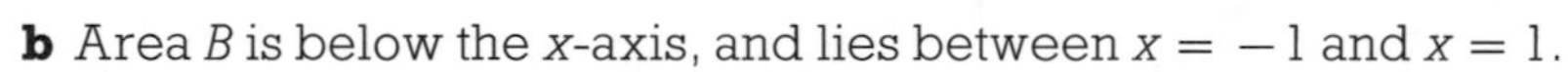

b Area B is below the x-axis, and lies between $x = -1$ and $x = 1$.

$$\int_{-1}^{1} (x^2 - 1)\,dx = \left[\tfrac{1}{3}x^3 - x\right]_{-1}^{1} = (\tfrac{1}{3} - 1) - (-\tfrac{1}{3} + 1) = -1\tfrac{1}{3}.$$

The definite integral is negative, but the area has measure $1\frac{1}{3}$.
The total shaded area is $1\frac{1}{3} + 1\frac{1}{3} = 2\frac{2}{3}$.

Note Integration will give negative values for areas under the x-axis since, for the approximating strips, $f(x)$ is <0 while $\Delta x > 0$.

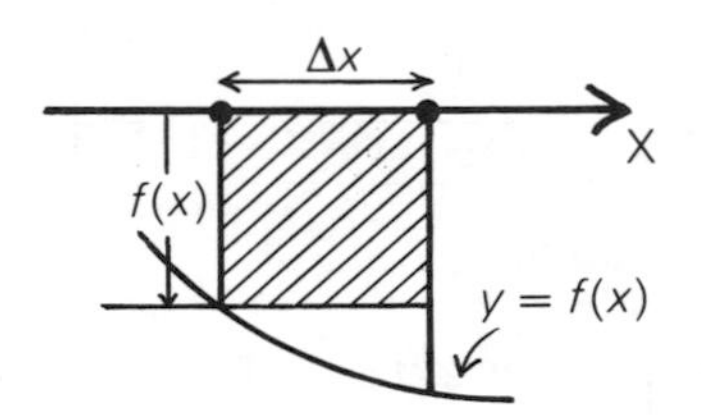

Exercise 7

In each question make a quick sketch of the curve, and shade the region defined by the integral.

1 For the above example evaluate $\int_{-1}^{2} (x^2 - 1)\,dx$. Are you surprised at the answer? Explain.

2 Calculate the area bounded by $y = x^2$, the x-axis and the lines:
a $x = 1$ and $x = 2$ **b** $x = -2$ and $x = 1$.

3 Calculate the area enclosed by $y = 4x^3$, the x-axis and the lines:
a $x = 1$ and $x = 2$ **b** $x = -1$ and $x = 2$.

4 Calculate the area of the region between the line $y = 4(x - 1)$, the x-axis, $x = -1$ and $x = 2$.

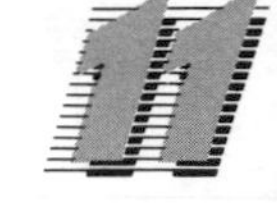

5 Calculate the area of the region bounded by the x-axis and:
a $y = x^2+3$ for $0 \leqslant x \leqslant 2$ **b** $y = 2x-1$ for $\frac{1}{2} \leqslant x \leqslant 3$
c $y = \dfrac{1}{x^3}$ for $1 \leqslant x \leqslant 2$ **d** $y = \sin x$ for $0 \leqslant x \leqslant \pi$.

6 Calculate the total shaded area in each diagram:

a

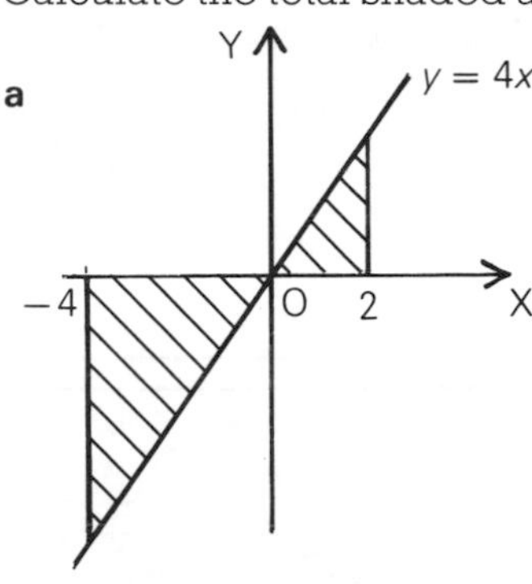

b

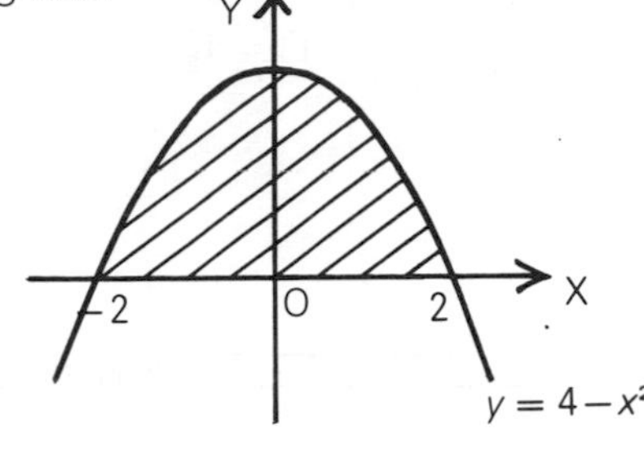

7 Evaluate the area enclosed by $y = \cos x$, the x-axis, $x = 0$ and $x = \dfrac{2\pi}{3}$.

8 f is defined on R by the formula $f(x) = x^3-4x$.
a Find the coordinates and nature of the stationary points of f.
b Find the intersections with the x-axis.
c Sketch the graph of f.
d Calculate the area of the region between the graph and the x-axis.
e Explain why $\int_{-2}^{2} f(x)\,dx = 0$.

Evaluate:

a $\int_{-1}^{1} \sqrt{(1-x^2)}\,dx$ **b** $\int_{0}^{2} \sqrt{(4-x^2)}\,dx$ **c** $\int_{-1}^{1} \sqrt{(4-x^2)}\,dx$.

Think geometrically!

AREAS BETWEEN CURVES

Example Calculate the area enclosed by:
a the parabola $y = x^2$ and the line $y = 2x$
b the parabola $y = x^2$, line $y = 2x$ and line $x = -1$.

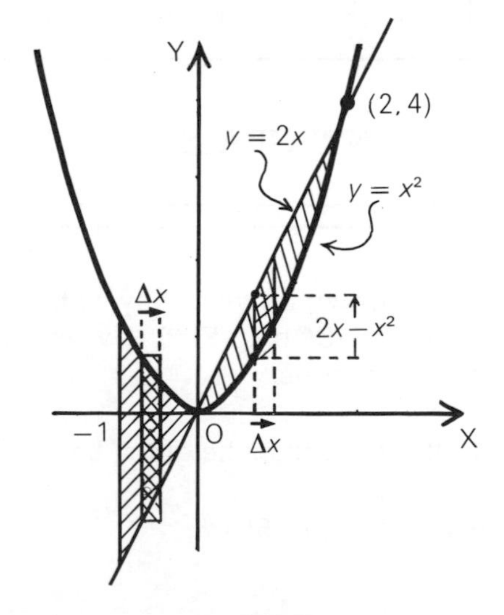

$y = x^2$ meets $y = 2x$ where $x^2 = 2x$, i.e. $x(x-2) = 0$, and so where $x = 0$ or 2.

Check that the graphs are as shown in the sketch.

a Area of strip $\doteqdot (2x-x^2)\Delta x$.

So area $= \int_{0}^{2} (2x-x^2)\,dx$

$= \left[x^2-\frac{1}{3}x^3\right]_0^2 = (4-\frac{8}{3})-(0-0) = 1\frac{1}{3}$.

b Area of strip $\doteqdot (x^2-2x)\Delta x$.

So area $= \int_{-1}^{0} (x^2-2x)\,dx$

$= \left[\frac{1}{3}x^3 - x^2\right]_{-1}^{0} = (0-0)-(-\frac{1}{3}-1) = 1\frac{1}{3}$.

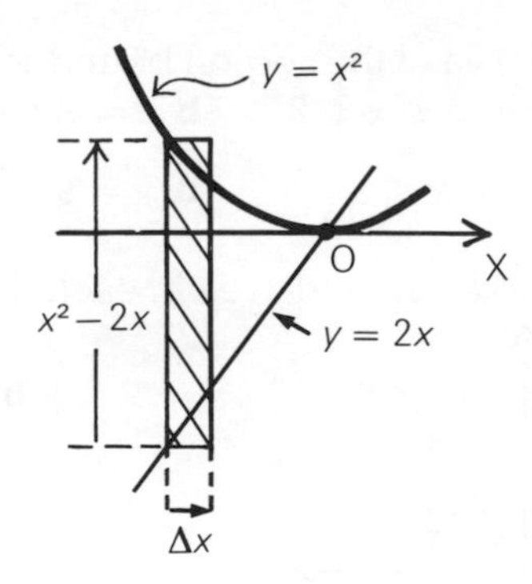

Exercise 8

In each question draw a sketch, and shade the region.

1 Calculate the area enclosed by each parabola and line:
a $y = 2x^2$, $y = 4x$ **b** $y = x^2$, $y = -x$
c $y = x^2$, $y = x+2$ **d** $y = 4-x^2$, $y = 4-2x$.

2 Calculate the area enclosed between the curve $y = x^3$ and the line $y = x$.

3 **a** Find the points of intersection of the parabola $y = 2+x-x^2$ and the line $y = 3x-1$.
b Show the two curves in the same sketch.
c Calculate the area bounded by the two curves.

4 **a** Sketch the parabolas $y = 4x-x^2$ and $y = x^2-6$.
b Calculate the area between them.

5 Find the area bounded by $y = \sin x$ and $y = \frac{2}{\pi}x$ for $0 \leqslant x \leqslant \frac{\pi}{2}$.

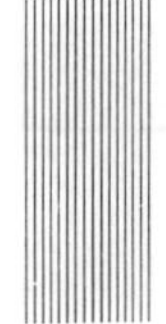

Calculate the area bounded by $y = \cos x$ and $y = \cos\frac{x}{2}$ for $0 \leqslant x \leqslant \frac{4\pi}{3}$.

DEFINITE INTEGRALS IN ACTION

Exercise 9

1 A ball is thrown up vertically so that its velocity at time t seconds is $v(t) = 20t-5t^2$ m/s. Find (by integration) the height gained in the first 2 seconds.

Hint If $h(t)$ denotes height at time t, $\frac{dh}{dt} = 20t-5t^2$ and the height gained is $\int_0^2 (20t-5t^2)\,dt$.

2 A rocket weighs 120 grams. As it rises it burns up fuel, and loses weight. Its weight $w(s)$ at height s metres is $w(s) = 120 - 2s$. The work done in lifting the rocket is $\int w\,ds$.
Calculate the work done (in gram metres):
a in the first 20 metres
b between 20 and 25 metres.

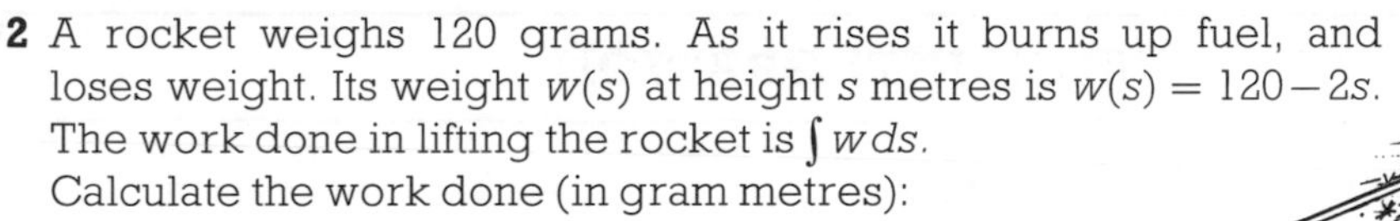

3 The monthly profit $P(x)$ (in £) of a manufacturer of fishing rods (profit in producing x rods) is such that $\dfrac{dP}{dx} = 50 - 0{\cdot}6x$. Find his profit in producing the first 100 rods.

4 A bobsleigh heads down the Vesta Run—length 1500 metres. Its velocity after travelling s metres is modelled by $v = \sqrt{(100 + 4s)}$. The time t (in seconds) taken between any two distances can be computed using

$$t = \int_{s_1}^{s_2} \frac{1}{v}\,ds.$$

Calculate the time taken for:
a the first 100 metres **b** the last 100 metres.

5 A hemisphere is formed by rotating a quarter circular disc about the x-axis. The volume generated by the shaded rectangular strip is $\pi y^2\,dx$. Why?
Calculate the volume of the hemisphere. ($V = \int \pi y^2\,dx$)

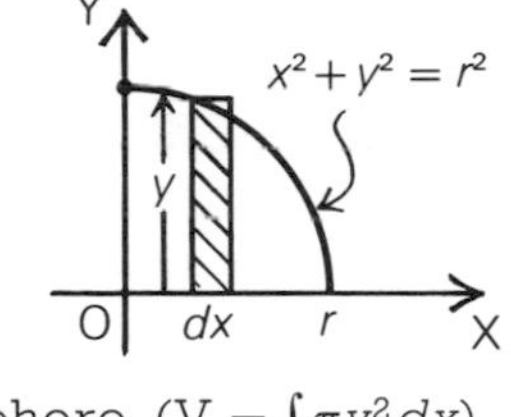

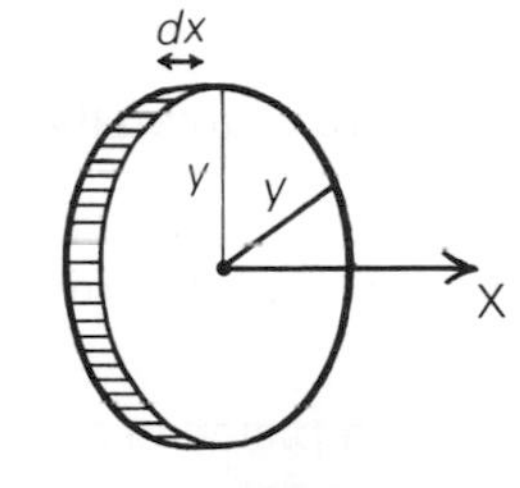

6 The horizontal velocity of the midpoint of the FASTA's rear window wiper is modelled by the equation $\dfrac{dx}{dt} = 2 - 2\sin t$.
Find the horizontal distance travelled in the first $\dfrac{\pi}{3}$ seconds.

7 When designing a bridge, the area presented to the wind must be calculated to forecast the wind-force the bridge will have to withstand.

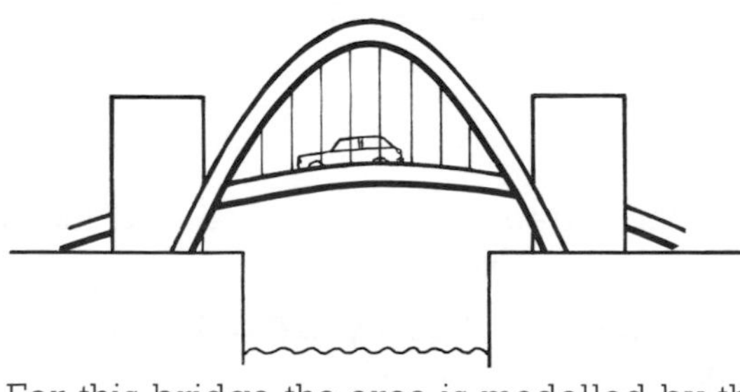

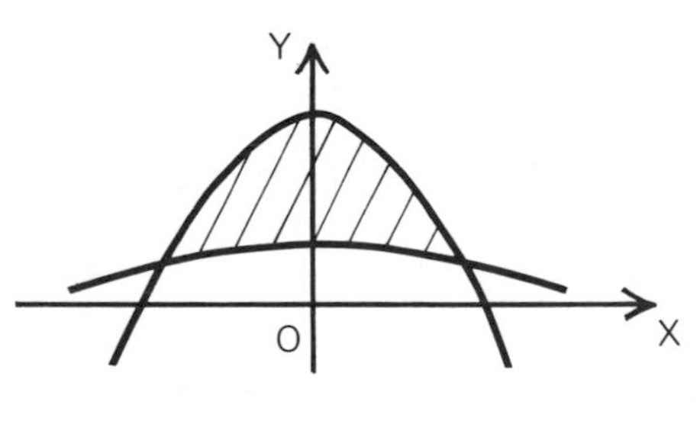

For this bridge the area is modelled by the region between the parabolas $y = 10 - x^2$ and $y = 19 - 2x^2$.
a Sketch the parabolas.
b Calculate the area bounded by them.

CHECK-UP ON **INTEGRATION**

1 Integrate with respect to the relevant variable:

a $2-x^3$ **b** $(4x+1)^3$ **c** $\frac{1}{\sqrt{t}}-6t^2$ **d** $\sin 5u-\cos 3u$ **e** $\left(x-\frac{1}{x}\right)^2$.

2 Find the following anti-derivatives:

a $\int(3+4x-3x^2)\,dx$ **b** $\int\cos(3t+2)\,dt$ **c** $\int\sqrt{(2u-3)}\,du$ **d** $\int x^4\left(3x-\frac{2}{x^3}\right)dx$.

3 Find $f(x)$ where:

a $f'(x)=3x^2-4x$ **b** $f'(x)=2+\cos 4x$ **c** $f'(x)=6x^2-2\sin 6x$ **d** $f'(x)=\frac{2-x^4}{x^2}$.

4 Find the general solution of:

a $\frac{dy}{dx}=3-6x^2$ **b** $\frac{dy}{dx}=3-2\sin 2x$ **c** $\frac{dy}{dx}=4+2\cos 4x-3\sin 2x$.

5 For each of these differential equations find the particular solution satisfying the given condition.

a $\frac{dy}{dx}=3-2x+6x^2$, given that $y=5$ when $x=1$

b $\frac{dy}{dx}=2\cos 4x-3\sin 6x$, given that $y\left(\frac{\pi}{12}\right)=\frac{1}{4}\sqrt{3}$.

6 The velocity of a piston in a cylinder is modelled by the equation $\frac{dx}{dt}=6\cos 2t$.

If the piston starts from the point A(2, 0), find its displacement $x(t)$ at time t (seconds).

7 Evaluate:

a $\int_0^1(4+2x)\,dx$ **b** $\int_{-1}^4\frac{dt}{\sqrt{(3t+4)}}$ **c** $\int_{\pi/6}^{\pi/3}\sin 3u\,du$ **d** $\int_1^2\frac{2v^2+3}{v^4}\,dv$.

8 Calculate the area bounded by the x-axis and:

a $y=8-2x^2$, $x=-1$, $x=2$ **b** $y=6\cos 3x$, $x=-\frac{\pi}{18}$, $x=\frac{\pi}{6}$.

(A rough sketch will help in each case.)

9 a Calculate the area of the region bounded by the x-axis and the curve $y=x(x^2-1)$ between $x=0$ and $x=1$.

b Explain geometrically why $\int_0^{\pi}\cos 2x\,dx=0$.

10 Calculate the area of the region bounded by the parabola $y=3+2x-x^2$ and the line $y=3(x-1)$.

11 A meteor streaked across the sky. Its velocity was modelled by $v(t)=6+0{\cdot}006t$ km/s, where t is measured from the instant of detection.

If the meteor burned out in 3 seconds, calculate how far it travelled in the atmosphere.

$\left(v(t)=\frac{dx}{dt}\right)$

3 DIMENSIONS

We all live in a 3-dimensional world—a world of length, breadth and height. Even this sheet of paper has 3 dimensions, although its thickness is very small in comparison with its length and breadth.

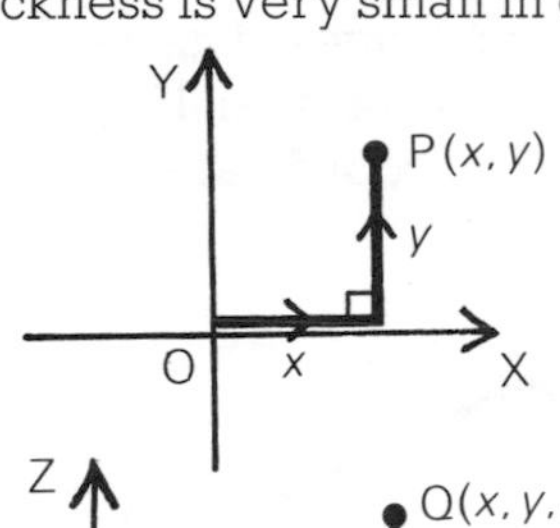

To fix the position of a point on the plane you need two axes OX and OY, and two coordinates x and y. P is the point (x, y).

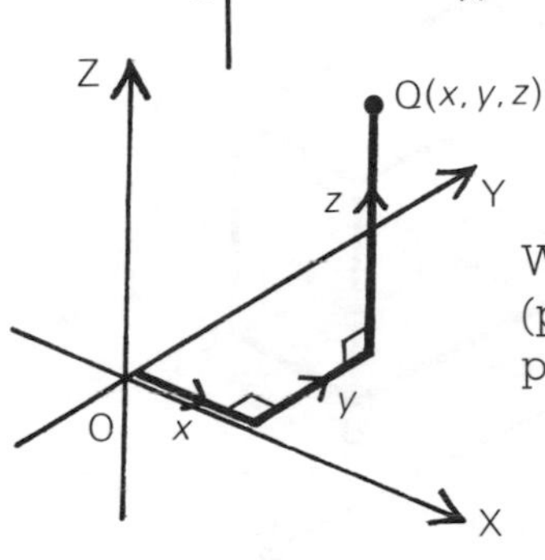

When you move out of that plane all you need is one other axis OZ (perpendicular to OX and OY), and one other coordinate z, to fix the position of $Q(x, y, z)$.

Exercise 1

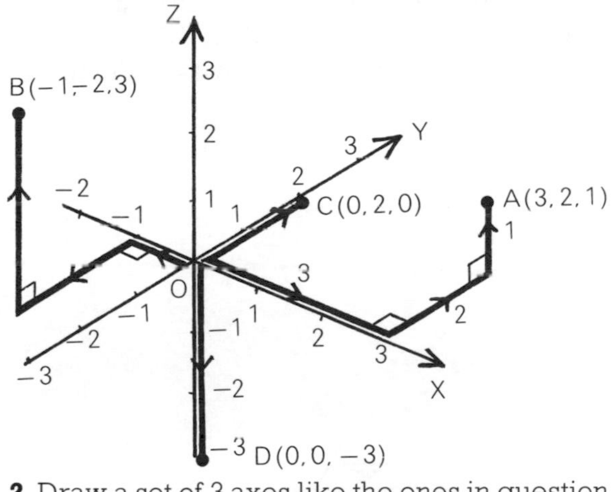

1 Check the positions of the points A(3, 2, 1), B(−1, −2, 3), C(0, 2, 0) and D(0, 0, −3) by following the trail from O in the x-direction, then in the y-direction, then in the z-direction.

2 Draw a set of 3 axes like the ones in question **1**, and mark the scales 1, 2, 3, etc. Show the x, y and z-components in red leading to the points P(3, 1, 1), Q(−1, −1, 2), R(2, −1, 0), S(0, 2, 1), T(0, 0, 3) and U(0, −2, −3).

3 The cuboid sits in the OX, OY, OZ corner. P is the point (2, 1, 3). Write down the coordinate triples for the other seven vertices of the cuboid (x-direction, then y-direction, then z-direction).

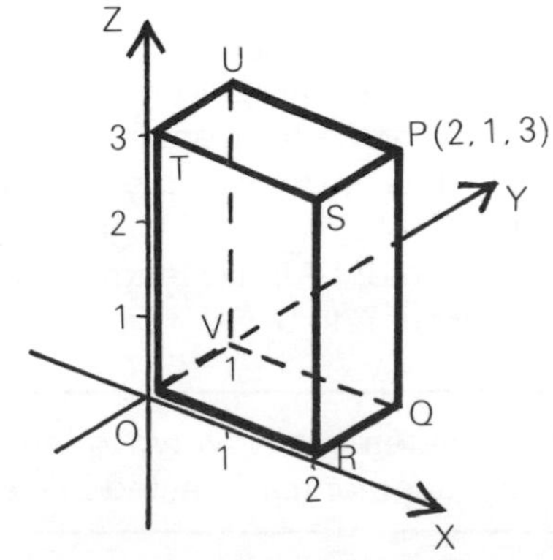

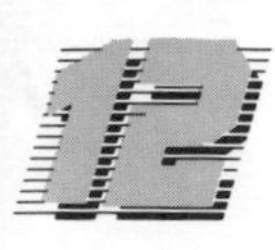

4

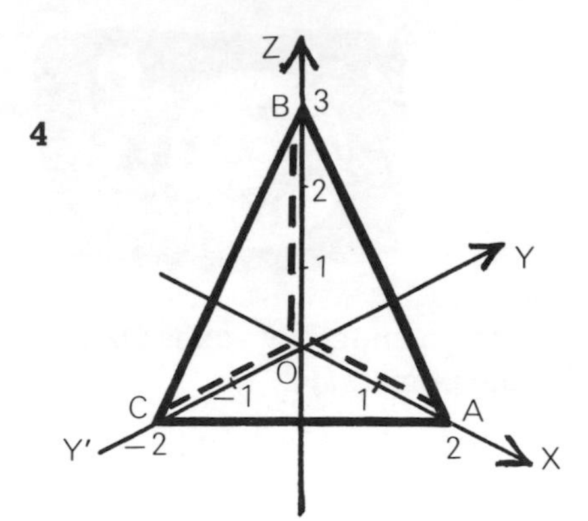

The triangular prism OABC sits in the OX, OY′, OZ corner.

a Write down the coordinates of its vertices.

b If the prism is turned through 180° about OB, what will the coordinates of the vertices be?

5 The edges of the cuboid are parallel to the axes. AB = 4 units, BC = 2 units and CG = 3 units. A is the point $(-1, 2, 0)$. Write down the coordinates of the other vertices.

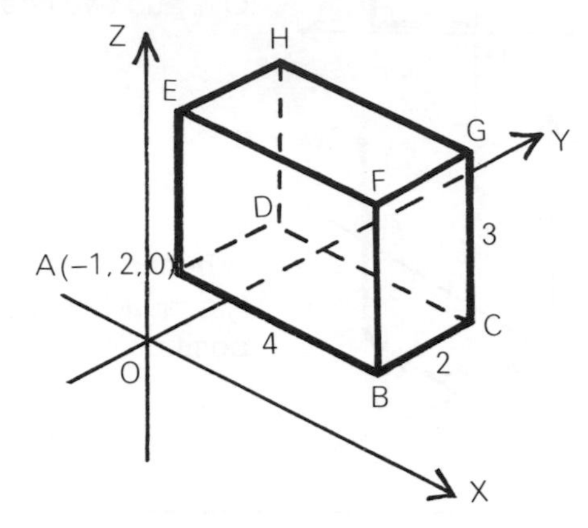

6

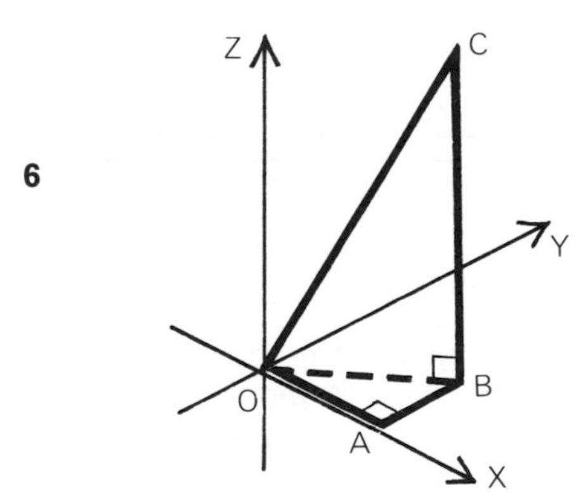

A is (4, 0, 0), B is (4, 3, 0) and C is (4, 3, 12).

a Calculate the length of: (i) OB (ii) OC.

b If C is (x, y, z), write down a formula for OC^2.

7 Find the coordinates of the images of:

a A(2, 2, 2) when reflected in the XOY plane

b B(−1, 5, 7) reflected in the YOZ plane

c C(1, 2, 4) reflected in the origin O.

VECTORS

(i) The car is travelling due east at 45 mph.

(ii) The force F has magnitude 5 units in a direction 30° to the horizontal.

(iii) The crate has a displacement of 10 m from P to Q.

(iv) The line segment $\overrightarrow{AB}$ has length AB and direction from A to B.

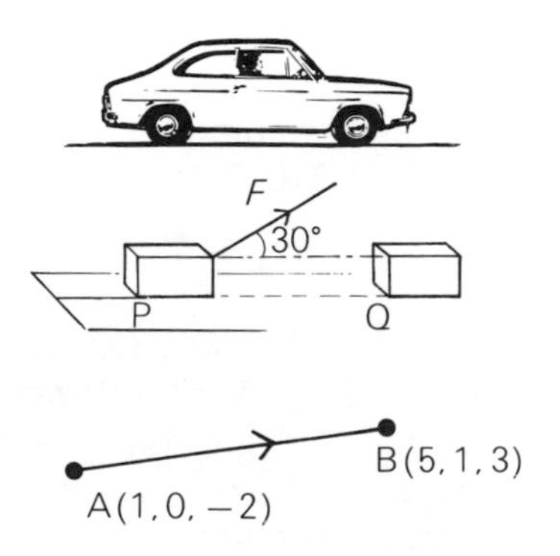

These four statements involve quantities with *magnitude* and *direction*—a velocity, a force, a displacement, a directed line segment. These, and similar objects, are *vectors*.

Vectors can be represented by directed line segments

If $\overrightarrow{AB}$ represents $\boldsymbol{u}$, we write $\boldsymbol{u} = \overrightarrow{AB}$.

> If $\boldsymbol{u} = \overrightarrow{AB} = \overrightarrow{DC}$, then
> ABCD is a parallelogram,
> or A, B, C, D are collinear.

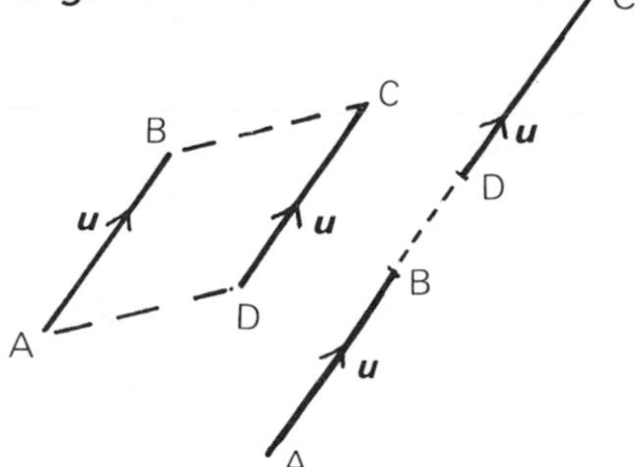

Components of a vector

The cuboid has its edges parallel to the axes. $\overrightarrow{AB}$ is a space diagonal. The step AP parallel to $OX = x_B - x_A = 5 - 1 = 4$.
This is the x-component of $\overrightarrow{AB}$.
The y-component of $\overrightarrow{AB} = y_B - y_A = 6 - 3 = 3$.
The z-component of $\overrightarrow{AB} = z_B - z_A = 4 - 2 = 2$.

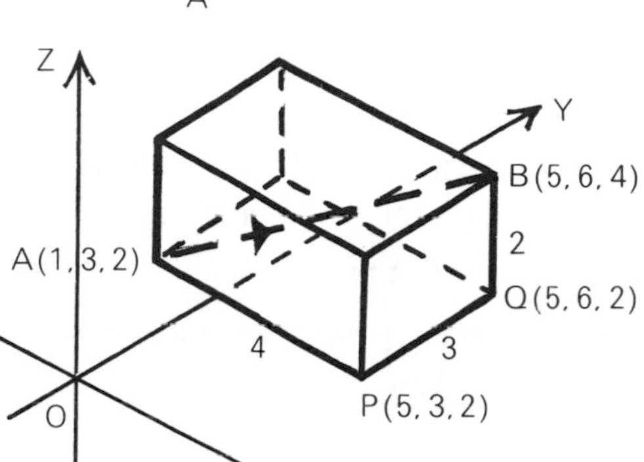

In component form, $\overrightarrow{AB} = \begin{pmatrix}4\\3\\2\end{pmatrix} = \begin{pmatrix}x\text{-component}\\y\text{-component}\\z\text{-component}\end{pmatrix}$.

If $\boldsymbol{u} = \overrightarrow{AB}$, then $\boldsymbol{u} = \begin{pmatrix}4\\3\\2\end{pmatrix}$.

> In general, $\overrightarrow{AB} = \begin{pmatrix}x_B - x_A\\y_B - y_A\\z_B - z_A\end{pmatrix}$.

Example If P is (6, 2, 0) and Q is (8, −3, 1), then

$$\overrightarrow{PQ} = \begin{pmatrix}8-6\\-3-2\\1-0\end{pmatrix} = \begin{pmatrix}2\\-5\\1\end{pmatrix}, \text{ a column vector.}$$

Magnitude of a vector

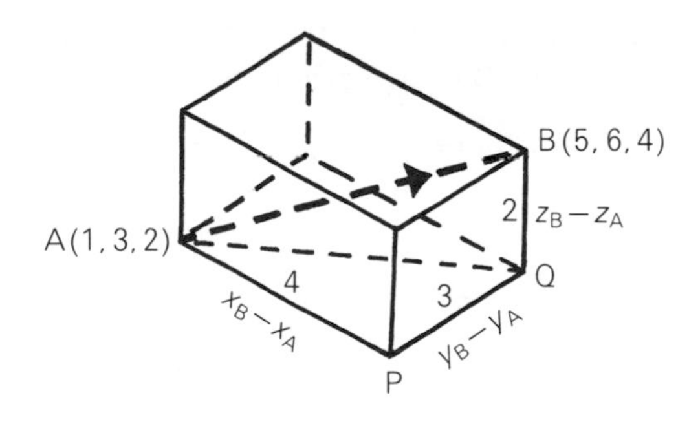

$$\begin{aligned} AB^2 &= AQ^2 + BQ^2 \text{ (Pythagoras' Theorem)} \\ &= (AP^2 + PQ^2) + BQ^2 \\ &= (x_B - x_A)^2 + (y_B - y_A)^2 + (z_B - z_A)^2 \\ &= 4^2 + 3^2 + 2^2 \\ &= 29 \end{aligned}$$

If $\boldsymbol{u} = \overrightarrow{AB}$, the magnitude of $\boldsymbol{u}$, $|\boldsymbol{u}| = AB = \sqrt{29}$.

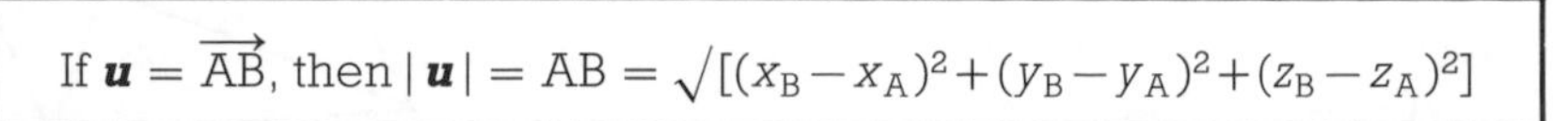

If $\boldsymbol{u} = \overrightarrow{AB}$, then $|\boldsymbol{u}| = AB = \sqrt{[(x_B - x_A)^2 + (y_B - y_A)^2 + (z_B - z_A)^2]}$

This is the distance formula in 3 dimensions.

VECTORS

Exercise 2

1 Which of these quantities is a vector and which is a *scalar* (i.e. a quantity with magnitude only, independent of direction)?
a speed **b** acceleration **c** a real number
d electric current **e** area **f** temperature.

2 Write down the components of $\overrightarrow{AB}$ as a column vector:
a A(0, 1, −2) B(2, 1, 3) **b** A(1, 1, 1) B(2, 4, −5)
c A(1, 2, 4) B(1, 2, 6) **d** A(0, 2, 0) B(−3, 1, −2)
e A(7, 5, −1) B(2, 7, −3) **f** A(−2, 1, 6) B(5, −2, 3)
g A(−6, −4, 2) B(1, 3, 5) **h** A(a, b, c) B(p, q, r).

3 Calculate the length of AB in **a–d** of question **2**.

4 A is (x_A, y_A, z_A) and B is (x_B, y_B, z_B).
a Write down in column form the components of: (i) $\overrightarrow{AB}$ (ii) $\overrightarrow{BA}$.
b If $\overrightarrow{PQ} = \begin{pmatrix} u \\ v \\ w \end{pmatrix}$, what is $\overrightarrow{QP}$ in component form?

5 A vector has only one set of components, so $\begin{pmatrix} a \\ b \\ c \end{pmatrix} = \begin{pmatrix} d \\ e \\ f \end{pmatrix} \Rightarrow \begin{matrix} a = d \\ b = e. \\ c = f \end{matrix}$

If $\begin{pmatrix} x+2 \\ 5 \\ -2 \end{pmatrix} = \begin{pmatrix} 2y+1 \\ x+y \\ -2 \end{pmatrix}$, find x and y.

6 $\overrightarrow{PQ}$ and $\overrightarrow{RS}$ both represent vector $\boldsymbol{v}$, and are not in the same straight line. Draw a diagram, and explain why PQSR is a parallelogram.

7 P is the point (6, 4, 2), Q is (8, 6, 4) and R is (2, 2, 2). Prove that OPQR is a parallelogram, where O is the origin.

8 Show that A(4, 11, 2), B(2, −3, 10) and C(0, 5, −4) lie on the surface of a sphere with centre S(2, 3, 4).

9 △LMN has vertices L(4, 7, −3), M(5, 7, −5) and N(2, 7, −4). Prove that the triangle is:
(i) isosceles (ii) right-angled.

10 $\overrightarrow{AB} = \begin{pmatrix} x_B - x_A \\ y_B - y_A \\ z_B - z_A \end{pmatrix}$. If $\overrightarrow{AB} = \begin{pmatrix} 7 \\ 5 \\ 3 \end{pmatrix}$, and A is (3, 2, −1), find B.

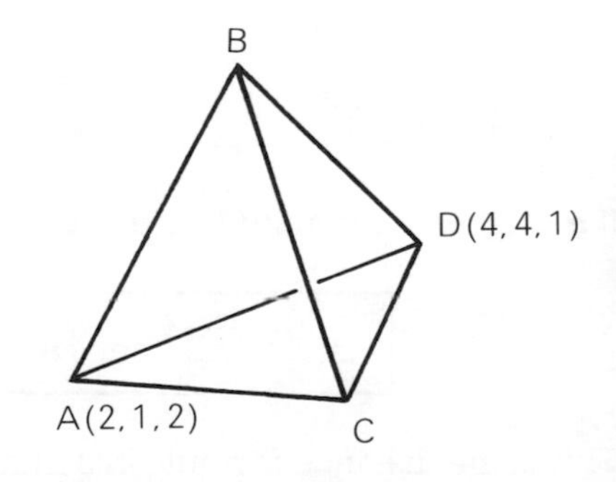

11 **a** Express $\overrightarrow{AD}$ in component form.
b Calculate the length of AD.
c If $\overrightarrow{AC} = \begin{pmatrix} 3 \\ 1 \\ -1 \end{pmatrix}$ find C.

d $\overrightarrow{DB} = \begin{pmatrix} -1 \\ -1 \\ 3 \end{pmatrix}$. Find B.

e Express $\overrightarrow{BC}$ in component form.

f Calculate the length of BC.

12 When lightning strikes the old tower, it is directed safely to ground by the lightning conductor. A is (1, 1, 6), B(0, 0, 3), E(2, 2, 0).

a Write down the coordinates of C and D.

b Express $\overrightarrow{AB}$ in component form, and calculate $|\overrightarrow{AB}|$.

c Find the total length of the conductor.

d Find $\overrightarrow{AE}$ in components and magnitude.

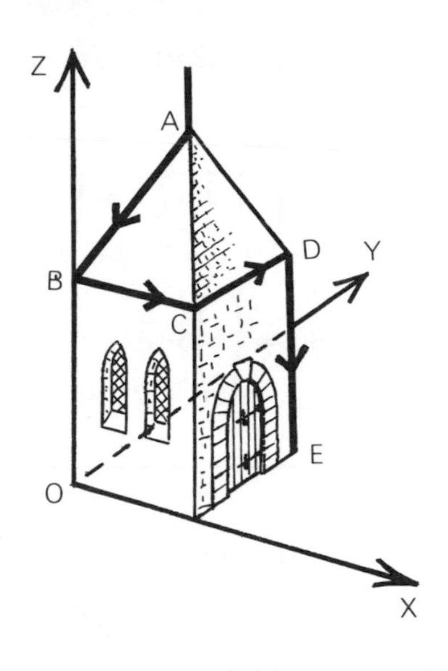

ADDITION OF VECTORS

Personal stereo	*£25*
Batteries	*2·50*
Total	*£27·50*

Temperature 10pm	*= −1°C*
Temperature 2am	*= −5°C*
Fall in temperature	*= 4°C*

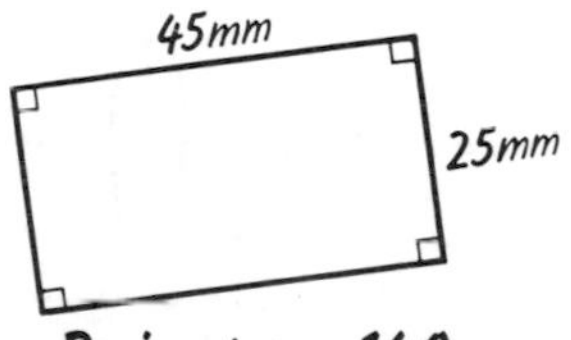

Perimeter = 140mm

It's easy to add and subtract scalars. But what about vectors, which have *direction* as well as *size, or magnitude*?

400 km/h
E
50 km/h
S
?

For example, the aircraft is flying at 400 km/h in an easterly direction, but is subject to a wind from the north of 50 km/h.

Or, two tugs are towing the liner with equal force, acting at an angle of 30°. How can these be combined?

A common sense solution

Concorde is pinpointed at A(1, 1, 1), then B(3, 3, 3), then C(4, 6, 7).
Instead of flying from A to B, then from B to C, it could have flown directly from A to C.
In this sense, $\overrightarrow{AB} + \overrightarrow{BC} = \overrightarrow{AC}$.

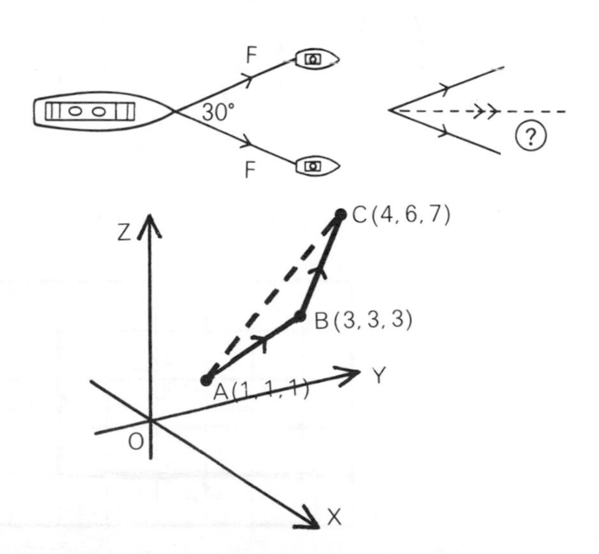

Vectors are, in fact, added like this, 'nose-to-tail'.

$\overrightarrow{AB} = \begin{pmatrix}2\\2\\2\end{pmatrix}$, $\overrightarrow{BC} = \begin{pmatrix}1\\3\\4\end{pmatrix}$ and $\overrightarrow{AC} = \begin{pmatrix}3\\5\\6\end{pmatrix}$.

Check $\overrightarrow{AB}+\overrightarrow{BC} = \begin{pmatrix}3\\5\\6\end{pmatrix} = \overrightarrow{AC}$.

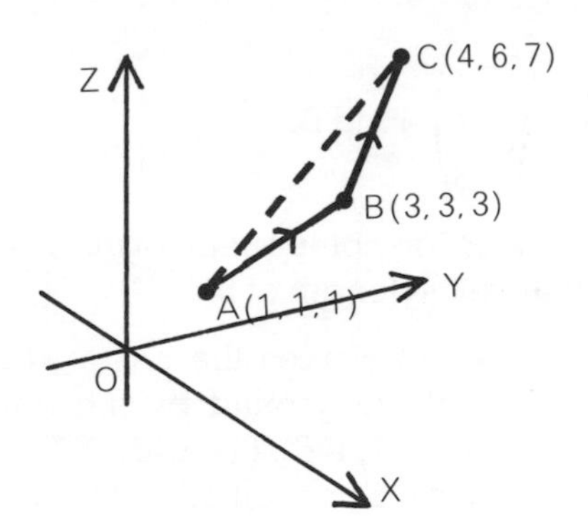

Triangle rule for adding vectors:

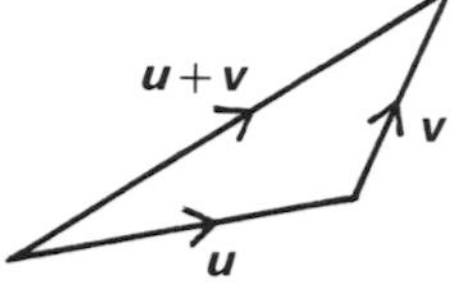

Components $\begin{pmatrix}a\\b\\c\end{pmatrix}+\begin{pmatrix}p\\q\\r\end{pmatrix}=\begin{pmatrix}a+p\\b+q\\c+r\end{pmatrix}$

Example $\boldsymbol{u} = \begin{pmatrix}1\\1\\1\end{pmatrix}$ and $\boldsymbol{v} = \begin{pmatrix}2\\-3\\0\end{pmatrix}$. Calculate: (i) $\boldsymbol{u}+\boldsymbol{v}$ (ii) $|\boldsymbol{u}+\boldsymbol{v}|$

(i) $\boldsymbol{u}+\boldsymbol{v} = \begin{pmatrix}1\\1\\1\end{pmatrix}+\begin{pmatrix}2\\-3\\0\end{pmatrix}=\begin{pmatrix}1+2\\1-3\\1+0\end{pmatrix}=\begin{pmatrix}3\\-2\\1\end{pmatrix}$.

(ii) $|\boldsymbol{u}+\boldsymbol{v}|^2 = 3^2+(-2)^2+1^2 = 14$, so $|\boldsymbol{u}+\boldsymbol{v}| = \sqrt{14}$.

Exercise 3

1 Calculate: **a** $\boldsymbol{u}+\boldsymbol{v}$ **b** $|\boldsymbol{u}+\boldsymbol{v}|$ for:

(i) $\boldsymbol{u} = \begin{pmatrix}2\\1\\-1\end{pmatrix}$, $\boldsymbol{v} = \begin{pmatrix}-2\\1\\3\end{pmatrix}$ (ii) $\boldsymbol{u} = \begin{pmatrix}3\\4\\5\end{pmatrix}$, $\boldsymbol{v} = \begin{pmatrix}0\\-4\\5\end{pmatrix}$ (iii) $\boldsymbol{u} = \begin{pmatrix}-1\\-2\\-3\end{pmatrix}$, $\boldsymbol{v} = \begin{pmatrix}5\\-2\\1\end{pmatrix}$.

2 On squared paper construct vector triangles to add the pairs of vectors: (i) $\boldsymbol{p}+\boldsymbol{q}$ (ii) $\boldsymbol{p}+\boldsymbol{r}$ (iii) $\boldsymbol{p}+\boldsymbol{s}$ (iv) $\boldsymbol{q}+\boldsymbol{r}$ (v) $\boldsymbol{q}+\boldsymbol{s}$ (vi) $\boldsymbol{r}+\boldsymbol{s}$.

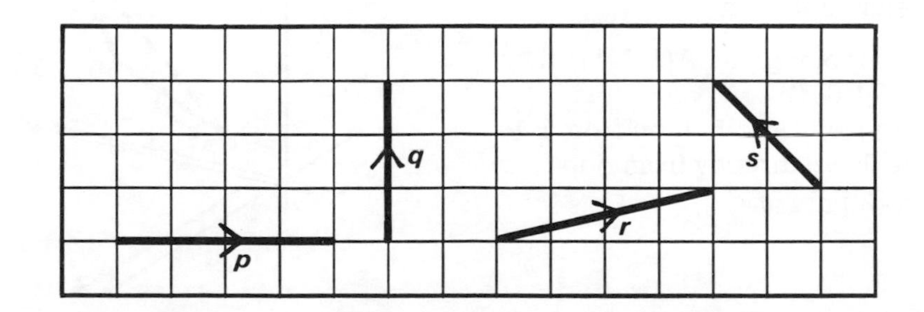

3 *The negative of a vector*

a Calculate $\boldsymbol{u}+\boldsymbol{v}$:

(i) $\boldsymbol{u}=\begin{pmatrix}4\\2\\1\end{pmatrix}, \boldsymbol{v}=\begin{pmatrix}-4\\-2\\-1\end{pmatrix}$ (ii) $\boldsymbol{u}=\begin{pmatrix}5\\0\\-1\end{pmatrix}, \boldsymbol{v}=\begin{pmatrix}-5\\0\\1\end{pmatrix}$ (iii) $\boldsymbol{u}=\begin{pmatrix}-1\\3\\2\end{pmatrix}, \boldsymbol{v}=\begin{pmatrix}1\\-3\\-2\end{pmatrix}$.

You should have obtained $\begin{pmatrix}0\\0\\0\end{pmatrix}$ each time. $\begin{pmatrix}0\\0\\0\end{pmatrix}$ is the zero vector, $\boldsymbol{0}$.

In each case, $\boldsymbol{v}$ is called the *negative of* $\boldsymbol{u}$, and written $\boldsymbol{v}=-\boldsymbol{u}$.
So $\boldsymbol{u}+(-\boldsymbol{u})=\boldsymbol{0}$.

b Write down the negative of each of these vectors:

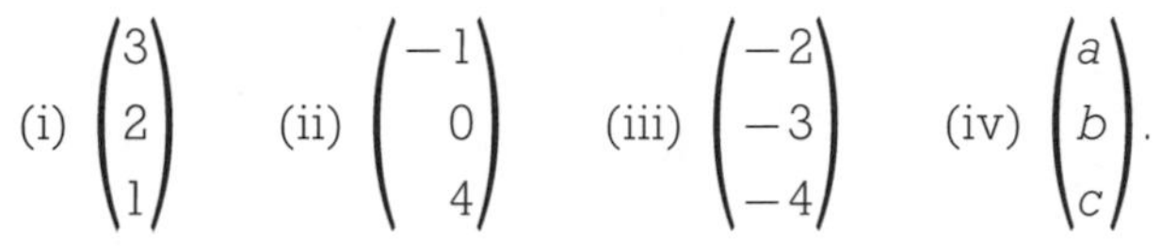

(i) $\begin{pmatrix}3\\2\\1\end{pmatrix}$ (ii) $\begin{pmatrix}-1\\0\\4\end{pmatrix}$ (iii) $\begin{pmatrix}-2\\-3\\-4\end{pmatrix}$ (iv) $\begin{pmatrix}a\\b\\c\end{pmatrix}$.

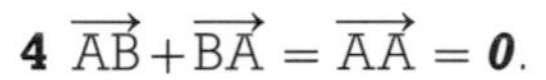

4 $\overrightarrow{AB}+\overrightarrow{BA}=\overrightarrow{AA}=\boldsymbol{0}$.
$\overrightarrow{BA}$ is the negative of $\overrightarrow{AB}$, i.e. $\overrightarrow{BA}=-\overrightarrow{AB}$.
What can you say about: **a** the length **b** the direction of $\overrightarrow{BA}$ compared to $\overrightarrow{AB}$?

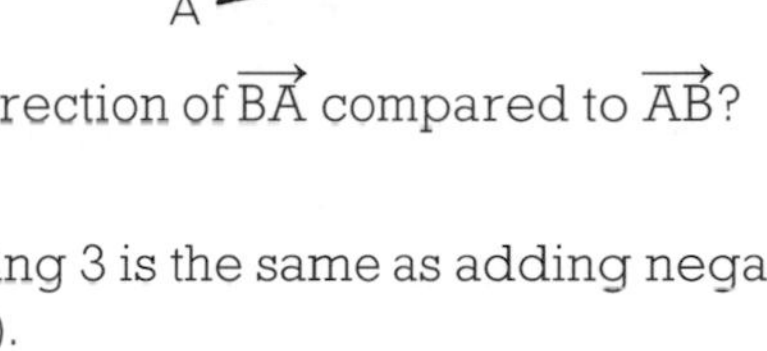

5 *Subtraction of vectors*
You'll remember that $5-3=5+(-3)$, i.e. subtracting 3 is the same as adding negative 3.
The same idea is used for vectors. $\boldsymbol{u}-\boldsymbol{v}=\boldsymbol{u}+(-\boldsymbol{v})$.

For example, if $\boldsymbol{u}=\begin{pmatrix}1\\2\\-5\end{pmatrix}$ and $\boldsymbol{v}=\begin{pmatrix}1\\-2\\3\end{pmatrix}$, then $\boldsymbol{u}-\boldsymbol{v}=\begin{pmatrix}1\\2\\5\end{pmatrix}+\begin{pmatrix}-1\\2\\3\end{pmatrix}=\begin{pmatrix}0\\4\\-8\end{pmatrix}$.

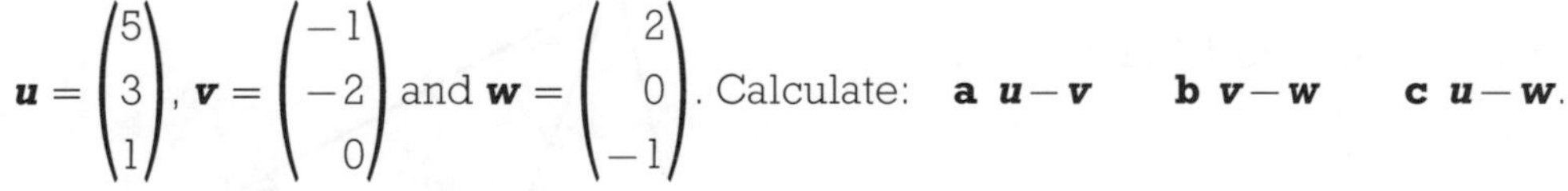

$\boldsymbol{u}=\begin{pmatrix}5\\3\\1\end{pmatrix}, \boldsymbol{v}=\begin{pmatrix}-1\\-2\\0\end{pmatrix}$ and $\boldsymbol{w}=\begin{pmatrix}2\\0\\-1\end{pmatrix}$. Calculate: **a** $\boldsymbol{u}-\boldsymbol{v}$ **b** $\boldsymbol{v}-\boldsymbol{w}$ **c** $\boldsymbol{u}-\boldsymbol{w}$.

6 The diagram shows the construction of $\boldsymbol{u}+\boldsymbol{v}$, and also $\boldsymbol{u}-\boldsymbol{v}$ based on $\boldsymbol{u}+(-\boldsymbol{v})$.
Use the vectors in question **2** to construct vector triangles on squared paper for:
(i) $\boldsymbol{p}-\boldsymbol{q}$ (ii) $\boldsymbol{p}-\boldsymbol{r}$ (iii) $\boldsymbol{p}-\boldsymbol{s}$
(iv) $\boldsymbol{q}-\boldsymbol{r}$ (v) $\boldsymbol{q}-\boldsymbol{s}$ (vi) $\boldsymbol{r}-\boldsymbol{s}$.

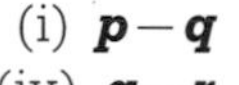

7

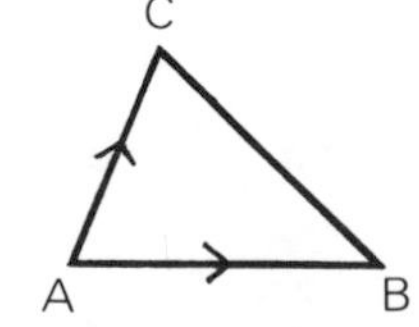

$\overrightarrow{AB}-\overrightarrow{AC}=\overrightarrow{AB}+\overrightarrow{CA}=\overrightarrow{CA}+\overrightarrow{AB}=\overrightarrow{CB}$.
Simplify $\overrightarrow{BA}-\overrightarrow{BC}$ in the same way.

8 Rain runs off the parasol along the ribs and edges as shown. State a single directed line segment equal to:

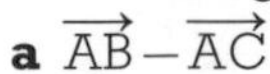

a $\overrightarrow{AB}-\overrightarrow{AC}$

b $\overrightarrow{AE}-\overrightarrow{AC}+\overrightarrow{ED}$.

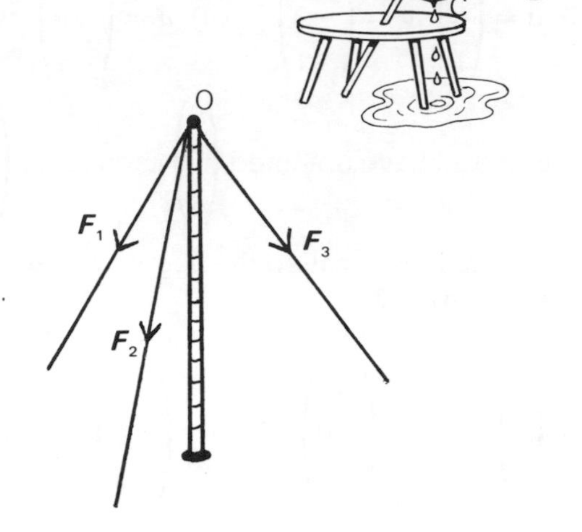

9 A circus tent support pole is held in position by three guy ropes, exerting

forces $\mathbf{F}_1=\begin{pmatrix}-3\\-2\\-5\end{pmatrix}$, $\mathbf{F}_2=\begin{pmatrix}4\\-1\\-5\end{pmatrix}$, $\mathbf{F}_3=\begin{pmatrix}7\\3\\-5\end{pmatrix}$.

Calculate:

a the components of the force equivalent to this system of three forces

b the magnitude of the resultant force.

10 Ken aims his canoe at the opposite bank of the fast-flowing river. He paddles at 12 mph, while the river flows at 5 mph. Find his actual speed and direction (to the nearest degree) across the river.

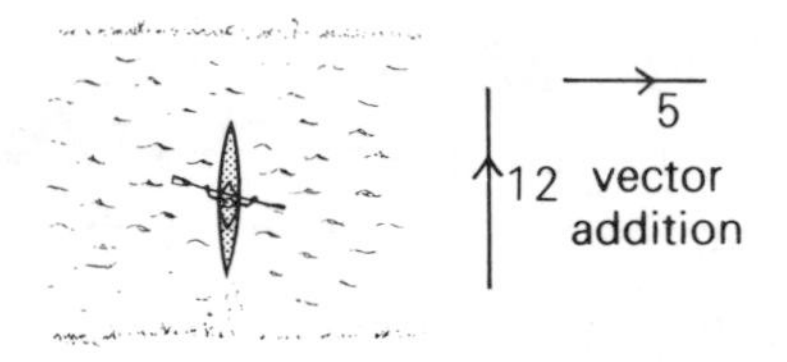

11

The Tiger Moth has set a northerly course at 120 km/h. A 60 km/h wind is blowing from the south-west. Calculate the actual velocity of the plane (i.e. speed and direction) to the nearest unit. (You'll need to use the cosine rule and the sine rule.)

12 Find a single directed line segment equal to:

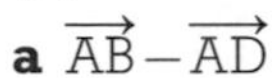

a $\overrightarrow{AB}-\overrightarrow{AD}$

b $\overrightarrow{BD}-\overrightarrow{BC}$

c $\overrightarrow{AD}-\overrightarrow{BD}-\overrightarrow{AC}$

d $\overrightarrow{AC}-\overrightarrow{AB}-\overrightarrow{BC}-\overrightarrow{CD}$.

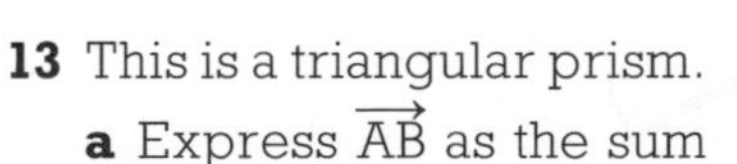

13 This is a triangular prism.

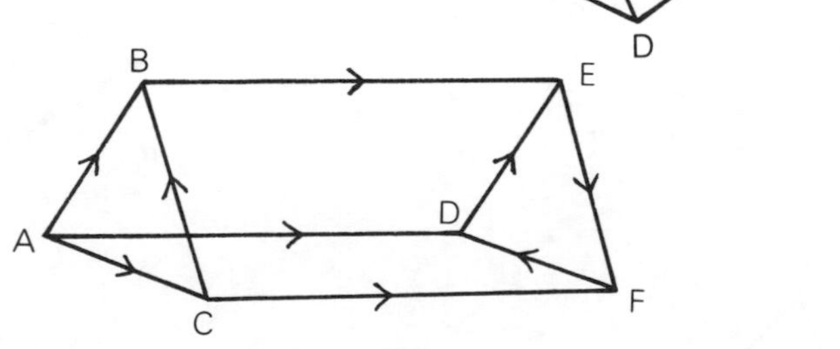

a Express $\overrightarrow{AB}$ as the sum of:

(i) two other directed line segments

(ii) four other directed line segments (there are two possibilities, watch directions!)

b Express $\overrightarrow{BE}$ as the sum of:

(i) three directed line segments in two ways

(ii) four directed line segments in four ways.

MULTIPLICATION OF A VECTOR BY A NUMBER (SCALAR MULTIPLICATION)

John thinks that his father is driving too slowly. 'Double your speed', John tells him.
If he does, what happens to the car's velocity, $\boldsymbol{v}$?
The direction doesn't change, but the speed goes to $2|\boldsymbol{v}|$. The new velocity is $2\boldsymbol{v}$.

Given vector $\boldsymbol{v}$, and number k, then $k\boldsymbol{v}$ has the same direction as $\boldsymbol{v}$ if $k > 0$, and the opposite direction if $k < 0$.

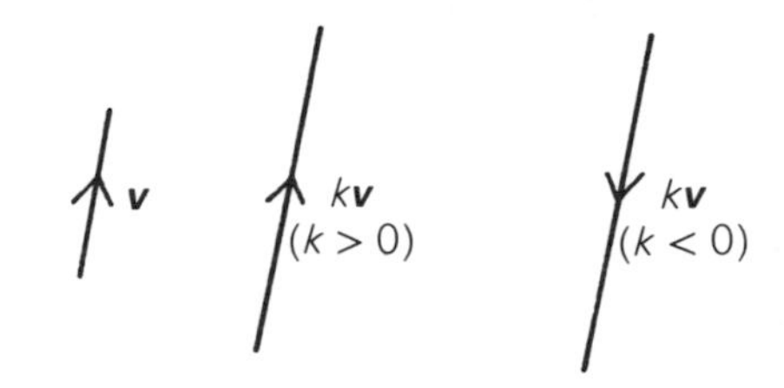

In component form, $k\begin{pmatrix} a \\ b \\ c \end{pmatrix} = \begin{pmatrix} ka \\ kb \\ kc \end{pmatrix}$.

> Vectors $\boldsymbol{u}$ and $\boldsymbol{v}$ are parallel $\Leftrightarrow \boldsymbol{u} = k\boldsymbol{v}$, k a real number.

Example Prove that the points A(2, −3, 4), B(8, 3, 1) and C(12, 7, −1) are collinear, and find the ratio $\frac{\text{AB}}{\text{BC}}$, i.e. the ratio in which B divides AC.

$$\overrightarrow{\text{AB}} = \begin{pmatrix} 8 \\ 3 \\ 1 \end{pmatrix} - \begin{pmatrix} 2 \\ -3 \\ 4 \end{pmatrix} = \begin{pmatrix} 6 \\ 6 \\ -3 \end{pmatrix} = 3\begin{pmatrix} 2 \\ 2 \\ -1 \end{pmatrix}$$

$$\overrightarrow{\text{BC}} = \begin{pmatrix} 12 \\ 7 \\ -1 \end{pmatrix} - \begin{pmatrix} 8 \\ 3 \\ 1 \end{pmatrix} = \begin{pmatrix} 4 \\ 4 \\ -2 \end{pmatrix} = 2\begin{pmatrix} 2 \\ 2 \\ -1 \end{pmatrix}$$

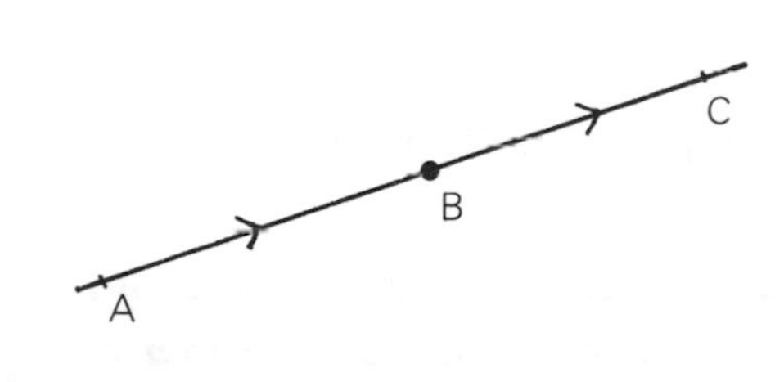

So $\overrightarrow{\text{AB}} = \frac{3}{2}\overrightarrow{\text{BC}}$, and $\overrightarrow{\text{AB}}$ is parallel to $\overrightarrow{\text{BC}}$.
But B is a common point, so A, B, C are collinear, and $\frac{\text{AB}}{\text{BC}} = \frac{3}{2}$.

Exercise 4

1 For each pair of vectors, write down the components of:
(i) $2\boldsymbol{u}$ (ii) $3\boldsymbol{v}$ (iii) $2\boldsymbol{u}+3\boldsymbol{v}$.

a $\boldsymbol{u} = \begin{pmatrix} 1 \\ -1 \\ 3 \end{pmatrix}$, $\boldsymbol{v} = \begin{pmatrix} 2 \\ 4 \\ -2 \end{pmatrix}$ **b** $\boldsymbol{u} = \begin{pmatrix} 0 \\ 1 \\ 2 \end{pmatrix}$, $\boldsymbol{v} = \begin{pmatrix} 6 \\ -2 \\ 0 \end{pmatrix}$.

2 $\boldsymbol{p} = \begin{pmatrix} 0 \\ 1 \\ 0 \end{pmatrix}$ and $\boldsymbol{q} = \begin{pmatrix} -2 \\ -1 \\ 4 \end{pmatrix}$. Calculate $|\boldsymbol{p} - 2\boldsymbol{q}|$.

3 $\boldsymbol{u} = \begin{pmatrix} 2 \\ -3 \\ 4 \end{pmatrix}$, $\boldsymbol{v} = \begin{pmatrix} 1 \\ 0 \\ 5 \end{pmatrix}$, $\boldsymbol{w} = \begin{pmatrix} -3 \\ 2 \\ -1 \end{pmatrix}$. Find $\boldsymbol{p}$, given $\boldsymbol{p} + \boldsymbol{v} = 2\boldsymbol{u} + 3\boldsymbol{w}$.

4 A is (1, 4, 5), B(2, 7, 8), C(0, 12, 15) and D(3, 21, 24).

a Calculate the components of: (i) $\overrightarrow{AB}$ (ii) $\overrightarrow{CD}$.

b Prove that $\overrightarrow{CD} = 3\overrightarrow{AB}$. What does this tell you about $\overrightarrow{CD}$ and $\overrightarrow{AB}$?

5 P is (0, 1, 1), Q(1, 1, 0), R(−5, −4, −5) and S(−3, −4, −7).

a Show that $\overrightarrow{RS} = k\overrightarrow{PQ}$ for some value of k.

b How are $\overrightarrow{RS}$ and $\overrightarrow{PQ}$ related?

6 Prove that D(1, 3, 4), E(5, 5, 6) and F(17, 11, 12) are collinear.

7 In each case show that A, B and C are collinear, and find $\frac{AC}{CB}$:

a A(1, 3, −4), B(9, −1, 0), C(3, 2, −3) **b** A(3, −2, 0), B(2, 1, 5), C(0, 7, 15).

8 $\boldsymbol{u} = \begin{pmatrix} -1 \\ 0 \\ 3 \end{pmatrix}$ and A is the point (2, 0, −3).

Find the coordinates of the points B and C such that $\overrightarrow{AB}$ and $\overrightarrow{AC}$ are parallel to $\boldsymbol{u}$, and both have twice the length of $\boldsymbol{u}$.

9 Prove that:

a A, B, C are collinear

b D, E, F are collinear.

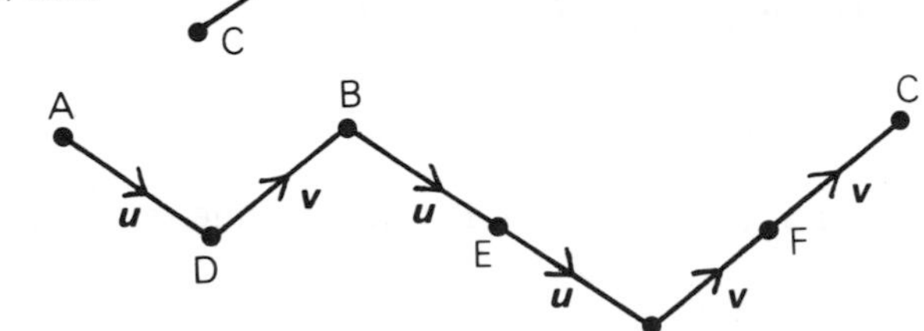

a Use these diagrams to illustrate why:

(i) $\boldsymbol{u} + \boldsymbol{v} = \boldsymbol{v} + \boldsymbol{u}$

(ii) $(\boldsymbol{u} + \boldsymbol{v}) + \boldsymbol{w} = \boldsymbol{u} + (\boldsymbol{v} + \boldsymbol{w})$.

b Draw a suitable diagram to illustrate why: $k(\boldsymbol{u} + \boldsymbol{v}) = k\boldsymbol{u} + k\boldsymbol{v}$ (k a number).

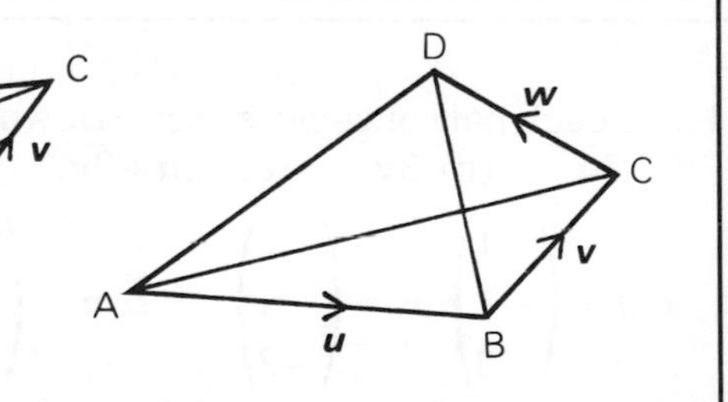

POSITION VECTORS

Unit vectors

A unit vector has magnitude 1.

$\boldsymbol{u} = \overrightarrow{AB} = \begin{pmatrix} a \\ b \\ c \end{pmatrix}$ is a unit vector $\Leftrightarrow a^2 + b^2 + c^2 = 1$.

The unit vectors in the directions OX, OY, OZ are denoted by:

$\boldsymbol{i} = \begin{pmatrix} 1 \\ 0 \\ 0 \end{pmatrix}, \boldsymbol{j} = \begin{pmatrix} 0 \\ 1 \\ 0 \end{pmatrix}, \boldsymbol{k} = \begin{pmatrix} 0 \\ 0 \\ 1 \end{pmatrix}$.

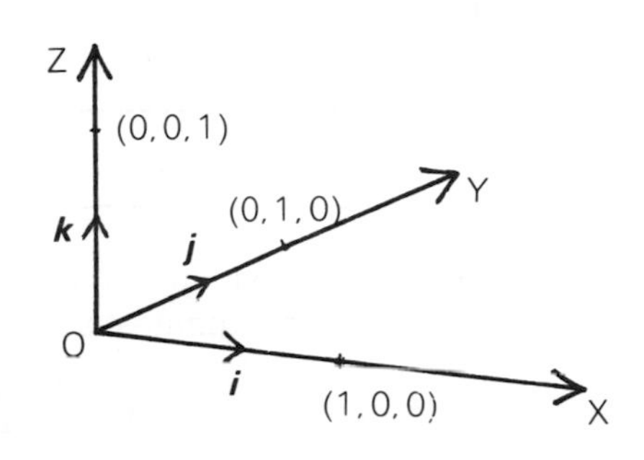

The vectors $\boldsymbol{i}$, $\boldsymbol{j}$, $\boldsymbol{k}$ form a *basis* for 3-dimensional space.

The position vector of a point

$\overrightarrow{OP}$ is the position vector of P(x, y, z).

In component form, $\overrightarrow{OP} = \begin{pmatrix} x \\ y \\ z \end{pmatrix}$.

Also, $\overrightarrow{OP} = \overrightarrow{OK} + \overrightarrow{KP} = \overrightarrow{OL} + \overrightarrow{LK} + \overrightarrow{KP}$
$= \overrightarrow{OL} + \overrightarrow{OM} + \overrightarrow{ON}$
$= x\boldsymbol{i} + y\boldsymbol{j} + z\boldsymbol{k}$

$$\boldsymbol{p} = \overrightarrow{OP} = \begin{pmatrix} x \\ y \\ z \end{pmatrix} = x\boldsymbol{i} + y\boldsymbol{j} + z\boldsymbol{k}.$$

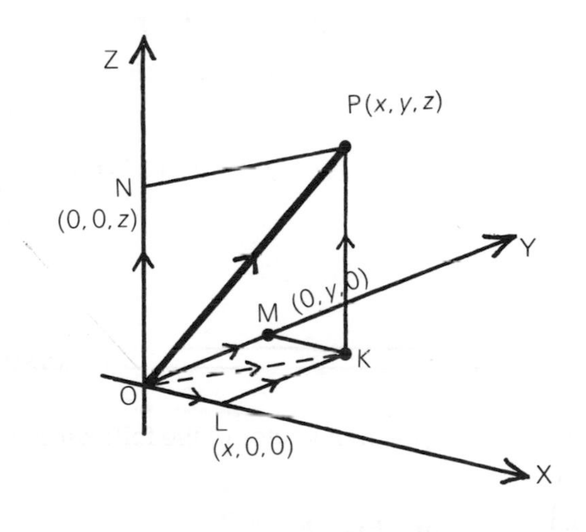

Two useful facts

(i) $\overrightarrow{OB} = \overrightarrow{OA} + \overrightarrow{AB}$,
$\overrightarrow{AB} = \overrightarrow{OB} - \overrightarrow{OA}$,

so $\boxed{\overrightarrow{AB} = \boldsymbol{b} - \boldsymbol{a}}$,

where $\boldsymbol{a}$, $\boldsymbol{b}$ are the position vectors of A, B.

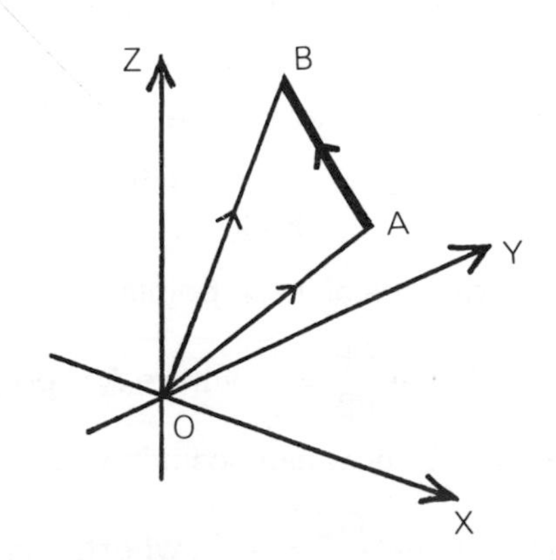

(ii) M is the midpoint of AB, and $\boldsymbol{a}$, $\boldsymbol{m}$ and $\boldsymbol{b}$ are the position vectors of A, M and B.

$$\overrightarrow{AM} = \overrightarrow{MB}$$

So $\boldsymbol{m} - \boldsymbol{a} = \boldsymbol{b} - \boldsymbol{m}$

$$\boldsymbol{m} = \tfrac{1}{2}(\boldsymbol{a} + \boldsymbol{b}).$$

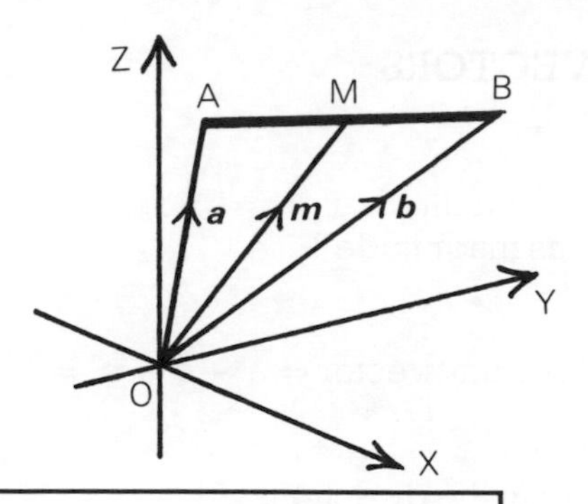

> The midpoint of AB has position vector $\boldsymbol{m} = \tfrac{1}{2}(\boldsymbol{a} + \boldsymbol{b})$.

Example P divides AB in the ratio 3 : 2. Find the coordinates of P.

A (2, −3, 4)
P
3
2
B(12, 7, −1)

The vector form of $\dfrac{AP}{PB} = \dfrac{3}{2}$ is $\overrightarrow{AP} = \dfrac{3}{2}\overrightarrow{PB}$.

$$\text{So } 2(\boldsymbol{p} - \boldsymbol{a}) = 3(\boldsymbol{b} - \boldsymbol{p})$$

$$5\boldsymbol{p} = 2\boldsymbol{a} + 3\boldsymbol{b}$$

$$\boldsymbol{p} = \frac{1}{5}(2\boldsymbol{a} + 3\boldsymbol{b}) = \frac{1}{5}\left[\begin{pmatrix}4\\-6\\8\end{pmatrix} + \begin{pmatrix}36\\21\\-3\end{pmatrix}\right] = \begin{pmatrix}8\\3\\1\end{pmatrix}.$$

P is the point (8, 3, 1).

Exercise 5

1 Write down the position vector of the following in the form:

(i) $\begin{pmatrix}a\\b\\c\end{pmatrix}$ (ii) $a\boldsymbol{i} + b\boldsymbol{j} + c\boldsymbol{k}$

a A(1, 2, 0) **b** B(−1, −5, 2) **c** C(6, 0, −7) **d** D(0, 8, −11).

2 a Calculate the length of the vector: (i) $\boldsymbol{u} = 3\boldsymbol{i} - 2\boldsymbol{j} + 6\boldsymbol{k}$ (ii) $\boldsymbol{v} = 5\boldsymbol{i} - 2\sqrt{2}\boldsymbol{j} - 4\sqrt{3}\boldsymbol{k}$.
b Find the two unit vectors of the form $a\boldsymbol{i} + \tfrac{1}{2}\boldsymbol{j} - \tfrac{1}{2}\boldsymbol{k}$ by calculating the values of a.

3 Find the coordinates of the midpoints of:
a A(4, 2, 3) B(6, 8, 1) **b** C(−2, 3, −1) D(−8, −7, 1).

4 Find the coordinates of these points:

a P on AB such that $\dfrac{AP}{PB} = \dfrac{1}{3}$, where A is point (2, 2, 3) and B(2, −2, −1).

Hint $\overrightarrow{AP} = \tfrac{1}{3}\overrightarrow{PB}$; now use position vectors.

b Q on CD such that $\dfrac{CQ}{QD} = -\dfrac{2}{1}$, where C is point (−1, 3, 2) and D(1, 4, 4).

5 a Find the coordinates of:
(i) P, the midpoint of AB
(ii) Q, the midpoint of BC.

b Show that $\overrightarrow{PQ} = \frac{1}{2}\overrightarrow{AC}$.

c Deduce the geometrical relationship between PQ and AC.

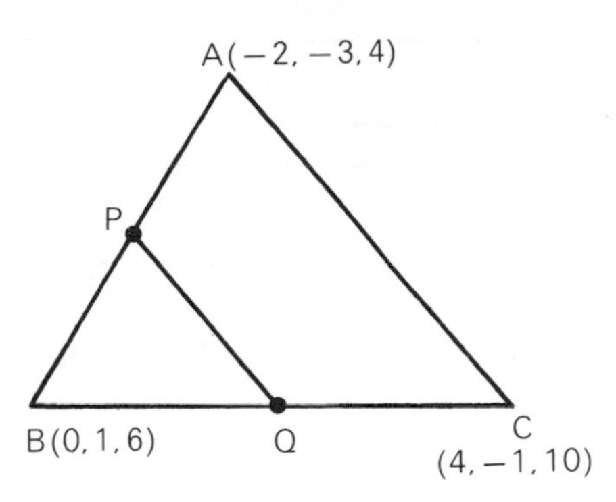

6 A is the point $(-2, -2, -2)$, B is $(1, -2, 1)$ and C is $(1, 0, -1)$. D divides AB in the ratio 2 : 1.

a Find the coordinates of D.

b Prove that $\angle BDC = 90°$.

7 There are 3 stations for the chairlift to the ski slopes:
A(base) has grid reference 575637 and is 0·09 units above sea-level (A is point (57·5, 63·7, 0·09)).
B(nursery slope) has reference 581645 and is 0·14 above sea-level.
C(main slope) has reference 593661 and is 0·24 above sea-level.

a Obtain the components of:
(i) $\overrightarrow{AB}$ (ii) $\overrightarrow{BC}$.

b Calculate the magnitudes of:
(i) $\overrightarrow{AB}$
(ii) $\overrightarrow{BC}$ (to 1 decimal place).

c Prove that A, B, C are collinear.

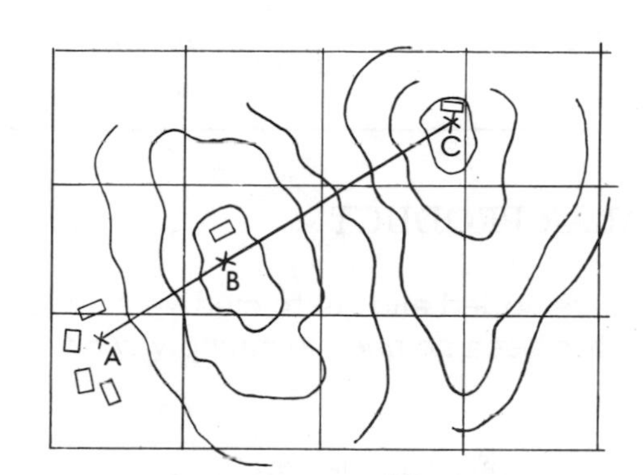

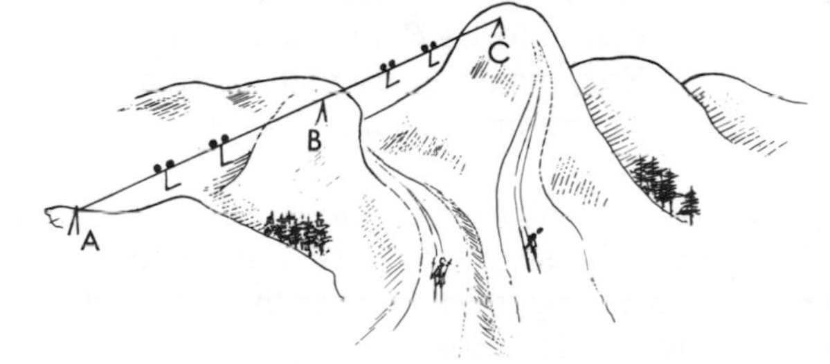

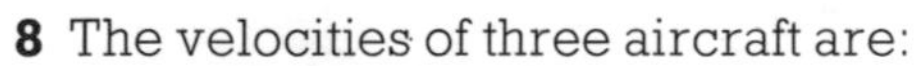

8 The velocities of three aircraft are:
$\boldsymbol{v}_A = 3\boldsymbol{i}+2\boldsymbol{j}-3\boldsymbol{k}$,
$\boldsymbol{v}_B = 2\boldsymbol{i}-\boldsymbol{j}+2\boldsymbol{k}$,
$\boldsymbol{v}_C = 3\boldsymbol{i}+2\boldsymbol{j}-\boldsymbol{k}$.
The 'velocity of B relative to A' is $\boldsymbol{v}_B - \boldsymbol{v}_A$.
Calculate the velocity and its magnitude of:

a B relative to A

b A relative to B

c A relative to C.

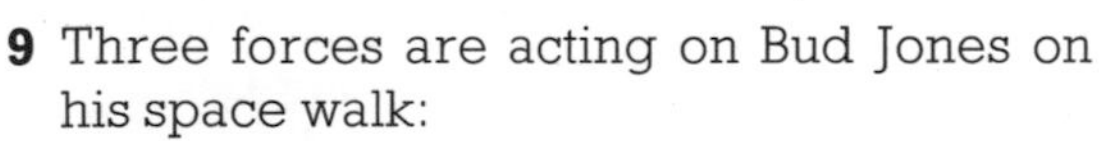

9 Three forces are acting on Bud Jones on his space walk:
$\boldsymbol{e} = 2\boldsymbol{i}+\boldsymbol{j}+\boldsymbol{k}$
$\boldsymbol{m} = 3\boldsymbol{i}+\boldsymbol{j}+2\boldsymbol{k}$
$\boldsymbol{s} = \boldsymbol{i}+\boldsymbol{j}-\boldsymbol{k}$.

a Find the resultant force.

b Calculate its magnitude.

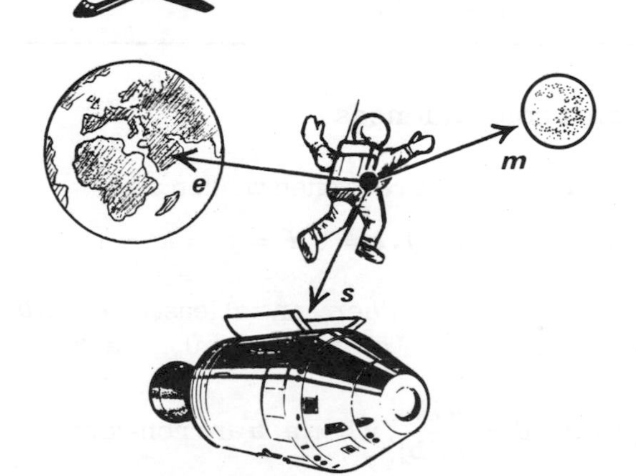

Section formula

Find a formula for the position vector, $\boldsymbol{p}$, of P on AB such that $\frac{\text{AP}}{\text{PB}} = \frac{m}{n}$ in terms of $\boldsymbol{a}$, $\boldsymbol{b}$, the position vectors of A, B.

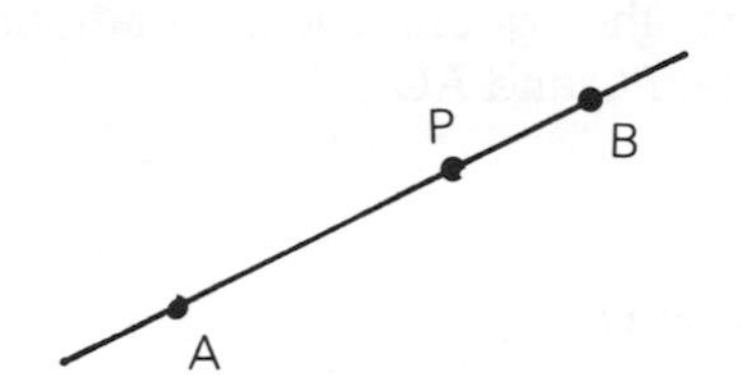

a Write down the vector equivalent of $\frac{\text{AP}}{\text{PB}} = \frac{m}{n}$.

b Prove that $n(\boldsymbol{p}-\boldsymbol{a}) = m(\boldsymbol{b}-\boldsymbol{p})$.

c Show that $\boldsymbol{p} = \frac{1}{m+n}(m\boldsymbol{b}+n\boldsymbol{a})$.

SCALAR PRODUCT

You have added and subtracted vectors, and multiplied vectors by real numbers. But so far there has been no need to multiply two vectors together. Here is an example of how this can arise:

Rashid is pulling the crate across the factory floor with force $\boldsymbol{F}$ (in magnitude and direction). He moves it through the displacement $\boldsymbol{x}$ (also in magnitude and direction).

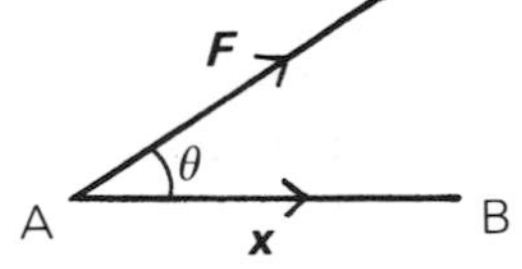

Physicists will tell you that the work done is $|\boldsymbol{F}|\,|\boldsymbol{x}|\cos\theta$. In a sense they are 'multiplying' two vectors together, even though the product is a real number.

$|\boldsymbol{F}|\,|\boldsymbol{x}|\cos\theta$ is called the *scalar product* $\boldsymbol{F}.\boldsymbol{x}$ of the vectors $\boldsymbol{F}$ and $\boldsymbol{x}$.

Definition The scalar product of $\boldsymbol{a}$ and $\boldsymbol{b}$, denoted by $\boldsymbol{a}.\boldsymbol{b}$ ('$\boldsymbol{a}$ dot $\boldsymbol{b}$'), is $\boldsymbol{a}.\boldsymbol{b} = |\boldsymbol{a}|\,|\boldsymbol{b}|\cos\theta$, neither $\boldsymbol{a}$ nor $\boldsymbol{b}$ being zero.

Some consequences

(i) $\boldsymbol{a}.\boldsymbol{a} = |\boldsymbol{a}|^2$ ($\boldsymbol{a}.\boldsymbol{a}$ is often written $\boldsymbol{a}^2$).

(ii) $\boldsymbol{i}.\boldsymbol{j} = \boldsymbol{i}.\boldsymbol{k} = \boldsymbol{j}.\boldsymbol{k} = 0$; $\boldsymbol{i}^2 = \boldsymbol{j}^2 = \boldsymbol{k}^2 = 1$.

(iii) $\boldsymbol{a}.\boldsymbol{b} = 0 \Leftrightarrow \begin{cases} \textit{either} & \text{one at least of } \boldsymbol{a}, \boldsymbol{b} \text{ is } \boldsymbol{0} \\ \textit{or} & \cos\theta = 0 \text{, i.e. } \boldsymbol{a}, \boldsymbol{b} \text{ are perpendicular.} \end{cases}$

(iv) $\cos\theta = \frac{\boldsymbol{a}.\boldsymbol{b}}{|\boldsymbol{a}|\,|\boldsymbol{b}|}$, when $\boldsymbol{a}$, $\boldsymbol{b}$ are non-zero.

Component form of a . b

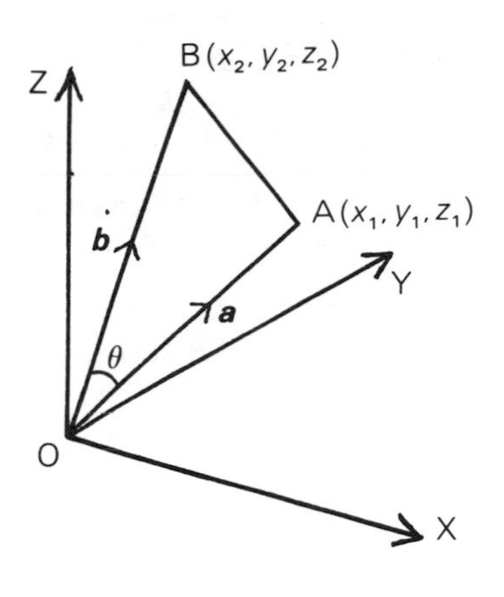

Use the representatives shown for **a**, ***b***.

$AB^2 = OA^2 + OB^2 - 2\,OA\,OB\cos\theta$ (cosine rule)

$= (x_1^2 + y_1^2 + z_1^2) + (x_2^2 + y_2^2 + z_2^2) - 2|\mathbf{a}|\,|\boldsymbol{b}|\cos\theta \ldots (1)$

But $AB^2 = (x_1 - x_2)^2 + (y_1 - y_2)^2 + (z_1 - z_2)^2$ (distance formula)

$= (x_1^2 + y_1^2 + z_1^2) + (x_2^2 + y_2^2 + z_2^2) - 2(x_1x_2 + y_1y_2 + z_1z_2) \ldots (2)$

From (1) and (2), and the fact that $\mathbf{a}\,.\,\boldsymbol{b} = |\mathbf{a}|\,|\boldsymbol{b}|\cos\theta$,

$$\mathbf{a}\,.\,\boldsymbol{b} = x_1x_2 + y_1y_2 + z_1z_2.$$

Notes (i) This formula holds whether or not **a**, ***b*** are ***0***.

(ii) When **a** and ***b*** are non-zero. $$\cos\theta = \frac{x_1x_2 + y_1y_2 + z_1z_2}{|\mathbf{a}|\,|\boldsymbol{b}|}.$$

Example $\mathbf{a} = \begin{pmatrix}1\\2\\1\end{pmatrix}$ and $\boldsymbol{b} = \begin{pmatrix}2\\1\\-1\end{pmatrix}$. Calculate the angle between the vectors.

$$\cos\theta = \frac{x_1x_2 + y_1y_2 + z_1z_2}{|\mathbf{a}|\,|\boldsymbol{b}|} = \frac{2+2-1}{\sqrt{6}\sqrt{6}} = \frac{3}{6} = \frac{1}{2}. \; (|\mathbf{a}|^2 = 1^2 + 2^2 + 1^2 = 6).$$

So $\theta = \frac{\pi}{3}$.

Exercise 6

1 Calculate **a . *b*** for:

a $|\mathbf{a}| = 2, |\boldsymbol{b}| = 1, \theta = \frac{\pi}{6}$ **b** $|\mathbf{a}| = 1, |\boldsymbol{b}| = 5, \theta = \frac{\pi}{4}$

c $|\mathbf{a}| = 1, |\boldsymbol{b}| = 5, \theta = 135°$ **d** $\theta = 90°$.

2 Calculate **a . *b*** for each pair of vectors.

a $\mathbf{a} = \begin{pmatrix}2\\-1\\-3\end{pmatrix}$ $\boldsymbol{b} = \begin{pmatrix}1\\0\\-2\end{pmatrix}$ **b** $\mathbf{a} = \begin{pmatrix}1\\2\\4\end{pmatrix}$ $\boldsymbol{b} = \begin{pmatrix}-3\\-2\\2\end{pmatrix}$ **c** $\mathbf{a} = \begin{pmatrix}1\\3\\-5\end{pmatrix}$ $\boldsymbol{b} = \begin{pmatrix}-2\\-1\\0\end{pmatrix}$

d $\mathbf{a} = \begin{pmatrix}3\\-2\\1\end{pmatrix}$ $\boldsymbol{b} = \begin{pmatrix}-2\\-4\\3\end{pmatrix}$ **e** $\mathbf{a} = \begin{pmatrix}a\\b\\c\end{pmatrix}$ $\boldsymbol{b} = \begin{pmatrix}1\\0\\0\end{pmatrix}$ **f** $\mathbf{a} = \begin{pmatrix}a\\b\\c\end{pmatrix}$ $\boldsymbol{b} = \begin{pmatrix}0\\0\\1\end{pmatrix}$.

3 Calculate $\boldsymbol{a}\,.\,\boldsymbol{b}$ for:

a $\boldsymbol{a} = \boldsymbol{i} - 2\boldsymbol{j} + 3\boldsymbol{k}$, $\boldsymbol{b} = 2\boldsymbol{i} - 3\boldsymbol{j} + 2\boldsymbol{k}$ **b** $\boldsymbol{a} = 3\boldsymbol{i} - \boldsymbol{j}$, $\boldsymbol{b} = 2\boldsymbol{j} + \boldsymbol{k}$

c $\boldsymbol{a} = \boldsymbol{i} + \boldsymbol{j} + \boldsymbol{k}$, $\boldsymbol{b} = \boldsymbol{j} + \boldsymbol{k}$ **d** $\boldsymbol{a} = 3\boldsymbol{i} + \boldsymbol{j} - 2\boldsymbol{k}$, $\boldsymbol{b} = \boldsymbol{i} - \boldsymbol{k}$.

4 A triangle has vertices A(2, 1, 5), B(3, −2, 4) and C(−3, 4, 1).

a Obtain: (i) $\overrightarrow{AB}$ (ii) $\overrightarrow{AC}$ in component form.

b Calculate $\overrightarrow{AB}\,.\,\overrightarrow{AC}$.

c Calculate AB and AC (leaving the answers in exact form).

d Calculate ∠BAC correct to 1 decimal place (in degrees).

e Similarly calculate ∠CBA and ∠ACB.

5 a Calculate the angle between $\boldsymbol{u} = 2\boldsymbol{i} + \boldsymbol{j} - \boldsymbol{k}$ and $\boldsymbol{v} = 3\boldsymbol{i} - 2\boldsymbol{j} + 2\boldsymbol{k}$, in degrees correct to 1 decimal place.

b Show that $\boldsymbol{p} = 2\boldsymbol{i} + 3\boldsymbol{j} - \boldsymbol{k}$ and $\boldsymbol{q} = 4\boldsymbol{i} - 2\boldsymbol{j} + 2\boldsymbol{k}$ are perpendicular.

6 Verify that the triangle with vertices A(2, −3, 4), B(5, −8, 13) and C(4, 0, 5) is right-angled at A, using a suitable scalar product.

7 PQRS is a quadrilateral with vertices P(−2, 0, −5), Q(1, 6, −8), R(7, 9, 4), S(7, 3, 16).

a Find the point T on PR such that $\frac{PT}{TR} = \frac{5}{4}$.

b Show that Q, T, S are collinear.

c Calculate the acute angle between the diagonals of PQRS in degrees correct to 1 decimal place.

8 $\boldsymbol{u} = \begin{pmatrix} 8 \\ 0 \\ 4 \end{pmatrix}$ and $\boldsymbol{v} = \begin{pmatrix} 4 \\ 0 \\ 1 \end{pmatrix}$. Calculate the angle between the vectors $\boldsymbol{u} + \boldsymbol{v}$ and $\boldsymbol{u} - \boldsymbol{v}$, in degrees correct to 1 decimal place.

9 P is the point (−2, −1, 1), Q is (−1, 1, 2) and O is the origin.

a Calculate: (i) $|\boldsymbol{p}|$ (ii) $|\boldsymbol{q}|$ (iii) ∠POQ.

b What kind of triangle is POQ?

10 A force $\boldsymbol{F}$ is given by $\boldsymbol{F} = 2\boldsymbol{i} + 3\boldsymbol{j} + \boldsymbol{k}$. Calculate, in exact form, the component of the force in the direction of the vector $\boldsymbol{a} = \boldsymbol{i} + \boldsymbol{j} + \boldsymbol{k}$.

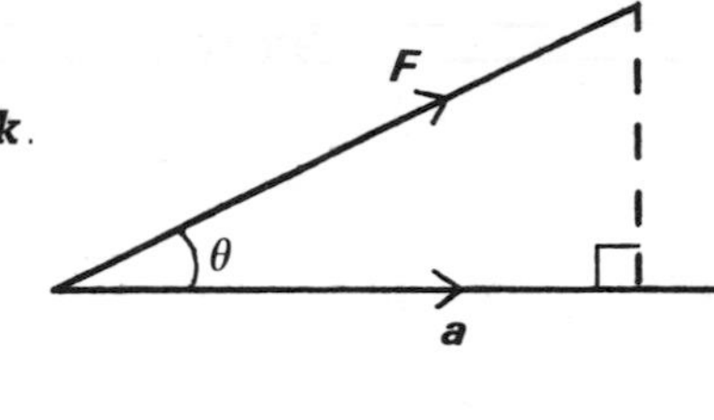

11 A potash alum crystal is shaped like an octahedron. In a suitable coordinate system its vertices are A(2, −4, 4), B(7, −2, 6), C(8, 2, 2), D(3, 0, 0), E(−3, 10, 12), F(13, −12, −6).

a Find the coordinates of the midpoints of AC, EF.

b Show that these space diagonals, AC and EF, bisect each other at right angles.

c Show that ABCD is a rhombus ($\overrightarrow{AB} = \overrightarrow{DC}$ and AB = AD).

d Find the exact value of cos BAD.

12 A 'space' quadrilateral has vertices A, B, C, D as shown. M_1, M_2, M_3, M_4 are the midpoints of the sides.

a Prove that $M_1M_2M_3M_4$ is a parallelogram.

b Calculate the acute angles in $M_1M_2M_3M_4$ in degrees correct to 1 decimal place.

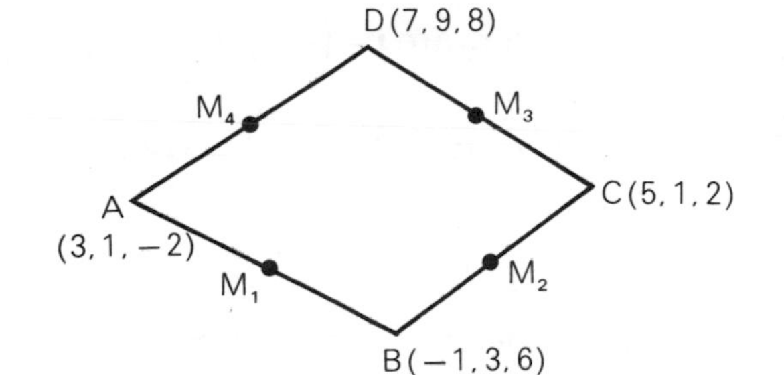

Using the component form of the vectors, prove that, for all vectors **a**, **b**, **c**, $\mathbf{a}.(\mathbf{b}+\mathbf{c}) = \mathbf{a}.\mathbf{b}+\mathbf{a}.\mathbf{c}$, i.e. the distributive law holds for scalar products.

2-DIMENSIONAL VECTORS

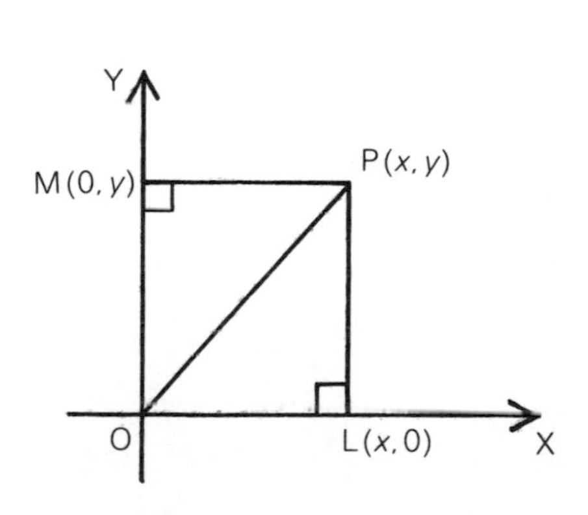

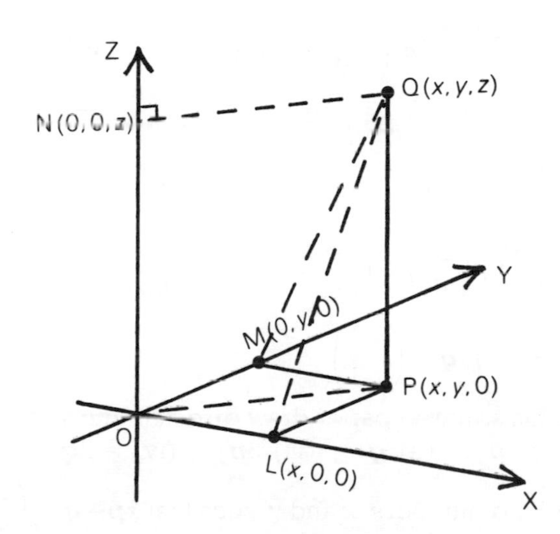

If you project down onto the *x*, *y*-plane, the *z*-coordinates are all 0. You can omit these 0s and return to the normal two coordinates *x*, *y*. Vectors are now in the 2-dimensional *x*, *y*-plane. Everything that you have done for 3 dimensions can be adapted immediately for 2 dimensions.

Exercise 7

1 Copy and complete. For 2 dimensions, on the XOY plane:

a $\overrightarrow{AB} = \begin{pmatrix} x_B - x_A \\ \ldots\ldots \end{pmatrix}$.

b If $\boldsymbol{u} = \overrightarrow{AB}$, then $|\boldsymbol{u}| = AB = \sqrt{[(\ldots)^2+(\ldots)^2]}$.

c $\begin{pmatrix} a \\ b \end{pmatrix} = \begin{pmatrix} c \\ d \end{pmatrix} \Rightarrow a = \ldots, b = \ldots$.

d $\boldsymbol{u}+\boldsymbol{v}$ and $k\boldsymbol{v}$ have the same definitions as before.

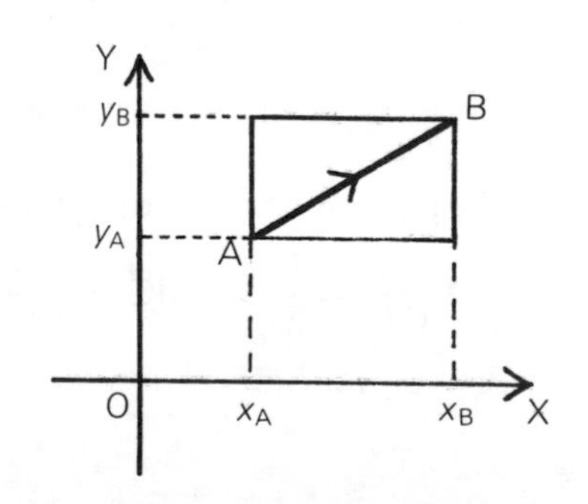

e Basis vectors are $\boldsymbol{i} = \begin{pmatrix} \dots \\ \dots \end{pmatrix}$, $\boldsymbol{j} = \begin{pmatrix} \dots \\ \dots \end{pmatrix}$.

f Position vector $\overrightarrow{OP} = \begin{pmatrix} \dots \\ \dots \end{pmatrix} = x\boldsymbol{i} + \dots$.

g $\overrightarrow{AB} = \boldsymbol{b} - \dots$.

h Midpoint of AB is $\frac{1}{2}(\dots + \dots)$.

i $\boldsymbol{a}.\boldsymbol{b}$ has the same definition, $\boldsymbol{a}.\boldsymbol{b} = \dots\dots\dots\dots$

j $\boldsymbol{a}.\boldsymbol{b} = \dots\dots\dots$, where $\boldsymbol{a} = \begin{pmatrix} x_1 \\ y_1 \end{pmatrix}$, $\boldsymbol{b} = \begin{pmatrix} x_2 \\ y_2 \end{pmatrix}$.

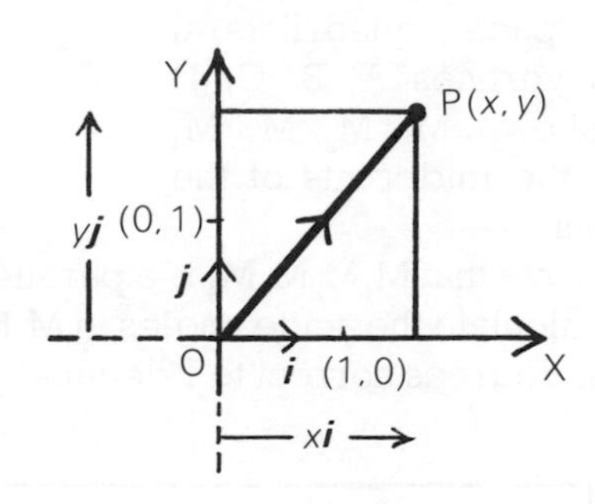

2 Find these vectors in component form:

a $\overrightarrow{AB}$, A(3, 1) B(4, 3) **b** $\overrightarrow{PQ}$, P(5, 1) Q(−3, −2)
c $\overrightarrow{MN}$, M(−3, 5) N(3, −2) **d** $\overrightarrow{KL}$, K(−6, 6) L(−6, −6).

3 Calculate in exact form the magnitude of each of these vectors:

a $\begin{pmatrix} 3 \\ -4 \end{pmatrix}$ **b** $\begin{pmatrix} -6 \\ 0 \end{pmatrix}$ **c** $\begin{pmatrix} -5 \\ -12 \end{pmatrix}$ **d** $\begin{pmatrix} 0 \\ 4 \end{pmatrix}$ **e** $\begin{pmatrix} 1{\cdot}5 \\ 2 \end{pmatrix}$.

4 Find a and b such that:

a $\begin{pmatrix} a \\ b \end{pmatrix} + \begin{pmatrix} 3 \\ 4 \end{pmatrix} = \begin{pmatrix} -1 \\ 3 \end{pmatrix}$ **b** $\begin{pmatrix} a \\ 3 \end{pmatrix} + \begin{pmatrix} 6 \\ b \end{pmatrix} = \begin{pmatrix} -1 \\ -1 \end{pmatrix}$

c $\begin{pmatrix} a \\ 2 \end{pmatrix} + \begin{pmatrix} 2a \\ 3b \end{pmatrix} = \begin{pmatrix} -6 \\ 8 \end{pmatrix}$ **d** $\begin{pmatrix} 3a \\ 4 \end{pmatrix} - \begin{pmatrix} 4a \\ b \end{pmatrix} + \begin{pmatrix} -2a \\ 3b \end{pmatrix} = \begin{pmatrix} -3 \\ -4 \end{pmatrix}$.

5 $\boldsymbol{p} = \begin{pmatrix} 1 \\ 3 \end{pmatrix}$, $\boldsymbol{q} = \begin{pmatrix} -2 \\ 1 \end{pmatrix}$.

a On squared paper draw directed line segments to represent:
(i) $\boldsymbol{p}$ (ii) $\boldsymbol{q}$ (iii) $3\boldsymbol{p}$ (iv) $-2\boldsymbol{q}$ (v) $4\boldsymbol{p} - 3\boldsymbol{q}$.

b Find numbers, x and y, such that $x\boldsymbol{p} + \boldsymbol{q} = \begin{pmatrix} 1 \\ y \end{pmatrix}$.

6 The position vectors of A, B and C are $\begin{pmatrix} -2 \\ -1 \end{pmatrix}$, $\begin{pmatrix} -5 \\ 4 \end{pmatrix}$ and $\begin{pmatrix} 3 \\ 2 \end{pmatrix}$ respectively.

a Express in component form: (i) $\overrightarrow{AB}$ (ii) $\overrightarrow{AC}$ (iii) $\overrightarrow{BC}$.
b By using AB, AC and BC, show that the triangle is right-angled.

7 For triangle ABC, L, M and N are the midpoints of the sides.

a Find the position vectors $\boldsymbol{a}$, $\boldsymbol{b}$, $\boldsymbol{m}$ of A, B and M.

b Find the position vector of G on MC such that $\frac{MG}{GC} = \frac{1}{2}$, i.e. $\overrightarrow{MG} = \frac{1}{2}\overrightarrow{GC}$.

c Prove that A, G, N are collinear, and find the ratio $\frac{NG}{GA}$.

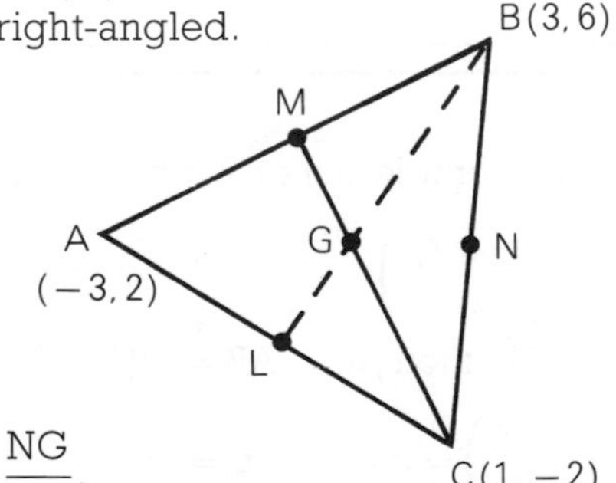

d Similarly, prove B, G, L collinear, and find $\frac{LG}{GB}$.

e Show that the medians of $\triangle$ABC are concurrent at G (the centroid).

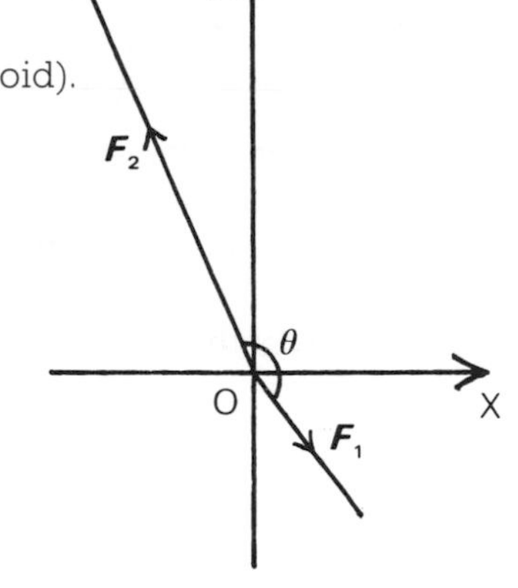

8 Forces $\boldsymbol{F}_1 = 3\boldsymbol{i} - 4\boldsymbol{j}$ and $\boldsymbol{F}_2 = -5\boldsymbol{i} + 12\boldsymbol{j}$ act at the origin O.

a Calculate $\cos\theta$.

b Find the magnitude and direction of the resultant of this system of forces, i.e. of $\boldsymbol{F}_1 + \boldsymbol{F}_2$, giving the angle made with OX in degrees correct to 1 decimal place.

9 P, Q and R have position vectors $\begin{pmatrix}4\\3\end{pmatrix}$, $\begin{pmatrix}3\\4\end{pmatrix}$ and $\begin{pmatrix}1\\1\end{pmatrix}$ respectively.

a Calculate: (i) cos POR (ii) cos ROQ, where O is the origin.

b Prove that the line $y = x$ bisects the angle between the lines $y = \frac{3}{4}x$ and $y = \frac{4}{3}x$.

Investigate the concurrency properties of the triangle, i.e. the concurrency of:
a the medians (at the centroid)
b the altitudes (at the orthocentre)
c the perpendicular bisectors of the sides (at the centre of the circumcircle)
d the bisectors of the angles (at the centre of the inscribed circle).
Start with vertices with numerical coordinates, and use vector methods. Then try to generalise with literal coordinates.

CHECK-UP ON **VECTORS**

1 a Write down the component form of $\overrightarrow{AB}$ for:
(i) A(−1, 2, −3) B(3, −2, 5) (ii) A(2, 4, −1) B(3, −2, −2).

b Calculate the distance AB for each pair of points.

2 $\begin{pmatrix}3a+1\\a^2\\1-b\end{pmatrix} = \begin{pmatrix}b-2\\9\\-4a-5\end{pmatrix}$. Find a and b.

3 For these pairs of vectors write down the components of:
(i) $\boldsymbol{u} + \boldsymbol{v}$ (ii) $3\boldsymbol{u}$ (iii) $-2\boldsymbol{v}$ (iv) $3\boldsymbol{u} - 2\boldsymbol{v}$:

a $\boldsymbol{u} = \begin{pmatrix}2\\1\\-3\end{pmatrix}$ $\boldsymbol{v} = \begin{pmatrix}3\\0\\-5\end{pmatrix}$ **b** $\boldsymbol{u} = \begin{pmatrix}-4\\0\\1\end{pmatrix}$ $\boldsymbol{v} = \begin{pmatrix}2\\-3\\-1\end{pmatrix}$.

4 $\boldsymbol{u} = \begin{pmatrix}1\\0\\2\end{pmatrix}$ and A is the point A(3, 1, −2).

Find the coordinates of the point B such that $\overrightarrow{AB}$ is parallel to $\boldsymbol{u}$ and has three times the length of $\boldsymbol{u}$.

5 The forces $\boldsymbol{F}_1 = \begin{pmatrix} 3 \\ 1 \\ 1 \end{pmatrix}$, $\boldsymbol{F}_2 = \begin{pmatrix} -1 \\ 0 \\ 2 \end{pmatrix}$, $\boldsymbol{F}_3 = \begin{pmatrix} 2 \\ 3 \\ -1 \end{pmatrix}$ act at O.

a Find the components of the resultant force $\boldsymbol{F}_1 + \boldsymbol{F}_2 + \boldsymbol{F}_3$.
b Find the magnitude of the resultant.
c Find the cosine of the angle made with: (i) OX (ii) OY (iii) OZ.

6 Prove that the points A(3, −1, 0), B(5, 2, 5) and C(9, 8, 15) are collinear, and find the ratio $\frac{\text{AB}}{\text{BC}}$.

7 Find the two unit vectors of the form $a\boldsymbol{i} + \frac{1}{3}\boldsymbol{j} - \frac{1}{3}\boldsymbol{k}$.

8 Find the coordinates of the point P on AB such that $\frac{\text{AP}}{\text{PB}} = \frac{3}{2}$, where A is the point (2, 0, −3) and B is (12, 10, 7).

9 In △ABC, $\frac{\text{AP}}{\text{PB}} = -\frac{3}{2}$, $\frac{\text{BQ}}{\text{QC}} = \frac{2}{3}$ and R is the midpoint of CA.
a Find the coordinates of:
(i) P (ii) Q (iii) R.
b Show that P, Q, R are collinear, and find $\frac{\text{PQ}}{\text{QR}}$.

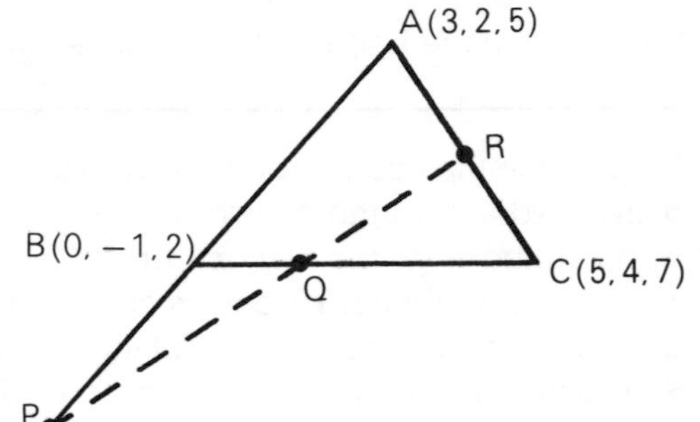

10 Calculate $\boldsymbol{a} \cdot \boldsymbol{b}$ for these pairs:

a $\boldsymbol{a} = \begin{pmatrix} 1 \\ 0 \\ -3 \end{pmatrix}$ $\boldsymbol{b} = \begin{pmatrix} 2 \\ -1 \\ 5 \end{pmatrix}$ **b** $\boldsymbol{a} = \begin{pmatrix} 2 \\ 3 \\ 1 \end{pmatrix}$ $\boldsymbol{b} = \begin{pmatrix} -1 \\ 2 \\ 4 \end{pmatrix}$.

11 a Calculate the angle between $\boldsymbol{u} = 2\boldsymbol{i} + \boldsymbol{j} - 2\boldsymbol{k}$ and $\boldsymbol{v} = \boldsymbol{i} + 2\boldsymbol{j} + \boldsymbol{k}$ in degrees correct to 1 decimal place.
b Show that $\boldsymbol{p} = 2\boldsymbol{i} + 3\boldsymbol{j} - \boldsymbol{k}$ and $\boldsymbol{q} = 3\boldsymbol{i} - \boldsymbol{j} + 3\boldsymbol{k}$ are perpendicular.

12 Show that triangle ABC is obtuse-angled at A, and calculate the angle in degrees correct to 1 decimal place.

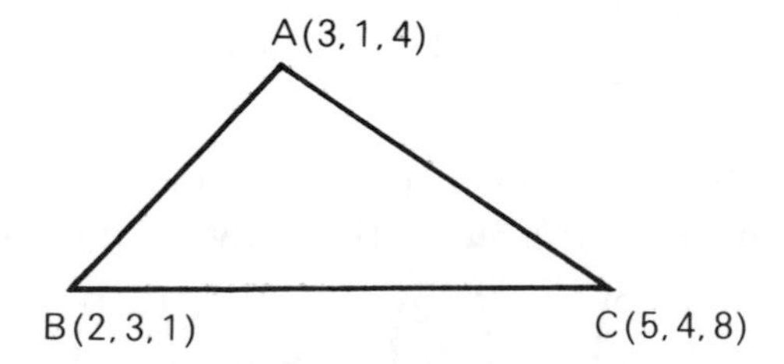

13 Find a and b such that:

a $\begin{pmatrix} a \\ b \end{pmatrix} - \begin{pmatrix} 2 \\ 3 \end{pmatrix} = \begin{pmatrix} 1 \\ 5 \end{pmatrix}$ **b** $\begin{pmatrix} a \\ 4 \end{pmatrix} + \begin{pmatrix} 5 \\ -b \end{pmatrix} = \begin{pmatrix} 4 \\ 3 \end{pmatrix}$.

14 Forces $\boldsymbol{F}_1 = 3\boldsymbol{i} + 4\boldsymbol{j}$ and $\boldsymbol{F}_2 = 4\boldsymbol{i} + 3\boldsymbol{j}$ act in the x, y-plane at the origin O.
a Calculate $\cos\theta$, where θ is the angle between $\boldsymbol{F}_1$ and $\boldsymbol{F}_2$.
b Find the magnitude and direction of the resultant of the system of forces, giving the angle made with OX.

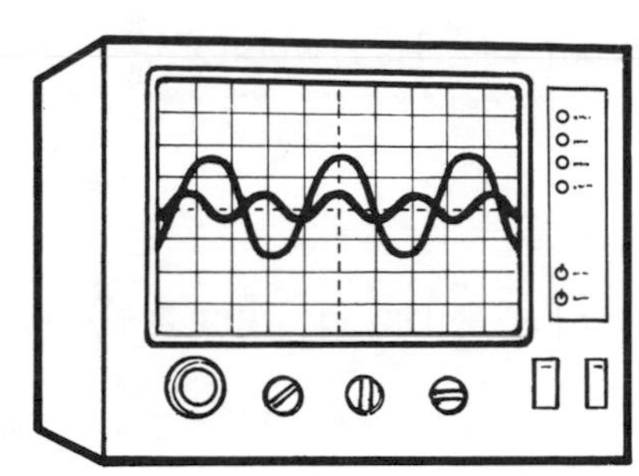

A ship at sea, a current passing through an oscilloscope, a singer in full voice—these all involve patterns of waves, mostly very complex. The study of waves is based on their graphs. Do you remember these graphs, for $0 \leqslant x \leqslant 2\pi$ (radians)?

(i) $y = \cos x$ (ii) $y = 2\cos x$ (iii) $y = \cos 2x$

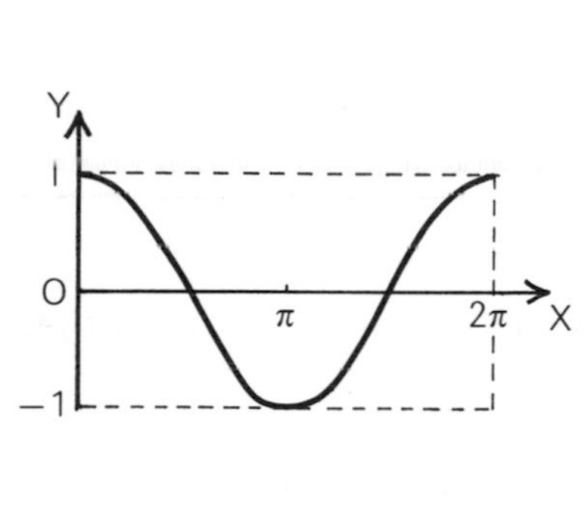

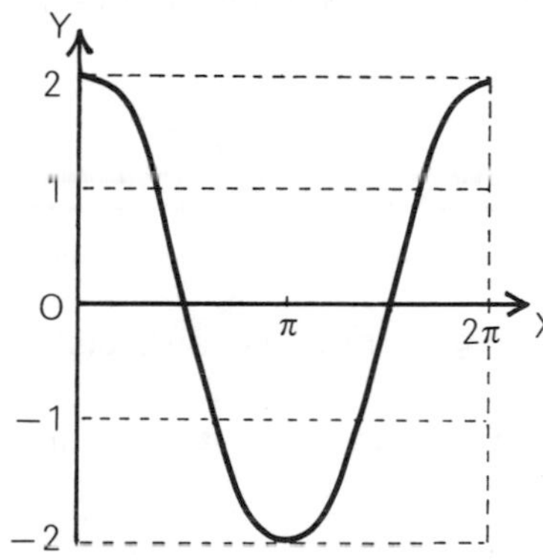

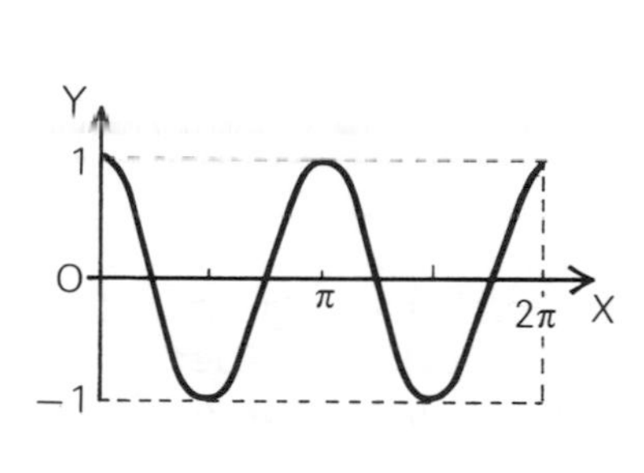

The *amplitude* of a wave is its maximum height.
The amplitudes and periods of the graphs above are:
(i) $y = \cos x$, 1 and 2π (ii) $y = 2\cos x$, 2 and 2π (iii) $y = \cos 2x$, 1 and π.

Example Sketch the graphs $y = \cos\left(x - \dfrac{\pi}{4}\right)$ and $y = 2\cos\left(x - \dfrac{\pi}{4}\right)$, for $0 \leqslant x \leqslant 2\pi$, and give their amplitudes and periods.

Intersections with x-axis

When $y = 0$, $\cos\left(x - \dfrac{\pi}{4}\right) = 0$

$$x - \frac{\pi}{4} = \frac{\pi}{2}, \frac{3\pi}{2}, \ldots$$

$$x = \frac{3\pi}{4}, \frac{7\pi}{4}, \ldots$$

Intersections with y-axis

When $x = 0$, $y = \cos\left(-\dfrac{\pi}{4}\right) = \cos\dfrac{\pi}{4} = \dfrac{1}{\sqrt{2}} \doteqdot 0{\cdot}7$.

$y = 2\cos\left(x - \dfrac{\pi}{4}\right)$

$y = \cos\left(x - \dfrac{\pi}{4}\right)$

Maximum and minimum values

The maximum value of $\cos\left(x-\frac{\pi}{4}\right)$ is 1, when $x-\frac{\pi}{4} = 0, 2\pi, \ldots$

$$x = \frac{\pi}{4}, \frac{9\pi}{4}, \ldots.$$

The minimum value of $\cos\left(x-\frac{\pi}{4}\right)$ is -1, when $x-\frac{\pi}{4} = \pi, \ldots$

$$x = \frac{5\pi}{4}, \ldots.$$

The amplitude of the wave is 1, and the period is 2π.

The graph of $y = \cos\left(x-\frac{\pi}{4}\right)$ is the same as that of $y = \cos x$, but moved $\frac{\pi}{4}$ to the right.

$y = 2\cos\left(x-\frac{\pi}{4}\right)$ is like $y = \cos\left(x-\frac{\pi}{4}\right)$, but with amplitude 2.

Exercise 1

1 *Sketch* these graphs, for $0 \leqslant x \leqslant 2\pi$ radians, and state their amplitudes and periods. (No working needed.)
a $y = \sin x$ **b** $y = 2\sin x$ **c** $y = \sin 2x$ **d** $y = 2\sin 2x$.

2 Follow the steps in the worked example to sketch the graphs $y = \sin\left(x-\frac{\pi}{2}\right)$ and $y = 3\sin\left(x-\frac{\pi}{2}\right)$, for $0 \leqslant x \leqslant 2\pi$, and to find their amplitudes and periods.

3 What is the amplitude and period of each of these graphs?
a $y = \cos x$ **b** $y = 4\cos x$ **c** $y = 3\cos 2x$ **d** $y = 2\cos 3x$.

4 *Sketch* the graphs $y = 3\sin x$ and $y = \sin 3x$, for $0 \leqslant x \leqslant 2\pi$, and give their amplitudes and periods.

5 a Follow the steps in the worked example to sketch the graphs $y = \sin\left(x+\frac{\pi}{4}\right)$ and $y = 2\sin\left(x+\frac{\pi}{4}\right)$, for $0 \leqslant x \leqslant 2\pi$.
b Explain how to get these graphs from $y = \sin x$.

6 Write down the equation of each graph (they are sines or cosines):

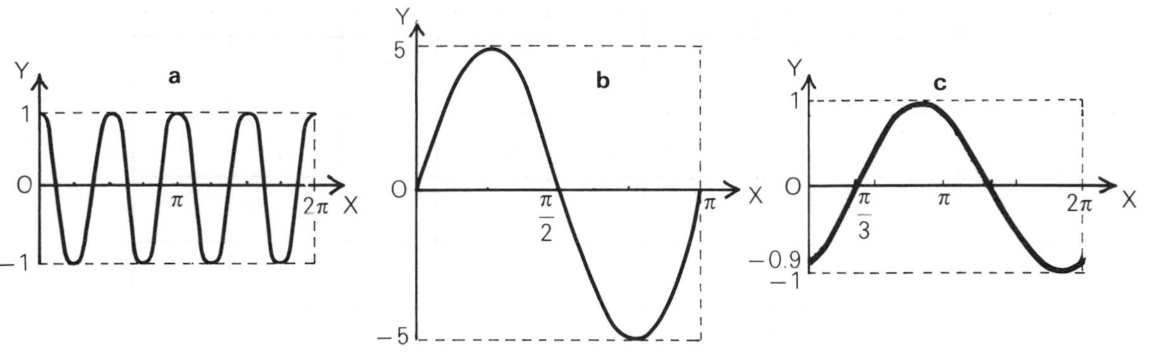

7 No working is necessary to sketch these graphs, for $0 \leqslant x \leqslant 2\pi$:

a $y = 2\sin\left(x+\dfrac{\pi}{2}\right)$ **b** $y = 3\cos\left(x-\dfrac{\pi}{6}\right)$ **c** $y = 4\cos\left(x+\dfrac{\pi}{6}\right)$ **d** $y = 5\sin\left(x+\dfrac{\pi}{3}\right)$.

8 a One of these graphs is $y = R\cos(x-\alpha)$ and the other $y = R\sin(x-\alpha)$. Which is which?

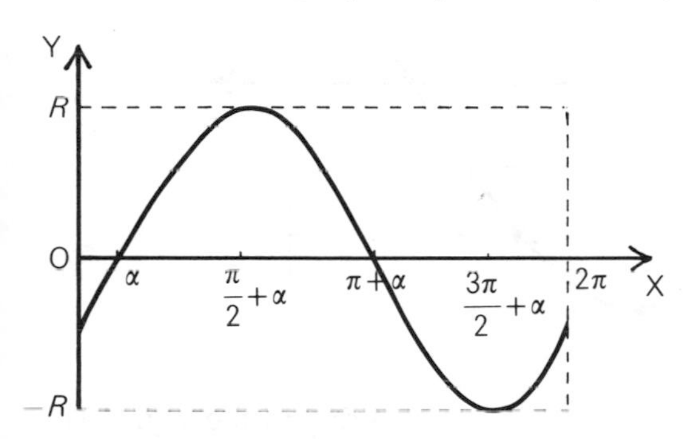

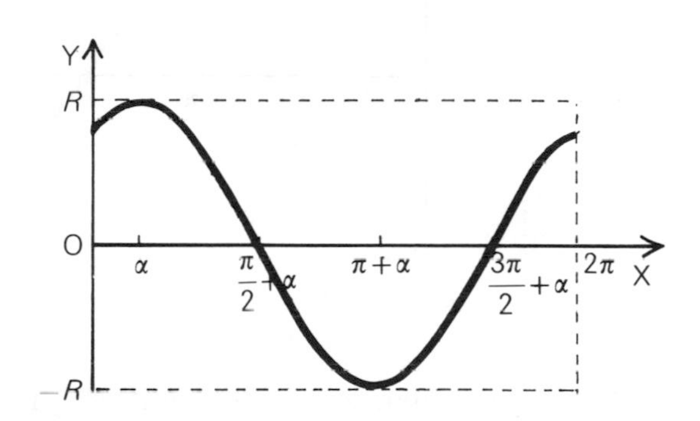

b For each graph, state:
  (i) its intersections with the x-axis and the y-axis
  (ii) its maximum and minimum values, and the corresponding values of x.

COMBINING WAVES

In speech, music, electronic circuits, vibrations, etc, wave forms can be very complicated since they may represent the sum of a large number of wave forms of different periods. A simple musical sound will normally consist of a basic note and its harmonics. Phenomena like these have complicated graphs.

In this chapter a method is developed for combining waves of the form $a\cos x$ and $b\sin x$ where a and b are constants.

Example Investigate the graph $y = \cos x + \sin x$ for $0 \leqslant x \leqslant 2\pi$.

x	0	$\frac{\pi}{4}$	$\frac{\pi}{2}$	$\frac{3\pi}{4}$	π	$\frac{5\pi}{4}$	$\frac{3\pi}{2}$	$\frac{7\pi}{4}$	2π
$\cos x$	1	$1/\sqrt{2}$	0	$-1/\sqrt{2}$	-1	$-1/\sqrt{2}$	0	$1/\sqrt{2}$	1
$\sin x$	0	$1/\sqrt{2}$	1	$1/\sqrt{2}$	0	$-1/\sqrt{2}$	-1	$-1/\sqrt{2}$	0
y	1	$2/\sqrt{2}$	1	0	-1	$-2/\sqrt{2}$	-1	0	1

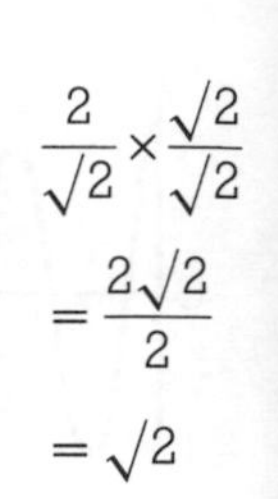

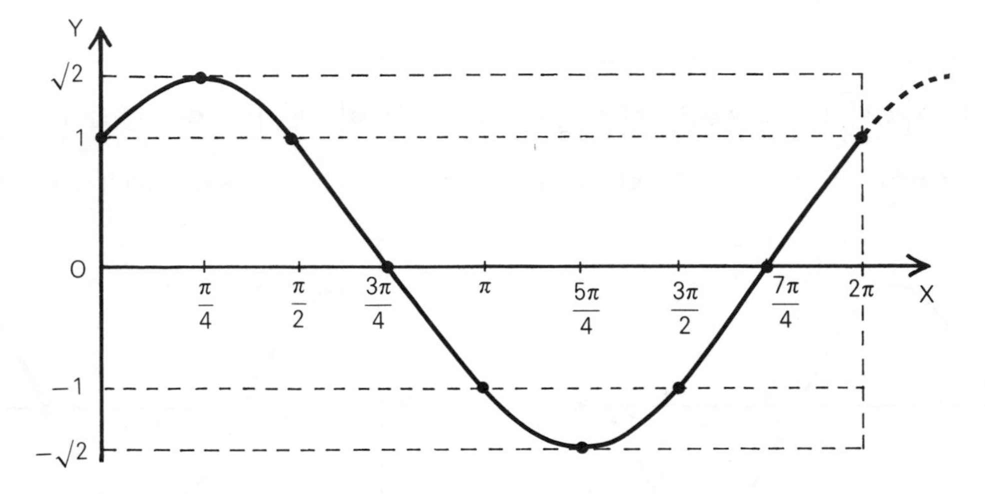

The graph is like $y = \cos x$, but its amplitude is $\sqrt{2}$, and it is displaced $\frac{\pi}{4}$ to the right. This suggests that its equation is $y = \sqrt{2} \cos\left(x - \frac{\pi}{4}\right)$.

Check $\sqrt{2} \cos\left(x - \frac{\pi}{4}\right) = \sqrt{2}\left(\cos x \cos\frac{\pi}{4} + \sin x \sin\frac{\pi}{4}\right)$

$= \cos x + \sin x$.

So $\cos x + \sin x = \sqrt{2} \cos\left(x - \frac{\pi}{4}\right)$,

and two waves, $y = \cos x$ and $y = \sin x$, have been combined into one wave,

$$y = \sqrt{2} \cos\left(x - \frac{\pi}{4}\right).$$

Investigate the graphs: **a** $y = \cos x - \sin x$, for $0 \leqslant x \leqslant 2\pi$
b $y = 3 \cos x° + 4 \sin x°$, for $0° \leqslant x° \leqslant 360°$.
Write their equations in the form $R \cos(x - \alpha)$, where R and α are constants.
Can you see how to obtain R and α from the given equations?

$a \cos x + b \sin x$ IN THE FORM $R \cos(x - \alpha)$, a, b NON-ZERO CONSTANTS

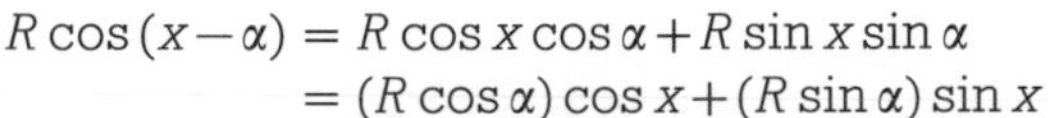

$$\begin{aligned} R\cos(x-\alpha) &= R\cos x\cos\alpha + R\sin x\sin\alpha \\ &= (R\cos\alpha)\cos x + (R\sin\alpha)\sin x \\ &= a\cos x + b\sin x, \text{ where } R\cos\alpha = a \text{ and } R\sin\alpha = b. \end{aligned}$$

Squaring and adding, $R^2(\cos^2\alpha + \sin^2\alpha) = a^2 + b^2$.
So $R = \sqrt{(a^2+b^2)}$, taking $R > 0$.

Also, $\tan\alpha = \frac{b}{a}$, and the quadrant of α is found from the signs of $\cos\alpha$ and $\sin\alpha$.

$$a\cos x + b\sin x = R\cos(x-\alpha), \text{ where } R = \sqrt{(a^2+b^2)} \text{ and } \tan\alpha = \frac{b}{a}.$$

The graph of the function is $y = a\cos x + b\sin x = R\cos(x-\alpha)$.

R is the *amplitude* of the function, the maximum height of a wave. α is the *phase angle*—the amount by which $y = R\cos x$ is moved to get $y = R\cos(x-\alpha)$.

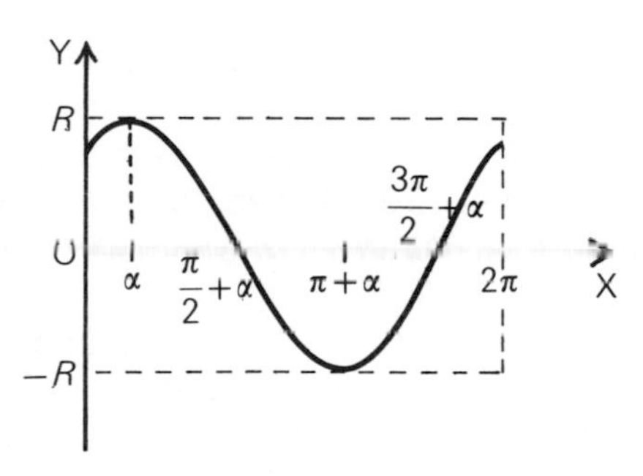

Example Express $\cos x - \sin x$ in the form $R\cos(x-\alpha)$, and sketch the graph of the function.
Let $\cos x - \sin x = R\cos(x-\alpha) = (R\cos\alpha)\cos x + (R\sin\alpha)\sin x$.

$\left.\begin{aligned} R\cos\alpha &= 1 \\ R\sin\alpha &= -1 \end{aligned}\right\}$ $R = \sqrt{(1^2+1^2)} = \sqrt{2}$, and α is in the fourth quadrant.

Sin +	All + ✓ α
Tan + ✓	$2\pi - \alpha$ Cos + ✓✓

$\tan\alpha = -1$, so $\quad \alpha = 2\pi - \frac{\pi}{4} = \frac{7\pi}{4}$.

So $\cos x - \sin x = \sqrt{2}\cos\left(x - \frac{7\pi}{4}\right)$.

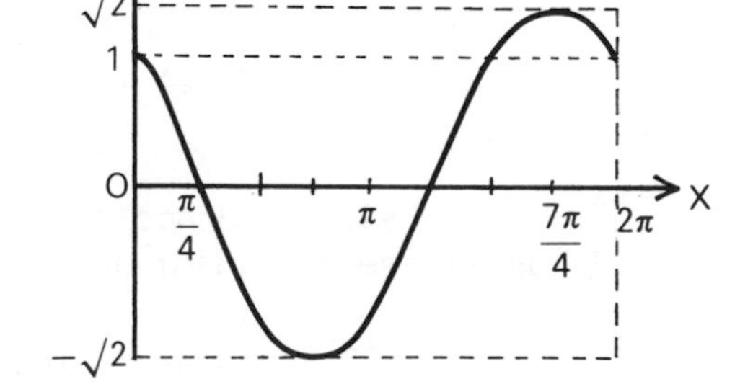

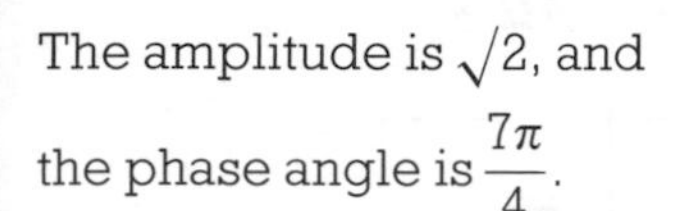

The amplitude is $\sqrt{2}$, and the phase angle is $\frac{7\pi}{4}$.

Exercise 2

In questions **1** and **2**:

(i) Express each function in the form $R\cos(x-\alpha)$, giving R exactly (e.g. $\sqrt{2}$) and α in radians, or correct to the nearest degree where necessary.

(ii) State the values of the amplitude R and the phase angle α.

(iii) Sketch the graph of the function.

1 For $0 \leqslant x \leqslant 2\pi$ radians:
a $\cos x+\sin x$ **b** $\cos x-\sin x$ **c** $\sqrt{3}\cos x+\sin x$ **d** $-\cos x+\sqrt{3}\sin x$.

2 For $0° \leqslant x° \leqslant 360°$:
a $3\cos x°+4\sin x°$ **b** $8\cos x°-6\sin x°$ **c** $\cos x°-\sqrt{3}\sin x°$ **d** $\cos x°+2\sin x°$.

3 By expanding $\sin(x+\alpha)$, show that $a\cos x+b\sin x$ can be expressed in the form $R\sin(x+\alpha)$, where R, α are constants and $R > 0$.

4 Express each of the functions in question **1** in the form $R\sin(x+\alpha)$.

5 Express each of these in the form $R\cos(kx-\alpha)$:
a $\cos 2x+\sin 2x$ (Let $\cos 2x+\sin 2x = R\cos(2x-\alpha)$.)
b $\sin 3x-\cos 3x$ **c** $3\cos\omega t-4\sin\omega t$.

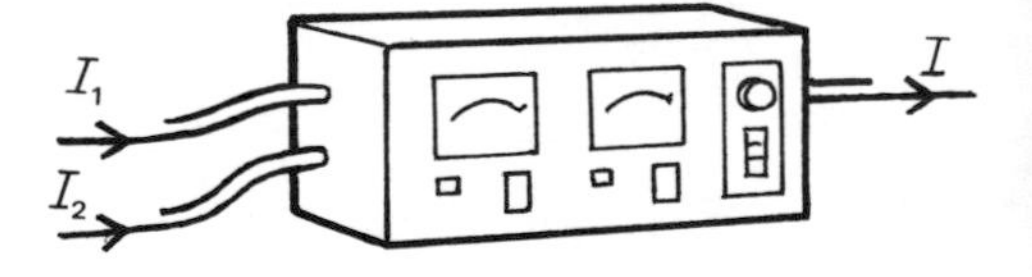

6 Electric currents $I_1 = \cos 300t°$ and $I_2 = \sqrt{3}\sin 300t°$ are fed into an ammeter. Express the resultant I_1+I_2 in the form $R\cos(300t-\alpha)°$.

7 The expression $10\cos 20t°+30\sin 20t°$ represents the displacement of a wave after t seconds.
a Express it in the form $R\sin(20t+\alpha)°$ where $R > 0$ and $0 \leqslant \alpha \leqslant 360$, giving R and α correct to 2 decimal places.
b Sketch the graph of the wave against t for $0 \leqslant t \leqslant 18$, showing where the graph cuts the t-axis and any stationary points.

MAXIMA AND MINIMA OF $a\cos x+b\sin x+c$

Example Find the maximum and minimum values of $f(x) = 12\sin x°-5\cos x°+10$, for $0° \leqslant x° \leqslant 360°$, and the corresponding values of x.

Let $12\sin x°-5\cos x° = R\cos(x-\alpha)° = (R\cos\alpha°)\cos x°+(R\sin\alpha°)\sin x°$.

$$\left.\begin{aligned}R\cos\alpha° &= -5\\ R\sin\alpha° &= 12\end{aligned}\right\} \begin{array}{l}R = \sqrt{(5^2+12^2)} = 13, \text{ and}\\ \alpha \text{ is in the second quadrant.}\end{array}$$

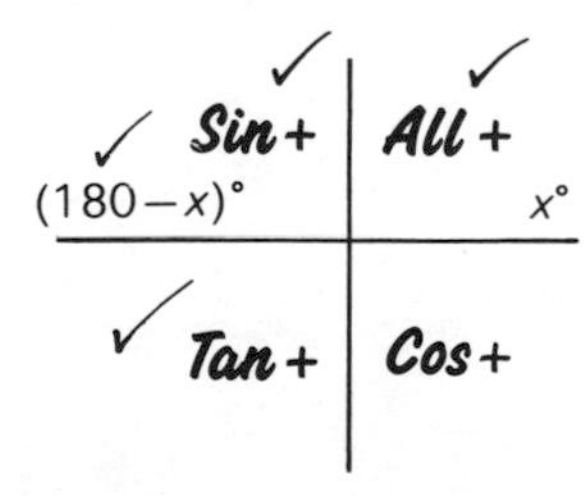

$$\tan\alpha° = -\frac{12}{5}, \text{ so } \alpha = 180-67 \text{ (to nearest degree)}$$
$$= 113.$$

So $f(x) = 13\cos(x-113)°+10$.

Its maximum value $= 13+10$
$= 23,$
when $\cos(x-113)^\circ = 1,$
$x-113 = 0$
$x = 113.$

Its minimum value $= -13+10$
$= -3,$
when $\cos(x-113)^\circ = -1$
$x-113 = 180$
$x = 293.$

Exercise 3

Find the maximum and minimum values of the functions in questions **1–4**, and the corresponding values of x.

1 For $0^\circ \leqslant x^\circ \leqslant 360^\circ$: **a** $\cos x^\circ + \sin x^\circ$ **b** $7\cos x^\circ - 24\sin x^\circ$ **c** $2\sin x^\circ - 3\cos x^\circ$.

2 a For $0 \leqslant x \leqslant 2\pi$ radians:
(i) $\sqrt{3}\sin x + \cos x + 2$ (ii) $\cos x - \sin x - 2$ (iii) $1 + \sqrt{3}\cos x - \sin x$.
b Sketch the graphs of the three functions.

3 For $0^\circ \leqslant x^\circ \leqslant 180^\circ$: **a** $4\cos 2x^\circ + 3\sin 2x^\circ + 5$ **b** $\cos 2x^\circ - 2\sin 2x^\circ + 8$.

4 For $0 \leqslant x \leqslant 2\pi$: **a** $4\cos^2 x + 4\sin x\cos x$ **b** $2\cos^2 x - 2\sin 2x$.
(Remember the formulae: $\cos 2x = 2\cos^2 x - 1$, and $\sin 2x = 2\sin x\cos x$.)

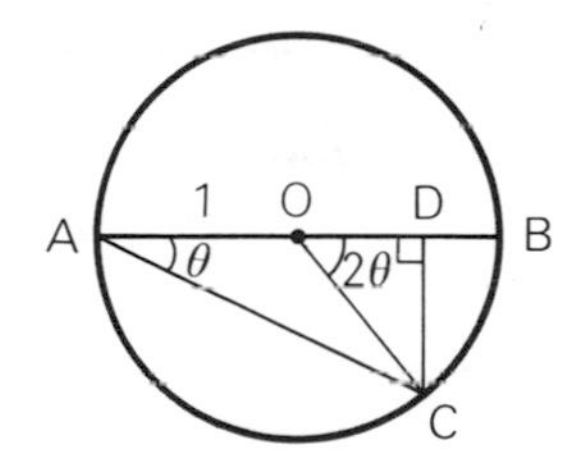

5 AB is a diameter of the circle centre O, radius 1; CD is perpendicular to AB.
a Show that the perimeter of $\triangle$COD is $P = 1 + \sin 2\theta + \cos 2\theta$.
b Find the maximum value of P, and the corresponding value of θ.

6 A bracing spar of length 2 m is placed horizontally between two perpendicular walls. For maximum strength it has to be placed so that it covers the maximum distance on the walls, i.e. so that OA + OB is a maximum.

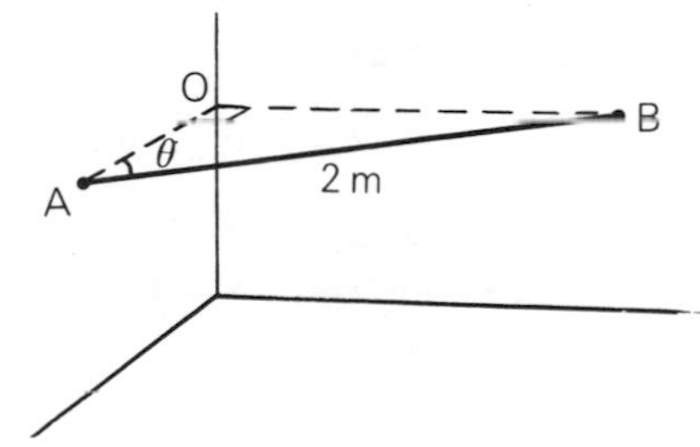

a Express OA + OB in terms of θ.
b Find the value of θ for which OA + OB is a maximum, and find the maximum value.

7 A rectangular car park is being planned. It consists of a square part ABCD and a rectangular part OADE, where OD = 50m. The problem is to choose the value of θ that makes the area S m² of the car park a maximum.
a Show that $S = 2500(\sin\theta\cos\theta + \cos^2\theta)$.
b Express S in the form $R\cos(2\theta - \alpha) + c$.
Hint See the reminder in question **4**.
c Find the maximum value of S, and the corresponding value of θ.

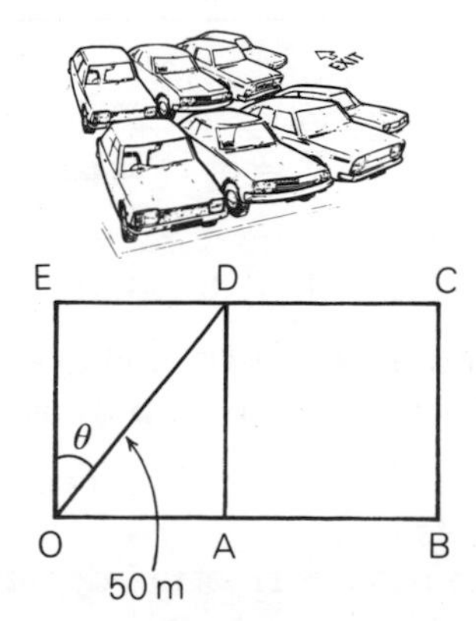

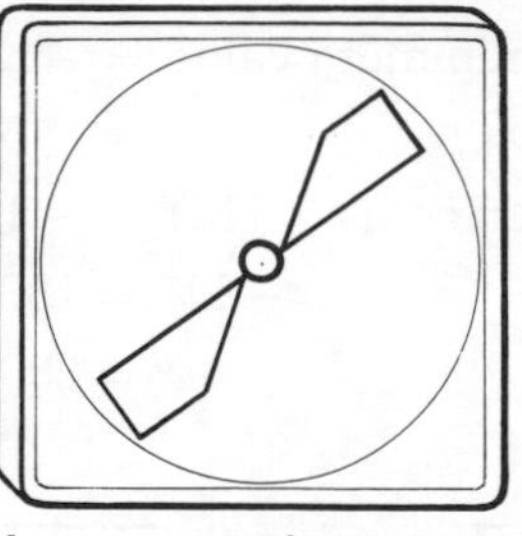

8 A new blade for a rotary lawnmower is constructed with the shape OMNQP shown, where MNQP is a square. (Two of these form a blade.) OMNQP has to have a maximum area.

a Show that its area is $A = \sin^2\theta + \frac{1}{4}\sin 2\theta$.

b Express this in the form $R\cos(2\theta - \alpha) + c$.

c Find the maximum value of A, and the corresponding value of θ.

SOLVING LINEAR TRIGONOMETRIC EQUATIONS

Example Solve the equation $\sin 2x° + 3\cos 2x° + 1 = 0$, for $0° \leqslant x° \leqslant 180°$.
Let $\sin 2x° + 3\cos 2x° = R\cos(2x - \alpha)° = (R\cos\alpha°)\cos 2x° + (R\sin\alpha°)\sin 2x°$.

$$\left.\begin{array}{l} R\cos\alpha° = 3 \\ R\sin\alpha° = 1 \end{array}\right\} \begin{array}{l} R = \sqrt{(3^2+1^2)} = \sqrt{10}, \text{ and} \\ \alpha \text{ is in the first quadrant} \end{array}$$

Sin + ✓	All + ✓✓
Tan +	Cos + ✓

$\tan\alpha° = \frac{1}{3}$, so $\alpha = 18$, to the nearest degree.

The equation is $\sqrt{10}\cos(2x - 18)° = -1$

$$\cos(2x-18)° = -\frac{1}{\sqrt{10}}$$

$$2x - 18 = 180 - 72 \text{ or } 180 + 72$$

$$2x = 126 \text{ or } 270$$

$$x = 63 \text{ or } 135 \text{ (to nearest degree).}$$

Sin + ✓ $(180-x)°$	All + $x°$
$(180+x)°$ Tan + ✓	Cos +

Exercise 4

1 Solve these equations, for $0° \leqslant x° \leqslant 360°$, to the nearest degree:
a $3\cos x° + 4\sin x° = 5$ **b** $9\sin x° - 12\cos x° = 10$ **c** $2\cos x° + 3\sin x° = -1$.

2 Solve these equations, for $0 < x \leqslant 2\pi$ radians:
a $\cos x - \sin x = 1$ **b** $\sqrt{3}\sin x + \cos x = 1$ **c** $\sqrt{2}\cos x + \sqrt{2}\sin x - 1 = 0$.

3 Solve, for $0° \leqslant x° \leqslant 180°$, giving solutions correct to 1 decimal place:
a $\cos 2x° - \sin 2x° = 1$ **b** $3\cos 2x° - 4\sin 2x° + 1 = 0$.

4 a Investigate the relation between a, b and c for the equation $a\cos x° + b\sin x° = c$ to have real roots.
b Find the roots correct to 1 decimal place, for $0° \leqslant x° \leqslant 360°$, if $a = 4$, $b = -5$, $c = 6$.

The sun is setting behind the lone rectangular standing-stone, casting a shadow with parallel sides.

a Show that the width of the shadow, $D = 5\sin\theta° + 2\cos\theta°$.

b Find θ (to the nearest degree) when $D = 4$.

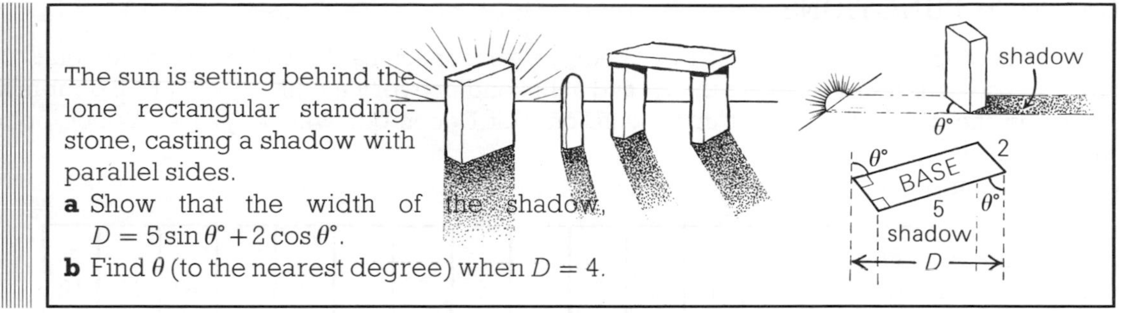

CHECK-UP ON **THE WAVE FUNCTION $a\cos x + b\sin x$**

1 Find the intersections of the graph $y = 2\cos\left(x - \frac{\pi}{6}\right)$ with the x- and y-axes ($0 \leqslant x \leqslant 2\pi$), and its maximum and minimum values. Sketch the graph and give its amplitude and period.

2 Express each function in the form $R\cos(x-\alpha)$, state the amplitude and phase angle of each (to the nearest degree), and sketch the graph of each function:
a for $0 \leqslant x \leqslant 2\pi$: (i) $\sqrt{3}\cos x - \sin x$ (ii) $-\cos x - \sin x$
b for $0° \leqslant x° \leqslant 360°$: (i) $-3\cos x° + 4\sin x°$ (ii) $5\sin x° + 2\cos x°$.

3 In an electric circuit the current is given by the formula $I = 6\cos 250t° + 3\sin 250t°$. Express this in the form $R\sin(250t + \alpha)°$, with α to the nearest degree. State the amplitude and phase angle of the current.

4 Find the maximum and minimum values of these functions, and the corresponding values of x, for $0° \leqslant x° \leqslant 360°$, to the nearest degree, where appropriate:
a $\sin x° + \cos x° + 3$ **b** $2\cos x° - 5\sin x° - 3$.

5

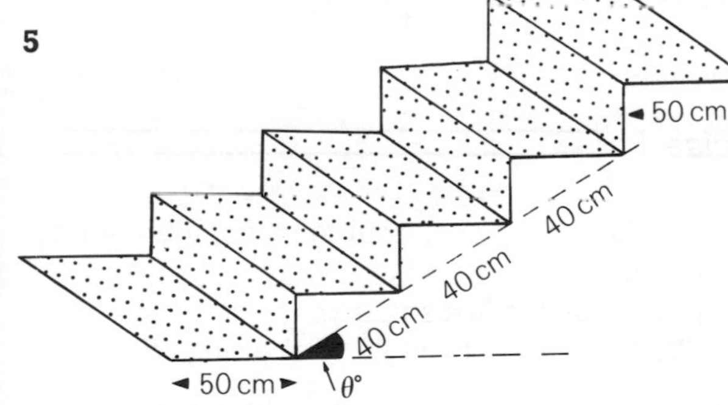

There are 50 cm lengths of carpet at the top and bottom of the stairs. The staircase contains 4 risers and 3 treads on the steps.

a Show that the total length of carpet is $L = 120\cos\theta° + 160\sin\theta° + 100$ cm.

b Find the maximum value of L, and the value of θ when L is a maximum.

6 Solve these equations for $0° \leqslant x° \leqslant 360°$, giving solutions correct to 1 decimal place:
a $3\cos x° - \sin x° = 2$ **b** $\sin 2x° + 2\cos 2x° = 1$.

7 A right-angled triangular plot of ground has 40 metres of fencing round it.

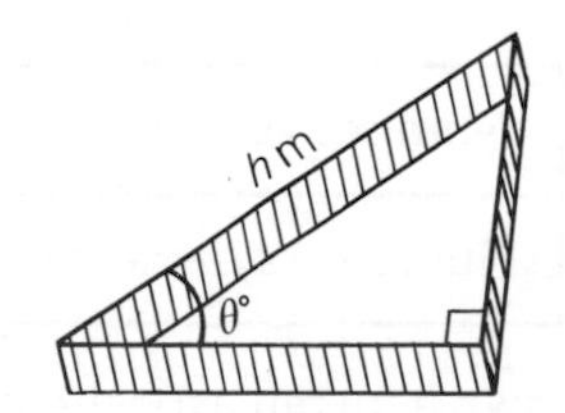

a Prove that $h = \dfrac{40}{\cos\theta° + \sin\theta° + 1}$.

b Find the minimum value of h, and the corresponding value of θ.

GROWTH FUNCTIONS

Mary Love was 18 on 1st January. Her father, an economist, said 'I will either give you £2000 now, or else I'll put £1 in the safe today, and at the end of every month for a year I'll double the money in the safe. The amount in the safe on 31st December will be yours!' Which option should she choose?
Copy and complete:

Number of months in safe (x)	0	1	2	3	4	5	6	7	8	9	10	11	12
Sum in safe	1	2	4										

There should be no doubt in her mind about her choice. How much more will she have on 31st December if she chooses the 'growth' option rather than the £2000? In fact, the doubling function f is given by the formula $f(x) = 2^x (x \geqslant 0)$. (Note that the starting sum is $2^0 = 1$.)

Plot the points in the table up to $x = 6$, and join them with a smooth curve. (1 unit = 2 cm on the x-axis, 1 unit = 2 mm on the y-axis.)
You now have part of the graph $y = 2^x$ for which $x \geqslant 0$.

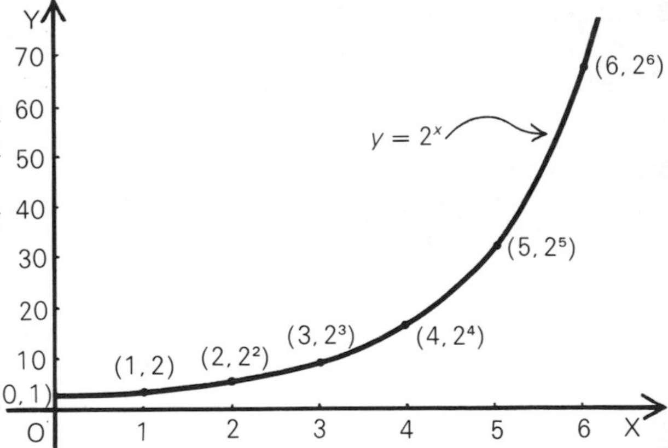

Exercise 1

1 Take a chess board (64 squares). Put a 1p coin on the 1st square, double that amount on the 2nd square, double this again on the 3rd square, and so on.
a How many 1p coins would you have to try to pile up on the last square?
b You have set up another doubling function f. Write down its formula. (Take care! $f(1) = 1$.)

2 A culture starting with 1 bacterium trebles the number of bacteria every 24 hours.
a Copy and complete:

Number of days (x)	1	2	3	4	5	6
Number of bacteria	3					

b Write down a formula for the function that models this situation.
c Draw the graph of the function (1 unit = 2 cm on x-axis, 10 units = 2 mm on y-axis).

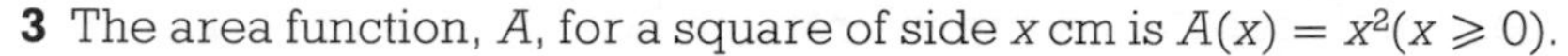

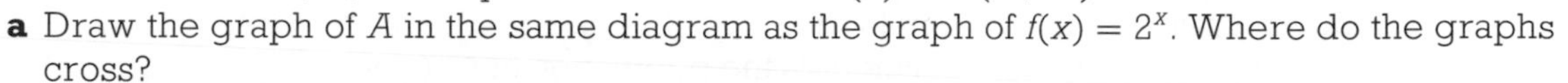

3 The area function, A, for a square of side x cm is $A(x) = x^2 (x \geqslant 0)$.

a Draw the graph of A in the same diagram as the graph of $f(x) = 2^x$. Where do the graphs cross?

b $y = 2^x$ and $y = x^2$ look alike, but their behaviour for x large is totally different. Which would you expect to be called a growth function?

4 The mathematical model for question **2** is the growth function with formula $f(x) = 3^x$. The volume function, V, for a cube of edge x is $V(x) = x^3$.
Investigate the graphs of these for $x \geqslant 0$ as in question **3**. (Check that one of the points of intersection has a value of x between 2·4 and 2·5.)

5 a On the same diagram, sketch the graphs for $x \geqslant 0$ of the functions defined by $f(x) = a^x$ where: (i) $a = 2$ (ii) $a = 3$ (iii) $a = 4$ (iv) $a = 5$.

a Explain why the graph of a function of the form a^x always passes through the point (0, 1).

6 Find the functions that model these facts:

a A culture of 1000 bacteria doubles its size by the end of every day.

b An investment starting with £200 is trebled every year.

c A population of cells, starting with 500, doubles its number every hour.

Example

Leave £100 with us, said the Northern Bank, and each year you'll see your deposit grow by 12%.

Amount in the bank, £A(n), after n years:

initially: $A(0) = 100$

after 1 year: $A(1) = 100 \times \left(1 + \frac{12}{100}\right) = 100(1{\cdot}12)$

after 2 years: $A(2) = A(1) \times (1{\cdot}12) = 100(1{\cdot}12)^2$

after 3 years: $A(3) = A(2) \times (1{\cdot}12) = 100(1{\cdot}12)^3$

. . . .

after n years: $A(n) = 100(1{\cdot}12)^n$.

Exercise 2

1 The above example involved a growth function, $A(n) = 100(1{\cdot}12)^n$. Using your calculator, copy and complete the table (giving the amounts correct to the nearest £).

Number of years (n)	0	10	20	30	40	50
Amount in £s in bank A(n)	100	311				

2 The Southern Bank, down the road from the Northern Bank, offers a growth factor (compound interest) of 15%.

a Construct the growth function for an initial sum of £100.

b Construct a table similar to that in question **1**.

3 Ulanda is troubled by a very high population growth rate, in fact 6% per year. The 1975 census showed a population of 100 million.

a Construct the growth function that models the population after t years.

b Estimate the population in the year 2000.

4 Certain types of laboratory mice breed very fast—their population increases by 30% per month. Starting with 400 mice:

a construct the function that models the population after t months

b calculate the population after:
(i) 10 months (ii) 1 year.

5 The 'Sprinter' cabbage in its most active growing period increases the radius of its heart by 50% each day. Taking the initial radius as 1 unit:

a construct the function that models the radius after x days

b estimate the radius after: (i) 8 days (ii) 14 days.

6 Some cells reproduce very rapidly. One population of 500 cells increases its number by 80% each hour.

a Construct the growth function that models the population after t hours.

b Estimate the population after: (i) 6 hours (ii) 1 day.

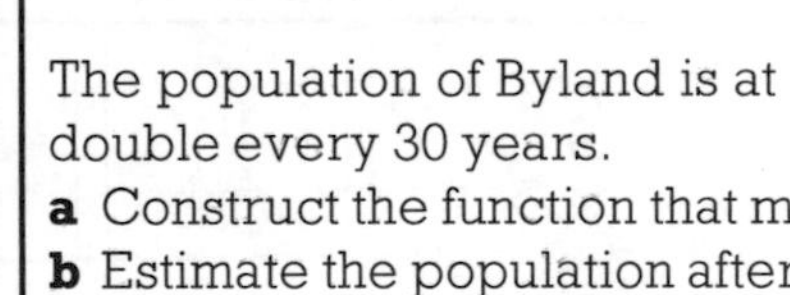

The population of Byland is at present 50 million and is increasing so fast that it will double every 30 years.

a Construct the function that models the population after t years.

b Estimate the population after: (i) 15 years (ii) 45 years.

DECAY FUNCTIONS

Leaving the cap off the 5 litre can of petrol is a big mistake. The petrol evaporates at the rate of 20% per week.

At the end of Week 1, 0·2 of the petrol is gone, so 5(0·8) litres are left.

At the end of Week 2, $5(0{\cdot}8)^2$ litres are left.

At the end of Week 3, $5(0{\cdot}8)^3$ litres are left.

A decay function D is set up, given by the formula $D(t) = 5(0{\cdot}8)^t (t \geqslant 0)$.
Copy and complete (giving the number of litres correct to 2 decimal places):

Number of weeks (t)	0	1	2	3	4	5	6
Petrol in can D(t) (litres)	5	4	3·2				

Plot the points in the table and join them with a smooth curve. The graph shows clearly that D(t) decays as t increases.

Use your calculator to check that after 40 weeks only the smell of petrol will be left!

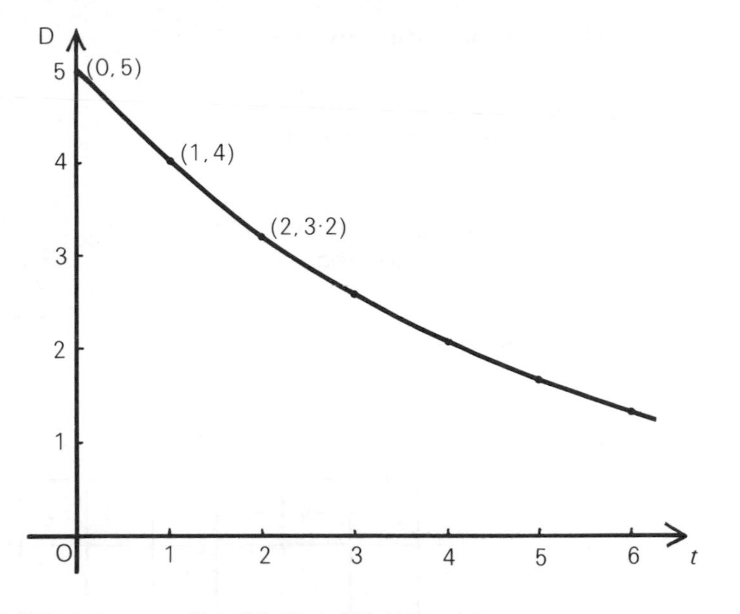

Exercise 3

1 A volatile fluid evaporates at the rate of 60% per week. An open can is filled with 10 litres of the fluid.

a Construct the decay function that models the fluid left after t weeks.
b Draw up a table as above.
c Draw the graph of the function.
d Estimate how long the fluid will take to evaporate.

2 The car salesman assured Mr Jamieson that the FASTA car, priced at £4000, would depreciate at the low rate of 15% per year over the first 6 years. He could not guarantee a depreciation rate after that.

a Construct the decay function that models the car's value over the 6 years.
b Calculate the estimated value (to the nearest £) of the car after 6 years.

3 Radium has a 'half-life' of 1600 years—if you start with a given amount (in grams), then after 1600 years, 50% of that amount will be left, the remainder decaying into other elements. Start with 5 grams.

a Explain why the decay function that models the amount present after t years is $D(t) = 5(0{\cdot}5)^{t/1600}$.
b Estimate (correct to 2 decimal places) the amount left after 600 years.
c After 20 000 years barely 1 mg will be left. Do you agree?

4 Some radioactive materials have a much shorter half-life than radium. One has a 20 year half-life.

a Construct the decay function if you start with 5 g of this material.
b Use your calculator to estimate how many years would pass before only 1 mg remained of the original 5 g.

5 Plutonium 239 has a half-life of 24 400 years.

a Starting with 1 kg of 239, construct the decay function that models the amount left after t years.
b Estimate the amount left after: (i) 50 000 years (ii) 100 000 years.

6 The half-life of C^{14}, the radioactive form of carbon, is 5720 years.

a If 1 g is originally present, construct the function that models the amount left after t years.

b Bones discovered in an archaeological dig contain 25% of their original C^{14} content. Why are the bones about 11 440 years old?

THE EXPONENTIAL FUNCTIONS

In dealing with growth and decay you met functions like $f(x) = 2^x$ and $g(x) = (0{\cdot}5)^x = (\frac{1}{2})^x = 2^{-x}$ for $x \geqslant 0$. Suppose you now investigate their graphs $y = 2^x$ and $y = 2^{-x}$ for $x \in R$.
Check this table:

x	-5	-4	-3	-2	-1	0	1	2	3	4	5
2^x	$\frac{1}{32}$	$\frac{1}{16}$	$\frac{1}{8}$	$\frac{1}{4}$	$\frac{1}{2}$	1	2	4	8	16	32
2^{-x}	32	16	8	4	2	1	$\frac{1}{2}$	$\frac{1}{4}$	$\frac{1}{8}$	$\frac{1}{16}$	$\frac{1}{32}$

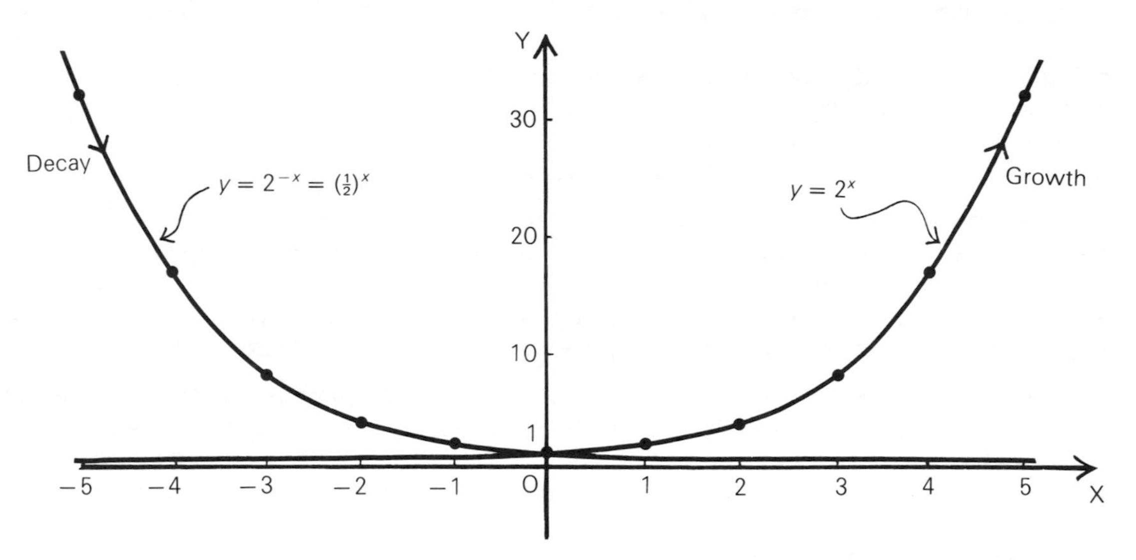

In the same way, $y = a^x$ and $y = a^{-x}$, for each $a > 1$, are mirror images of each other in the y-axis.

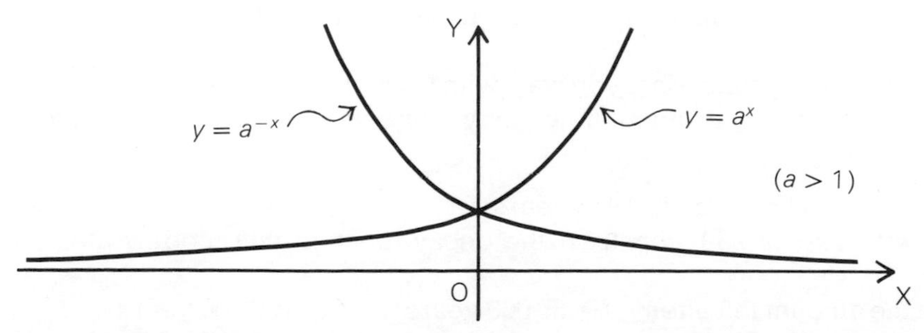

For a a positive constant, the formula

$$f(x) = a^x (x \in R)$$

defines the *exponential function with base a.*

Note For $a > 1$: (i) a^x is a growth function
(ii) a^{-x} is a decay function.

A special exponential function—the number e

The Northern Bank is making a very special offer. If you put in £1 they will pay you 100% interest per year. In addition, they will pay the interest 1, 2, 3, . . . , or n times a year.

With $n = 1$, the £1 becomes £$\left(1+\frac{100}{100}\right) =$ £2.

With $n = 2$, the £1 becomes £$\left(1+\frac{100}{2(100)}\right)^2 =$ £$\left(1+\frac{1}{2}\right)^2 =$ £2·25.

With $n = 3$, the £1 becomes £$\left(1+\frac{100}{3(100)}\right)^3 =$ £$\left(1+\frac{1}{3}\right)^3 =$ £2·44 . . . .

For n, the £1 gives rise in 1 year to £$\left(1+\frac{1}{n}\right)^n$.

What happens to $\left(1+\frac{1}{n}\right)^n$ as n increases?

Use your calculator to check this table:

n	1	2	4 (quarterly)	12 (monthly)	52 (weekly)	100
$\left(1+\frac{1}{n}\right)^n$	2	2·25	2·441406 . . .	2·613035 . . .	2·692596 . . .	2·704813 . . .

1000	10 000	100 000	1 000 000
2·716923 . . .	2·718145 . . .	2·718268 . . .	2·718280 . . .

It seems that as n gets larger and larger, $\left(1+\frac{1}{n}\right)^n$ gets closer and closer to 2·71828 . . .

> In fact, $\lim_{n \to \infty} \left(1 + \frac{1}{n}\right)^n$ does exist, and is denoted by e.

Use your calculator to check that $e = 2{\cdot}7182818$, correct to 8 significant figures. e is an important number in mathematics and its applications.
Your £1 will never exceed £e by the end of the year.
The exponential function with base e is often denoted by *exp*, so

$$\exp(x) = e^x \, (x \in R).$$

Exercise 4

Use your calculator where necessary in these questions.

1 Calculate, correct to 3 decimal places:

a e^5 **b** $e^{1\cdot2}$ **c** $e^{2\cdot3}$ **d** e^{-2} **e** $e^{-3/2}$ **f** $\frac{1}{\sqrt{e}}$.

2 The intensity of sunlight reaching a depth x metres in a muddy lake decreases exponentially with x according to the formula $I(x) = I_0 e^{-0\cdot5x}$, $(x \geqslant 0)$, where I_0 is the intensity at the water surface.
a Calculate $e^{-0\cdot5x}$ for $x = 0, 1, 2, 3$ and 4.
b Write down the value of x, to the nearest whole number, for which, correct to 1 decimal place, $I(x)/I_0 =$ (i) 1 (ii) 0·6 (iii) 0·1.

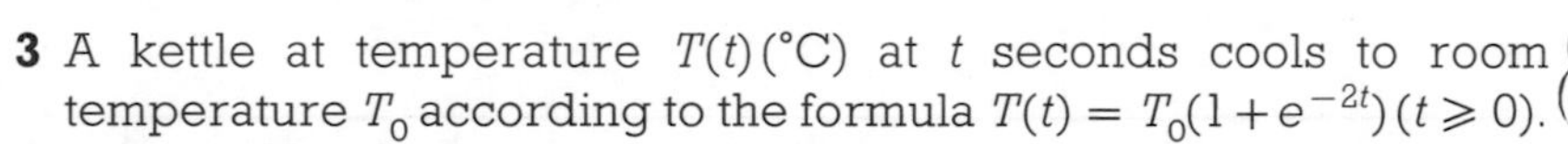

3 A kettle at temperature $T(t)$ (°C) at t seconds cools to room temperature T_0 according to the formula $T(t) = T_0(1 + e^{-2t})\,(t \geqslant 0)$.
a Calculate e^{-2t} for $t = 0, 1, 2$ and 3.
b Write down the value of t, to the nearest whole number, for which, correct to 1 decimal place, $T(t)/T_0 =$ (i) 1·1 (ii) 2.

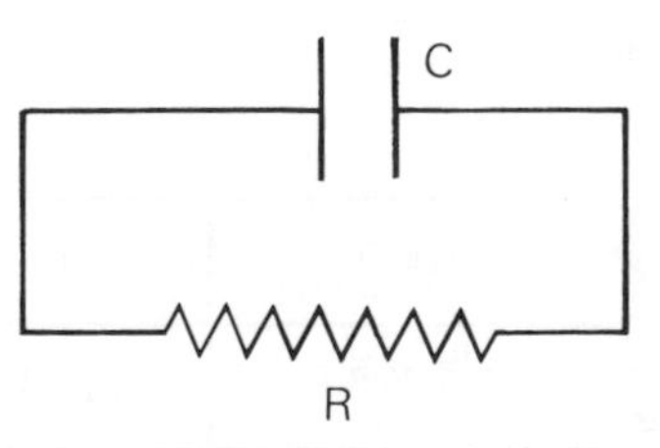

4 A charge q_0 (coulombs) on a condenser of capacity C (farads) is allowed to discharge through a resistor of R ohms. The amount of charge on the condenser at time t (seconds) is $q(t)$, where $q(t) = q_0 e^{-t/RC}$.
a Indicate in a 'rough' sketch how $q(t)$ varies with t.
b Find values of t of the form kRC (k constant) after which the charge is less than:
(i) $\frac{1}{2}q_0$ (ii) $\frac{1}{10}q_0$.

5 This electric circuit contains a resistor (R ohms), an inductance (L Henrys) and a source of electromotive force of E volts. If $i(t)$ (amperes) is the current flowing in the circuit at time t seconds and i_0 is the initial current (at $t = 0$), then

$$i(t) = \frac{E}{R} + \left(i_0 - \frac{E}{R}\right)e^{-(R/L)t}.$$

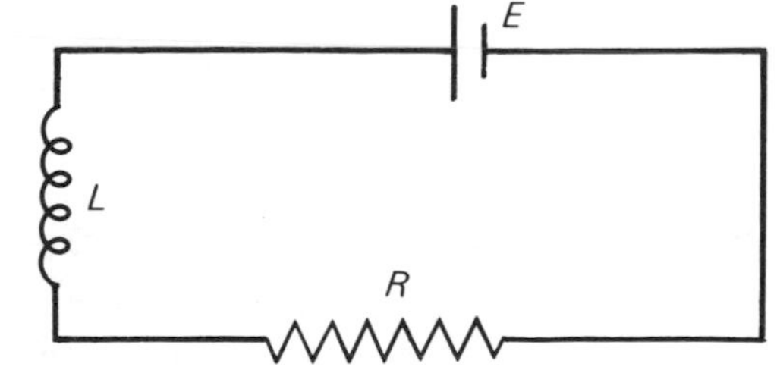

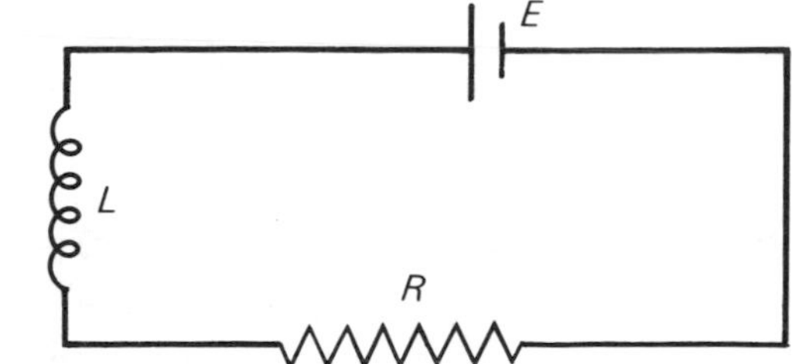

a Check that $i(0) = i_0$.
b What happens to $i(t)$ as $t \to \infty$?
c Indicate in a 'rough' sketch how $i(t)$ varies with t.

A new key to open a new door

Find the e^x and ln keys on your calculator, then copy and complete:

x	0	0·5	1	1·25	1·5	1·75
e^x						

x	1	2	3	4	5	6
$\ln x$						

Draw the graphs $y = e^x$ and $y = \ln x$ on the same sheet of 2 mm squared paper, using a scale of 2 cm to 1 unit on each axis.
Draw the line of symmetry. What is its equation?
From Chapter **6**, Functions, the symmetry should tell you something about the functions e^x and $\ln x$. Investigate the connection between them.

THE LOGARITHMIC FUNCTIONS

In Chapter **6** you saw that, if a function f sets up a one-to-one correspondence between its domain A and range B, then it has an inverse function f^{-1} from B to A. Also $y = f(x) \Leftrightarrow x = f^{-1}(y)$.

For the exponential function $f(x) = a^x (a > 1)$, the domain is R (the x-axis) and the range is $\{y \in R : y > 0\}$, the positive y-axis, often denoted by R^+.

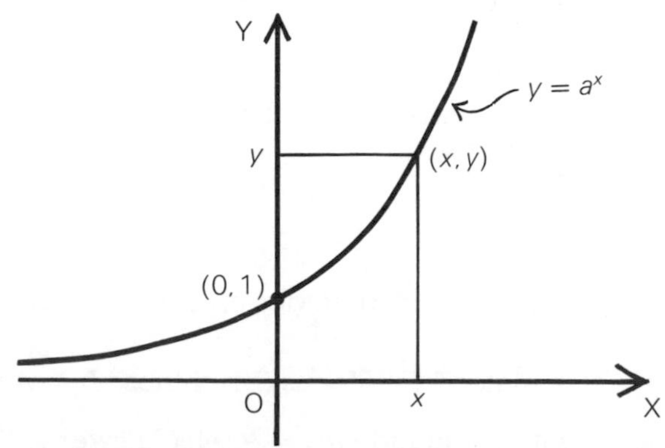

For each $y \in R^+$ there is exactly one $x \in R$ with $y = a^x$. $f(x) = a^x$ has an inverse function. It is denoted by $\log_a$ and read as 'log (or logarithm) to the base a'. So

$$y = a^x \Leftrightarrow x = \log_a y.$$

Note $1 = a^0$, so $\log_a 1 = 0$; $a = a^1$, so $\log_a a = 1$.

$\log_e x$ is often denoted by $\ln x$ and called the *natural logarithm* of x $(x > 0)$.
Before calculators and computers became readily available many calculations were carried out using the function $\log_{10} x$, i.e. logarithms with the decimal base 10. Many years were spent constructing tables of logarithms for these calculations.
When $\log x$ is used without mention of a base you should always know what the base is or be aware that the result involved is independent of the base.
It is worth remembering that a logarithm is an *index*.

Exercise 5

Making use of $y = a^x \Leftrightarrow x = \log_a y$.

1 Write in logarithmic form. *For example,* $2^4 = 16 \Leftrightarrow \log_2 16 = 4$.
a $3^4 = 81$ **b** $10^3 = 1000$ **c** $9^{1/2} = 3$ **d** $5^{-2} = \frac{1}{25}$ **e** $7^0 = 1$.

2 Simplify. *For example,* $\log_2 8 = x \Leftrightarrow 2^x = 8 = 2^3$, so $x = 3$.
a $\log_2 4$ **b** $\log_2 16$ **c** $\log_3 9$ **d** $\log_5 125$ **e** $\log_3 3$
f $\log_4 16$ **g** $\log_8 2$ **h** $\log_2(\frac{1}{2})$ **i** $\log_3(\frac{1}{9})$ **j** $\log_{25}(\frac{1}{5})$.

3 Change to logarithmic form. *For example,* $s = a^3 \Leftrightarrow \log_a s = 3$.
a $y = 10^2$ **b** $x = e^2$ **c** $v = e^{-1}$ **d** $u = 10^{-10}$
e $V = x^3$ **f** $p = q^{1/2}$ **g** $c = a^b$ **h** $x = y^{-4}$.

4 Change to exponential form. *For example,* $2x = \log_a u \Leftrightarrow u = a^{2x}$.
a $x = \log_e y$ **b** $x = \log_{10} p$ **c** $t = \log_5 w$ **d** $2u = \log_a v$.

5 If two earthquakes have energy levels E_1 and E_2, and Richter numbers M_1 and M_2, then $\log_{10} \dfrac{E_1}{E_2} = 1{\cdot}5\,(M_1 - M_2)$.
The San Francisco earthquake of 1906 (energy level E_1) had Richter number 8·3. The Whittier earthquake of 1987 (energy level E_2) had Richter number 6·1. Use the formula to calculate $\dfrac{E_1}{E_2}$.

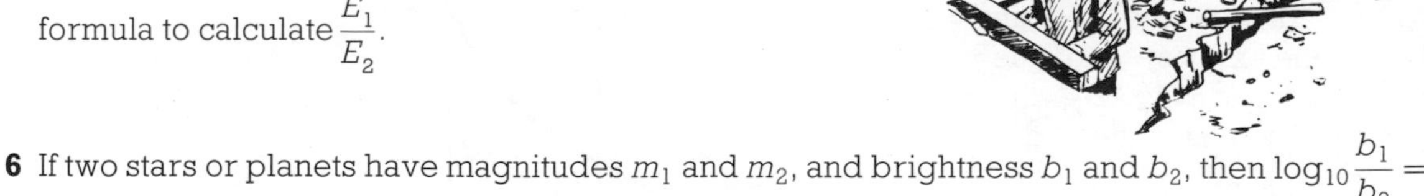

6 If two stars or planets have magnitudes m_1 and m_2, and brightness b_1 and b_2, then $\log_{10} \dfrac{b_1}{b_2} = 0{\cdot}4\,(m_2 - m_1)$. Calculate $\dfrac{b_1}{b_2}$ for Venus (brightness b_1), with magnitude $m_1 = -4{\cdot}4$, and Sirius (brightness b_2), with magnitude $m_2 = -1{\cdot}4$. (Answer to the nearest whole number.)

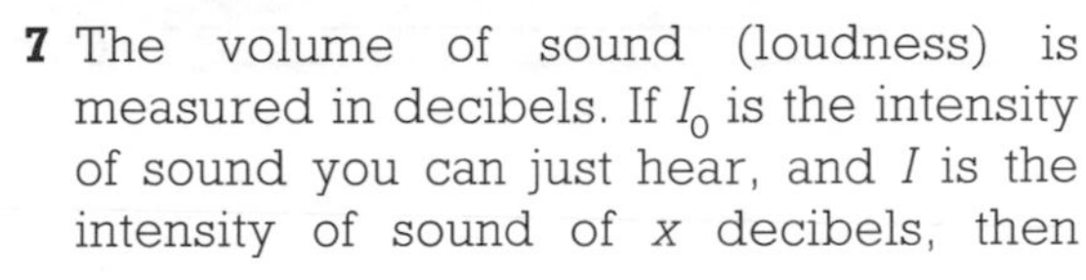

7 The volume of sound (loudness) is measured in decibels. If I_0 is the intensity of sound you can just hear, and I is the intensity of sound of x decibels, then

$$x = 10\log_{10}\frac{I}{I_0}.$$

The Big Blaster disco boasts 100 decibels. Compare its sound intensity (I) with I_0.

PROPERTIES OF LOGARITHMS

The graph $y = \log_a x$ is the mirror image of $y = a^x$ in the line $y = x$. From this you can answer the following questions:

a Is $\log_a x$ increasing?

b What happens to $\log_a x$ as:

(i) $x \to \infty$

(ii) $x \to 0+$ (i.e. $x \to 0$ through positive values)?

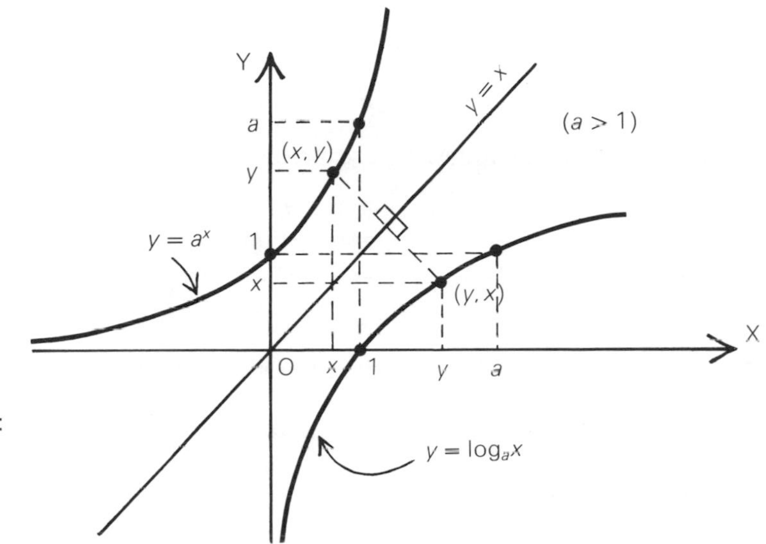

The exponential function a^x has a lot of index properties.
For example, $a^m \times a^n = a^{m+n}$.
What property of $\log_a x$ corresponds to this?
Follow this argument:
$\log_a(xy) = \log_a(a^m \times a^n) = \log_a a^{m+n} = m+n = \log_a x + \log_a y$.

For any base a, $\boxed{\log(xy) = \log x + \log y.}$

Using similar lines of argument, show that, for any base a, you also have:

$$\log\left(\frac{x}{y}\right) = \log x - \log y$$

$$\log(x^n) = n\log x \; (n \in R).$$

Exercise 6

1 Simplify each of these. *For example,*

$\log_2 4 + \log_2 6 - \log_2 3 = \log_2\left(\frac{4 \times 6}{3}\right) = \log_2 8 = \log_2 2^3 = 3.$

a $\log_3 3 + \log_3 3$ **b** $\log_{10} 5 + \log_{10} 2$ **c** $\log_5 7 + \log_5 \frac{1}{7}$
d $\log_2 12 - \log_2 6$ **e** $\log_2 20 - \log_2 5$ **f** $\log_{10} 8 + \log_{10} 5 - \log_{10} 4$.

2 Simplify each. *For example,*

$2\log_2 4 - 3\log_2 2 = \log_2 4^2 - \log_2 2^3 = \log_2 \frac{16}{8} = \log_2 2 = 1.$

a $2\log_6 2 + 2\log_6 3$ **b** $2\log_{10} 2 + 2\log_{10} 5$ **c** $2\log_5 10 - 2\log_5 2$
d $\log_3 12 - 2\log_3 2$ **e** $3\log_3 3 + \frac{1}{2}\log_3 9$ **f** $2\log_4 6 + \log_4 10 - \log_4 45$.

3 Solve for x:
a $\log_a x + \log_a 2 = \log_a 10$ **b** $\log_a x - \log_a 5 = \log_a 20$ **c** $\log_a x + 3\log_a 3 = \log_a 9$.

4 Find the flaw in the following argument:

$$\frac{1}{16} < \frac{1}{8}$$
$$\Rightarrow (\tfrac{1}{2})^4 < (\tfrac{1}{2})^3$$
$$\Rightarrow 4\log\tfrac{1}{2} < 3\log\tfrac{1}{2} \quad \text{(for any base)}$$
$$\Rightarrow 4 < 3.$$

5 Take logarithms to base 10 to find the least positive integer n for which $5^n > 10^{12}$.

(Start $5^n > 10^{12} \Rightarrow n\log_{10} 5 > 12\log_{10} 10 \Rightarrow n > \frac{12}{\log_{10} 5}$ and use the **log** key on your calculator for this division.)

6 Find the least positive integer n for which:
a $7^n > 10^{42}$ **b** $3^n > 10^{35}$.

7 How many years will £1000 take to grow to £2000 at 10% compound interest, paid annually?

8 The sum of n terms of the geometric series
$3 + 3 \times 2 + 3 \times 2^2 + 3 \times 2^3 + \ldots + 3 \times 2^{n-1}$ is $3(2^n - 1)$.
How many terms must be added for the sum to exceed one million?

Logarithms and the experimenter

In experimental work, data can often be modelled by equations of the form $y = ax^b$ and $y = ab^x$, with graphs like this:

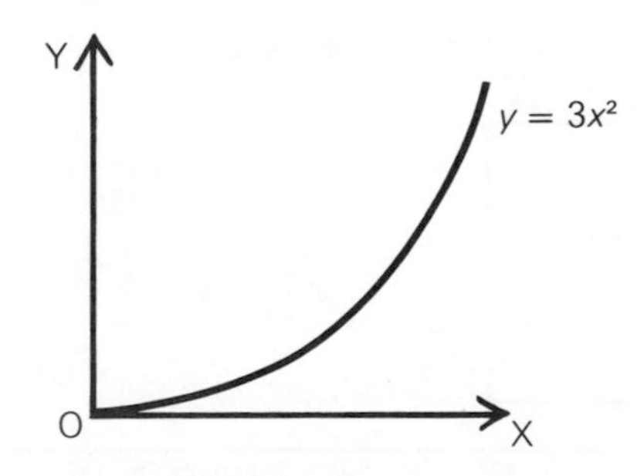

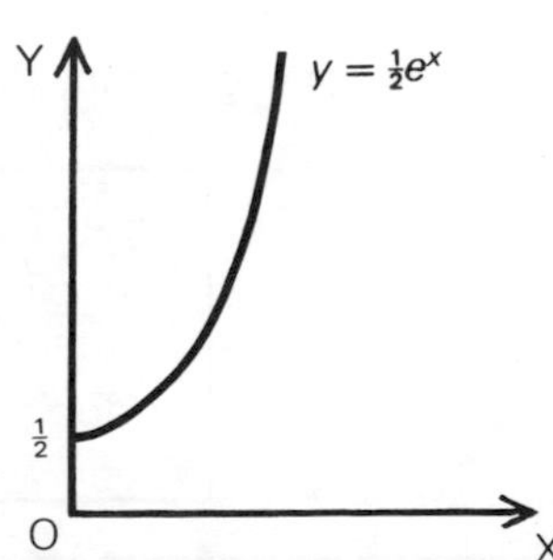

Sometimes it is difficult to know which model to choose.
One way to decide is to take logarithms.

For $y = ax^b$,	For $y = ab^x$,
$\log y = \log a + b \log x$	$\log y = \log a + x \log b$
'$Y = bX + c$'	'$Y = mx + c$'.

The graph of each is a straight line. By drawing the line, b and c, or m and c, can be found, and hence the equations obtained.

Example

Cynthia injects doses of a new stimulant into the leg muscles of a frog. She measures the extension of the muscle, y cm, against dose x and draws up this table of data.

x	1·1	1·2	1·3	1·4	1·5	1·6
y	2·06	2·11	2·16	2·21	2·26	2·30

She suspects that the data can be modelled by an equation of the form $y = ax^b$ with a, b positive constants. She tests this conjecture by taking logarithms to base 10.
$y = ax^b \Rightarrow \log_{10} y = \log_{10} a + b \log_{10} x \Rightarrow Y = bX + c$, where $X = \log_{10} x$, $Y = \log_{10} y$ and $c = \log_{10} a$.
Cynthia has now the much simpler problem of fitting a straight line $Y = bX + c$ to a new table of data of Y against X.

$X = \log_{10} x$	0·04	0·08	0·11	0·15	0·18	0·20
$Y = \log_{10} y$	0·31	0·32	0·33	0·34	0·35	0·36

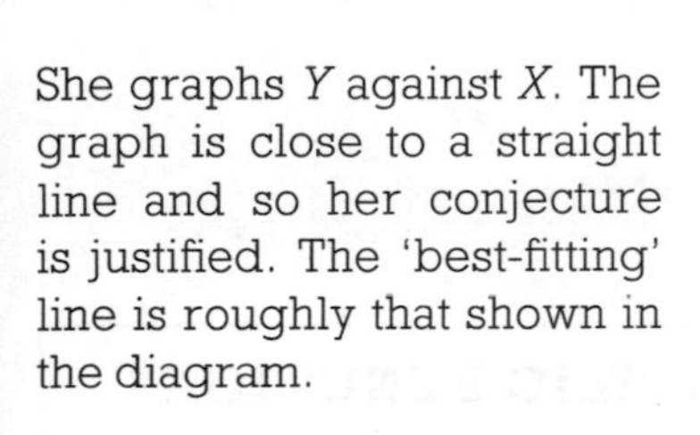

She graphs Y against X. The graph is close to a straight line and so her conjecture is justified. The 'best-fitting' line is roughly that shown in the diagram.

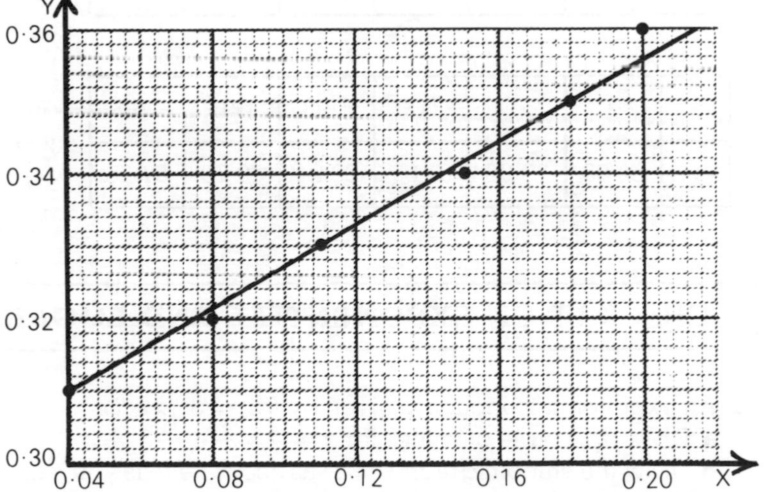

To find the constants b, c in $Y = bX + c$ she takes two points on the best-fitting line. She chooses $X = 0{\cdot}04$, $Y = 0{\cdot}31$ and $X = 0{\cdot}18$, $Y = 0{\cdot}35$.

$$\left.\begin{aligned} 0{\cdot}31 &= 0{\cdot}04b + c \\ 0{\cdot}35 &= 0{\cdot}18b + c \end{aligned}\right\} 0{\cdot}14b = 0{\cdot}04, \text{ so } b = 0{\cdot}29.$$

Then $c = 0{\cdot}30$.
Finally, $c = \log_{10} a = 0{\cdot}30$, so $a = 10^{0{\cdot}30} = 2{\cdot}00$ (correct to two decimal places).
Cynthia's model is approximately $y = 2x^{0{\cdot}29}$.

Exercise 7

1 By taking logarithms, express each equation in the form '$Y = a + bX$'.
a $y = x^6$ **b** $y = 3x^4$ **c** $y = 2x^{1/2}$ **d** $y = 1{\cdot}2x^{-4}$.

2 Express each equation in the form '$Y = a + bx$'.
a $y = 2e^x$ **b** $y = 3 \times 10^x$ **c** $y = 1{\cdot}3e^{-x}$ **d** $y = ke^{2x}$.

In questions **3–6**, find models of the form shown for the given data:

3 ; $y = x^b$

x	2·00	2·01	2·02	2·03	2·04
y	2·83	2·85	2·87	2·89	2·92

4 ; $y = b^x$

x	3·1	3·3	3·5	3·7	3·9	4·1
y	9·97	11·57	13·42	15·57	18·06	20·95

5 ; $y = ax^b$

x	10	20	30	40	50	60
y	9·49	13·42	16·43	18·98	21·21	23·24

6 ; $y = ab^x$

x	2·15	2·13	2·00	1·98	1·95	1·93
y	83·33	79·93	64·89	62·24	59·70	57·26

CHECK-UP ON **EXPONENTIAL AND LOGARITHMIC FUNCTIONS**

1 An exponential function is given by $f(x) = 3^x (x \in R)$.
a Draw up a table of values for $x = -5, -4, -3, -2, -1, 0, 1, 2, 3, 4, 5$.
b Draw the smooth graph $y = 3^x$ as accurately as you can.
c Use the graph to estimate: (i) $3^{2{\cdot}3}$ (ii) $\sqrt{3}$ (iii) $3^{-0{\cdot}3}$, and check by calculator.
d What happens to 3^x as: (i) $x \to \infty$ (ii) $x \to -\infty$?

2 The world population in 1987 was 5 billion (5×10^9). If we assume that it doubles every 80 years, show that the function $P(t) = (5 \times 10^9)2^{t/80} (t \geqslant 0)$ models the population growth from 1987 (t in years). Use the model to:
a estimate the population in the year 2027
b find the year in which the population will reach 6 billion.

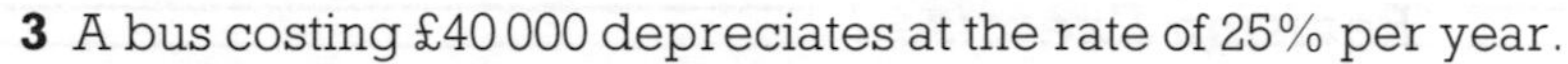

3 A bus costing £40 000 depreciates at the rate of 25% per year.
a Find the function that models its value after t years.
b How old is it when its value is for the first time $< £4000$?

4 Simplify: **a** $\log_2 16$ **b** $\log_3 27$ **c** $\log_9 3$ **d** $\log_{16} \frac{1}{4}$.

5 Convert these equations to equations involving logarithms with suitable bases:
a $y = 5^3$ **b** $u = e^{-2}$ **c** $w = v^{3/4}$ **d** $V = 10^{-t^2}$.

6 Convert these statements to statements involving suitable exponential functions:
a $2 = \log_e y$ **b** $3v = \log_{10} p$ **c** $\log_\alpha(2w) = \frac{1}{2}$ **d** $x^3 = \log_a y$.

7 Simplify:
a $\log_3 4 + \log_3 6 - \log_3 8$ **b** $\log_5 50 - \log_5 2$ **c** $2\log_3 12 - 4\log_3 2$.

8 Solve these equations for y in terms of x:
a $\log_a 3 + \log_a y = x$ **b** $\log_e y - 2\log_e x = 1$.

9 Find the least positive integer n for which $9^n > 10^{30}$.

10 The atmospheric pressure at different heights above sea-level is given by $P = P_0 e^{-0{\cdot}034x}$, where P_0 is the pressure at sea-level and x is the height in thousands of feet.
Find the heights at which P equals: **a** $\frac{1}{2}P_0$ **b** $\frac{1}{4}P_0$ **c** $\frac{1}{10}P_0$.

11 Find a model of the form $y = ab^x$ for this set of data:

x	1·5	2	2·5	3	3·5
y	18·44	34·99	66·39	125·97	239·01

Revision Exercise 1

1 A circle has equation $x^2+y^2-2x+4y=0$.
a Give the coordinates of its centre, and the length of its radius.
b What are the coordinates of the points where the circle cuts the x- and y-axes?
c Show the circle in a coordinate sketch, along with the line $x+y=2$.
d Find the coordinates of the points where this line cuts the circle.

2 Find the real factors of $2x^3-3x^2+x-6$.

3 A rubber ball, when dropped on smooth concrete, always rebounds 90% of the distance it falls. It is dropped from a height of 6 metres.
a What is its height at the top of its third bounce?
b Write down an expression (unsimplified) for the distance it has travelled to that point.

4 If $y(x)$ is such that $\dfrac{dy}{dx}=5-6x^2$ and $y(-1)=4$, find $y(x)$.

5 Find the equation of the tangent to the curve $y=x+\dfrac{2}{\sqrt{x}}$ at the point where $x=4$.

6 Interpret $\int_0^1 \sqrt{(1-x^2)}\,dx$ as the area of a region in the x, y-plane, and hence state the value of the integral.

7 For each graph $y=f(x)$, draw separate sketches to show the graphs of:
a $y=f(x)+1$
b $y=-f(x)$.

(i)
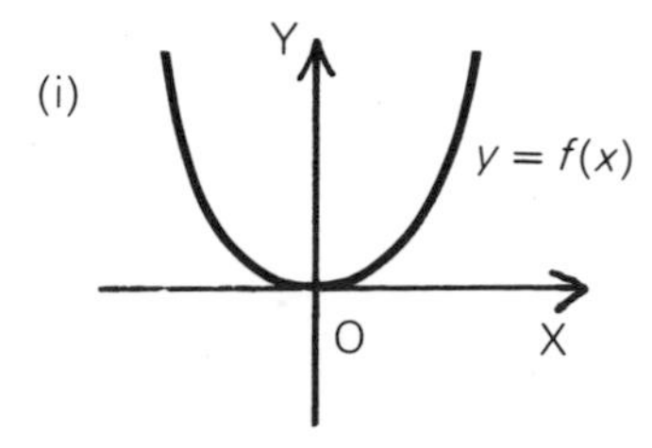

(ii)
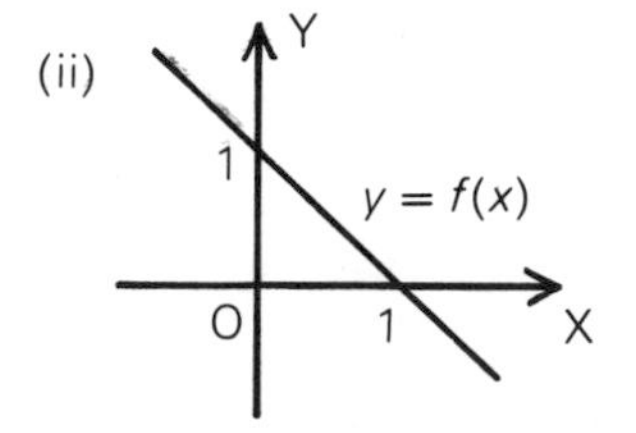

(iii)
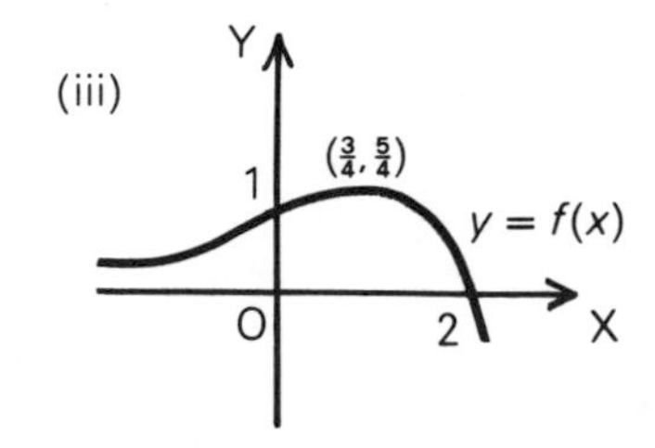

8 Write each of the following in the form $a(x-p)^2+b$, and then give its maximum or minimum value and the corresponding value of x.
a x^2+2x+5
b $2x^2-8x+1$
c $4-6x-x^2$.

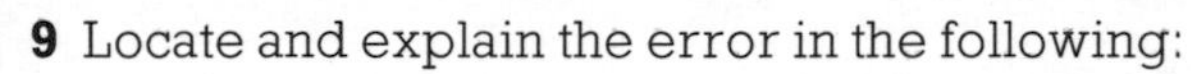

9 Locate and explain the error in the following:

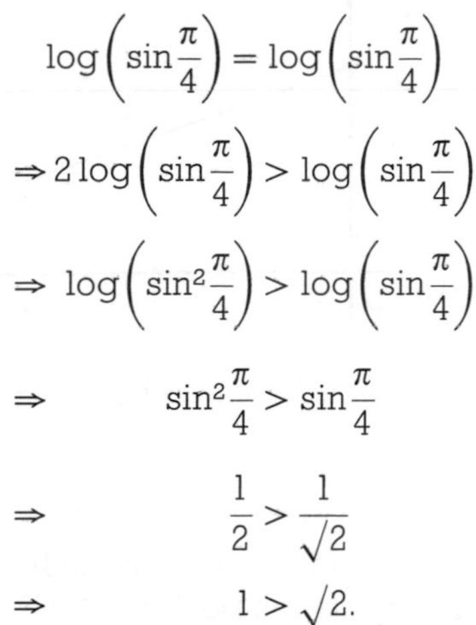

$$\log\left(\sin\frac{\pi}{4}\right) = \log\left(\sin\frac{\pi}{4}\right)$$

$$\Rightarrow 2\log\left(\sin\frac{\pi}{4}\right) > \log\left(\sin\frac{\pi}{4}\right)$$

$$\Rightarrow \log\left(\sin^2\frac{\pi}{4}\right) > \log\left(\sin\frac{\pi}{4}\right)$$

$$\Rightarrow \sin^2\frac{\pi}{4} > \sin\frac{\pi}{4}$$

$$\Rightarrow \frac{1}{2} > \frac{1}{\sqrt{2}}$$

$$\Rightarrow 1 > \sqrt{2}.$$

10 If $f(x) = 2x^2+1$ and $g(x) = 5-3x$, find formulae for the functions $f \circ g$ and $g \circ f$.

11 Assuming the formulae $\sin(A+B) = \sin A\cos B + \cos A\sin B$ and $\cos(A+B) = \cos A\cos B - \sin A\sin B$, prove that $\sin 3A = 3\sin A - 4\sin^3 A$.
Hint $3A = 2A + A$.

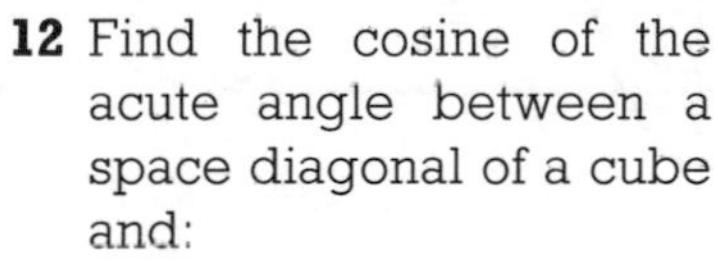

12 Find the cosine of the acute angle between a space diagonal of a cube and:
a one of its edges
b the diagonal of a face.

13 The points A(1, a, 2), B(3, 4, 5), C(2, −2, 3) and D(b, 2, −1) are such that AB is perpendicular to CD.
Find the relation between a and b.

14 For which values of x is the function $f(x) = 2x^3+3x^2-12x+1$:
a stationary **b** increasing **c** decreasing?

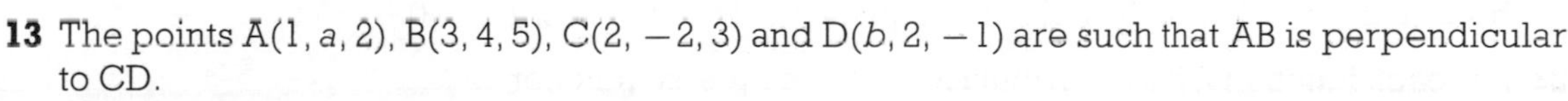

15 The current I in an electric circuit is given by $I = 3\sin 200t° - 4\cos 200t°$.
Express I in the form $R\cos(200t-\alpha)°$, giving R and α correct to 2 decimal places ($R > 0$).

16 △ABC has vertices A(−8, −4), B(−2, 4) and C(4, 2). By finding the equations of the perpendicular bisectors of the sides of the triangle, show that they are concurrent.

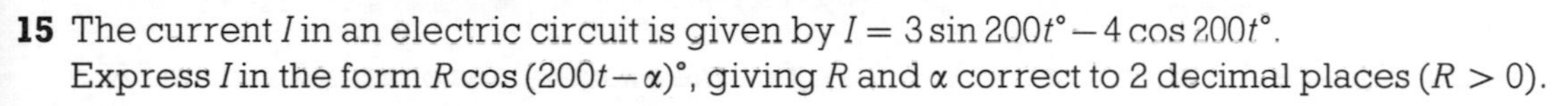

17 Differentiate $\sin(3x+1)+(2-3x)^{1/3}$.

18 Find the solutions in $[0, 2\pi]$ of the equation $3\sin\theta - 2\cos 2\theta + 1 = 0$, giving them correct to 2 decimal places, where appropriate.

19 Find the values of t for which $\mathbf{a} = t\mathbf{i}+\frac{1}{2}\mathbf{j}+\frac{1}{3}\mathbf{k}$ is a unit vector.

20 **a** Write down the expansions of $\cos(A+B)$ and $\cos(A-B)$ and show that $\cos(A-B)-\cos(A+B) = 2\sin A\sin B$.

b Evaluate $\displaystyle\int_0^{\pi/4} \sin x \sin 3x\, dx$.

WHAT DO YOU KNOW?

21 The graph illustrates the law $Y = kX^n$.
Find the values of k and n, correct to 1 decimal place.

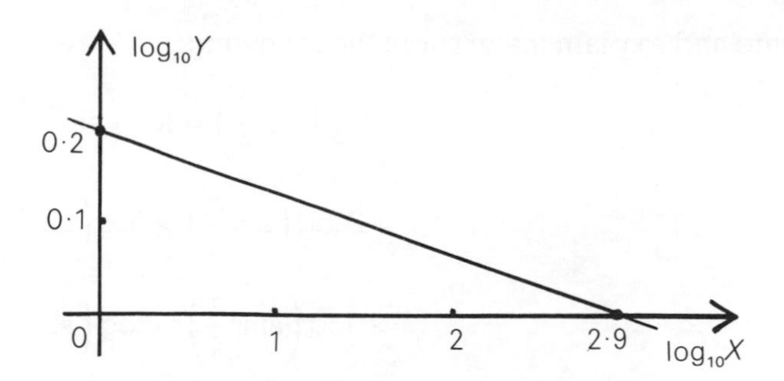

22 What is the domain of the real function f defined by $f(x) = 3+\sqrt{x}$, and what is the range of the function?

23 A population is 5 000 000 in 1980 (31st December) and increases by 500 000 each year.
a What will the population be at the end of the year 1998?
b After how many years will the population be 22 500 000?

24 Find the area of the shaded region bounded by the parabola $y = 3(x^2-1)$ and the line $y = 9$.

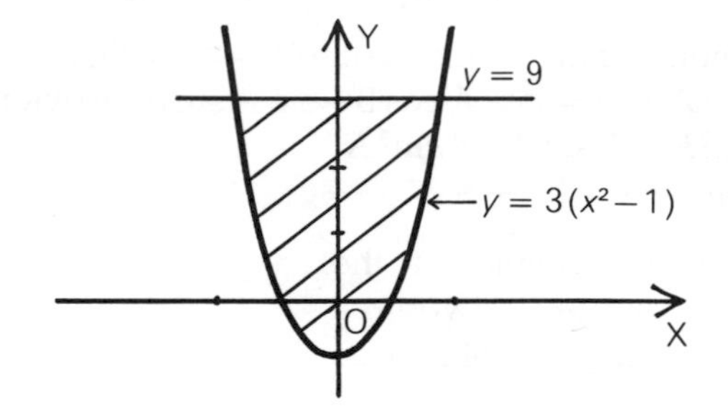

25 A function f is defined on the interval $x \geqslant 0$ by $f(x) = x^3 - 12x$. Find the minimum value of f on this interval.

26 **a** Express $\sqrt{3}\cos x - \sin x$ in the form $R\cos(x-\alpha)$ with R, α constants and $R > 0$.
b Find the solutions in $[0, 2\pi]$ of the equation $\sqrt{3}\cos x - \sin x + 1 = 0$.

27 For each function f, find a formula for f^{-1}, the inverse function of f.

a $f(x) = 3x-4$ **b** $f(x) = x^3+1$ **c** $f(x) = \dfrac{1}{x+1}$ **d** $f(x) = \dfrac{2x-3}{4x-5}$.

28 Find the values of a for which the line $y = x+a$ is a tangent to the circle $x^2+y^2 = 2$.

29 If the acute angle A is such that $\cos A = \dfrac{1}{2\sqrt{2}}$, find the exact values of $\cos 2A$ and $\sin 2A$.

30 A sphere is expanding in such a way that its radius is increasing at the rate of 1 cm/s. How fast is its volume expanding when its radius is 10 cm?

Hint Use the formula $V = \dfrac{4\pi}{3}r^3$.

Revision Exercise 2

1 **a** State the coordinates of the centre of the circle $x^2+y^2-4x+6y-37 = 0$.
b Verify that the point A(3, 4) lies on this circle, and find the equation of the tangent at A to the circle.
c Find also the equation of the parallel tangent.

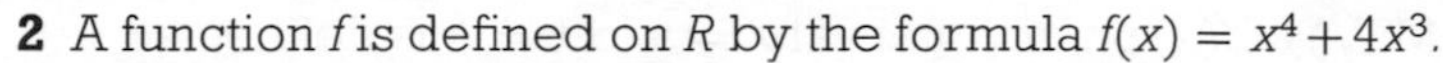

2 A function f is defined on R by the formula $f(x) = x^4 + 4x^3$.

a Determine the stationary values of f and the nature of each.

b Determine the points where the graph of $f(x)$ meets the x-axis.

c Indicate the shape of the graph as $x \to \infty$ and as $x \to -\infty$.

d Sketch the graph.

e For what values of a has the equation $x^4 + 4x^3 = a$ no real roots?

3 A triangle has vertices A(1, 3, 5), B(−4, −2, 0) and C(−1, −3, −3).

a Show that the triangle is obtuse-angled at B.

b Find the coordinates of the points P, Q, R on AB, BC, CA such that AP : PB = 2 : 3, BQ : QC = −3 : 2 and R is the midpoint of AC.

c Show that P, Q, R are collinear.

4 a If $S_n = 1 + \frac{1}{2} + \frac{1}{2^2} + \ldots + \frac{1}{2^{n-1}}$, then

$$\frac{1}{2}S_n = \qquad \frac{1}{2} + \frac{1}{2^2} + \ldots + \frac{1}{2^{n-1}} + \frac{1}{2^n}.$$

Deduce, by subtraction, that $S_n = 2 - \frac{1}{2^{n-1}}$.

b A sequence $\{u_n\}$ is defined by: $u_1 = 1$ and $u_{n+1} = 3 + \frac{1}{2}u_n$ $(n \geqslant 1)$.

(i) Write down unsimplified expressions for u_2, u_3, u_4.

(ii) Show that $u_n = 3\left(1 + \frac{1}{2} + \frac{1}{2^2} + \ldots + \frac{1}{2^{n-2}}\right) + \frac{1}{2^{n-1}}$.

(iii) Deduce that $u_n = 6 - \frac{5}{2^{n-1}}$ $(n \geqslant 1)$.

(iv) State the value of $\lim\limits_{n \to \infty} u_n$.

5 A large screen ABCDEF is being designed with a support strut AE of fixed length a cm. BCDE is a square and AF = $\frac{1}{2}$BE. The screen has to be constructed so that its area A cm² has maximum value.

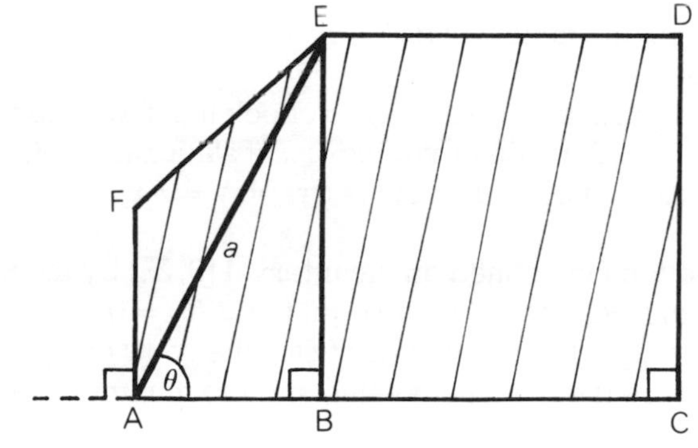

a Show that $A = a^2 \sin^2\theta + \frac{3}{4}a^2 \sin\theta \cos\theta$.

b Express A in the form $R\cos(2\theta - \alpha) + c$ with R, c constants, $R > 0$.

c Deduce the maximum value of A, and obtain the value of θ for which this is attained, correct to 1 decimal place.

6 A block of metal in the form of a cuboid has square ends of side x cm and length y cm. The volume of the block is 8 cm³.

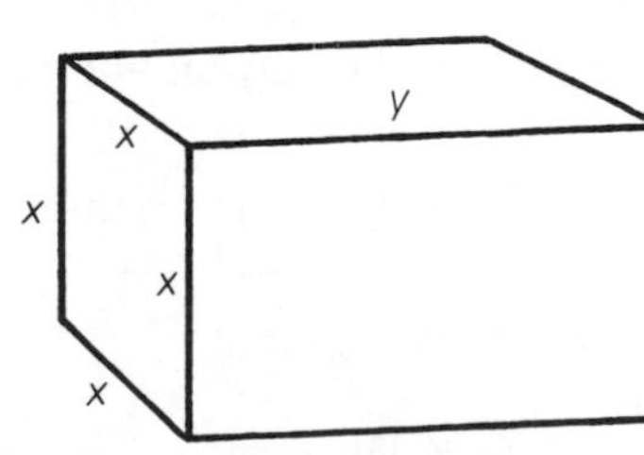

a If S cm² is the total surface area, express S in terms of x.

b Find the value of x for which S is a minimum.

c If, as an additional condition, $1 \leqslant x \leqslant 3$, find the largest possible value of S.

7 The root-mean-square value (RMSV) of $f(x)$ on the interval $x = a$ to $x = b$ is defined as

$$\sqrt{\left(\frac{1}{b-a}\int_a^b [f(x)]^2\,dx\right)}.$$

a Find the RMSV of $\cos x$ over the period $x = 0$ to $x = 2\pi$.
b Find the RMSV of $\sin x + \cos x$ over its period.
c The current, I amperes, in an electric circuit varies according to the relationship $I(t) = 3\sin 300t$, where t (seconds) denotes time from switch-on.
Calculate the RMSV of the current over the first $\frac{1}{150}$th of a second after switch-on, correct to 1 decimal place.

8 A tetrahedron ABCD has vertices A(4, 2, 5), B(2, 0, −1), C(0, −2, 3), D(6, −4, −1).
a Verify that B, C, D lie on the plane $x + y + z = 1$, but that A does not.
b Let M_1, M_2, M_3, M_4, M_5, M_6 be the midpoints of the sides AB, BC, CD, DA, AC, BD respectively. Prove that M_1M_3, M_2M_4 and M_5M_6 meet and bisect each other.
c Find the cosine of the acute angle between M_1M_3 and M_2M_4.

9 In the Richter scale, the intensity M of an earthquake is related to the energy E of the earthquake by the formula $\log_{10}E = 11{\cdot}4 + 1{\cdot}5M$ (E in ergs).
a If one earthquake has a thousand times the energy of another, how much larger is its Richter number M?
b What is the ratio of the energy of the San Francisco earthquake of 1906 ($M = 8{\cdot}3$) to that of the Eureka earthquake of 1980 ($M = 7$)?
c What is the Richter number (to 1 decimal place) of a 10 megaton H-bomb, i.e. an H-bomb whose energy is equivalent to that in 10 million tons of TNT? (Assume that 1 ton of TNT contains $4{\cdot}2 \times 10^6$ ergs.)

10 a Obtain the factors of $f(x) = x^4 + x^3 - 3x^2 - x + 2$, and deduce that the equation $f(x) = 0$ has a repeated root.
b Find $f'(x)$, and use it: (i) to check that the graph $y = f(x)$ has a stationary point at $x = 1$ (ii) to find the other values of x at stationary points.
c Make a rough sketch of the graph $y = f(x)$.

11 A function f is defined on the interval $[0, 2\pi]$ by the formula $f(\theta) = 8\sin^3\theta + 3\cos 2\theta$.
a Form $f'(\theta)$, and solve the equation $f'(\theta) = 0$.
b Draw up a table of values of f for the roots of $f'(\theta) = 0$.
c Deduce the maximum value and the minimum value of f.

12 a Check that the line $y = \dfrac{2}{\pi}x$ meets the curve $y = \sin^2 x$ at $x = 0, \dfrac{\pi}{4}$ and $\dfrac{\pi}{2}$.
b Sketch the line and the curve for $0 \leqslant x \leqslant 2\pi$.
c Find the area of the region enclosed by the line and the curve.

13 a If $S_n = 1 + r + r^2 + \ldots + r^{n-1}$, then
$rS_n = \quad r + r^2 + \ldots + r^{n-1} + r^n$.
Show by subtraction that

$$S_n = \frac{1-r^n}{1-r}\ (r \neq 1).$$

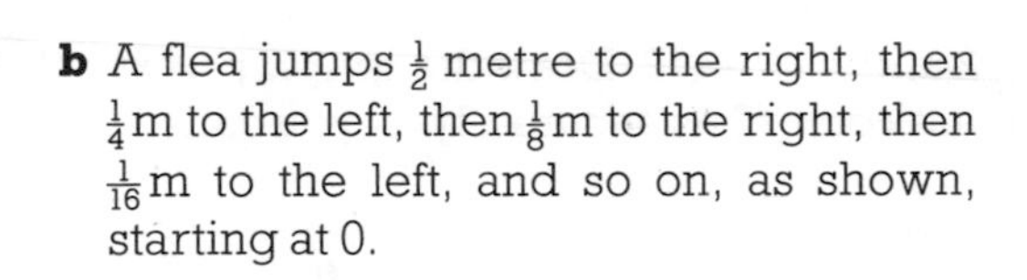

b A flea jumps $\frac{1}{2}$ metre to the right, then $\frac{1}{4}$m to the left, then $\frac{1}{8}$m to the right, then $\frac{1}{16}$m to the left, and so on, as shown, starting at 0.

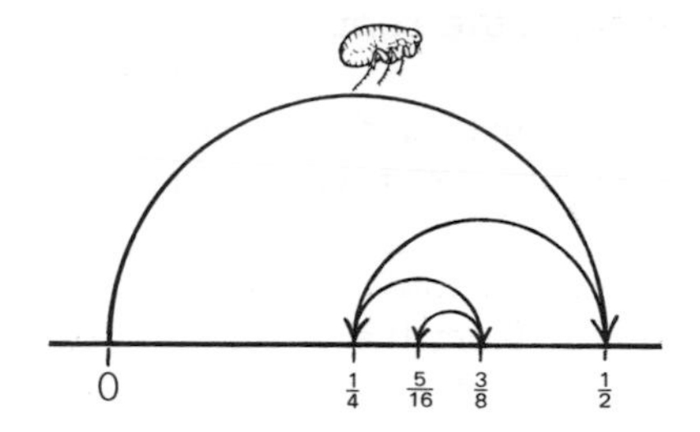

(i) Show that after n jumps, where n is odd, the flea is at the number $\frac{1}{3}\left(1+\frac{1}{2^n}\right)$.

(ii) Show that after n jumps, where n is even, the flea is at the number $\frac{1}{3}\left(1-\frac{1}{2^n}\right)$.

(iii) As the flea keeps on jumping, what single number does it keep jumping over?

14 The equation $\tan x = \frac{1}{x}\,(x > 0)$ occurs in the theory of vibrations.

a Draw a sketch of $y = \tan x$ and $y = \frac{1}{x}$ for $0 < x < \frac{5\pi}{2}$.

b How many roots has the equation?

c The first root lies in the interval $\left(0, \frac{\pi}{2}\right)$. Give intervals for the next two roots of the equation.

d Find an approximation for the first root correct to 2 decimal places.

15 If the selling price of an article is fixed at £p it is estimated that $y = 250 - p$ articles can be sold at that price ($0 \leqslant p \leqslant 250$). Each article costs £10 to manufacture and there is a fixed cost of £100 no matter how many articles are produced.

a Obtain an expression for P if £P is the profit in selling the y articles.

b What price should be charged for the article to obtain maximum profit?

c If the machine involved is such that, when switched on, it has to be set to make 150, 200 or 250 articles, what should the manufacturer do?

1 Getting at the roots

a The equation $x^2-2=0$ has two roots, $x=\sqrt{2}$ and $x=-\sqrt{2}$. Draw the graph $y=x^2-2$ of $f(x)=x^2-2$ on squared paper and estimate the root $\sqrt{2}$ as accurately as you can.

b Isaac Newton invented a more accurate method: (Note, from the graph, that $1<\sqrt{2}<2$.)

(i) Take $x_1=2$ as a first approximation to the root. The tangent at $P_1(2, 2)$ meets the x-axis at $M_2(x_2, 0)$. Use calculus to find the exact value of x_2.

(ii) The tangent at $P_2(x_2, f(x_2))$ meets the x-axis at $M_3(x_3, 0)$. Find the exact value of x_3.

(iii) Obtain the value of x_4 arising similarly from $P_3(x_3, f(x_3))$.

(iv) Explain why it is clear that the sequence $x_1, x_2, x_3, x_4, \ldots$ approaches $\sqrt{2}$. How accurate is x_4 as an approximation to $\sqrt{2}$?

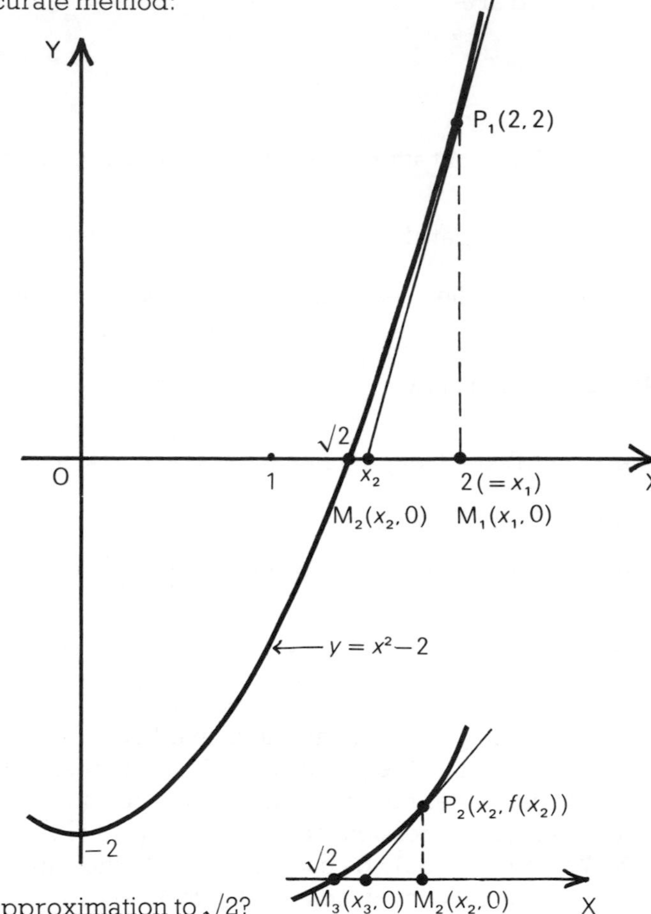

(v) Explain what happens if you start with a first approximation $x_1=1$.

c *Generalising*

x_1 is a first approximation to root a of equation $f(x)=0$. The tangent at $P_1(x_1, f(x_1))$ on the graph $y=f(x)$ meets the x-axis at $M_2(x_2, 0)$. Since P_1M_2 is this tangent, its gradient is $f'(x_1)$, so

$$f'(x_1)=\frac{y_{M_2}-y_{P_1}}{x_{M_2}-x_{P_1}}=\frac{0-f(x_1)}{x_2-x_1}=-\frac{f(x_1)}{x_2-x_1}.$$

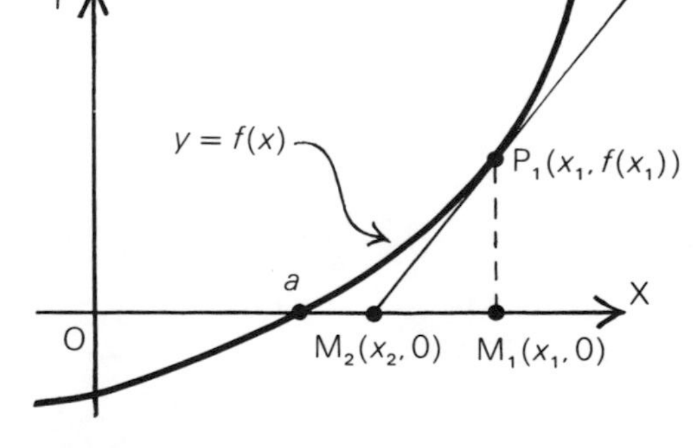

$x_2-x_1=-\dfrac{f(x_1)}{f'(x_1)}$, and $x_2=x_1-\dfrac{f(x_1)}{f'(x_1)}$ is a second approximation to a.

(i) Starting with x_2 you get a third approximation x_3. Write down the equation connecting x_2 and x_3.

(ii) Continuing in this way you get a sequence of approximations $x_1, x_2, x_3, x_4, \ldots$. Write down the recurrence relation that connects x_n and x_{n+1}.

(iii) Obtain the recurrence relation for the sequence in **b** that used $f(x) = x^2 - 2$.

(iv) Use the Newton method to find an approximation for $\sqrt{21}$ correct to 4 decimal places.

(v) If $f(x) = x^3 - x$ and you use $x_1 = \frac{1}{\sqrt{5}}$ as a first approximation for the root $x = 1$ (or $x = 0$) the method fails. Investigate why.

INVESTIGATIONS

2 Approximations for integrals, Simpson's Rule

a There is exactly one parabola $y = ax^2 + bx + c$ through three non-collinear points, for example $(-1, 1)$, $(0, 4)$, $(1, 3)$, no pair lying on a line parallel to the y-axis.

For,
$$1 = a - b + c$$
$$4 = c$$
$$3 = a + b + c,$$
and this system of equations can be solved for a, b, c. Find the parabola.

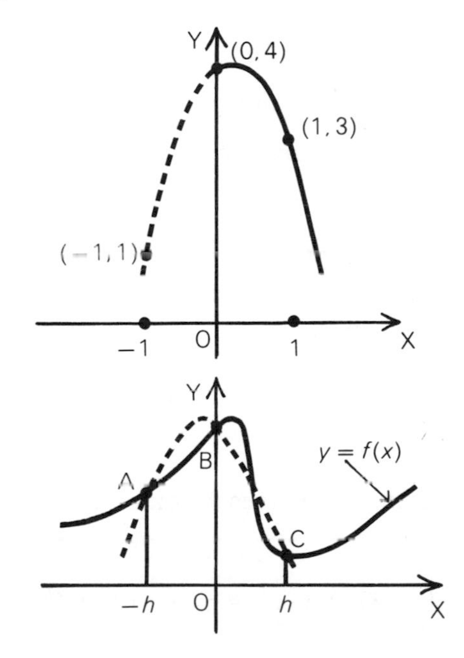

b *Parabolic approximation*

Take the points A$(-h, f(-h))$, B$(0, f(0))$, C$(h, f(h))$ on the graph $y = f(x)$ $(h > 0)$. Let $y = ax^2 + bx + c$ be the parabola through A, B, C.

The area under this parabola from $x = -h$ to $x = h$ is an approximation to the area under $y = f(x)$, i.e. to $\int_{-h}^{h} f(x)\,dx$.

The area under the parabola is $\int_{-h}^{h} (ax^2 + bx + c)\,dx$.

Show that this is $\frac{h}{3}(2ah^2 + 6c)$.

But
$$f(-h) = ah^2 - bh + c$$
$$f(0) = c$$
$$f(h) = ah^2 + bh + c.$$
Use these facts to deduce that

$$\int_{-h}^{h} f(x)\,dx \doteqdot \frac{h}{3}[f(-h) + 4f(0) + f(h)], \ldots (1)$$

c *Estimating* $\int_a^b f(x)\,dx$

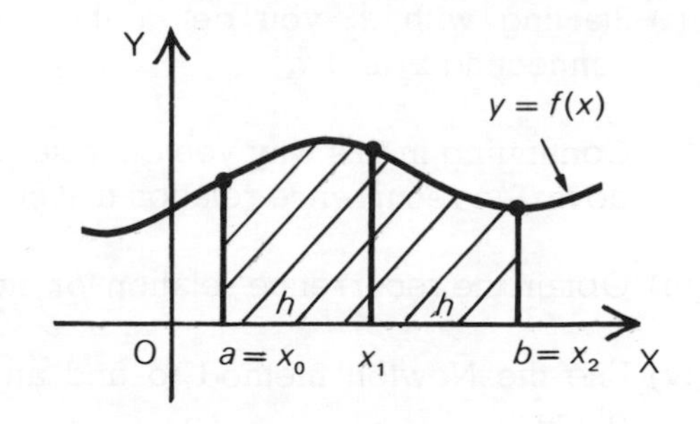

Divide the interval $[a, b]$ into two subintervals of equal length $h = \dfrac{b-a}{2}$, using points $x_0 = a$, $x_1 = a+h$, $x_2 = b$.
Using (1), with a shift of origin,

$$\int_a^b f(x)\,dx \doteqdot \frac{h}{3}[f(a)+4f(a+h)+f(b)].$$

With four subintervals of equal length,

$h = \dfrac{b-a}{4}$ and the points of division are

$x_0 = a$, $x_1 = a+h$,
$x_2 = a+2h$,
$x_3 = a+3h$,
$x_4 = b$.
Explain why

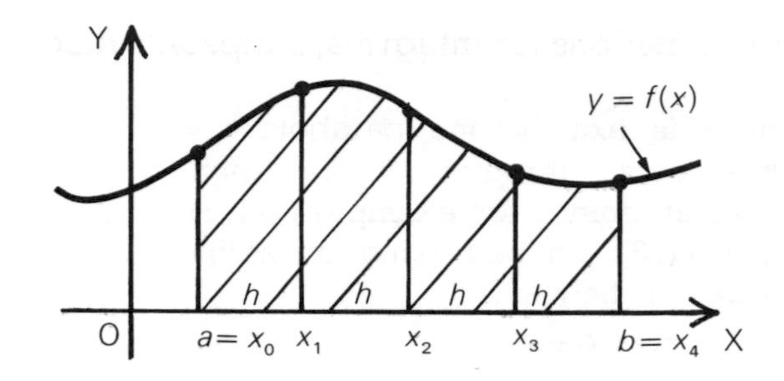

$$\int_a^b f(x)\,dx \doteqdot \frac{h}{3}[f(a)+4f(a+h)+2f(a+2h)+4f(a+3h)+f(b)].$$

Simpson's Rule is the extension of this to $2n$ equal subintervals. Here $h = \dfrac{b-a}{2n}$, and

$$\int_a^b f(x)\,dx \doteqdot \frac{h}{3}[f(a)+4f(a+h)+2f(a+2h)+4f(a+3h)+2f(a+4h)$$
$$+\ldots+2f(a+(2n-2)h)+4f(a+(2n-1)h)+f(b)].$$

d Estimate $\int_0^1 x^4\,dx$ using Simpson's Rule with $2n = 2, 4, 6, \ldots$, and discuss the accuracy of the approximations.

e Now consider $\int_0^1 \dfrac{dx}{x+1}$ $(= \log_e 2 \doteqdot 0{\cdot}6931472)$.

Investigate the number of strips needed in Simpson's Rule to obtain an approximation for the integral correct to 7 decimal places.

f The Simpson's Rule estimate is exact (for every n) for $f(x)$ a polynomial of degree $\leqslant 3$. Try to prove this. *Hint* See **b**.

3 Arithmetic progressions (APs)

The sequence 1,3,5,7,9,11,13 is an *arithmetic progression* (AP) of length 7, each term differing from the previous one by 2.
1,5,9,13 is a *subsequence* of the above and is also an AP, of length 4 and common difference 4.
In general an AP takes the form

$$a, a+d, a+2d, a+3d, \ldots,$$

where a is the first term and d is the common difference.

a *APs consisting of prime numbers*
3, 13, 23 is an AP consisting of primes only.
(i) Explain why you cannot have an AP of primes with more than two members if the first term is 2.
(ii) Investigate APs of primes < 100. What is the longest sequence you can find?
(iii) Can there be an AP of primes of infinite length?

Primes <100

2	3	5	7	11
13	17	19	23	29
31	37	41	43	47
53	59	61	67	71
73	79	83	89	97

b *Subsequences*
(i) Explain why the sequence of odd integers 1,3,5,7,9,11, . . . has lots of AP subsequences of infinite length.
(ii) The sequence $1, 2, 2^2, 2^3, 2^4, \ldots$ (of powers of 2) has no AP subsequence with more than two members. Can you prove this?

c *Greedy sequences*
An *n-greedy sequence* is a sequence of increasing integers starting with 1 which has *no subsequence* forming an AP with more than n members. When building up such a sequence each integer, starting at 1, should be considered in turn and rejected if it belongs to a subsequence with more than n members forming an AP. For example, the 2-greedy sequence can be found as follows:

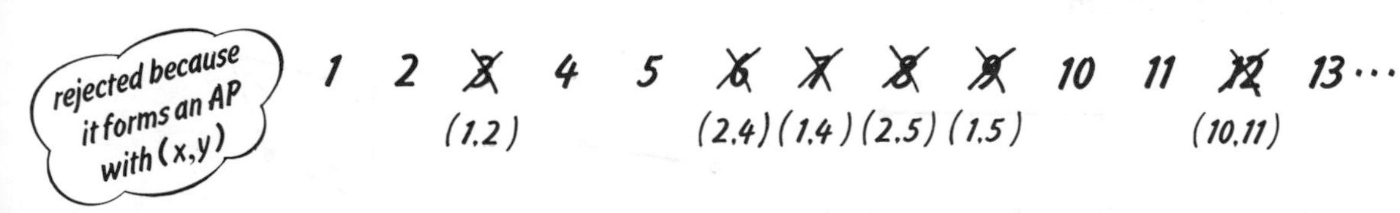

(i) Investigate the 2-greedy sequence further finding more terms and any rejection pattern.
(ii) The 3-greedy sequence starts 1,2,3,5,6,8,9, ... Investigate further.

4 Triangles with consecutive sides

The lengths of the sides of $\triangle ABC$ are three consecutive positive integers: $n-1$, n, $n+1$ units.

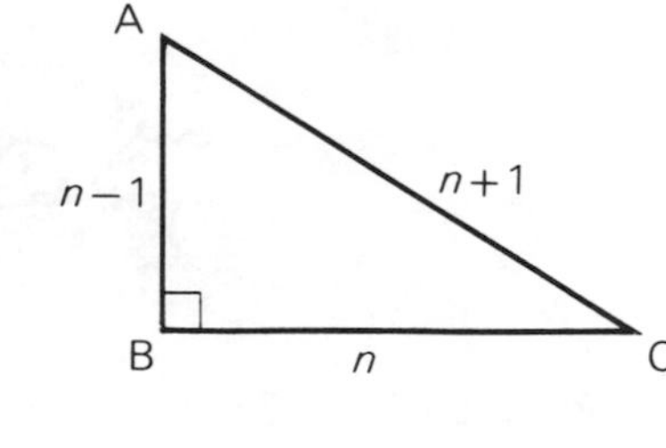

a Explain why $n \geqslant 3$.

b $\triangle ABC$ is right-angled. Use Pythagoras' Theorem to investigate this case.

c $\triangle ABC$ is obtuse-angled. Use the cosine rule to investigate this case.

$$\cos B = \frac{a^2 + c^2 - b^2}{2ac}$$

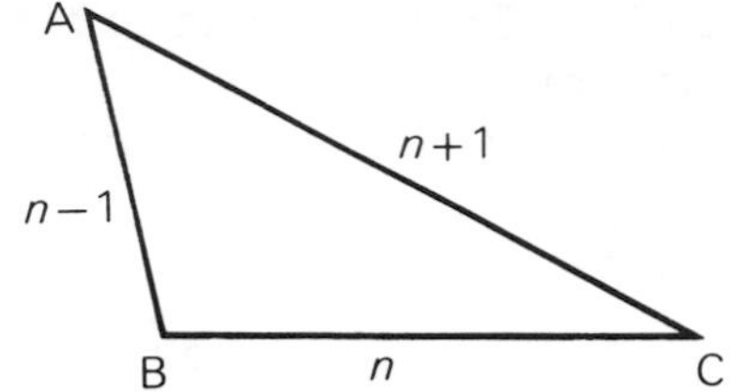

d $\triangle ABC$ is acute-angled.

(i) Explain why $n \geqslant 5$. (The formula for $\cos B$ may be helpful.)

(ii) Investigate what happens to $\angle B$ as $n \to \infty$.

(iii) For each triangle, (1) and (2), calculate $x-y$. (Pythagoras could be useful.) Investigate $x-y$ for other triangles of your choice.

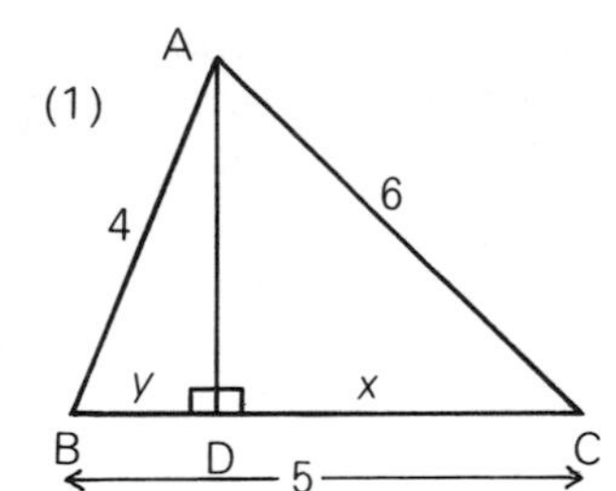

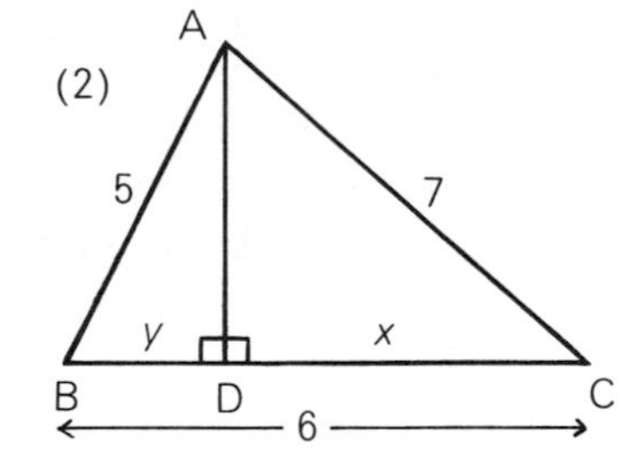

(iv) What result is suggested by (iii)? Prove it for the general $n-1$, n, $n+1$ triangle.

(v) Find an expression for the area ($\triangle$) of the $n-1$, n, $n+1$ triangle.

(vi) Investigate the existence of such triangles with $\triangle$ an integer.

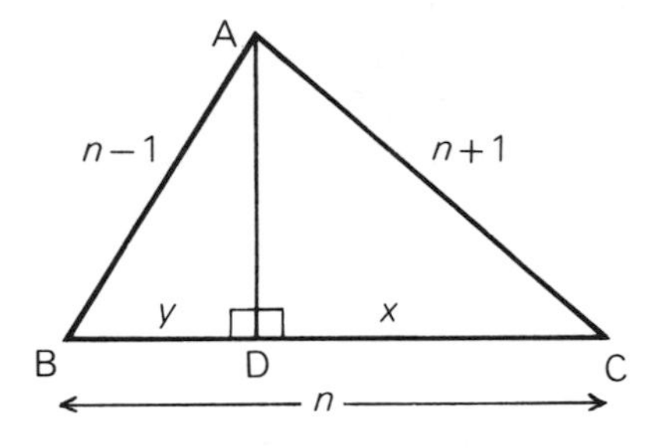

5 The pizza problem

a

2 objects produce 1 pair of objects,
3 objects produce 3 pairs,
4 objects produce 6 pairs.
Investigate the number of pairs of objects produced by n objects.

b A *plane graph* is a set of points (*vertices*) in a plane with some (or all) pairs joined by arcs (*edges*).

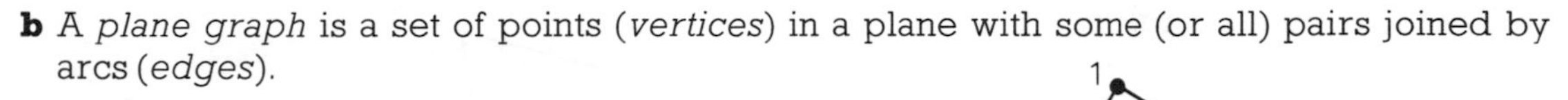

This graph has 4 vertices, 5 edges and 2 bounded *regions* (enclosed by edges and not overlapping). Vertices 1 and 3 can be joined by an arc as in (1), (2) or (3):

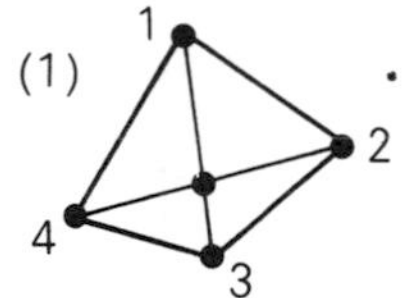

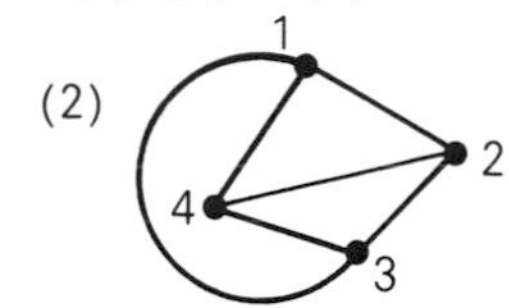

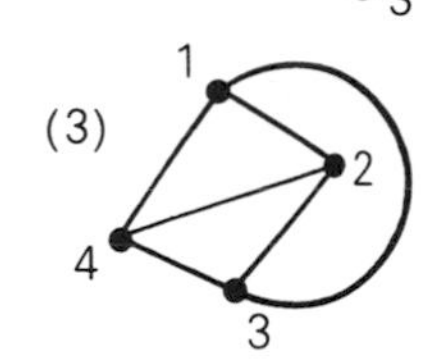

with at most 1 more vertex.

(i) For each of the above graphs and others of your own construction, count the numbers of vertices (V), edges (E) and regions (R).

(ii) In each case compare $V+R$ with E. Write down an equation involving V, E and R. This is called *Euler's Formula*—assume it is true for any plane graph.

c *Pieces of pizza*

A circular pizza is being cut into pieces with straight cuts. The problem is to find the *maximum number of pieces* P_n, from n cuts.

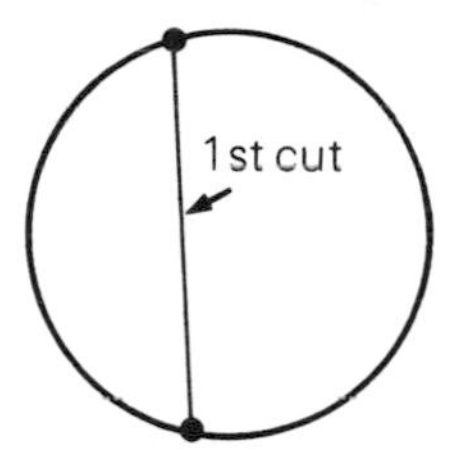

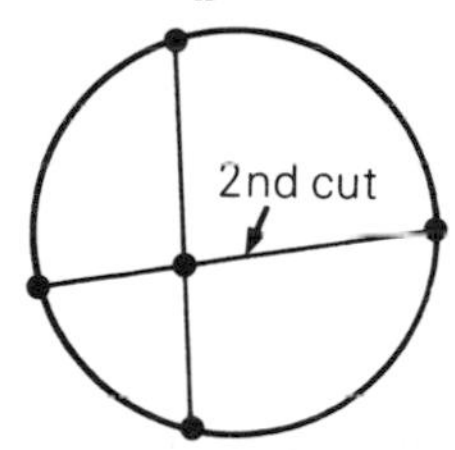

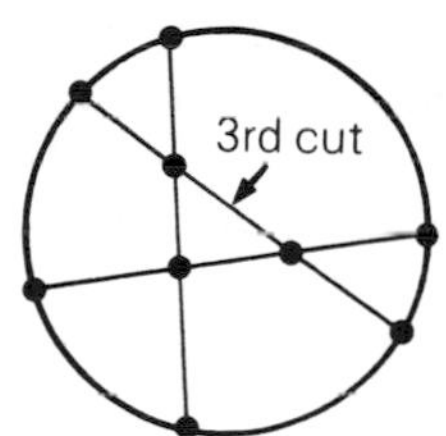

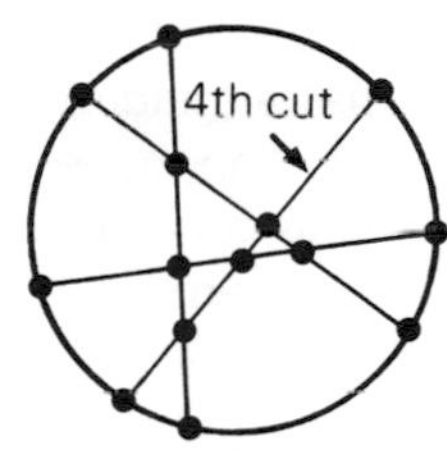

Each cutting pattern can be regarded as a plane graph, and P_n is the number of regions for n cuts.

To obtain P_n, each new cut must cross every previous cut and not pass through any existing vertex.

(i) State the values of P_1, P_2, P_3 and P_4.

(ii) Make a suitable fifth cut. Count V and E, and then find R from Euler's formula. Count P_5. Does this value agree with R?

(iii) Let V_n be the number of vertices when n cuts give P_n pieces.

(1) Find a formula for the number of these that are on the circumference (cases $n = 1, 2, 3, 4, 5$ should help).

(2) Examine the sequence of the number of internal vertices for cases $n = 1, 2, 3, 4, 5$. Find a formula for the P_n case (Part **a** may help.)

(3) Write down a formula for V_n in terms of n.

(iv) Let E_n be the number of edges when n cuts give P_n pieces.

(1) For $n = 1, 2, 3, 4, 5$, count 3 edges for each vertex on the circumference and count 4 edges for each internal vertex. How does the total count compare with E?

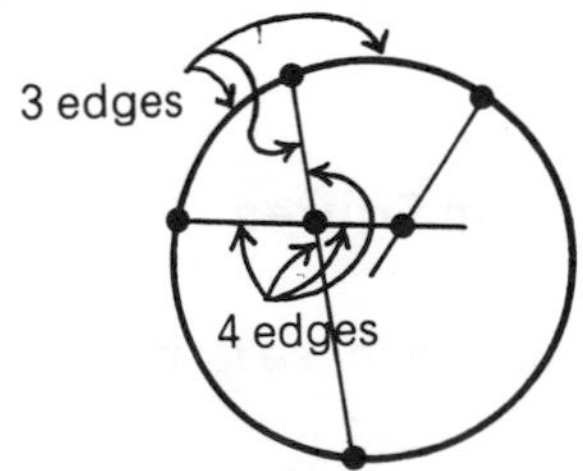

(2) Find a formula for E_n in terms of n.

(v) Use (iii), (iv) and Euler's formula to find a formula for P_n in terms of n.

6 Pythagorean triangles

These special right-angled triangles have sides with *integral* lengths.

They are *Pythagorean* triangles.

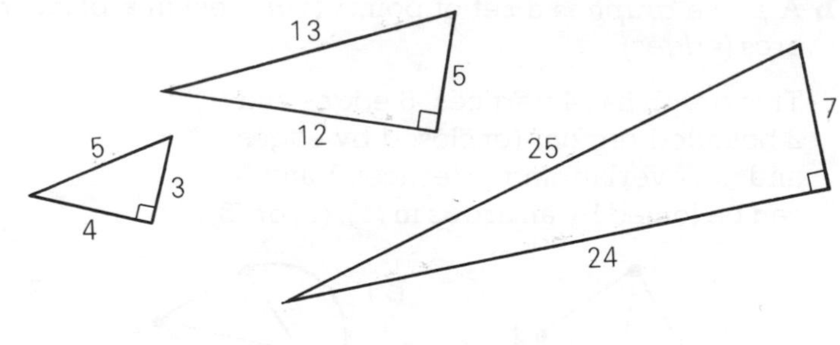

a Prove that this is a Pythagorean triangle (m and n are any positive integers with $m > n$).

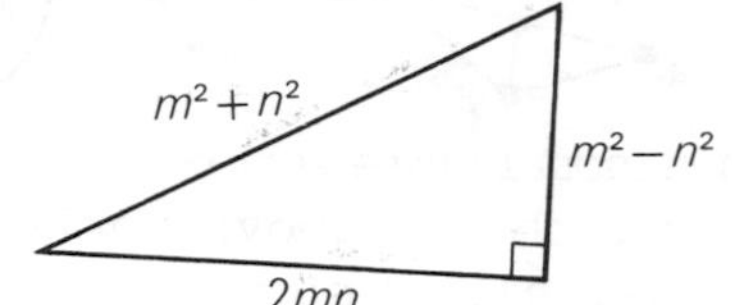

b In **a**, m and n are *generating numbers* for the triangle.
By using different pairs of generating numbers generate at least 10 different Pythagorean triangles.

c Show that if the two generating numbers have a factor (> 1) in common then so will the three sides.
Check your examples in **b**. Did this happen?

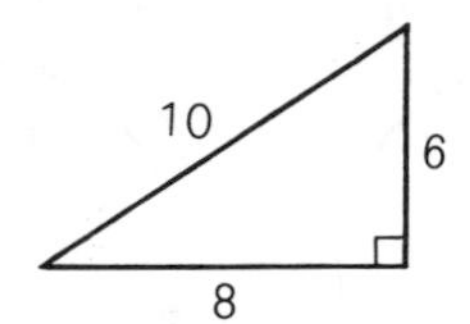

d The sides of a *primitive* Pythagorean triangle have no factor (> 1) in common. Which of your examples are primitive?
By considering whether m and n are odd or even, give conditions for m and n to generate a primitive triangle.

e Investigate primitive triangles whose hypotenuse differs by 1 from one of the other sides.
Try to obtain a condition on m and n for this to happen, and generate at least 5 examples.

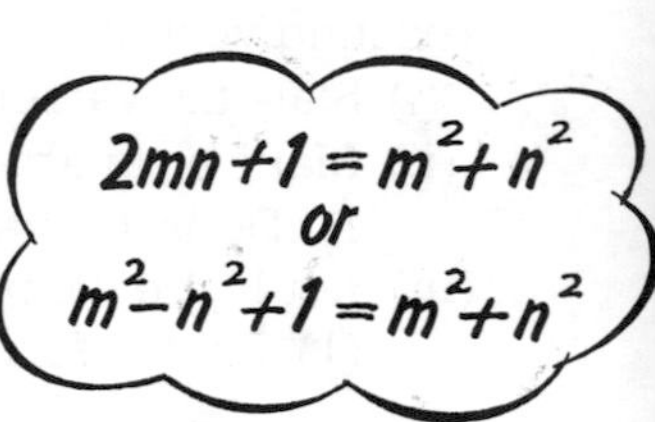

f Investigate triangles whose shorter sides differ by 1. Try to show that $2n^2+1$ or $2n^2-1$ has to be a square. Investigate the sequence of values of n that satisfy this condition, and so generate at least 10 examples.

7 Figures in figures

You will find these formulae useful:

$$1+2+3+\ldots+n=\tfrac{1}{2}n(n+1),$$
$$1^2+2^2+3^2+\ldots+n^2=\tfrac{1}{6}n(n+1)(2n+1).$$

a *Squares in a square*

A square is divided up into n^2 congruent squares (the case $n = 7$ is shown). Investigate the total number, S_n, of squares in the figure of all possible sides $n, n-1, \ldots, 1$.

(i) Calculate $S_2, S_3, S_4, S_5, \ldots$.

(ii) Write down a possible formula for S_n and try to prove the result.

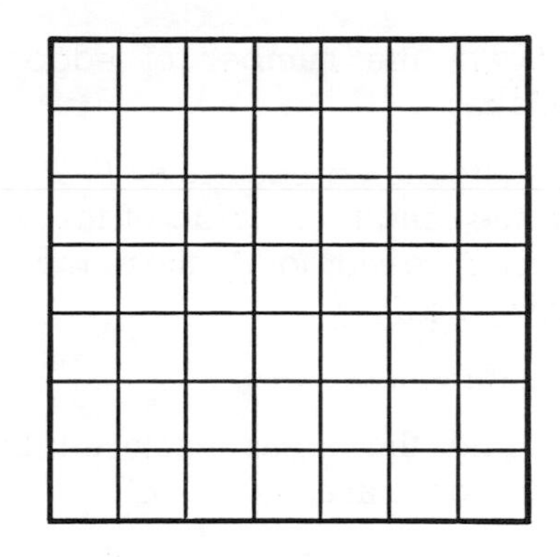

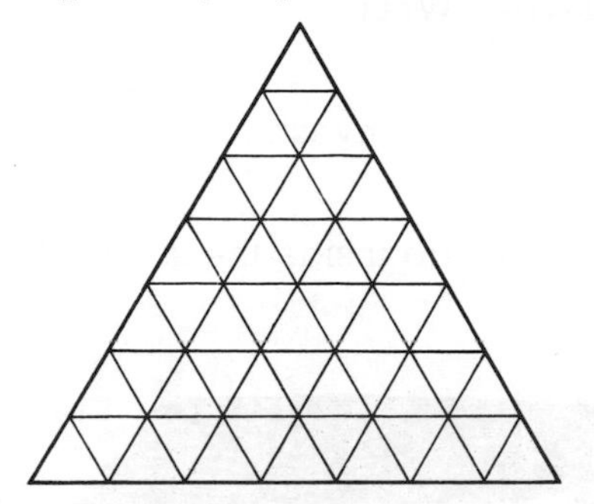

b *Triangles in a triangle*

Each side of an equilateral triangle is divided into n equal parts and the triangle is divided up into a network of congruent triangles (the case $n = 7$ is shown).

(i) Investigate the total number, T_n, of triangles in the figure pointing upwards, like the given triangle, of all possible sides $n, n-1, \ldots, 1$.

(1) Obtain $T_2, T_3, T_4, T_5, \ldots$.

(2) Suggest a possible formula for T_n and try to prove the result.

(ii) Investigate also the total number of 'upside down' triangles in the figure, U_n.

(1) Calculate $U_2, U_4, U_6, \ldots$, and $U_3, U_5, U_7, \ldots$.

(2) Try to explain why $U_n = \frac{1}{2}n(n-1) + \frac{1}{2}(n-2)(n-3) + \ldots + \frac{1}{2}.2.1$ when n is even.

(3) Investigate a similar formula for U_n when n is odd.

8 Snowflakes

Start with 'snowflake' S_0, an equilateral triangle.

Basic step Trisect each line segment and build an equilateral triangle on the centre section of each of these.

From S_0 you get the star snowflake S_1.

Repeat the basic step to get snowflake S_2.

Further repetition of the basic step produces more and more complex snowflakes $S_3, S_4, \ldots$.

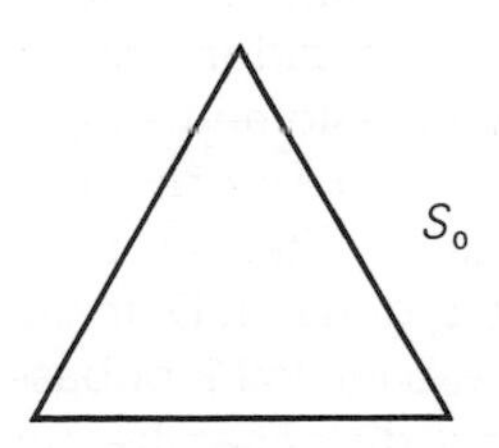

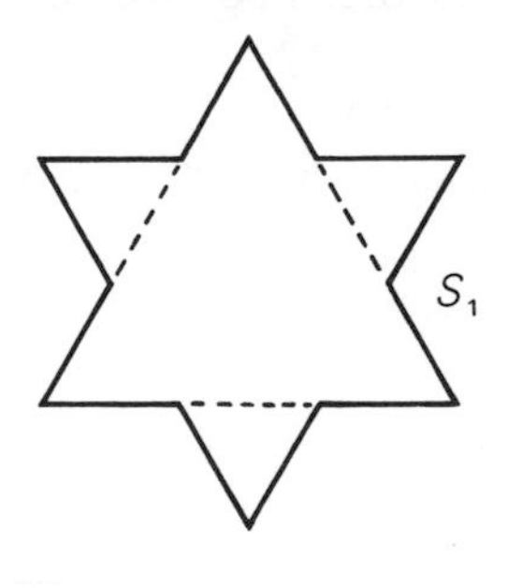

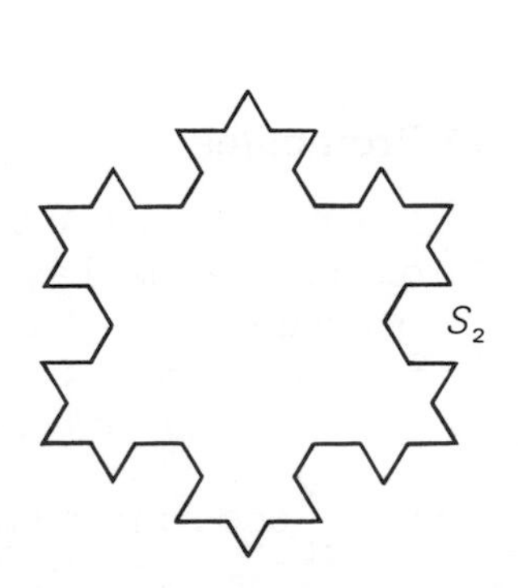

a Investigate the number of edges $N_0, N_1, N_2, N_3, \ldots$ of the snowflakes.

b Investigate the perimeters $P_0, P_1, P_2, P_3, \ldots$ of these snowflakes. (Take the length of a side of the original triangle as 1 unit.) Find an expression for P_n, the perimeter of snowflake S_n.

c What happens as $n \to \infty$?

d Investigate the areas enclosed by the snowflakes. What happens to the area as $n \to \infty$?

recurrence relations should help

e This 'fractal geometry' started with an equilateral triangle. Investigate what happens if you start with a square, a hexagon, etc.

f Investigate also what happens if the succeeding shapes go inside the snowflake rather than outside it.

9 Moon walks

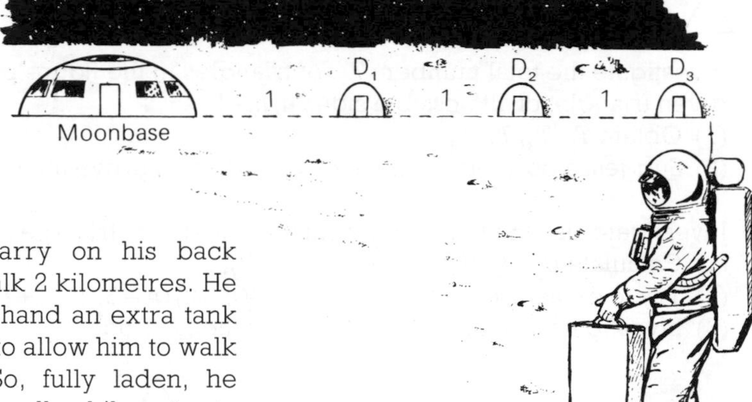

An astronaut can carry on his back enough oxygen to walk 2 kilometres. He can also carry in his hand an extra tank with enough oxygen to allow him to walk another kilometre. So, fully laden, he can leave Moonbase, walk a kilometre to depot D_1, leave the tank and return to base. He can repeat this as often as is necessary until enough oxygen is at D_1 to allow him to move a further kilometre to depot D_2, deposit a tank and still get back to Moonbase.

Investigate the increasing distances he has to walk in order to get $1, 2, 3, \ldots$ kilometres from Moonbase:

a if he must return to that base

b if the last depot reached is in fact another Moonbase and he can stay there.

10 Projectiles

If a projectile P moving on the x-axis has displacement $x(t)$ at time t from the origin O, then at time t its *velocity* is

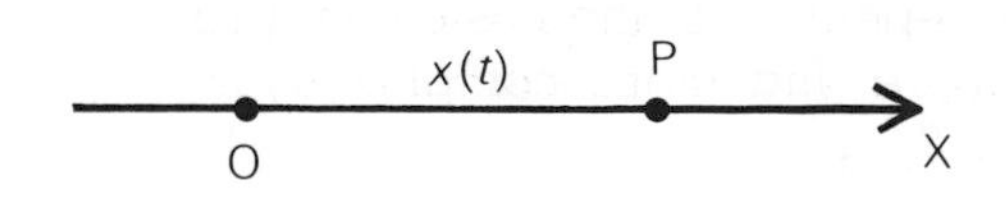

$v = \dfrac{dx}{dt}$ and its *acceleration* is $\dfrac{dv}{dt} = \dfrac{d}{dt}\left(\dfrac{dx}{dt}\right)$, which is written as $\dfrac{d^2x}{dt^2}$.

When a projectile is launched into space it is subject to gravity and other forces like air resistance. Concentrate on gravity only.

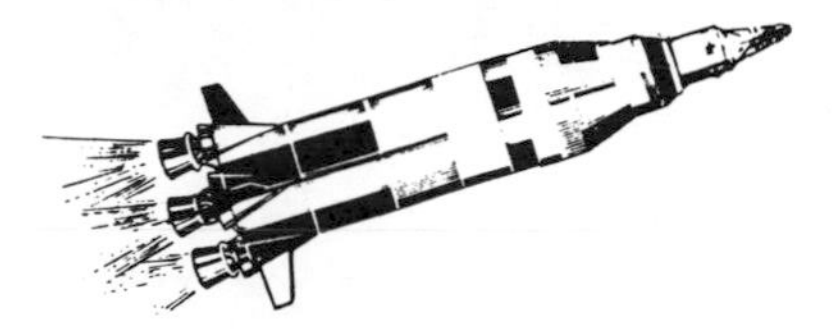

Gravity gives rise to a constant vertical acceleration, usually denoted by g; its value is approximately $10\,\text{m/s}^2$ (metres per second per second).

INVESTIGATIONS

Choosing suitable axes with OX horizontal and OY vertically upwards, a projectile is launched from point $(0, h)$ (h in metres) with speed V m/s at angle α with OX. Investigate its motion under gravity.

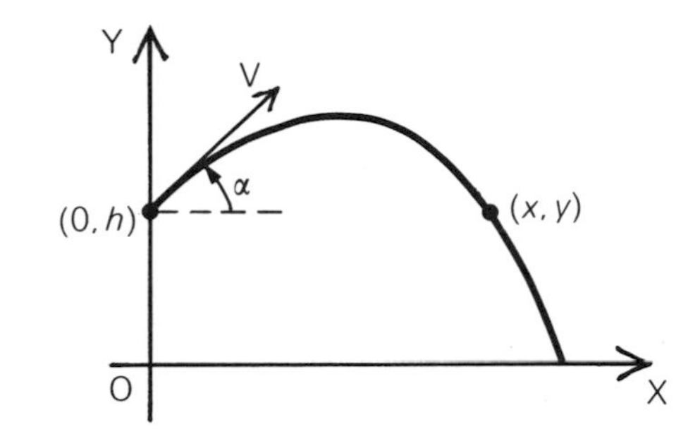

Initially (i.e. at $t = 0$),

$\dfrac{dx}{dt} = V\cos\alpha$ (initial horizontal velocity), $x = 0$,

and $\dfrac{dy}{dt} = V\sin\alpha$ (initial vertical velocity), $y = h$.

If the projectile is at position (x, y) at time t (seconds), then

$$\frac{d^2x}{dt^2} = 0,\ \frac{d^2y}{dt^2} = -g,$$

since the only acceleration is that due to gravity (downwards).

a Investigate the case $h = 0$, $\alpha = \dfrac{\pi}{3}$, $V = 100$, $g = 10$.

(i) State the initial conditions for the motion.

(ii) Solve the differential equations to obtain x and y, using the initial conditions for $\dfrac{dx}{dt}, \dfrac{dy}{dt}$ and then those for x, y.

(iii) Find out as much as you can about the flight path (e.g. maximum height reached, time to that point, point of impact, equation of path, etc).

b Investigate the case $h = 10$, $\alpha = \dfrac{\pi}{4}$, $V = 50$, $g = 10$.

c Investigate the general case (h, V, α, g).

d A target is placed at distance a from O. If the initial speed V is fixed, find a formula to determine the angle α at which the launching should take place to score a direct hit.

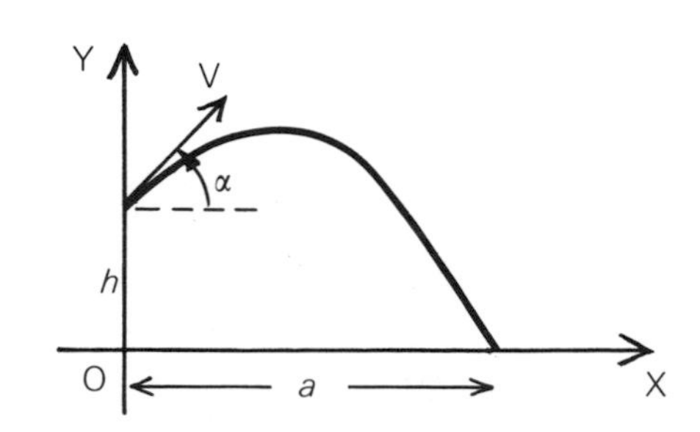

e If α is now fixed at $\dfrac{\pi}{4}$, at what speed should the projectile be launched to ensure a strike?

ANSWERS

Page 2 ***Exercise 1***

1a (i) $\frac{1}{3}$ (ii) 2 (iii) 1 (iv) $\frac{1}{3}$ (v) $-\frac{5}{3}$ (vi) 0 **b** AB and FG **c** MN
d (1) AB, OC, DE, FG (2) HK
2a (i) 3 (ii) -3 (iii) 1 (iv) $-\frac{3}{4}$ **b** (i) 72°, 108°, 45°, 143°
(ii)

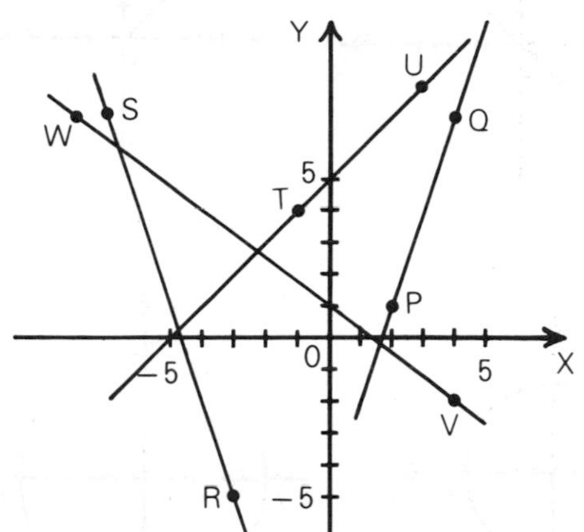

3 $m_{DG} = m_{EF} = -\frac{3}{2} \Rightarrow DG \parallel EF$
4 $m_{AB} = m_{CD} = 1,\ m_{AD} = m_{BC} = -6.$
$AB \parallel CD,\ AD \parallel BC \Rightarrow$ ABCD a parallelogram
5a $m_{PQ} = m_{QR} = \frac{1}{3}$, Q a common point $\Rightarrow$ P, Q, R collinear
b $m_{SQ} = m_{QT} = -1$, Q a common point $\Rightarrow$ S, Q, T collinear
c $m_{PS} = m_{TR} = 1 \Rightarrow PS \parallel TR$; by alternate angles, △s equiangular, so similar
6a 1 **b** −3·7 **c** −0·5
7a $\frac{1}{3}$, $3\frac{1}{4}$ **b** (i) 072° (ii) 017° **c** 126° **d** 504635
8 0·6 or −2·5 **9a** −0·3 **b** −0·1 **c** −0·5

Page 4 ***Exercise 2***

1a AB and IJ, EF and KL **b** AB and GH, CD and EF, CD and KL, GH and IJ
2a (i) 4 (ii) −1 (iii) $\frac{1}{3}$ (iv) $\frac{3}{4}$ (v) $-2\frac{1}{2}$
b (i) $-\frac{1}{4}$ (ii) 1 (iii) −3 (iv) $-\frac{4}{3}$ (v) $\frac{2}{5}$
3a $y = -\frac{1}{3}x$ **b** $y = \frac{1}{5}x$ **c** $y = -10x$ **d** $y = -\frac{3}{2}x$ **e** $y = 2x$
4a (iii), (iv) **b** (i), (ii)
5 $m_{CD} \times m_{DB} = \frac{1}{2} \times (-2) = -1 \Rightarrow \angle CDB = 90°$;
$m_{AC} \times m_{AB} = -2 \times \frac{1}{2} = -1 \Rightarrow \angle CAB = 90°$
6a $m_{PQ} = m_{RS} = -\frac{1}{3}$ and $m_{QR} = m_{SP} = 3$, so opposite sides are parallel; $m_{PQ} \times m_{PS} = -1$, so all angles are 90°
b 161·6°, 71·6°

Page 5 ***Exercise 3***

1a

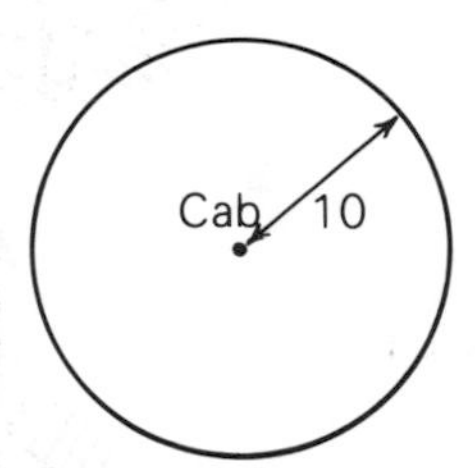

Within and on the circle

b

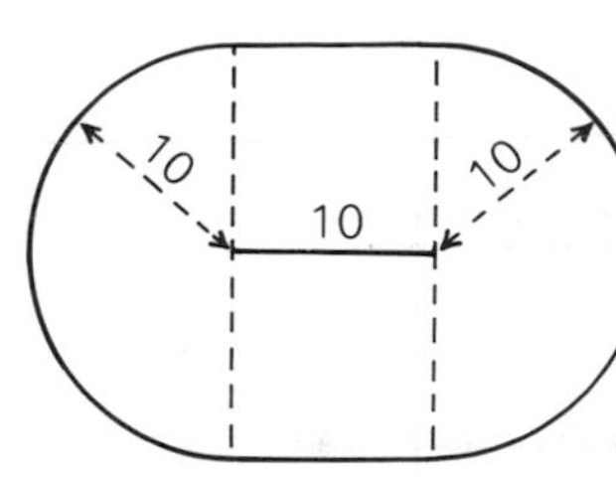

At any instant he can be heard within a circle, radius 10 miles, centre on his 10 mile route

2

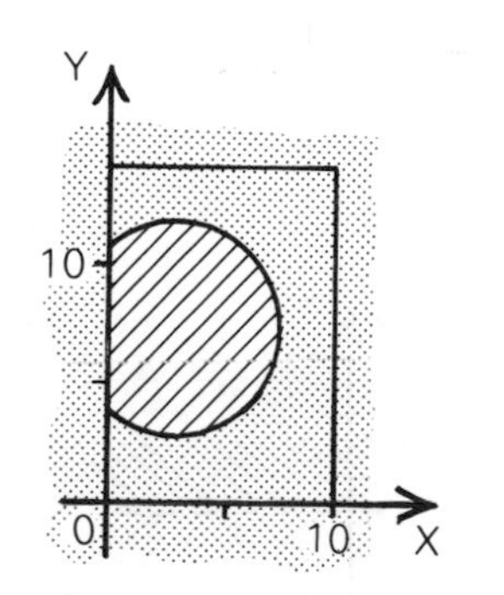

3

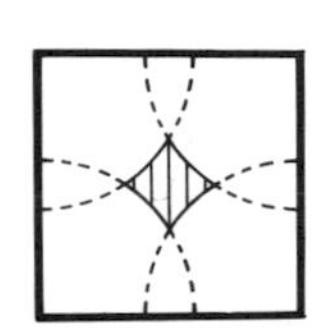

4a (i)

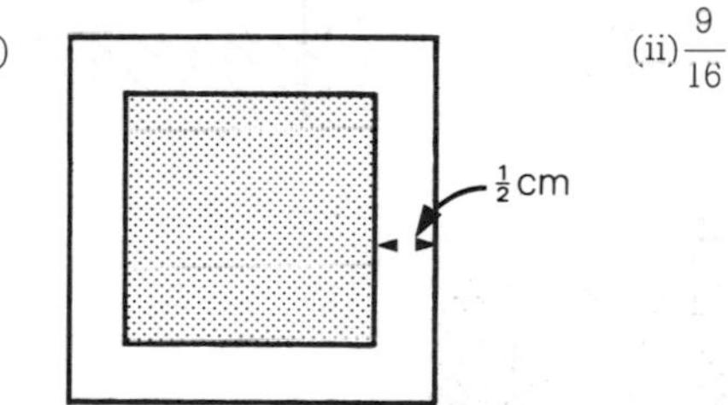

(ii) $\frac{9}{16}$

b (i) Ratio $= \frac{9}{16}$ (ii) Ratio $= \frac{(4-\sqrt{3})^2}{16}$

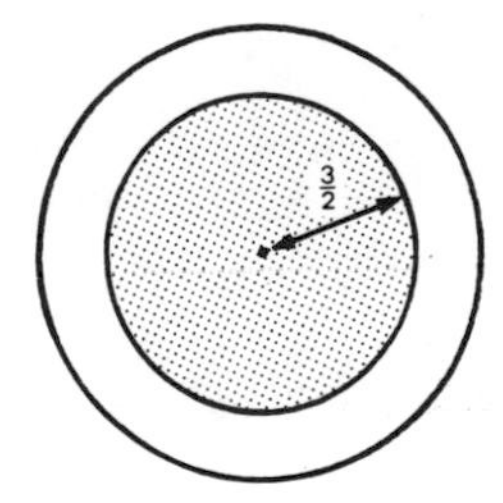

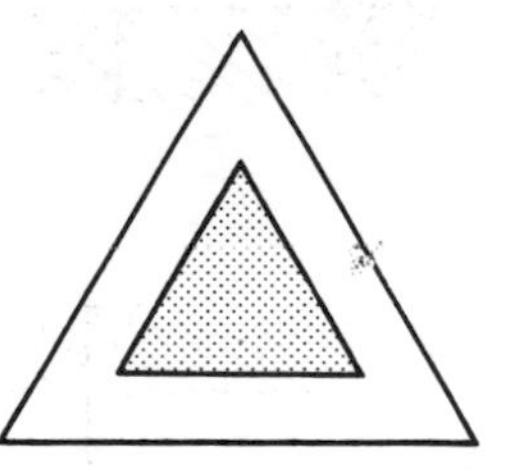

Side of inner △ = $(4-\sqrt{3})$ cm

Page 6 ***Investigation***

a Circle, centre intersection, radius $\frac{1}{2}$ mile **b** An ellipse with one or other road as major axis. One equation:
$$\frac{x^2}{(\frac{3}{4})^2} + \frac{y^2}{(\frac{1}{4})^2} = 1$$

Page 7 ***Exercise 4***

1

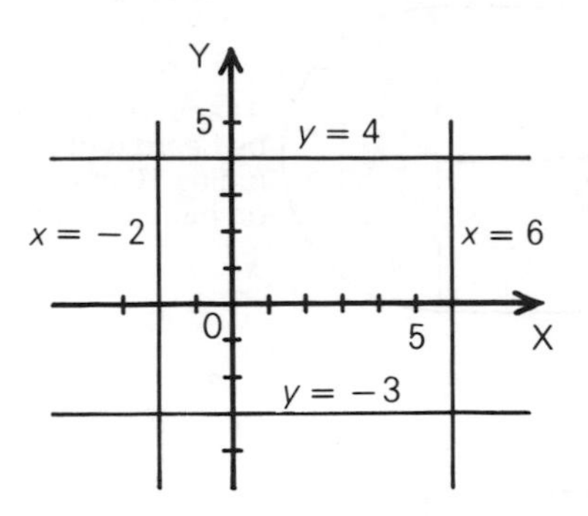

2 a **b**

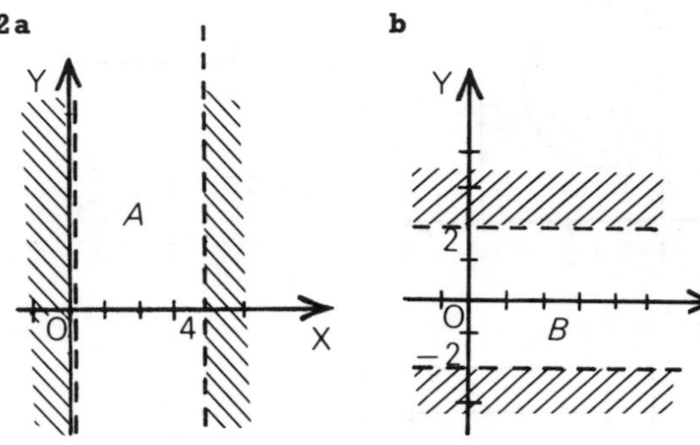

c

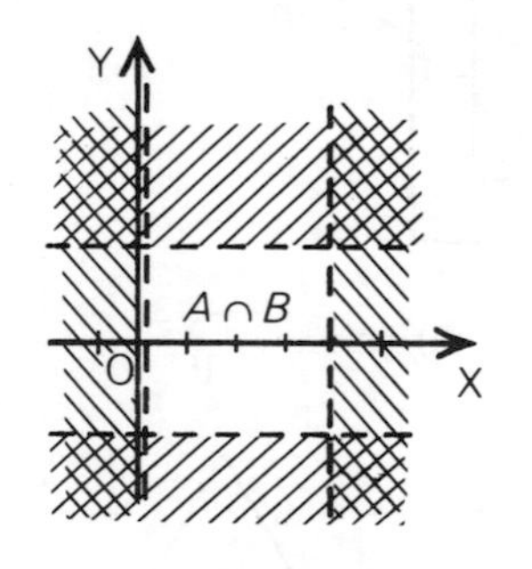

3 a

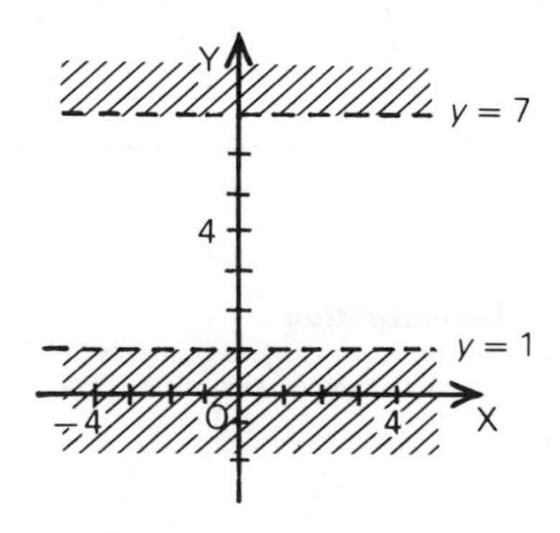

b $\{(x, y): 1 < y < 7\}$

4 a (i)

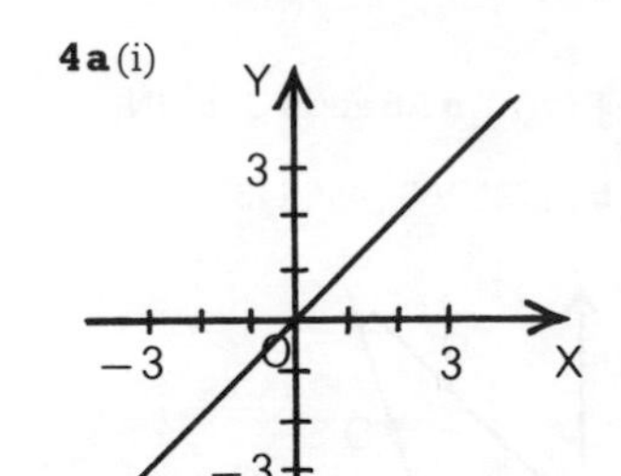

(ii) $y = x$

b (i)

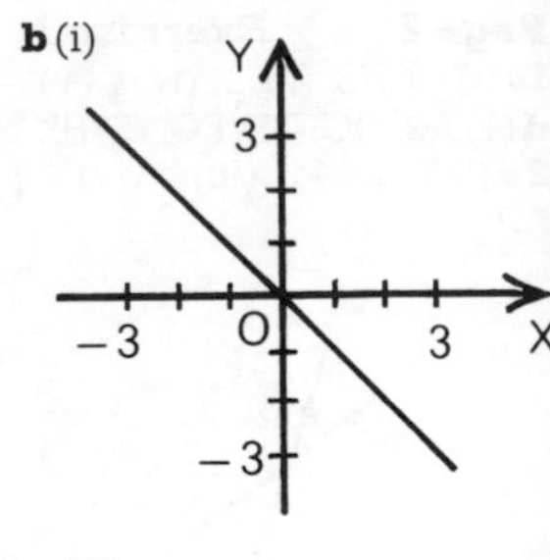

(ii) $y = -x$

5 a

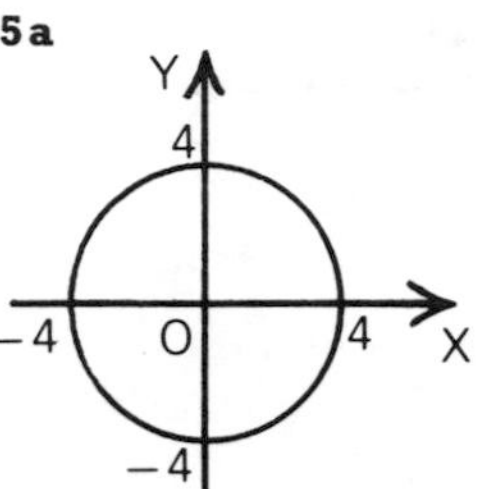

b

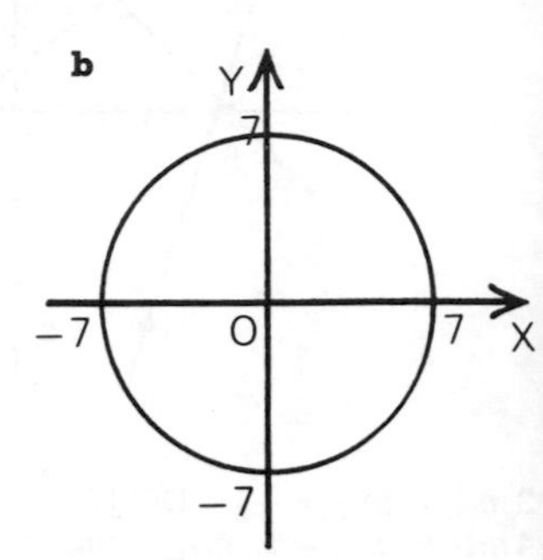

c

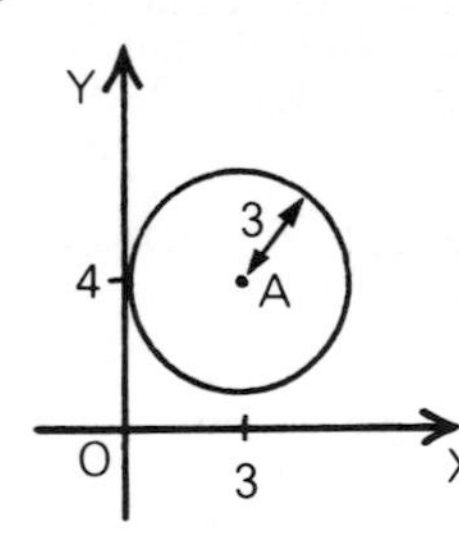

d

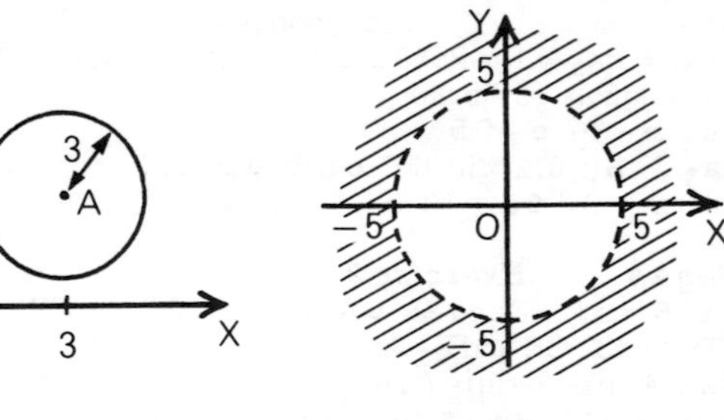

e

Y 6 2 −6 −2 O 2 6 X −2 −6

f

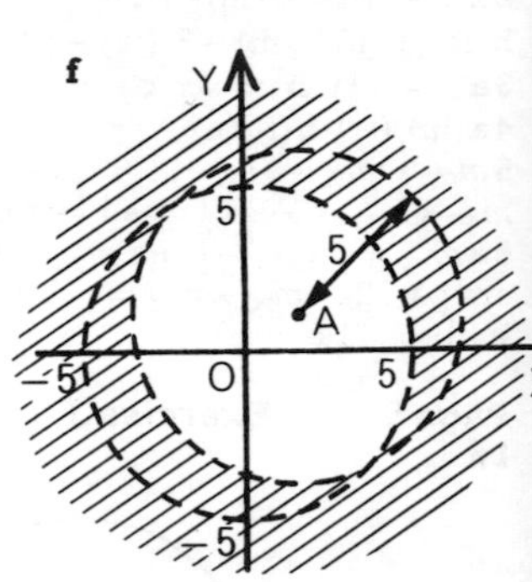

6 a

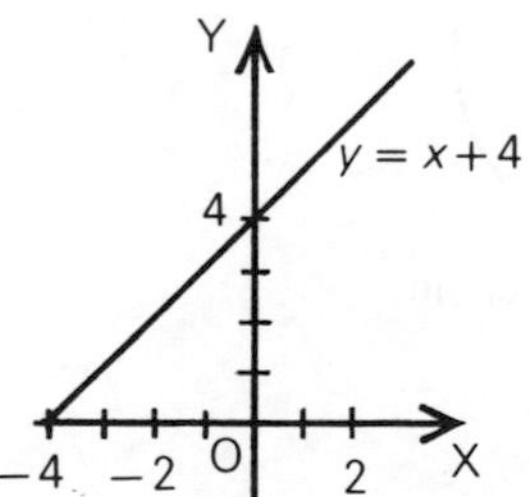

b

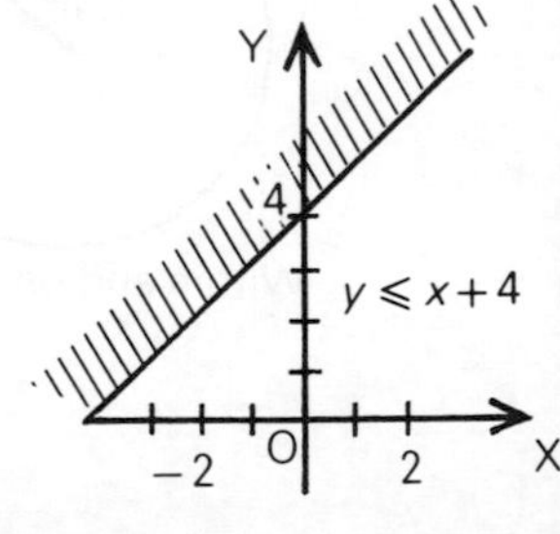

c (i)

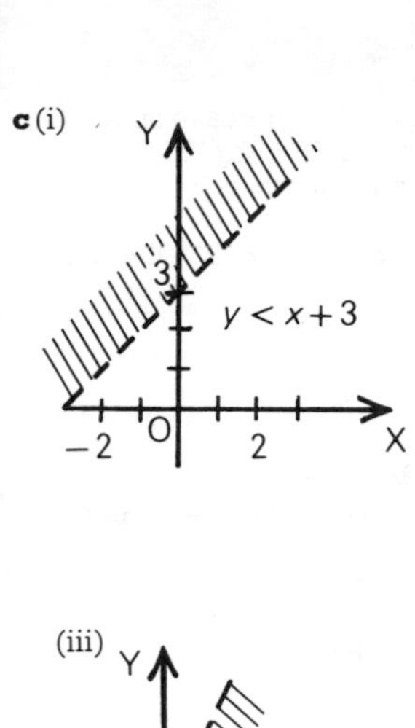

(ii)

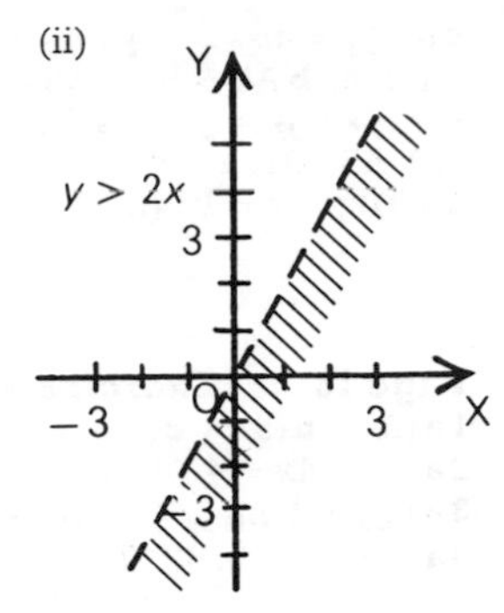

(iii)

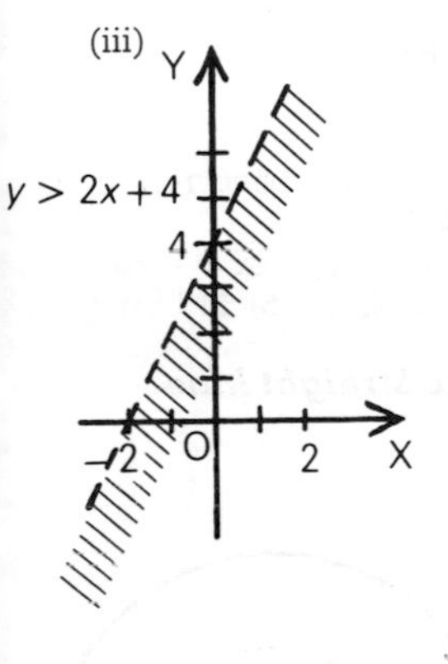

(iv)

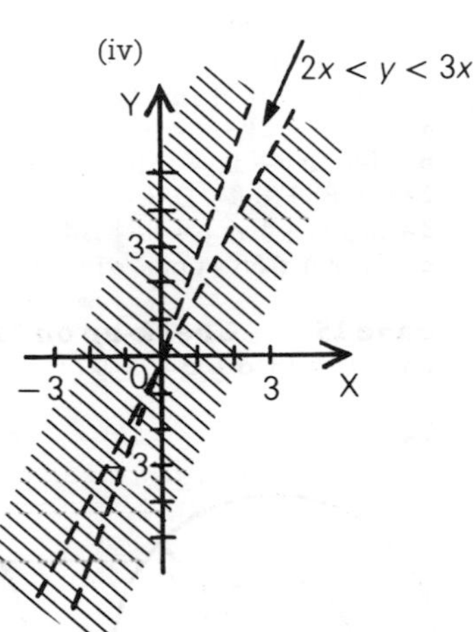

7

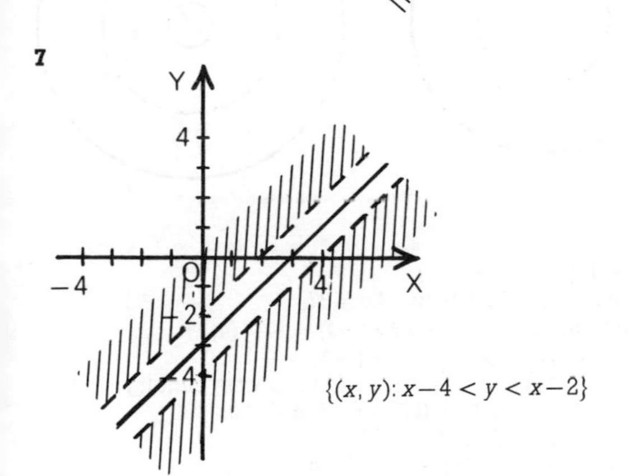

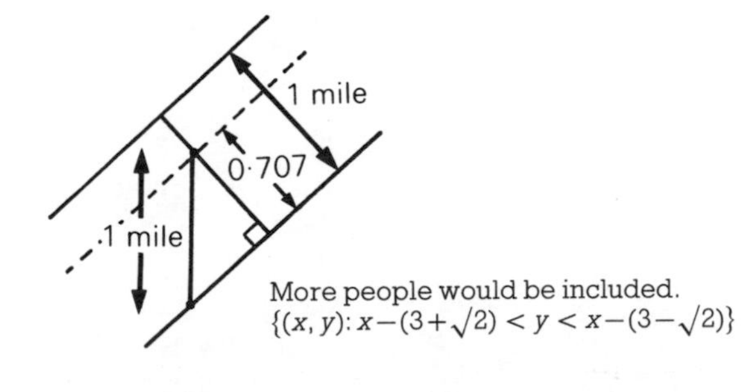

More people would be included.
$\{(x, y): x-(3+\sqrt{2}) < y < x-(3-\sqrt{2})\}$

Page 8 ***Exercise 5***

1 a (i) $y = 4x+2$ (ii) $y = 4x-1$ (iii) $y = 4x$ **b** (i) $y = -x+2$ (ii) $y = -x-5$ (iii) $y = -x$
2 a 1, 6, $y = x+6$ **b** $-\frac{1}{2}$, -3, $y = -\frac{1}{2}x-3$
3 a 1, $y = x+4$ **b** 2, $y = 2x+2$ **c** $-\frac{2}{3}$, $y = -\frac{2}{3}x+3$
d -4, $y = -4x-2$
4 a $y = x+2$; on **b** $y = -2x-1$; above **c** $y = \frac{1}{3}x-4$; below
d $y = -\frac{2}{3}x+2$; on **e** $y = \frac{4}{3}x-1$; below
5 a $y = 2x+4$, 2, 4 **b** $y = -3x+6$, -3, 6 **c** $y = \frac{2}{3}x+2$, $\frac{2}{3}$, 2
d $y = -x+5$, -1, 5 **e** $y = -\frac{3}{4}x+3$, $-\frac{3}{4}$, 3

6 a

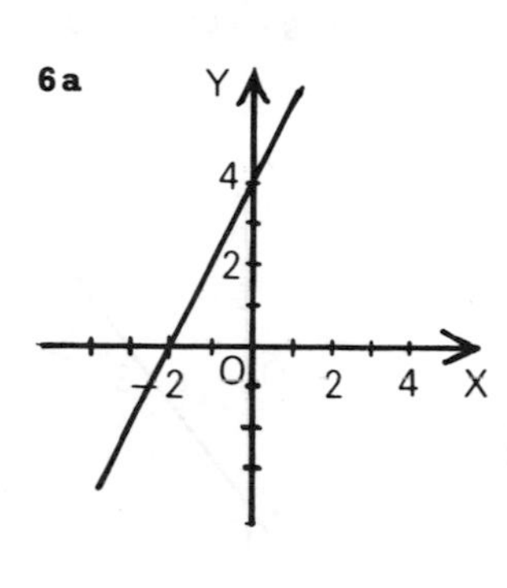

b

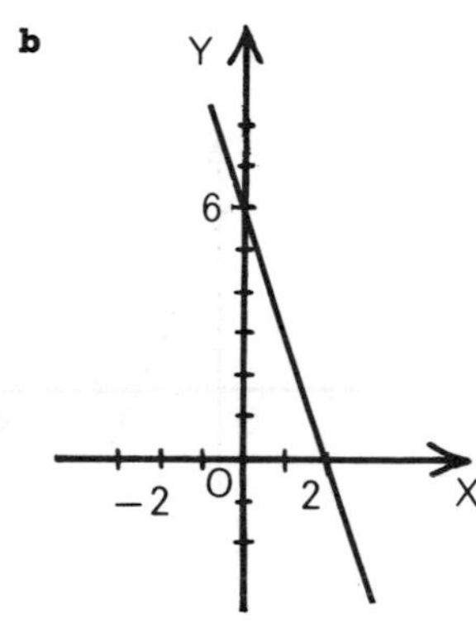

c

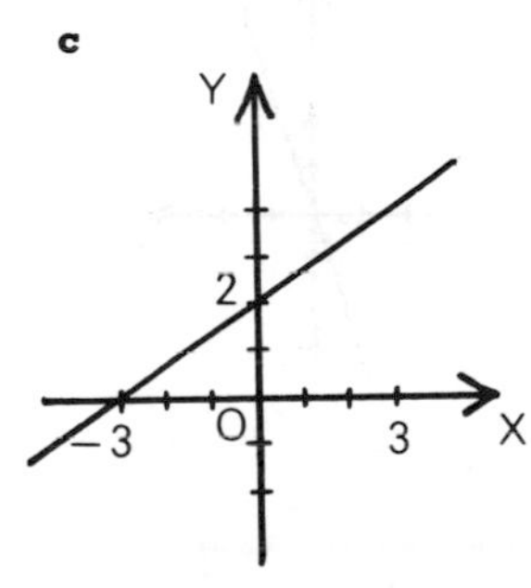

d

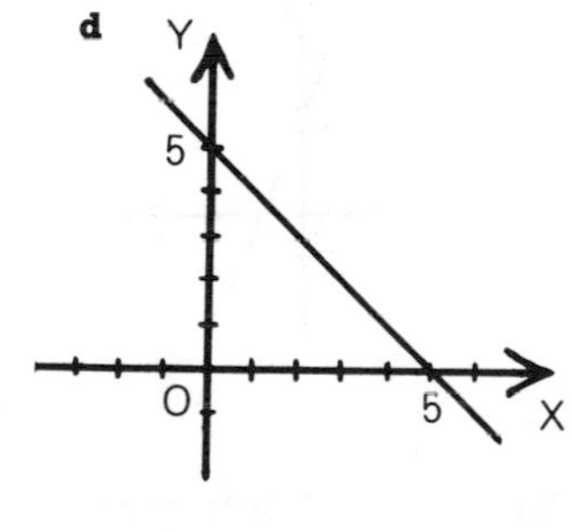

e

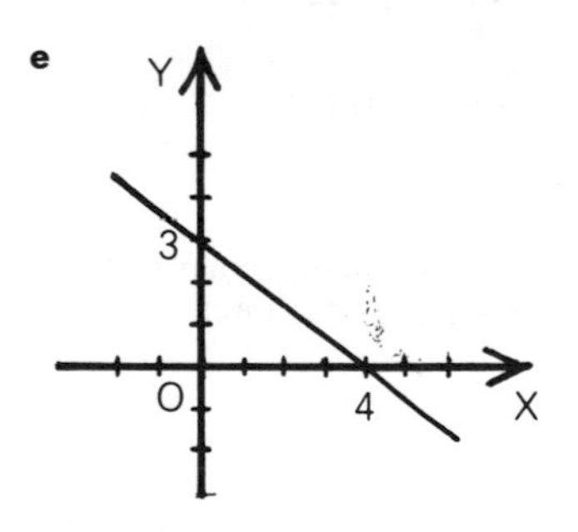

7 a $\frac{1}{4}$ **b** (i) $y = \frac{1}{4}x+5$ (ii) $y = -4x+5$
8 a $y = \frac{1}{2}x+5$; on **b** $y = 4x-1$; below

Page 9 ***Exercise 6***

1 a $y-2x = -2$ **b** $y-3x=3$ **c** $3y-x = 9$ **d** $5y+2x = 1$
2 a $\frac{2}{3}$, $3y-2x = 7$ **b** $-\frac{1}{4}$, $4y+x = 18$ **c** $-\frac{3}{2}$, $2y+3x = 13$
d $-\frac{1}{3}$, $3y+x = -5$
3 a $y-x = 1$ **b** $y-3x = -13$ **c** $y+x = -5$ **d** $2y-3x = -9$
4 a $2y+x = 4$ **b** $4y+x = 21$ **c** $y+2x = 10$ **d** $2y-3x = -15$
5 a (i) $4y-x = 13$ (ii) $4y-x = 13$
b Isosceles; median same as altitude

6 a $m_{ST} = m_{VU} = \frac{1}{2}$ and $m_{VS} = m_{UT} = \frac{7}{3}$, so opposite sides are parallel **b** ST: $2y-x=7$, VU: $2y-x=-4$, VS: $3y-7x=27$, UT: $3y-7x=-17$
7 a (i) $m_{AC} = -\frac{2}{3}$ (ii) $m_{BD} = \frac{3}{2}$ **b** $2y-3x=-11$ **c** $(5, 2)$
d (i) $2y-3x=-11$ with $5<x$ and $x \neq 9$
(ii) $2y-3x=-11$ with $x<5$ (iii) $(1, -4)$
8 a AS: $y+3x=22$, BL: $y+3x=52$ **b** $3y-x=26$ **c** $y=7$
d $3y+x=16$

Page 11 *Exercise 7*

1 a $2x-y-1=0$ **b** $-2x+y-1=0$ **c** $2x-3y-6=0$
d $4x+5y-5=0$ **e** $4y-3=0$ **f** $3x+2=0$ **g** $x+6y+4=0$
h $6x-4y-3=0$

2 a

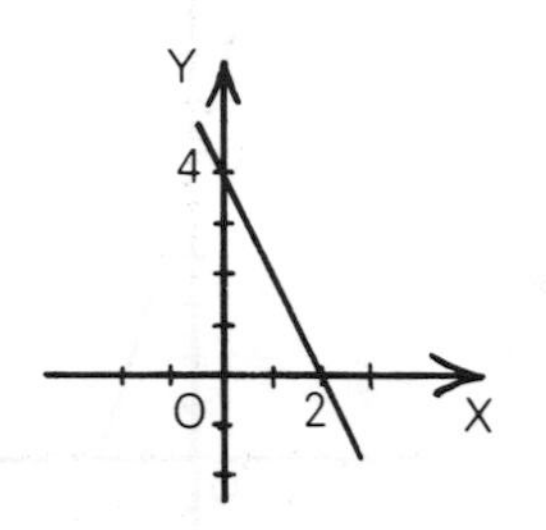

b

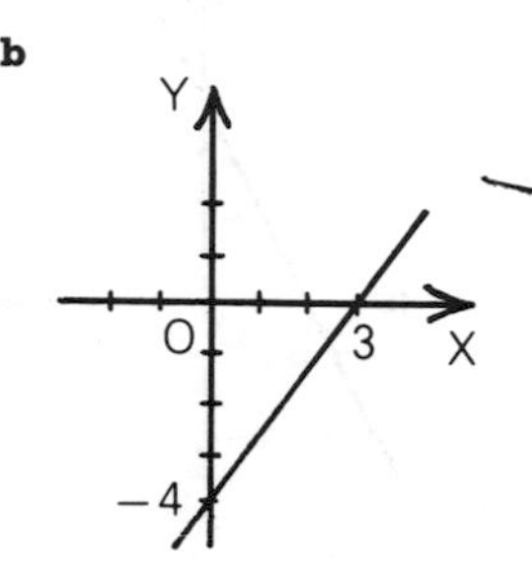

c

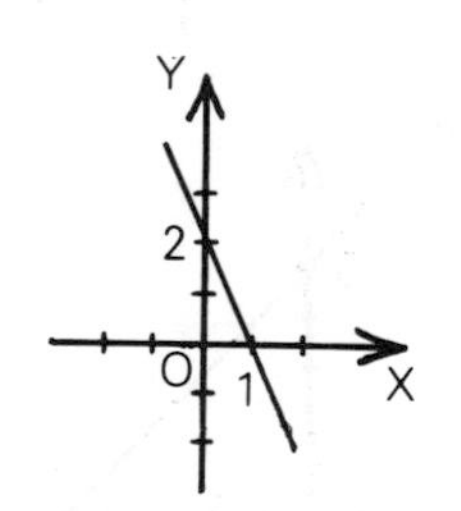

d

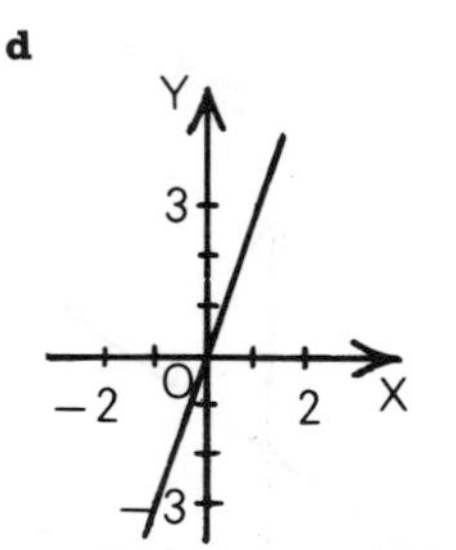

3 a $2x+y-5-0$ **b** $3x+4y+12=0$ **c** $-x+2y-7=0$
d $x+2y=0$ **e** $x+2y-12=0$ **f** $3x-4y-10=0$
4 a $4y+3x=33$ **b** $3y-4x=-19$
5 a (i) 1 (ii) $y=x+4$ **b** $y=-x-4$, $y=-x+8$
6 a $h=0$, $k=6$ **b** $3y+x=11$ **c** B$(-1, 4)$, D$(5, 2)$

Page 11 *Exercise 8*

1 a 10 **b** 13 **c** 25 **2** AB = BC = $5\sqrt{2}$
3 a $m_{RP} \times m_{RQ} = 2 \times (-\frac{1}{2}) = -1 \Rightarrow$ PR perpendicular to RQ
b $RQ^2 + RP^2 = PQ^2 = 100$ (converse of Pythagoras' Theorem)

4 a $m_{AB} = m_{CD} = -\frac{1}{4}$, $m_{AD} = m_{BC} = 4 \Rightarrow$ opposite sides parallel **b** AB = BC = CD = DA = $\sqrt{68} = 2\sqrt{17}$
c $m_{AB} \times m_{AD} = -1 \Rightarrow$ AB perpendicular to AD. ABCD is a square
5 a (i) $2y+x=10$ (ii) $y-2x=0$ **b** D(2, 4) **c** 027°, 4·47 km

Page 13 *Exercise 9*

1 a (2, 1) **b** (5, 0) **c** (3, −1) **d** (1, −1)
2 a (i) $y-4x=4$ (ii) $2y-x=8$ **b** V(0, 4)
3 a $m_{AB} = \frac{1}{3}$, $m_{BC} = -\frac{1}{2}$ **b** $y+3x=34$, $y-2x=-16$ **c** (10, 4)
4 a (i) A(−5, 3), B(5, 7), C(1, −5)
(ii) $4y+x=7$, $7y-8x=9$, $y+10x=5$
b Point of intersection $(\frac{1}{3}, \frac{5}{3})$. Medians are concurrent
5 a (i) $3y-2x=6$ (ii) $2y+3x=30$ **b** (i) $y=3$
(ii) E$(\frac{3}{2}, 3)$, C(8, 3) (iii) 3·61 units
6 a (i) $y=x$, $3y+x=-10$ (ii) K$(-\frac{5}{2}, -\frac{5}{2})$
b K lies on $y+3x=-10$. The altitudes are concurrent
7 a B(8, 8), F$(8, \frac{20}{3})$, D(8, 4), H(8, 0) **b** (8, 6)
8 a $m_{OA} = -\frac{4}{3}$, $m_{OB} = -\frac{3}{4}$ **b** $4y-3x=25$, $3y-4x=-25$
c L(25, 25) **d** 35·4 units **e** P(−15, −5), Q$(\frac{5}{2}, -5)$ **f** 17·5 units

Page 15 *Check-up on The Straight Line*

1 a $(1, \frac{1}{2})$ **b** 13 **c** $\frac{5}{12}$ **d** $\frac{5}{12}$, $-\frac{12}{5}$

2 a

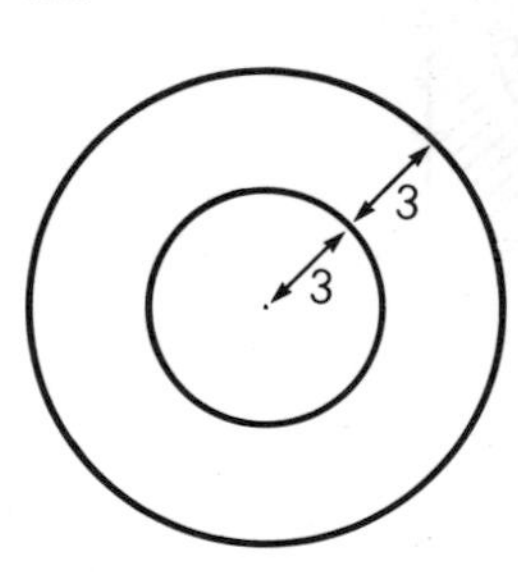

b

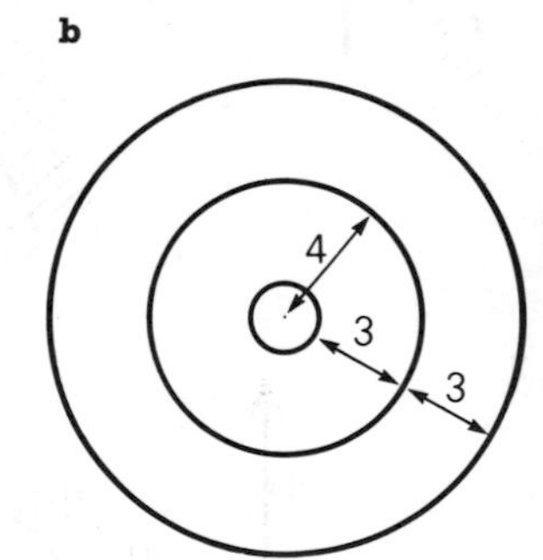

3 9·5°
4 a 3, 4 **b** $\frac{2}{3}$, −2 **c** $-\frac{4}{5}$, $-\frac{3}{5}$
5 a (i) $4y-3x=16$ (ii) $3y+4x=37$
b Gradients are $\frac{3}{4}$ and $-\frac{4}{3}$, $\frac{3}{4} \times (-\frac{4}{3}) = -1$; (4, 7)
6 a (−3, 1) **b** $m_{BD} \times m_{AC} = \frac{1}{3} \times (-3) = -1$.
Also (−3, 1) is mid pt. of AC ⇒ BD axis of symmetry
c AB = CB = 10, AD = DC = $5\sqrt{2}$ **d** V-kite
7 a (2, −3) **b** 33·7°, 31·0° 115·3°
8 a (i) $y-x=-2$ (ii) $y+2x=18$ **b** S$(\frac{20}{3}, \frac{14}{3})$.

Page 17 ***Exercise 1***

1 a 48·8 mph **b** (i) 47·5, 45·3, 44, 43·8, 42
(ii) Speeds are decreasing; between 40 and $41\frac{1}{2}$ possibly
2 a $-2{\cdot}7$°C per hour, temperature is decreasing
b (i) $-1{\cdot}1$, $-1{\cdot}6$, $-3{\cdot}5$, $-5{\cdot}1$, $-2{\cdot}6$, $-2{\cdot}2$° per hour
(ii) 10.00 to 11.00 pm

3 a $\frac{4-1}{2-1}=3$

b

x_Q	y_Q	x_Q-x_P	y_Q-y_P	average gradient m_{PQ}
1·8	3·24	0·8	2·24	2·8
1·6	2·56	0·6	1·56	2·6
1·4	1·96	0·4	0·96	2·4
1·3	1·69	0·3	0·69	2·3
1·2	1·44	0·2	0·44	2·2
1·1	1·21	0·1	0·21	2·1

(i) They get closer to 2 (ii) Possibly 2

Page 19 ***Exercise 2***

2

x_Q	y_Q	x_Q-x_P	y_Q-y_P	average gradient m_{PQ}
1·09	1·1881	0·09	0·1881	2·09
1·08	1·1664	0·08	0·1664	2·08
⋮	⋮	⋮	⋮	⋮
1·01	1·0201	0·01	0·0201	2·01
1·009	1·018...	0·009	0·018...	2·009
1·008	1·016...	0·008	0·016...	2·008
⋮	⋮	⋮	⋮	⋮

4 4 **5 a** 6 **b** $2a$ **6 a** 6 **b** $6a$ **7** 7 **8 a** 4 **b** 5 **c** 5

Page 21 ***Investigation***

1 a 2 **b** 10 **c** 12; $2+10=12$ **2 a** 3 **b** 10 **c** 13
To differentiate a sum, differentiate each part and add the results

Page 22 ***Formulae save time***

$f(x)$	x	x^2	x^3	x^4	x^5	x^6	x^{10}
$f'(x)$	1	$2x$	$3x^2$	$4x^3$	$5x^4$	$6x^5$	$10x^9$

Page 24 ***Exercise 3***

1 a 0 **b** 1 **c** $2x$ **d** $3x^2$ **e** $4x^3$ **f** $5x^4$ **g** $50x^{49}$ **h** $100x^{99}$
2 a $-2x^{-3}$ **b** $-10x^{-11}$ **c** $-15x^{-16}$ **d** $-\frac{1}{x^2}$ **e** $-\frac{3}{x^4}$ **f** $-\frac{5}{x^6}$
g $-\frac{8}{x^9}$ **h** $-\frac{20}{x^{21}}$
3 a $\frac{3}{2}x^{1/2}$ **b** $\frac{5}{2}x^{3/2}$ **c** $\frac{1}{3}x^{-2/3}$ **d** $-\frac{1}{2}x^{-3/2}$ **e** $-\frac{5}{2}x^{-7/2}$
f $-\frac{1}{2(\sqrt{x})^3}$ **g** $-\frac{2}{3x^{5/3}}$ **h** $-\frac{3}{4}x^{-7/4}$
4 a -6 **b** $-\frac{1}{16}$ **c** $\frac{20}{3}$
5 a $3a^2$ **b** $12a^{11}$ **c** $-\frac{3}{a^4}$ **d** $-4a^{-5}$ **e** $\frac{1}{3}a^{-2/3}$ **f** $-\frac{2}{5}a^{-7/5}$
g $-\frac{3}{4}a^{-7/4}$ **h** $\frac{4}{5}a^{-1/5}$
6 a 3 **b** 12 **c** 0 **d** $\frac{3}{4}$ **e** 3 **7 a** $\frac{1}{2}x^{-1/2}$ **b** $\frac{3}{2}$ **c** 4
8 a $\frac{1}{4}$ **b** 5 **c** 15 **d** 100 **e** 12 **f** $-\frac{1}{48}$ **g** $-\frac{1}{128}$ **h** $\frac{1}{320}$

9 a $f'(x)=2x$ **b** $f'(x)=3x^2$

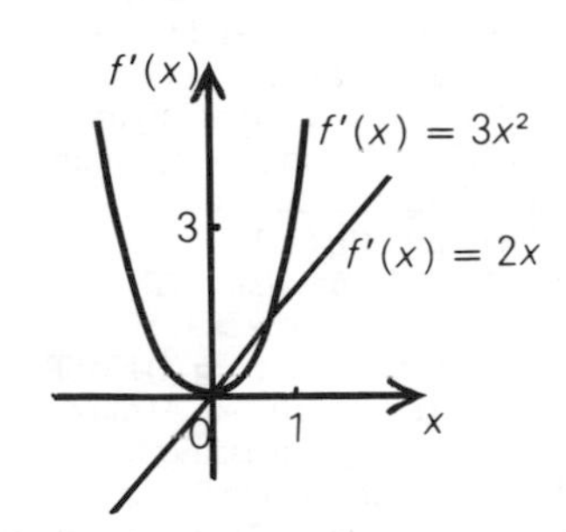

10 4 g/s **11 a** $2\frac{1}{2}$ km/h **b** $\frac{5}{16}$ km/h
12 a $\frac{1}{4}$ m/h **b** decreases. The soil becomes waterlogged

Page 25 ***Brainstormer***

a $m_{AB}=-\frac{1}{a^2}$ **c** $A(2a, 0)\,B\left(0, \frac{2}{a}\right)$. Constant area of 2 units²

Page 25 ***Two useful rules***

Example **2 a** Initial cost of machine etc

Page 26 ***Exercise 4***

1 a $2x$ **b** $6x^2$ **c** $12x^3-2$ **d** $2t+2$ **e** $6t^2-5$ **f** $6p-7$
g $3p^2-12p^5$ **h** $10x^4+11$ **i** $81x^2-9$ **j** $12x^3-6x^2+5$
2 a $2x-\frac{1}{x^2}$ **b** $6x^2-\frac{2}{x^3}$ **c** 1 **d** $12x^3-10x^4$ **e** $4-8x+3x^2$
f $-\frac{1}{3x^2}$ **g** $-\frac{6}{5x^3}$ **h** $6x^2-\frac{3}{2x^4}$ **i** $\frac{3}{2\sqrt{x}}-\frac{1}{6(\sqrt{x})^3}$ **j** $\frac{1}{\sqrt{x}}+\frac{1}{2(\sqrt{x})^3}$
3 a 4 **b** 9 **c** $-\frac{1}{9}$ **d** -13 **e** $-\frac{3}{4}$ **f** 1
4 a -51 **b** 1 **c** 1 **d** -10
5 2π. Rate of change of area of circle as the radius increases = circumference
6 a $4\pi r^2$ **b** Rate of change of volume of a sphere = surface area **7** 660 bacteria/second **8** $1\frac{1}{4}$ words/minute
9 a $x'(t)=4-2t$ **b** After 2 seconds. Train has stopped
c -2. Train is reversing

Page 28 ***Exercise 5***

1 a $4x$ **b** $9x^2-2$ **c** $4t^3-4t$ **d** $4p-4$
2 a $8x^3-9x^2+4$ **b** $3-\frac{1}{3x^2}$ **c** $8t^{-1/3}-\frac{4}{t^2}$ **d** $6u^2+\frac{3}{2u^4}$

3 a $y-12x=-12$ **b** $y-4x=-1$ **c** $y-4x=-3$
d $4y-x=4$ **e** $2y-9x=-27$ **f** $16y+3x=-8$
4 a $y-5x=-2$ **b** $9y+x=21$ **c** $y+13x=-8$
5 At A: $y-2x=2$. At B: $y-x=1$
6 $\frac{dT}{dL}=\frac{\pi}{\sqrt{(gL)}}, \frac{\pi}{4\sqrt{g}}$
7 a $8x-12$ **b** $9x^2+\frac{2}{x^2}$ **c** $-\frac{1}{2u^2}-\frac{1}{4}u^{-3/2}$
d $8t-12t^2+4t^3$ **e** $-\frac{2}{3}v^{-4/3}-2v^{-1/3}$
8 a 2 m **b** after 2 seconds
c 0 m/s; reached maximum height **d** 10 m/s, -10 m/s

Page 29 ***Brainstormer***
$a=-1$

Page 31 ***Exercise 6***
1 a $(1, -1)$ minimum TP **b** $(2, 9)$ maximum TP **c** $(0, 4)$ PI
2 a (i) $(0, 1)$ (ii) minimum TP (iii) decreasing $x<0$; increasing $x>0$ **b** (i) $(0, 3)$ (ii) maximum TP (iii) increasing $x<0$; decreasing $x>0$ **c** (i) $(0, 0)$ (ii) PI (iii) increasing $x<0$ and $x>0$ **d** (i) and (ii) $(-1, -2)$ minimum TP and $(1, 2)$ maximum TP (iii) increasing for $-1<x<1$; decreasing for $x<-1$ and $x>1$ **e** (i) and (ii) $(0, 0)$ PI and $(1, -1)$ minimum TP (iii) increasing for $x>1$; decreasing for $x<0$ and $0<x<1$ **f** (i) and (ii) $(-1, 4)$ and $(1, 4)$ minimum TPs; $(0, 5)$ maximum TP (iii) increasing for $-1<x<0$ and $x>1$; decreasing for $x<-1$ and $0<x<1$ **g** (i) No SPs (iii) always increasing **h** (i) and (ii) $(-1, 4)$ maximum TP; $(0, 2)$ PI; $(1, 0)$ minimum TP (iii) increasing for $x<-1$ and $x>1$; decreasing for $-1<x<0$ and $0<x<1$
i (i) and (ii) $(0, 1)$ PI; $(\frac{1}{2}, \frac{17}{16})$ maximum TP (iii) increasing for $x<0$ and $0<x<\frac{1}{2}$; decreasing for $x>\frac{1}{2}$
3 Rising: $0<t<2$ Falling: $2<t<4$
4 (i) Increasing: $0<t<35$ (ii) decreasing $35<t<60$
5 Decreasing for $0<x<10$; increasing for $x>10$. Minimum cost when $x=10$.

Page 32 ***Brainstormer***
a C_1: always increasing; C_2: always decreasing
b C: decreasing for $0<x<2$; increasing for $x>2$
c $(2, 4+a)$; minimum cost of £$(4+a)$ when radius is 2 feet

Page 33 ***Exercise 7***

1 a

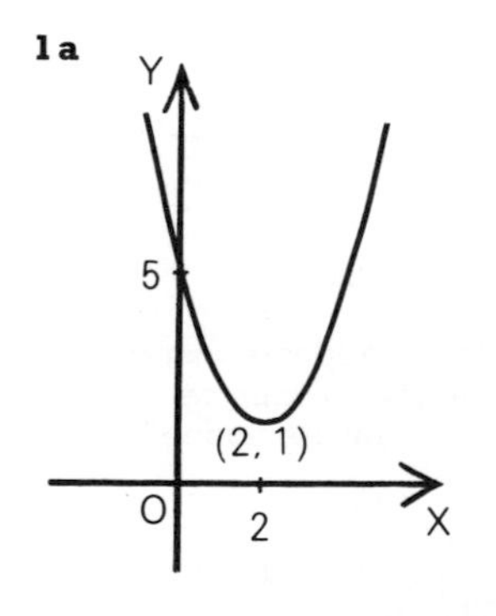

b

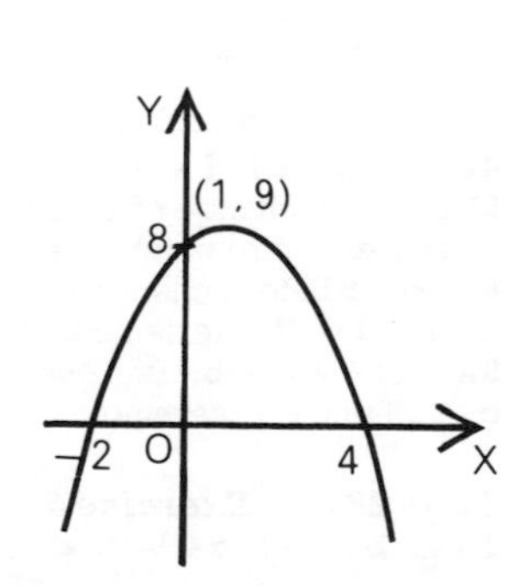

c

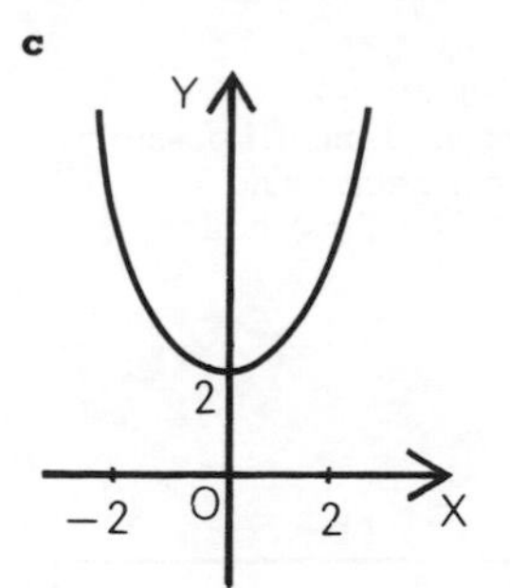

2 a

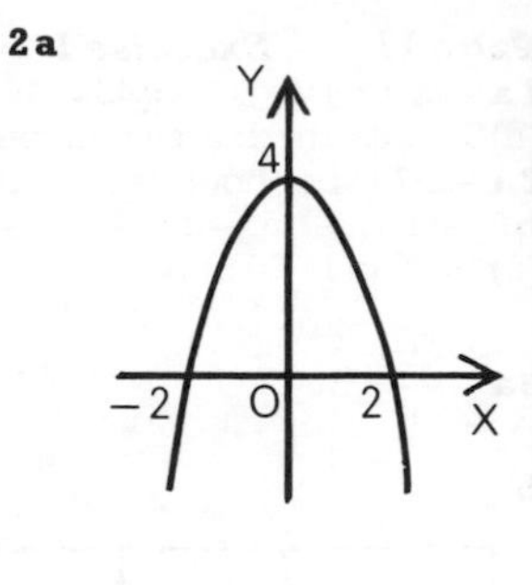

b

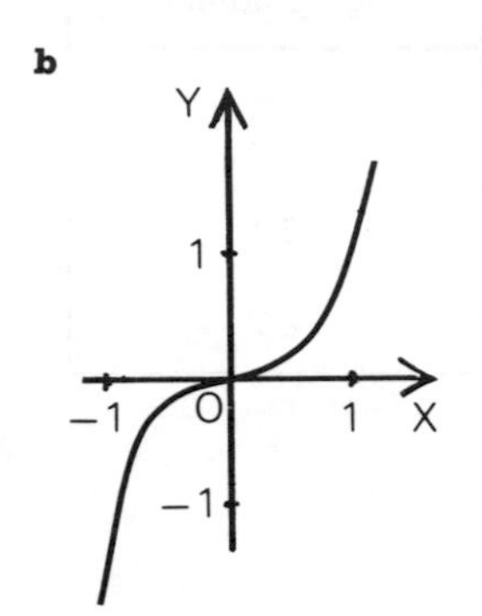

c

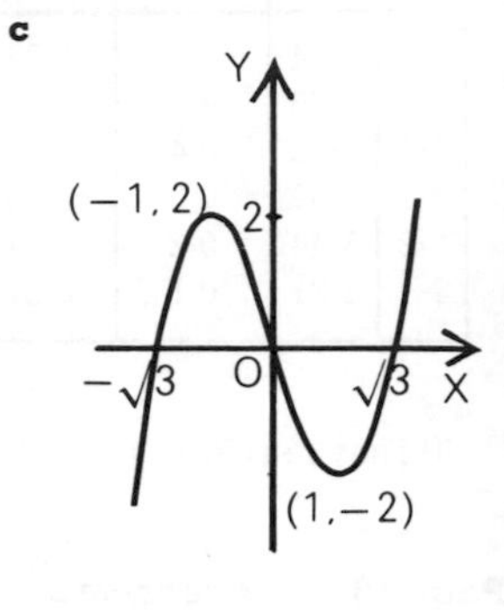

d

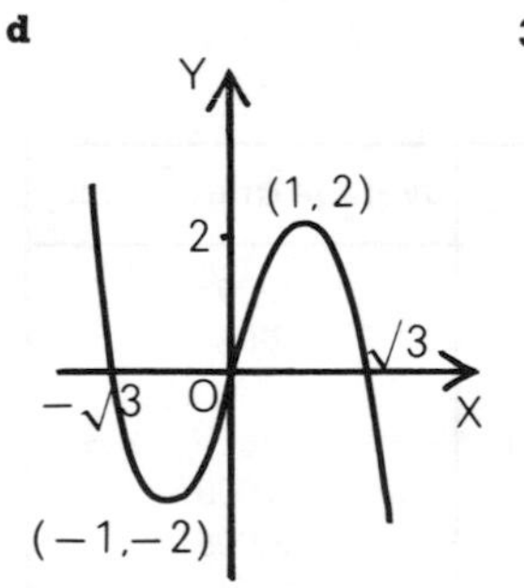

3 a

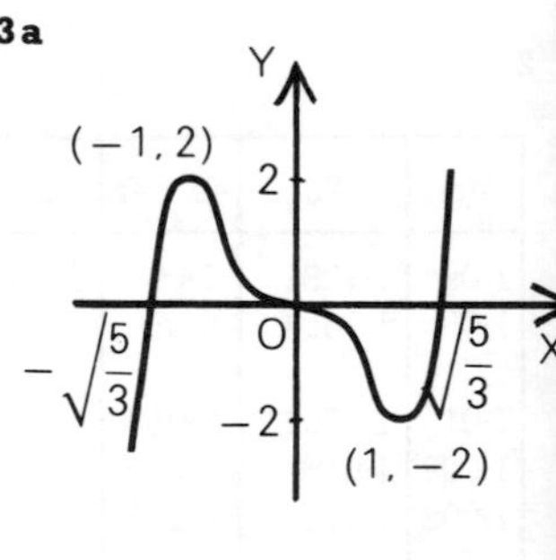

b

c

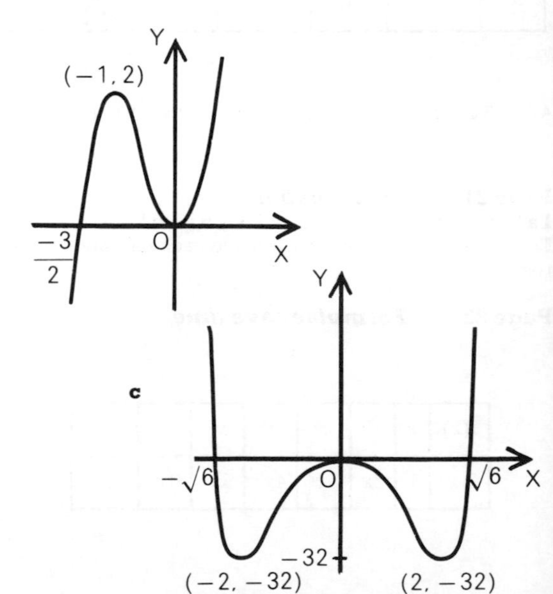

d

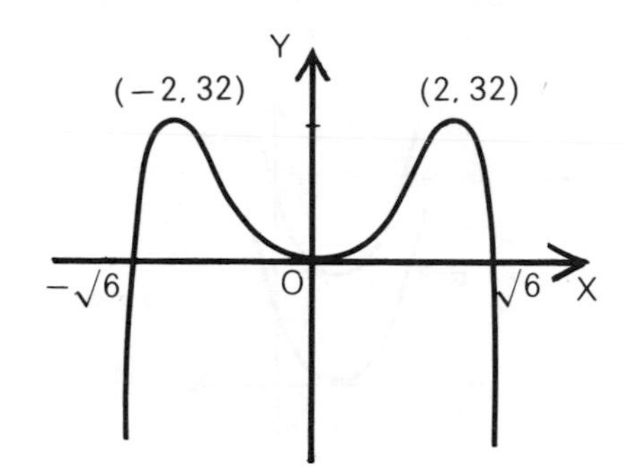

4 a **b**

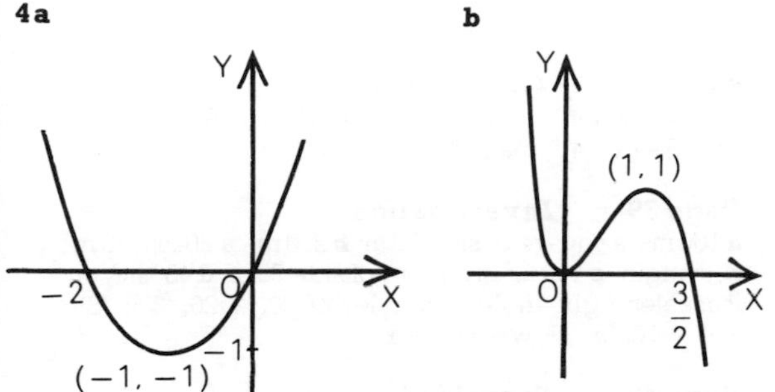

c

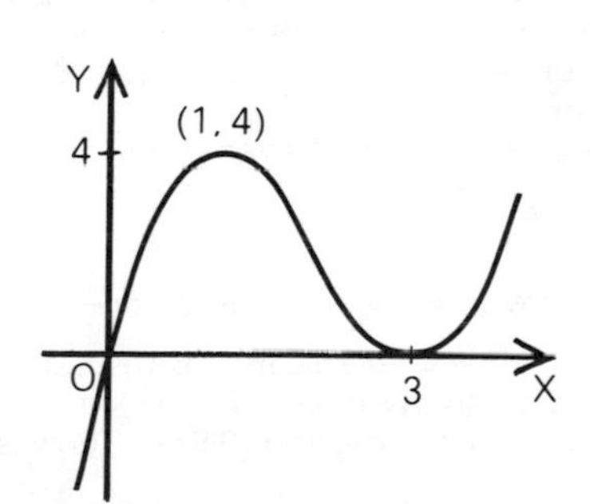

d

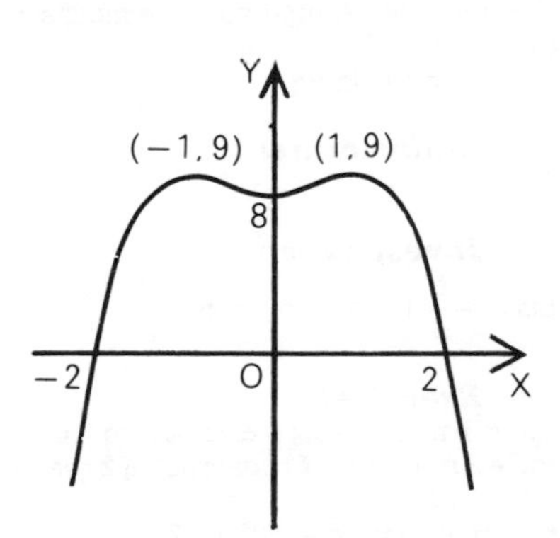

5

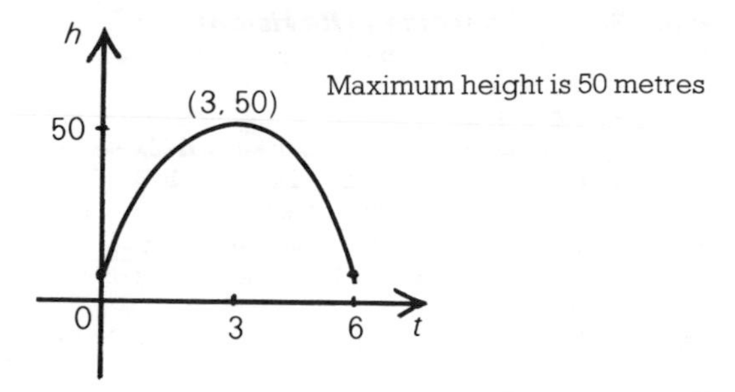

Maximum height is 50 metres

6 a

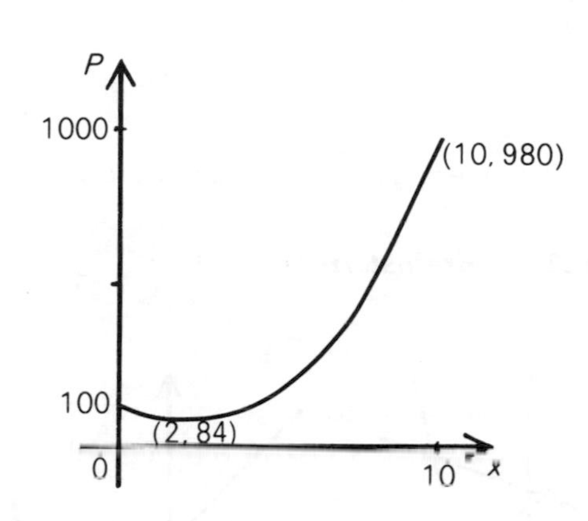

b Least profit: £84

Maximum profit: £980 for $0 \leqslant x \leqslant 10$

Page 34 *Check-up on Differentiation—1*

1 a 7 **b** 66·3 mph

2 a $f'(x) = 6x$ **b** (i) 4 (ii) 3

3 a $3x^2+5$ **b** $14x^6-12x^2$ **c** $\frac{1}{2}t^{-1/2}-1$ **d** $2+\frac{1}{t^3}$ **e** $\frac{3}{2}u^{1/2}-\frac{1}{2}u^{-3/2}$

f $v^{-2/3}+\frac{1}{9}v^{-4/3}$ **g** $8x^3-5x^{-2}$

h $-20u^{-6}+16u^{-5}-3u^{-4}$

4 $y+5x=-1$ **5** Rising: $0 < t < 4$; falling: $4 < t < 8$

6 a $f'(x) = 12x^3-12x^2$

b Decreasing for $x < 0$ and $0 < x < 1$; increasing for $x > 1$

c (0, 0) is a PI; (1, −1) is a minimum TP

d

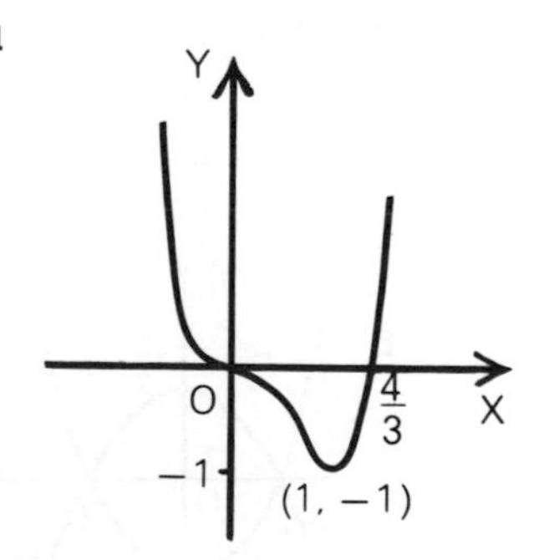

Page 36 ***Exercise 1 (Revision)***

1 a $a = 6, b = -7, c = 1$ **b** $a = -1, b = 2, c = 4$
c $a = -3, b = 0, c = 7$ **d** $a = 2, b = 0, c = 0$
e $a = -10, b = 4, c = 0$
2 a $(t+2)(t-1)$ **b** $(p-4)(p-2)$ **c** $(n-1)^2$ **d** $(2x+3)(x-1)$
e $(x-7)^2$ **f** $(2y-3)(y-2)$ **g** $(m-1)(m+1)$ **h** $(4-5r)(4+5r)$
3 a $t = -3$ or 1 **b** $x = -3$ or -2 **c** $p = -15$ or -1
d $y = -2$ or 4 **e** $x = \frac{1}{3}$ **f** $y = -3$ or 3 **g** $r = 0$ or $\frac{1}{2}$
h $q = -2$ or 0 **i** $x = -\frac{1}{2}$ or 3 **j** $d = \frac{3}{2}$ **k** $x = -3$ or 4
l $y = -5$ or 3
4 a Peter **b** $x^2+x-6 = 0$; $2x^2+2x-12 = 0$

Page 36 ***Exercise 2***

1 Line meets parabola where $(x-2)^2 = 0$; $(2, 4)$
2 $y = 6x-9$; $(3, 9)$
3 $(-2, 4)$ **4 a** $(2, 8)$ **b** $(1, 3)$ **c** $(-1, 4)$ **5** $(\frac{1}{2}, 2)$
6 $(1, 2)$ for $y = 4-2x$; $(\frac{1}{2}, 4)$ and $(-2, -1)$ for $y = 3+2x$
7 Tangent with point of contact $(-1, -2)$
8 a Circle, centre origin, radius $\sqrt{2}$ **b** Yes, with point of contact $(1, 1)$

Page 37 ***Brainstormer***

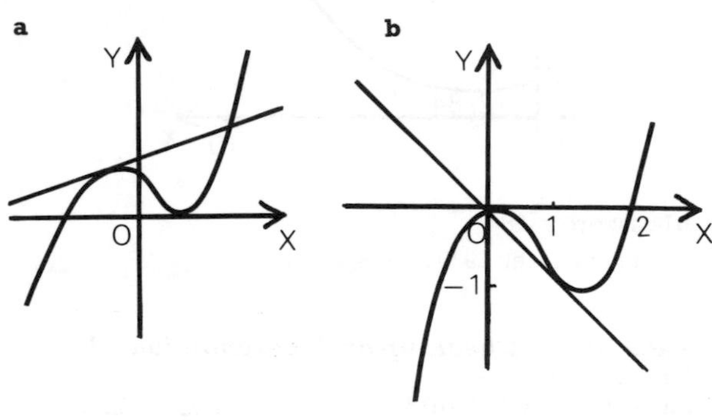

In **b**, points of intersection are (0. 0) and $(1, -1)$. When $x = 1$ gradient of $y = x^3-2x^2$ is -1, the same as $y = -x$

Page 38 ***Exercise 3***

1 a $(x+1)^2+4$ **b** $(y+3)^2-10$ **c** $(t-5)^2-22$ **d** $(k-1)^2-6$
e $2(x+2)^2-7$ **f** $3(y-1)^2-5$ **g** $2(t+1)^2+2$ **h** $5(p-2)^2-13$
i $(u+\frac{3}{2})^2-\frac{13}{4}$ **j** $(n-\frac{1}{2})^2+\frac{19}{4}$ **k** $2(c+\frac{1}{2})^2+\frac{5}{2}$ **l** $4(r-\frac{3}{2})^2-6$
m $8-(x+1)^2$ **n** $8-3(x+1)^2$ **o** $14-2(t+1)^2$ **p** $\frac{21}{2}-2(x-\frac{3}{2})^2$
2 c 1; 2 **3 c** -1; 1
4 a Concave up; $a = 1$ **b** Concave down; $a = -1$

5

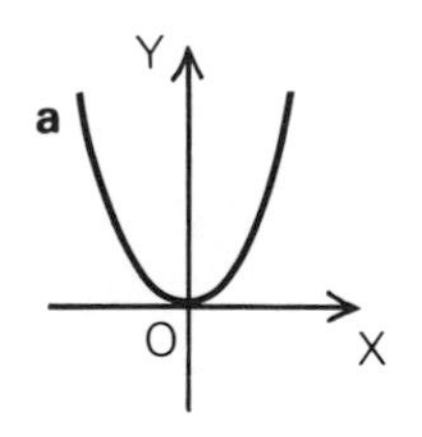

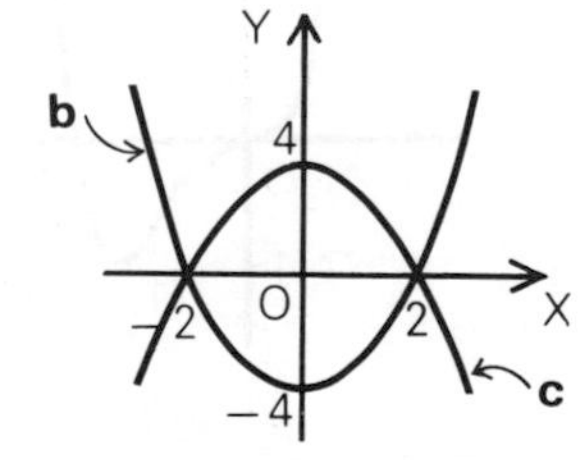

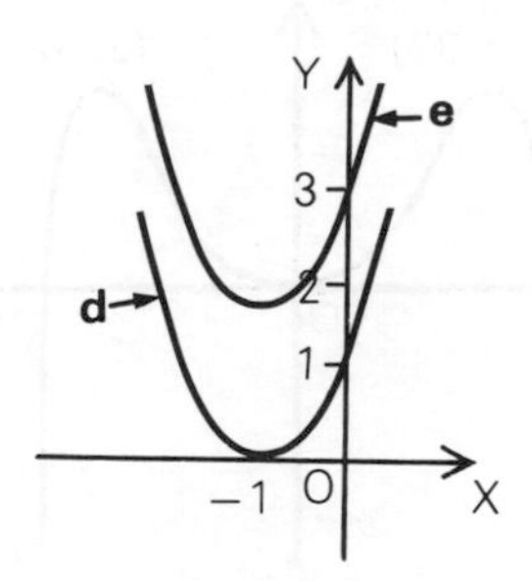

6 a, b (i) $(x+2)^2+6$; min. 6 when $x = -2$ (ii) $(x-3)^2-8$; min. -8 when $x = 3$ (iii) $9-(x+1)^2$; max. 9 when $x = -1$ (iv) $28-(x-4)^2$; max. 28 when $x = 4$

Page 39 ***Investigation***

a $100\,m^2$, a square of side 10 m **b** $200\,m^2$, a 20 m × 10 m rectangle **c** $400\,m^2$, a square of side 20 m **d** $400\,m^2$, an isosceles right-angled triangle, 40, $20\sqrt{2}$, $20\sqrt{2}$ m. (Show $4A^2 = 1600t-t^2$, where $t = x^2$)

Page 40 ***Exercise 4***

1 a $x = 1$ or 2 **b** $x = -1$ **c** $y = -2$ or 3
d $y = -1$ or $\frac{1}{2}$ **e** $x = -3{\cdot}24$ or $1{\cdot}24$ **f** $x = -4{\cdot}56$ or $-0{\cdot}44$
g $x = -2{\cdot}35$ or $0{\cdot}85$ **h** $x = 0{\cdot}18$ or $1{\cdot}82$
2 a $y = -1$ or $-\frac{2}{3}$ **b** $t = -3{\cdot}30$ or $0{\cdot}30$ **c** $x = 0$ or 1
d $x = 0{\cdot}76$ or $5{\cdot}24$
3 $x = 20, -25$, so 20 to break even ($x > 0$)
4 a $\frac{x}{1-x} = \frac{1}{x}$ **b** $\frac{\sqrt{5}-1}{2}$ ($x > 0$)
5 a Time upstream $= \frac{2}{12-v}$, downstream $\frac{2}{v}$
b $v = 10{\cdot}9$ or $1{\cdot}1$. $v = 10{\cdot}9$ means upstream speed 1·1 mph, downstream speed 10·9 mph
$v = 1{\cdot}1$ means upstream speed 10·9 mph, downstream speed 1·1 mph
6 a Time with wind $= \frac{40}{100+v}$, against $\frac{40}{100-v}$
b $v = \pm\sqrt{2000} \doteqdot \pm 44{\cdot}7$ (negative value means wind is in opposite direction)
7 Neil: 12 km/h; Tim: 15 km/h

Page 41 ***Brainstormer***

$x = \pm 1{\cdot}55$

Page 41 ***Investigation***

Sum of roots is $-\frac{b}{a}$; product of roots is $\frac{c}{a}$

Page 42 ***Exercise 5***

1 a 2 real roots **b** no real roots **c** no real roots
d 1 real root **e** 2 real roots **f** 1 real root **g** 2 real roots
h 2 real roots
2 a $-\frac{1}{2}, 2$ **b** no real roots **c** $-0{\cdot}32, 6{\cdot}32$
3 a (i) 2 roots (ii) equal roots (iii) no real roots
b (i) no real roots (ii) 2 roots (iii) equal roots
4 a $p < 1$ **b** $p = 1$ **c** $p > 1$
5 b $y = 2x, y = -2x$ **6 b** $c = -4$ **c** $(2, -2)$
7 b -7 **c** $(1, -2)$

Page 43 ***Exercise 6***

1 $x < -1$ or $x > 3$ **2** $x \leqslant -4$ or $x \geqslant -1$
3a -3 and 3 **b** $x < -3$ or $x > 3$ **c** $-3 \leqslant x \leqslant 3$
4a $x \leqslant 0$ or $x \geqslant 5$ **b** $1 < x < 4$ **c** $x < -1$ or $x > \frac{3}{2}$ **d** $\frac{5}{2} \leqslant x \leqslant \frac{7}{2}$
5a $-4 < x < 1$ **b** (i) $t \leqslant -4$ or $t \geqslant 1$ (ii) $p \leqslant -\frac{1}{2}$ or $p \geqslant 1$
6 (i) with **b** (ii) with **a** (iii) with **c** (iv) with **d**
7a $x \leqslant 0$ or $x \geqslant 1$ **b** $x < 1$ or $x > 1$ ($x \in R$ with $x \neq 1$)
c $x \leqslant -5$ or $x \geqslant 1$
8a $x(x+1) < 6$ **b** $-3 < x < 2$
9a (i) $-2 \leqslant x \leqslant 2$ (ii) $x < -2$ or $x > 2$
b (i) $x \leqslant -1$ or $x \geqslant \frac{5}{2}$ (ii) $-1 < x < \frac{5}{2}$
10 Limits lie between 15 m and 20 m in breadth

Page 44 ***Check-up on Quadratic Theory***

1a $(x+3)(x-1)$ **b** $(p+3)(p-4)$ **c** $(t+1)(3t-5)$
2a $(1, -3), (3, 1)$ **b** $(-2, -3)$ **c** $(2, 4), (-1, -8)$ It is a tangent
3a $(x-2)^2+7$ **b** $\frac{49}{8}-2(x+\frac{3}{4})^2$ **c** $3(t+1)^2-8$
4a $(x-1)^2-7$; minimum value -7 when $x = 1$
b $8-2(x-1)^2$; maximum value 8 when $x = 1$ **c** $(x+\frac{3}{2})^2-\frac{5}{4}$; minimum value $-\frac{5}{4}$ when $x = -\frac{3}{2}$

5a

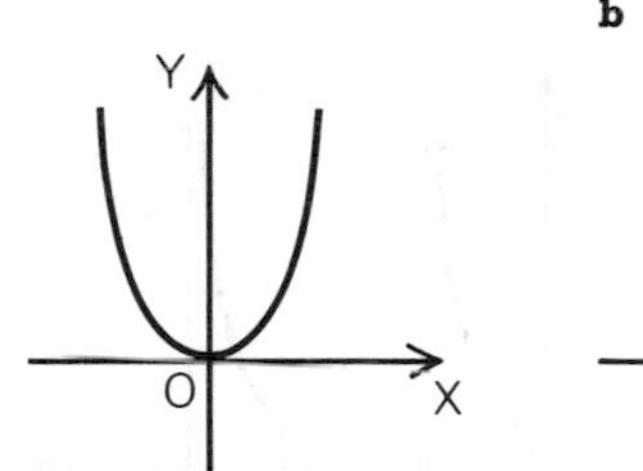

b

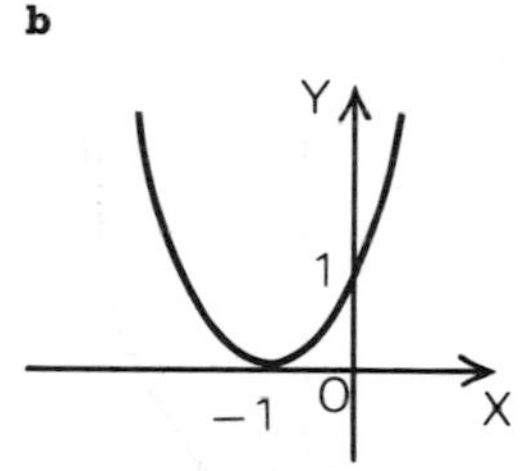

c

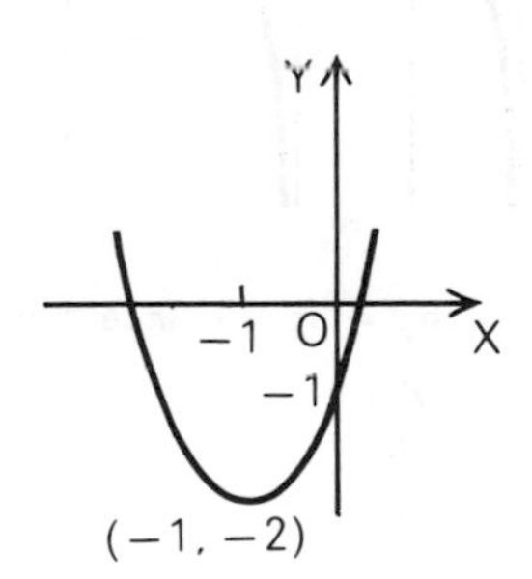

6a $-1{\cdot}24, 3{\cdot}24$ **b** $0{\cdot}68, 7{\cdot}32$ **c** $-0{\cdot}82, 1{\cdot}82$
7a (i) 1 (ii) 2 real roots **b** (i) -8 (ii) no real roots **c** (i) 0 (ii) equal roots **d** (i) 21 (ii) 2 real roots **e** (i) 1 (ii) 2 real roots **f** (i) 21 (ii) 2 real roots
8 Maximum profit of £900 for 100 toys
9b $t = \frac{7}{4}$, $(\frac{3}{2}, \frac{25}{4})$
10a $-1 \leqslant x \leqslant -\frac{1}{4}$ **b** for all $x \in R$.

Page 45 ***Exercise 1***

1a 28·7° **b** 14·2 **c** 6·3
2a 7 m **b** 10 m **3a** 256 m **b** 282 m
4 $\alpha = 75{\cdot}4$, $\beta = 37{\cdot}6$, $\theta = 37{\cdot}8$ No, since $\theta > 35$
5 5·0 cm **6a** 40 m **b** 37 m
7a 38·9° **b** 3051 m² **8** 131·1°
9a 93 km **b** 076°

Page 47 ***Exercise 2***

1a 10 cm **b** 26·6° **c** 39·8°
2a ∠TQP, ∠VQR, ∠RWS **b** ∠VQR or ∠WPS, ∠TQP or ∠WRS, ∠PWS or ∠QVR
3a 45° **b** 35·3°
4a ABCD is symmetrical about plane AMD
b 1·7 m, 0·6 m **c** 70·5°
5a (i) 9·4 m (ii) 42·0° **b** No. Eye moves 32·4°
6a 1·6 m **b** 39·8° **c** 1·3 m **d** 33·2° **e** 35·5°

Page 48 ***Brainstormer***

$$\operatorname{Sin} \alpha \cos \beta = \frac{AD}{AC} \cdot \frac{AC}{AB} = \sin \gamma$$

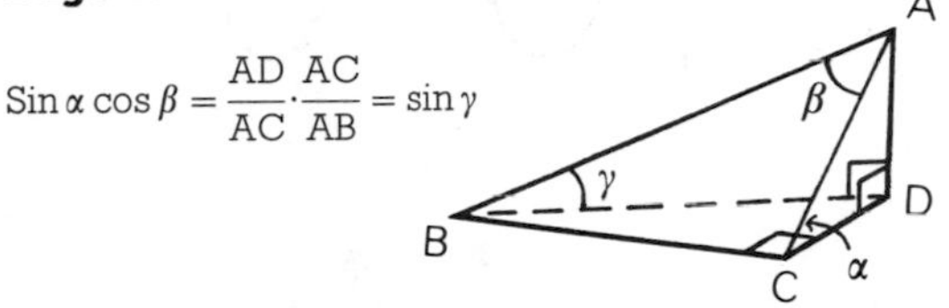

Page 49 ***Exercise 3***

1a $\frac{\pi}{6}$ **b** $\frac{\pi}{3}$ **c** $\frac{\pi}{2}$ **d** $\frac{2\pi}{3}$ **e** π **f** 2π
2a 45° **b** 90° **c** 120° **d** 270°

3a

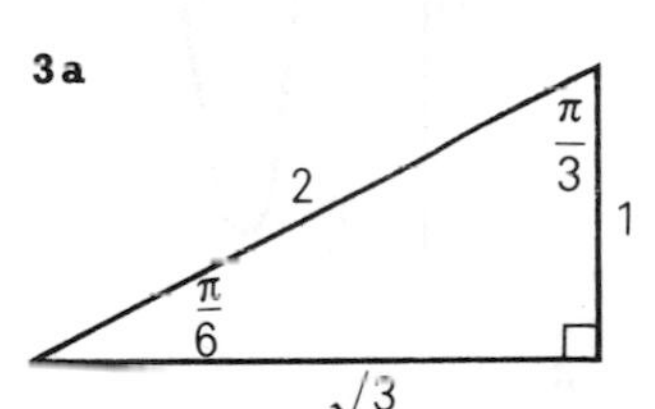

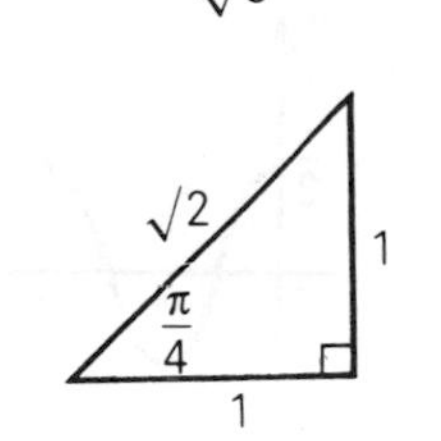

b

θ	$\frac{\pi}{6}$	$\frac{\pi}{4}$	$\frac{\pi}{3}$
$\sin\theta$	$\frac{1}{2}$	$\frac{1}{\sqrt{2}}$	$\frac{\sqrt{3}}{2}$
$\cos\theta$	$\frac{\sqrt{3}}{2}$	$\frac{1}{\sqrt{2}}$	$\frac{1}{2}$
$\tan\theta$	$\frac{1}{\sqrt{3}}$	1	$\sqrt{3}$

4a $\sin 30° = 0{\cdot}5$, $\cos 45° \doteqdot 0{\cdot}7071067$, $\tan 60° \doteqdot 1{\cdot}7320508$
5 2 revs/second = 12·6 radians/second

6a $$\frac{\theta}{2\pi} = \frac{\text{arc AB}\,(l)}{\text{circumference}\,(2\pi r)} = \frac{\text{area of sector AOB}}{\text{area of circle}\,(\pi r^2)}$$

7a 84 mm **b** 1470 mm²

Page 50 ***Brainstormer***

At least 57·3 cm long (to 1 decimal place)

Page 51 ***Exercise 4***

1a $\frac{3}{5}, \frac{4}{5}, \frac{3}{4}$ **b** $\frac{3}{5}, -\frac{4}{5}, -\frac{3}{4}$ **c** $-\frac{3}{5}, -\frac{4}{5}, \frac{3}{4}$ **d** $-\frac{3}{5}, \frac{4}{5}, -\frac{3}{4}$
3a Negative **b** negative **c** positive **d** positive **e** negative **f** positive

4

θ	0	$\frac{\pi}{2}$	π	$\frac{3\pi}{2}$	2π
$\sin\theta$	0	1	0	-1	0
$\cos\theta$	1	0	-1	0	1
$\tan\theta$	0		0		0

5

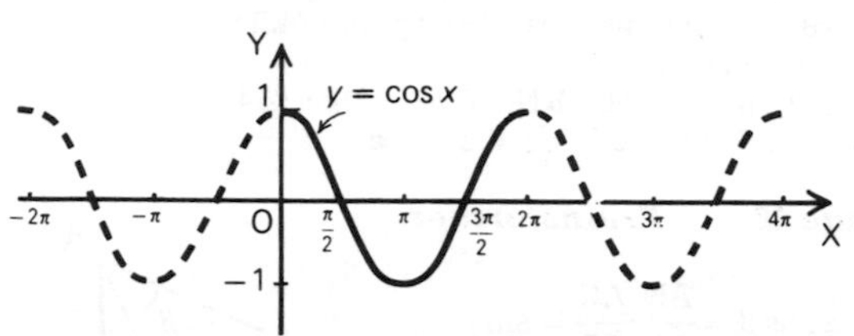

6a

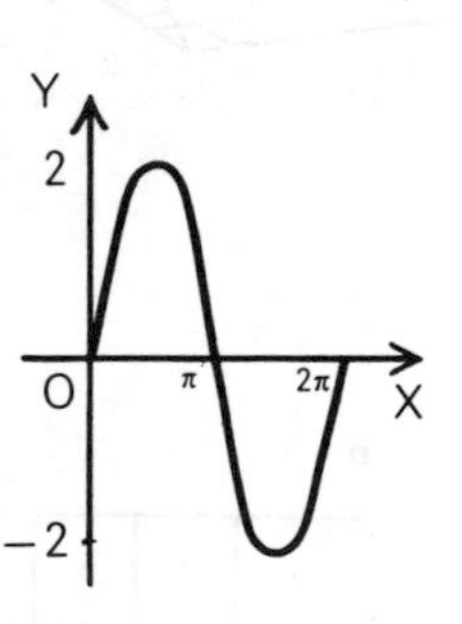

b

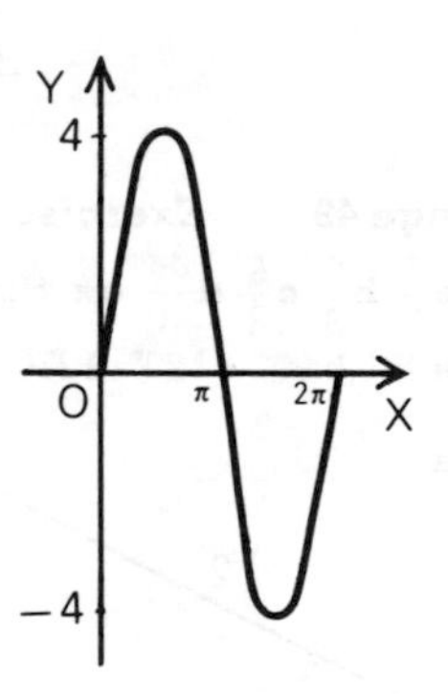

c

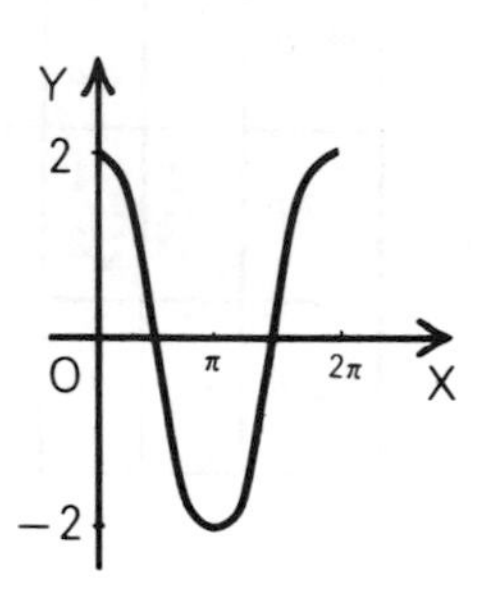

d

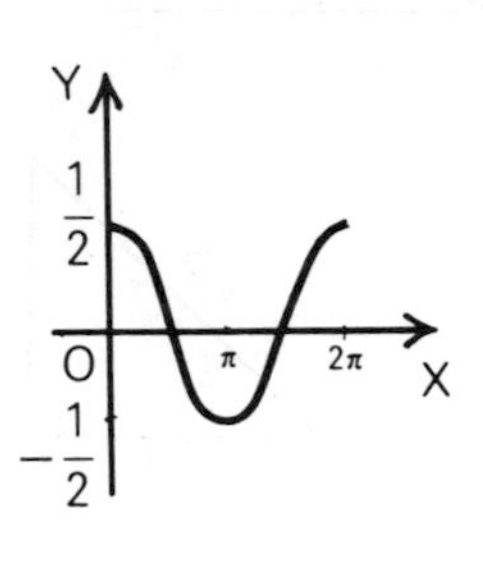

e

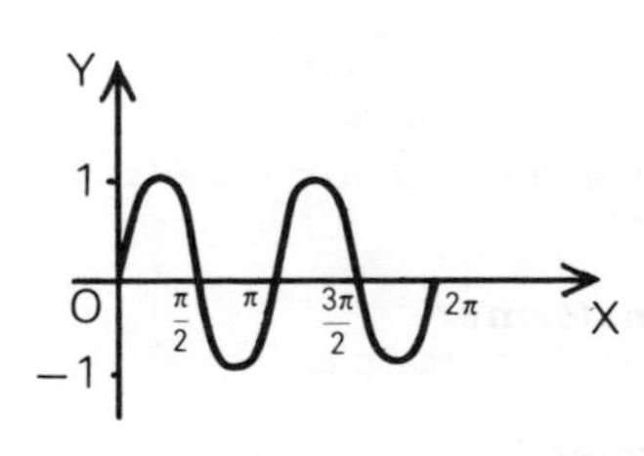

f

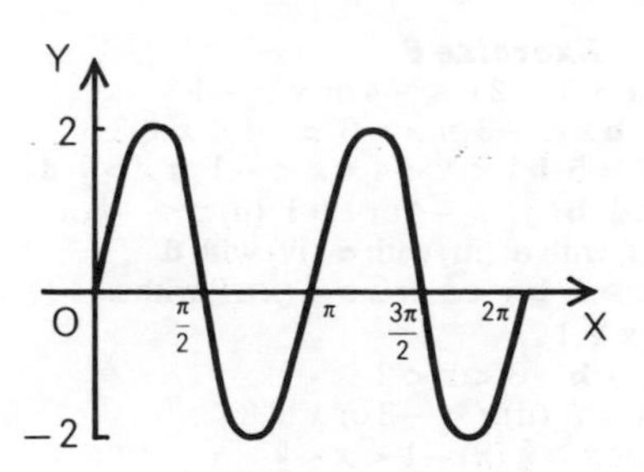

g

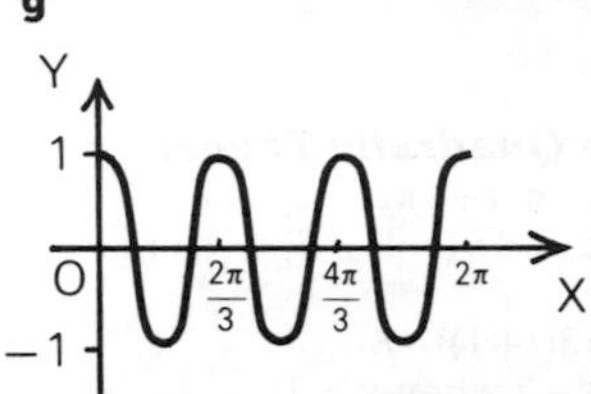

h

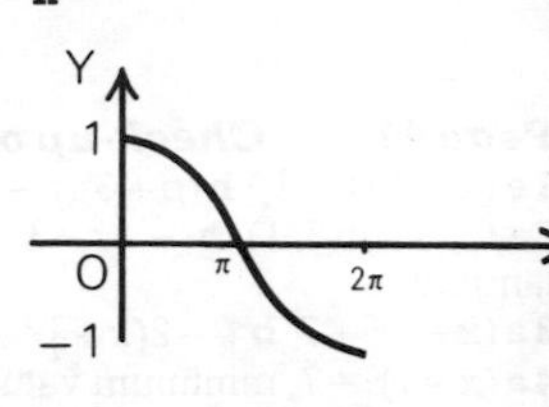

7 a $\tan\theta = \frac{y}{x}$, $\tan(\pi+\theta) = \frac{-y}{-x} = \frac{y}{x} = \tan\theta$ **b** π radians

c

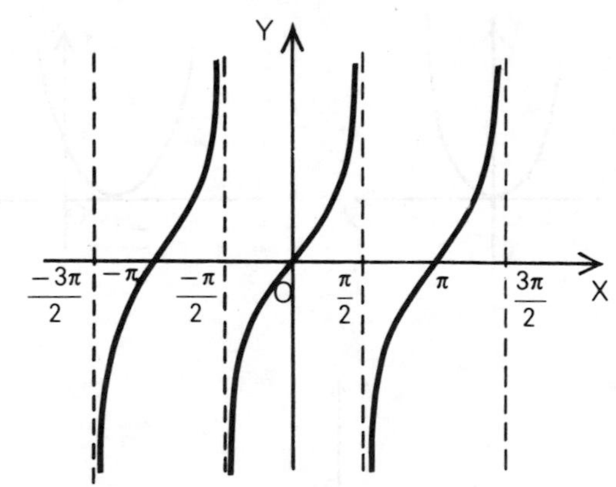

8 a $y = \sin 3x$; $\frac{2\pi}{3}$ radians **b** $y = 3\cos x$; 2π radians

9 a (i) 1 (ii) 1 (iii) 2 (iv) 3 (v) 6

b (i) $\frac{\pi}{2}$ (ii) 0 (iii) $\frac{\pi}{2}$ (iv) 0 (v) $\frac{\pi}{2}$

10 a 6 m **b**

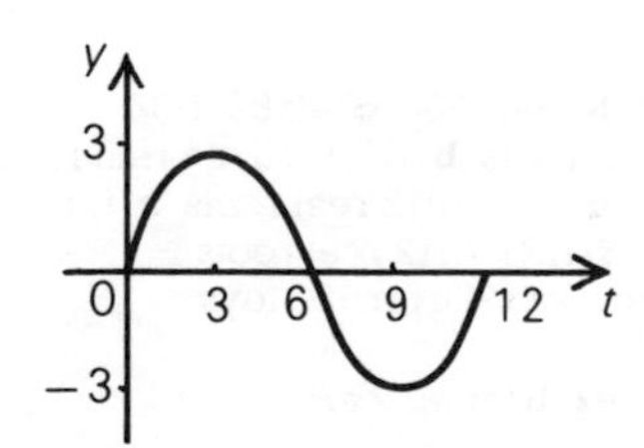

11 a $\frac{2}{3}$ m **b** $\frac{2\pi}{3}$ seconds

c

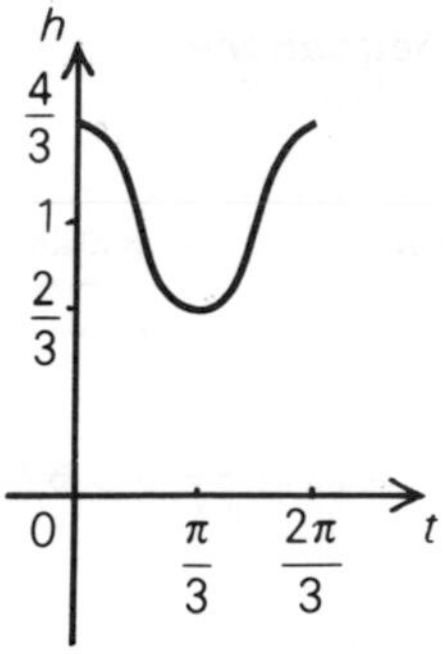

Page 54 *Exercise 5*

1 a $\frac{\pi}{2}$ **b** $0, 2\pi$ **c** $0, \pi, 2\pi$ **d** π

2 a $\frac{\pi}{4}, \frac{3\pi}{4}$ **b** $\frac{5\pi}{4}, \frac{7\pi}{4}$ **c** $\frac{2\pi}{3}, \frac{4\pi}{3}$ **d** $\frac{\pi}{6}, \frac{11\pi}{6}$

3 a 210, 330 **b** 60, 300 **c** 240, 300 **d** 45, 315

4 a $\frac{\pi}{4}, \frac{5\pi}{4}$ **b** $\frac{3\pi}{4}, \frac{7\pi}{4}$ **c** $\frac{\pi}{3}, \frac{4\pi}{3}$ **d** $\frac{5\pi}{6}, \frac{11\pi}{6}$

5 After $\frac{\pi}{9}$ seconds **6** At time 1 hour

Page 54 *Brainstormer*

[illegible]

a $\theta = \begin{cases} \frac{\pi}{6} + 2n\pi \\ \frac{5\pi}{6} + 2n\pi \end{cases}, n \in Z$ **b** $\theta = \pm\frac{\pi}{4} + 2n\pi, n \in Z$

c $\theta = \frac{3\pi}{4} + n\pi, n \in Z$

Page 55 *Check-up on Trigonometry*

1 a $r = \sqrt{(R^2 - d^2)}$ **b** $\sin\theta° = \frac{1}{R}\sqrt{(R^2 - d^2)}$; $\cos\theta° = \frac{d}{R}$; $\tan\theta° = \frac{1}{d}\sqrt{(R^2 - d^2)}$

2 a 71 m **b** 29° **c** 64 m

3 a $\frac{\pi}{3}$ **b** $\frac{2\pi}{3}$ **c** $\frac{5\pi}{6}$ **d** $\frac{3\pi}{2}$ **e** $\frac{7\pi}{4}$

4 a 90°/second **b** $\frac{\pi}{2}$ radians/second

5 a

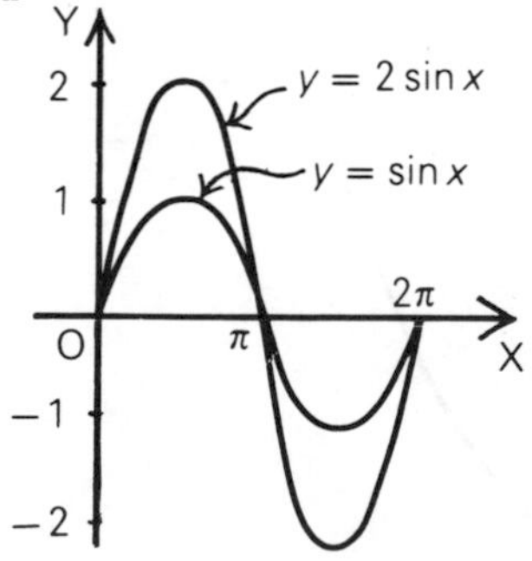

b

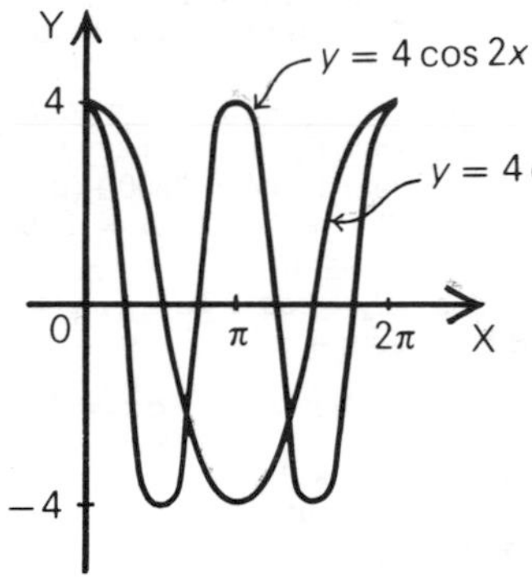

6 a $\frac{1}{2}$ m **b** $t = \frac{\pi}{18} + \frac{2}{3}n\pi$, $t = \frac{5\pi}{18} + \frac{2}{3}n\pi$ $(n \in Z)$

7 a $\frac{3\pi}{2}$ **b** $\frac{\pi}{2}, \frac{3\pi}{2}$ **c** $\frac{7\pi}{6}, \frac{11\pi}{6}$ **d** $\frac{5\pi}{6}, \frac{7\pi}{6}$.

Page 56 *Exercise 1*

1 a 6 **b** 8 **c** 5 **d** 11

2 a Sequences **c** and **d** finite; **a** and **b** infinite.

b Sequence **c**: 1, 2, 4, 5, 10, 20. Sequence **d**: 1, 2, 4, 11, 22, 44

3 a 2, 4, 6, 8, 10 **b** 4, 7, 10, 13, 16 **c** 2, 5, 10, 17, 26

d $\frac{1}{3}, \frac{1}{9}, \frac{1}{27}, \frac{1}{81}, \frac{1}{243}$ **e** 0, −7, −26, −63, −124 **f** $1\frac{1}{2}, 1\frac{3}{4}, 1\frac{7}{8}, 1\frac{15}{16}, 1\frac{31}{32}$

4 $\frac{1}{2}, \frac{\sqrt{3}}{2}, 1, \frac{\sqrt{3}}{2}, \frac{1}{2}, 0$

5 a $a = 3, b = 2$ **b** 11, 14, 17

6 a $0, \frac{1}{2}, \frac{2}{3}, \frac{3}{4}, \frac{4}{5}$ **b** 0·9, 0·99, 0·999, 0·9999, 0·999 99, 0·999 999

c 1 **d** $\frac{1}{n} \to 0$ as $n \to \infty$ so $1 - \frac{1}{n} \to 1$ as $n \to \infty$

7 a 0·49 g **b** 0·12 g **c** less than 0·65 g

8 a 0 **b** 2 **c** 3

9 a $\sin x°, \sin^2 x°, \sin^3 x°$ **b** $\sin^n x°$ **c** $OP_n \to 0$ as $n \to \infty$

d $\cos x°, \sin x° \cos x°, \sin^2 x° \cos x°$ **e** $\sin^{n-1} x° \cos x°$, 0

Page 57 *Investigation*

1 a $2n-1$ **b** $5n-7$ **c** $10-6n$

2 a Quadratic

n	1	2	3	4	5
u_n	$a+b+c$	$4a+2b+c$	$9a+3b+c$	$16a+4b+c$	$25a+5b+c$
1st differences	$3a+b$	$5a+b$	$7a+b$	$9a+b$	
2nd differences	$2a$	$2a$	$2a$		

b $\frac{1}{2}n^2 + \frac{1}{2}n$ or $\frac{1}{2}n(n+1)$ (i) n^2+n-5 (ii) $\frac{3}{2}n^2 - \frac{1}{2}$ or $\frac{1}{2}(3n^2-1)$

3 a Of the form $an^3 + bn^2 + cn + d$ (a, b, c, d constants)

b $\frac{1}{3}n^3 + \frac{1}{2}n^2 + \frac{1}{6}n$ or $\frac{1}{6}n(2n+1)(n+1)$

Page 59 *Exercise 2*

3, 4, 7, 8 $n \geqslant 1$; **5, 6** $n \geqslant 0$

1 a (i) 1, 2, 4, 8 (ii) $u_n = 2^{n-1} (n \geqslant 1)$ **b** (i) 3, 8, 13, 18 (ii) $u_n = 5n - 2 (n \geqslant 1)$ **c** (i) $6, 3, \frac{3}{2}, \frac{3}{4}$ (ii) $u_n = \frac{3}{2^{n-2}} (n \geqslant 1)$

d (i) 5, 1, −3, −7 (ii) $u_n = 9-4n (n \geqslant 1)$
2 a (i) Current salary is £2000 more than previous year's salary, $W_{n+1} = W_n + 2000$. (ii) $L_{n+1} = L_n + 1500, n \geqslant 1$
b $W_2 = 9500$; $W_3 = 11\,500$; $W_4 = 13\,500$ By year n, $n-1$ lots of 2000 have been added to the initial salary of 7500 since 2000 was not added in year 1. $W_n = 7500+(n-1)2000$
c $L_2 = 11\,500$; $L_3 = 13\,000$; $L_4 = 14\,500$; $L_n = 10\,000+(n-1)\times 1500, n \geqslant 1$ **d** In year 6
3 a $l_2 = \cos\theta$; $l_3 = \cos^2\theta$; $l_4 = \cos^3\theta$ **b** $l_{n+1} = l_n \times \cos\theta$
c $l_n = \cos^{n-1}\theta$ **d** $l_n \to 0$ as $n \to \infty$
4 a $\frac{r_n}{r_{n+1}} = \sin 30° = \frac{1}{2}$, etc **b** $r_n = 2^{n-1}$ **c** $x_n = 2^n \times \sqrt{3}$
5 a $A_{n+1} = \frac{9}{10}A_n$ **b** $A_n = (\frac{9}{10})^n \times A_0$
c 7 filters reduce light to about 48%
6 a $A_{n+1} = \frac{11}{10}A_n$ **b** $A_n = (\frac{11}{10})^n \times 1000$ **c** 8 years
7 a $P_n = (1{\cdot}045)^n \times 210$ **b** $M_n = (1{\cdot}08)^n \times 190$ **c** 3 years
8 a (i) $(0{\cdot}4)^n \times 200$ (ii) $(0{\cdot}1)^n \times 200$ (iii) $(0{\cdot}6)^n \times 200$
b (i) 4 weeks (ii) 2 weeks (iii) 6 weeks
9 a $f_{n+1} = 2f_n$ where f_n is the frequency of C_n **b** 65·4, 130·8, 261·6, 523·2, 1046·4, 2092·8, 4185·6 **c** $f_8 = 2^7 f_1 = 2^7(32{\cdot}7)$

Page 61 ***Brainstormer***
Recurrence relation $C_{n+1} = RC_n + 100$ where R is the fraction left after a spraying.

$C_n = R^n C_0 + 100(R^{n-1}+R^{n-2}+R^{n-3}+\ldots+R+1), (n \geqslant 1)$

For Tritox $R = 0{\cdot}4$; Pillary $R = 0{\cdot}1$; Nopest $R = 0{\cdot}6$

Page 62 ***Exercise 3***
1 a By adding 180° **b** $S_{n+1} = S_n + 180\ (S_3 = 180)\ n \geqslant 3$
c $S_n = (n-2)\times 180$
2 a $5, 5k, 5k^2, 5k^3, 5k^4$ **b** $5k^n\ (n \geqslant 0)$
3 a $m = 2, c = 1$ **b** 63
4 a (i) $a_2 = \frac{1}{2}a_1$ (ii) $a_3 = \frac{1}{2}a_2$ (iii) $a_4 = \frac{1}{2}a_3$ **b** $a_{n+1} = \frac{1}{2}a_n$, $a_1 = \frac{1}{2}$
c $a_n = \frac{1}{2^n}$ **d** $a_n \to 0$ as $n \to \infty$
5 a $t_n = 1 - \frac{1}{2^n}$ **b** $t_n \to 1$ as $n \to \infty$
c $t_{n+1} = 1 - \frac{1}{2^{n+1}} = \frac{1}{2}\left(1-\frac{1}{2^n}\right)+\frac{1}{2} = \frac{1}{2}t_n + \frac{1}{2}$.
6 b (i) $a = -1, d = 3$ (ii) $a = 10, d = -4$ **c** $S_{n+1} = S_n + a + nd$
7 b (i) $a = 3, r = 2$ (ii) $a = 1, r = -3$ **c** na
d $S_n \to \frac{1}{1-r}$ as $n \to \infty$

Page 63 ***Brainstormer***
a A is reduced to $A(\frac{3}{10})^n$ after n days and 300 to $300(\frac{3}{10})^r$ after r days.
b $N_n \to \frac{3000}{7} (\doteqdot 429)$ as $n \to \infty$

Page 64 ***Exercise 4***
1 a 1, 3, 4, 7, 11, 18, 29, 47, 76, 123 **b** −4, 5, 1, 6, 7, 13, 20, 33, 53, 86
2 a 239, 577, 1393, 3363, 8119, 19 601
b $b_{n+2} = 2b_{n+1} + b_n (n \geqslant 1)$; 169, 408, 985, 2378, 5741, 13 860
3 a $1+1\sqrt{2}, 3+2\sqrt{2}, 7+5\sqrt{2}, 17+12\sqrt{2}$
c $a_{n+1} + b_{n+1}\sqrt{2} = (a_n + 2b_n) + (a_n + b_n)\sqrt{2}$ and equate coefficients
4 a $\{c_n\}$ is $\{1, 4, 10, 28, 76, 208, \ldots\}$. $\{d_n\}$ is $\{1, 2, 6, 16, 44, 120, \ldots\}$ **b** $c_{n+1} = c_n + 3d_n$; $d_{n+1} = c_n + d_n$

Page 64 ***Check-up on Sequences***
1 a −2, −7, −12, −17 **b** 5, −1, −11, −25 **c** $1\frac{1}{4}, 1\frac{1}{16}, 1\frac{1}{64}, 1\frac{1}{256}$ or 1·25, 1·0625, 1·015625, 1·00390625; $u_n \to 1$ as $n \to \infty$
2 $u_n = 2n^2 + 3n - 5\ (n \geqslant 1)$
3 a $A_{n+1} = (1{\cdot}12)A_n\ (A_0 = 2000)$ **b** $A_n = (1{\cdot}12)^n \times 2000$
c After 4 years
4 a $P_{n+1} = (0{\cdot}6)P_n\ (P_0 = 21)$ **b** $P_n = (0{\cdot}6)^n \times 21$ **c** 6 weeks
5 a $m = -2, c = 3$ **b** $u_4 = -31$
6 a $a = 3, d = 5$ **b** 3, 8, 13, 18, 23, 28, 33
7 a Either $a = 2$ with $r = 3$ or $a = -2$ with $r = -3$ **b** 2, 6, 18, 54; −2, 6, −18, 54.

Page 66 ***Exercise 1***
1 b (i) 10 (ii) 90 (iii) 360 (iv) $2\frac{1}{2}$ **c** Not defined: $x = -2$ makes no sense as x is a length, $A(-2)$ has no meaning in this context
2 a (i) 7 (ii) 13 (iii) $5\frac{1}{8}$ **b** $2x^2+5$ **c** (ii)
3 a (i) 145 (ii) $102\frac{1}{2}$ (iii) $111\frac{1}{4}$ **b** Breadth 25 cm gives 100 cm minimum perimeter **c** 625 cm²

4 a

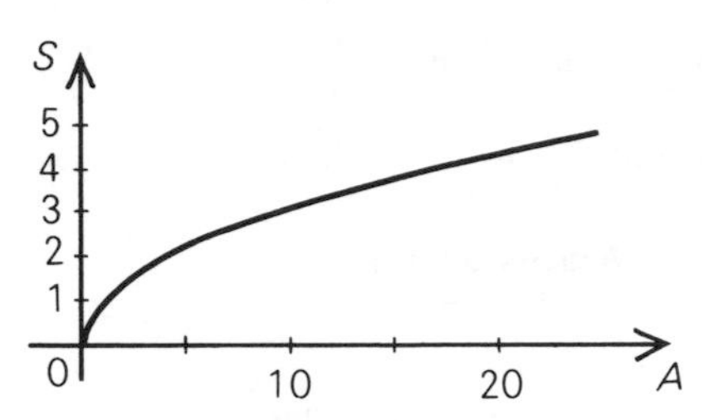

b Area is always positive or zero
5 a $6+c$ **b** $c = 4$

Page 68 ***Exercise 2***
1 a {5, 6, 7} **b** {−3, −1, 1} **c** $\{\frac{1}{3}, \frac{1}{2}, 1\}$ **d** {0, 25, 100}
e {0, 1, 4} **f** {0, 1, 4} **g** $\{x \in R : x \geqslant 0\}$ **h** $\left\{0, \frac{1}{\sqrt{2}}, 1\right\}$
i $\{-1, -\frac{1}{2}, 0, \frac{1}{2}, 1\}$ **j** $\{x \in R : -2 \leqslant x \leqslant 2\}$
2 a Domain = {1, 2, 3, 4, 5} Range = {−2, 0, 5, 6}
b Domain = {−3, −2, −1, 0, 1, 2, 3}
Range = {−5, 1, 5, 7, 9, 11}
3 a $\{u \in R : 0 < u \leqslant 70\}$ **b** $\{d \in R : 0 < d \leqslant 315\}$
4 a $\{x \in R : x \geqslant 0\}$ **b** $\{x \in R : x \geqslant 1\}$ **c** $\{x \in R : x \leqslant 3\}$
d $\{x \in R : x \leqslant -1 \text{ or } x \geqslant 1\}$ **e** $\{x \in R : x \neq 0\}$
f $\{x \in R : x \neq -1 \text{ and } x \neq 1\}$

Page 70 ***Exercise 3***
1 a Yes **b** No **c** No **d** Yes

2 a (i)

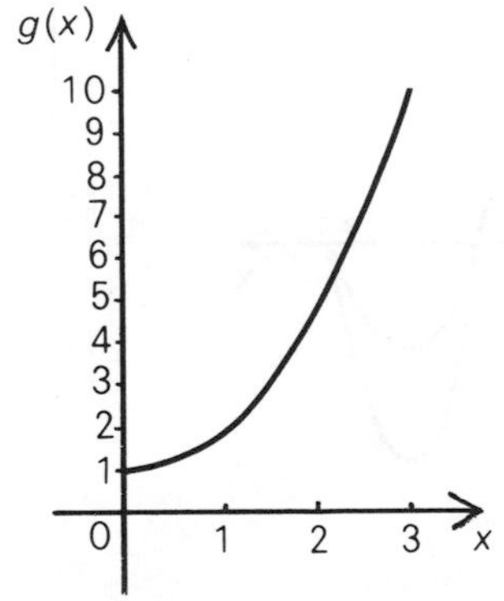

(ii) Range = $\{g \in R : g \geqslant 1\}$
b (i) The graph extends into the second quadrant, its reflection in g-axis
(ii) Range is the same
3 (i) **a**

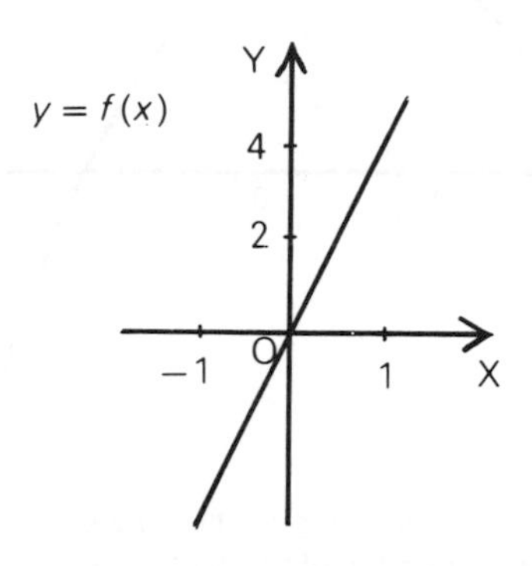

b

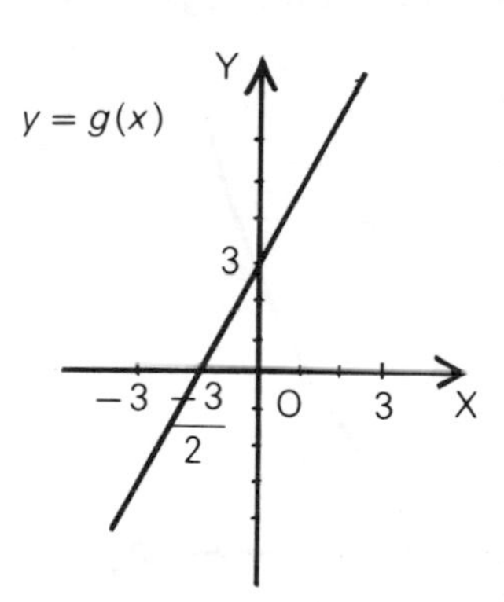

c

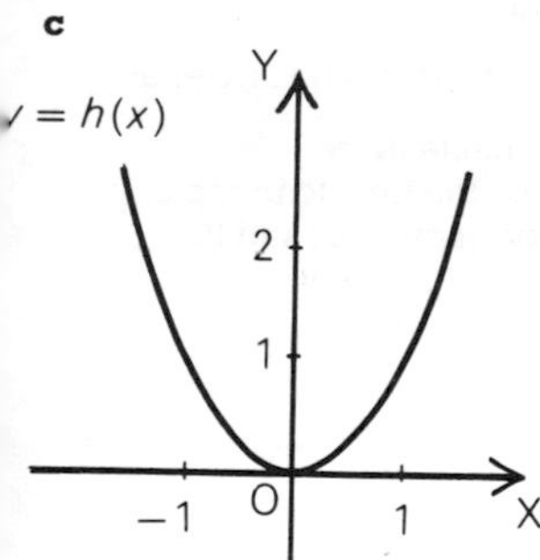

d

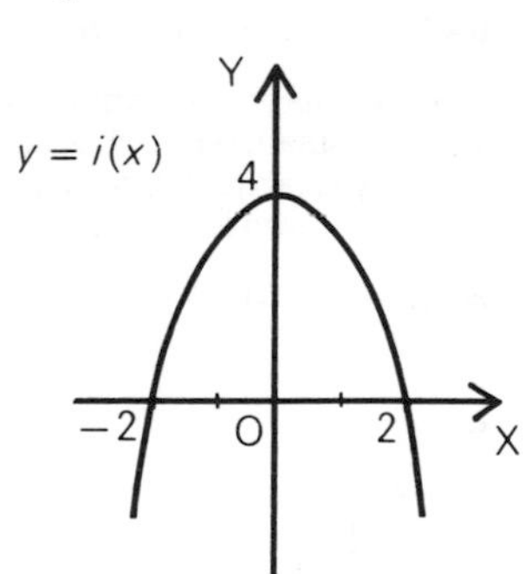

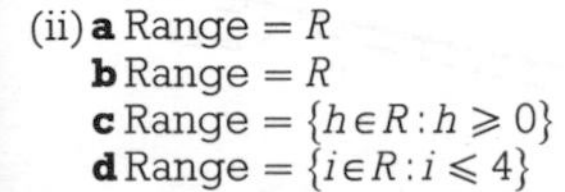

(ii) **a** Range = R
b Range = R
c Range = $\{h \in R : h \geqslant 0\}$
d Range = $\{i \in R : i \leqslant 4\}$

4 a

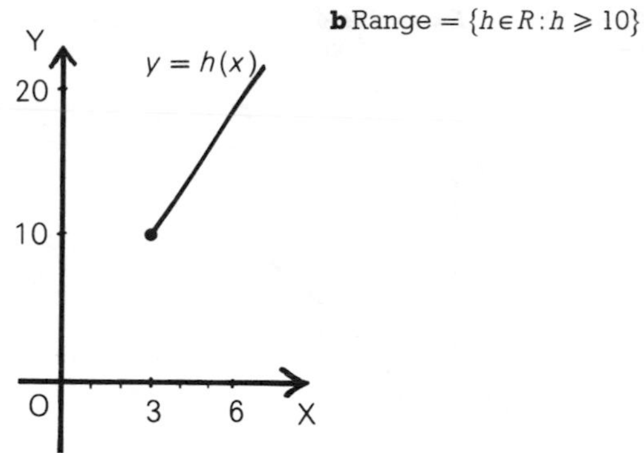

b Range = $\{h \in R : h \geqslant 10\}$

5 b The symbol ● indicates the end point is included. The symbol ○ indicates the end point is not included
c Range = $\{-1, 1\}$
6 a

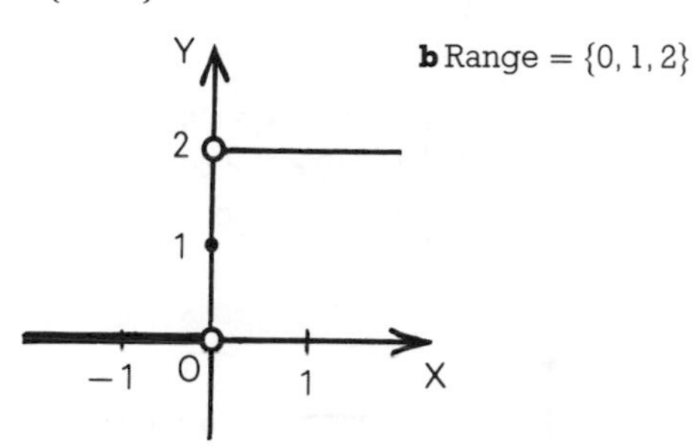

b Range = $\{0, 1, 2\}$

7 Here, **a** and **c** represent functions, since for *every* x in the domain there is *one and only one* $f(x)$. But, **b** and **d** cannot represent functions as there are some x in the domain which have two $f(x)$.
8 a

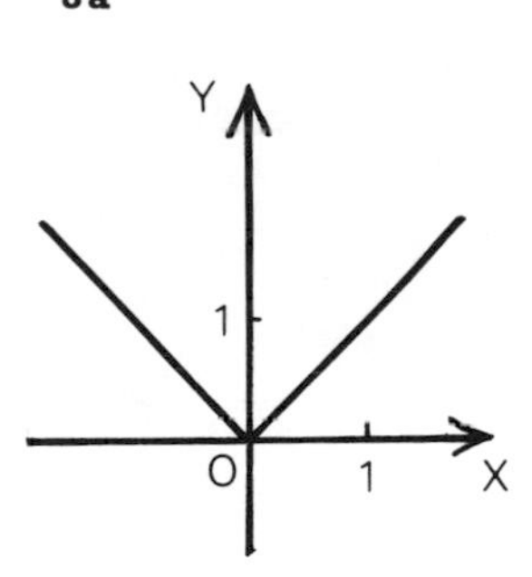

Range = $\{y \in R : y \geqslant 0\}$

b

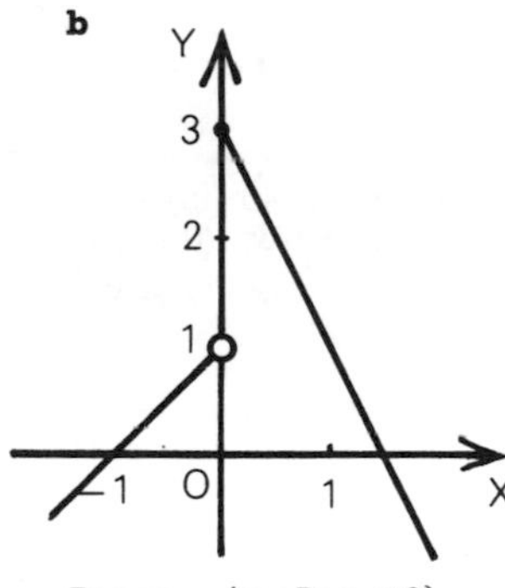

Range = $\{y \in R : y \leqslant 3\}$

c

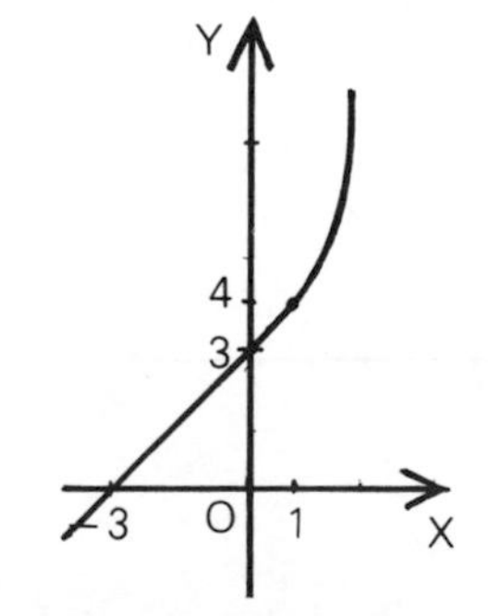

Range = R

Page 71 ***Exercise 4***

1 a $f(x) = \begin{cases} 7x\,(0 \leqslant x \leqslant 500) \\ 3x + 2000\,(x > 500) \end{cases}$

b

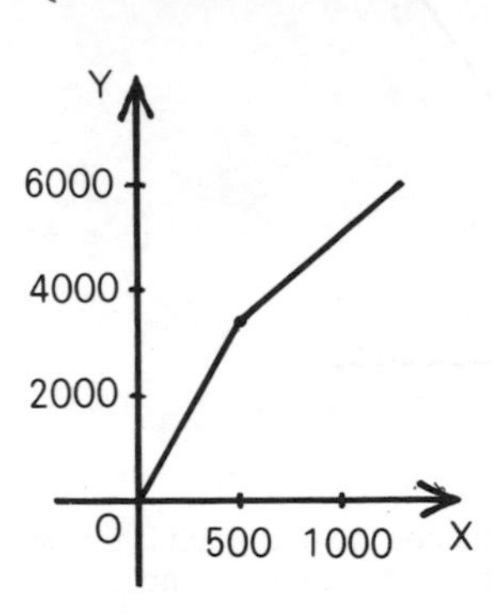

b Range $= \{f(x) \in W : f(x) \geqslant 0\}$

2 a

$$C(x) = \begin{cases} 0 & (0 < x < 10) \\ 1 & (10 \leqslant x < 50) \\ 5 & (x \geqslant 50) \end{cases}$$

b

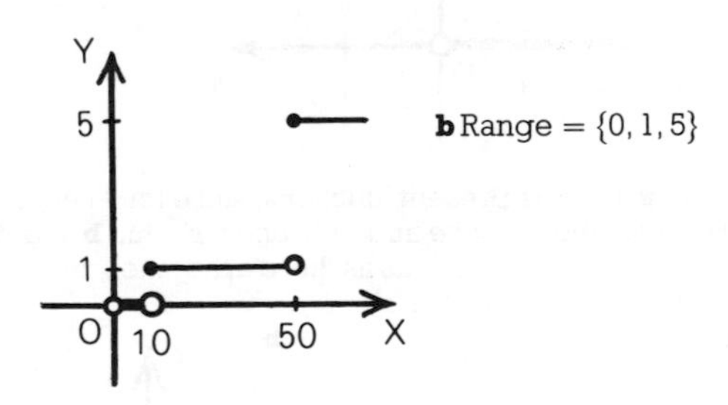

b Range $= \{0, 1, 5\}$

3 a (i) 20 (ii) 40 (iii) 54

b $B(x) = \begin{cases} 10x & (0 \leqslant x < 5) \\ 9x & (x \geqslant 5) \end{cases}$

c

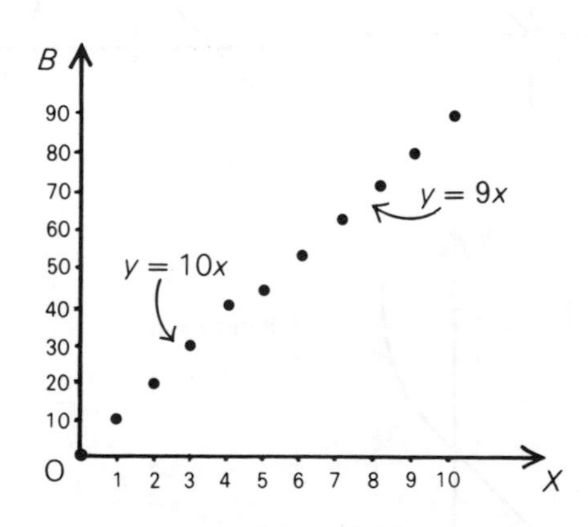

4 a

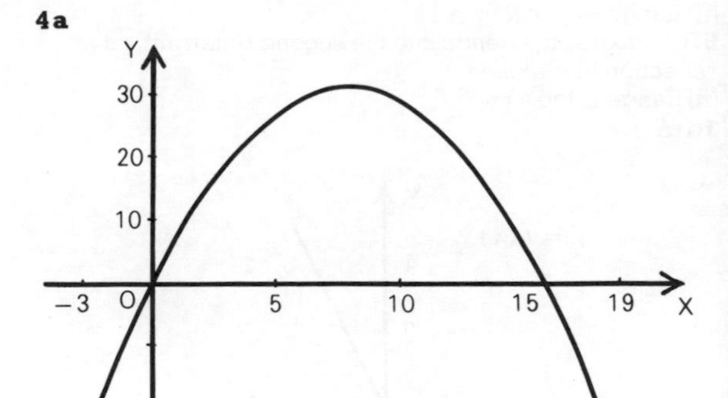

b Range $= \{y \in R : -28\frac{1}{2} \leqslant y \leqslant 32\}$ This gives maximum height of $60\frac{1}{2}$ units from top of bridge to river level.

5 a

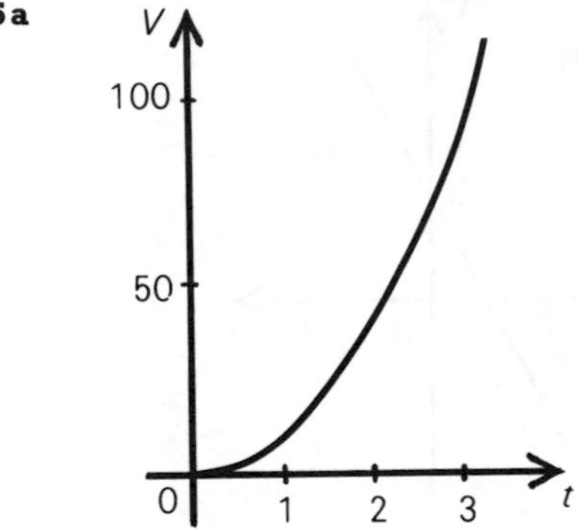

b $V \to \infty$ as $t \to \infty$

It does not correctly model 'reality'

6 a $F = \dfrac{k}{s}$, k a constant **b** They get larger and larger as $s \to 0$ **c** This would require an infinite force

7 a If too small, cost increases and the time to pump any quantity is large. If too large, cost increases and the pressure needed to pump water is too great

b (i) $c \doteqdot 2a$ (ii) $c \doteqdot \dfrac{200}{9a}$

c

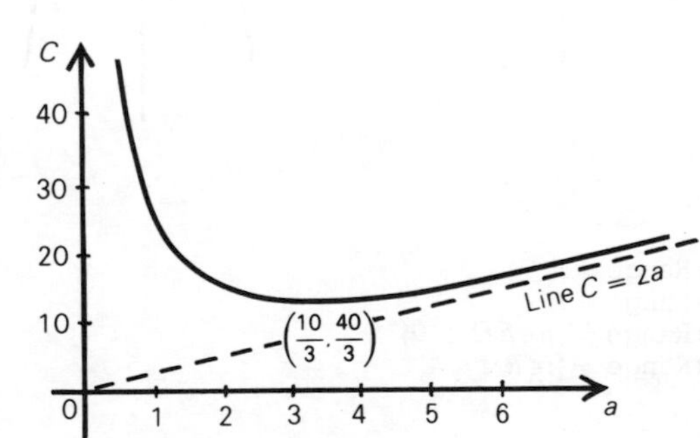

d Minimum cost is £13⅓ million

Page 73   *Exercise 5*
1 a 6 **b** 8 **c** 10 **d** $2x+6$
2 a Range $= \{0, 4\}$ **b** $4x^2$
3 a $h(x) = 2x+1$, $k(x) = 2x+2$ **b** Order is important (not commutative)
4 a (i) $x+5$ (ii) $x+5$ **b** (i) $6x+11$ (ii) $6x+7$ **c** (i) x^6 (ii) x^6
d (i) x^2-4x+4 (ii) x^2-2 **e** (i) $3x^3$ (ii) $27x^3$ **f** (i) $2x^2+11$
(ii) $4x^2+20x+28$ **g** (i) $2\sin x$ (ii) $\sin 2x$ **h** (i) $1-\cos^2 x$
(ii) $\cos(1-x^2)$
5 Two pairs (**a** and **c**) **a** $15x+5+a$, $15x+3a+1$ **b** $a=2$
6 a $V(s) = \dfrac{s}{10}$ **b** $s(x) = \sqrt{(x^2+1)}$ **c** $V(s(x)) = \dfrac{\sqrt{(x^2+1)}}{10}$
7 a $d(x) = 10\sin x°$ **b** $h(d) = d+12$ **c** $s(h) = h\tan 20°$
d (i) $s(d) = (d+12)\times\tan 20°$ (ii) $s(x) = (10\sin x°+12)\times\tan 20°$

Page 74   *Investigation*
1 In all cases, where the compositions are meaningful
$h\circ(g\circ f) = (h\circ g)\circ f$
First example: $(g\circ f)(x) = 2x+1$, $(h\circ g)(x) = x^2+2x+1$,
with $[h\circ(g\circ f)](x) = [(h\circ g)\circ f](x) = 4x^2+4x+1$
Second example: $(g\circ f)(x) = 6x+11$, $(h\circ g)(x) = 3x+6$,
with $[h\circ(g\circ f)](x) = [(h\circ g)\circ f](x) = 6x+15$
2 a 0 **b** 1 **c** $i(x) = x$ **e** Graph is $y = x$

Page 75   *Exercise 6*
3 No inverse
4 Tables **a** and **b** have no inverse; **c** and **d** have inverses.
5 c

y	2	3	4	5	6	7	8
$f^{-1}(y)$	1	2	3	4	5	6	7

d

y	10	10^2	10^3	10^4	10^5	10^6
$f^{-1}(y)$	1	2	3	4	5	6

6 For each graph there are different values of x for which $f(x)$ is the same

7 a For example, $f(-1) = f(1) = -3$. Thus cannot be 'reversed' to give a function.

−1   1   −3 f

b Range $= B = \{y\in R: y \geqslant -4\}$
c The right-hand 'half' of the parabola $y = x^2-4$. Each x in A has a unique $f(x)$ in B and so f can be 'reversed' to give the inverse function, $y = \sqrt{(x+4)}$
d $A_1 = \{x\in R: x \leqslant 0\}$

Page 77   *Exercise 7*
1 a $f^{-1}(x) = \frac{1}{2}x$ **b** $g^{-1}(x) = \frac{1}{4}(x+2)$ **c** $h^{-1}(x) = \frac{1}{3}(x-7)$
d $i^{-1}(x) = \frac{1}{6}(x-2)$ **e** $j^{-1}(x) = \frac{1}{5}(1-2x)$ **f** $k^{-1}(x) = 2(x-3)$
g $m^{-1}(x) = x^{1/3}$ **h** $n^{-1}(x) = (x-1)^{1/5}$ **i** $p^{-1}(x) = \dfrac{1-x}{x}$
j $q^{-1}(x) = \dfrac{2+3x}{x}$ **k** $r^{-1}(x) = \dfrac{-2x}{x-1}$ **l** $s^{-1}(x) = \dfrac{3x-2}{2x}$

2 a For each $x\in A$ there is only one $y\in B$, and for each $y\in B$ there is only one $x\in A$ **b** $f^{-1}(x) = \sqrt{(x+1)}$

Page 78   *Investigation*

a

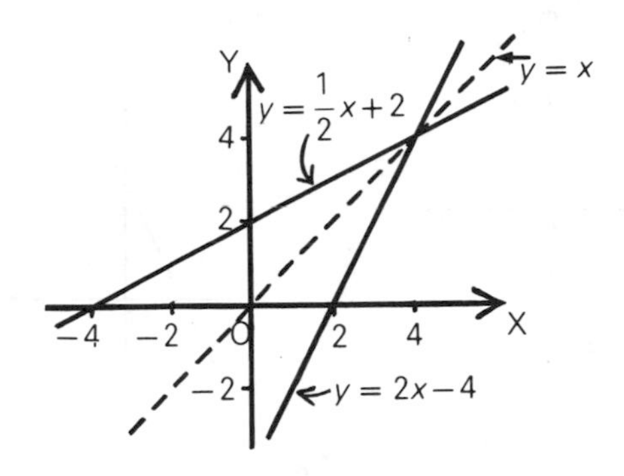

b

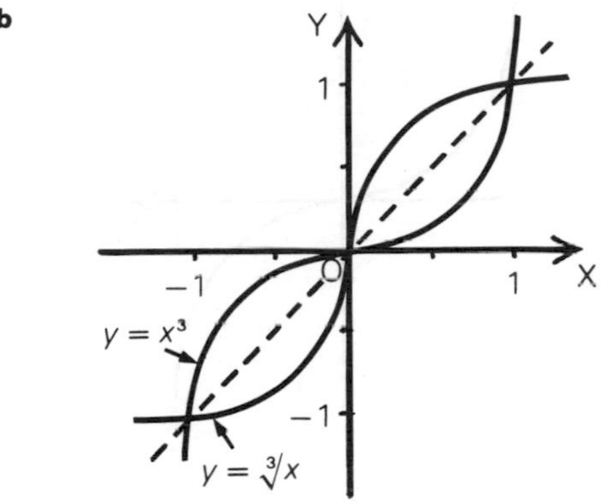

c Reflection in the line $y = x$ of the graph $y = f(x)$ will give $y = f^{-1}(x)$

Page 78   *Check-up on Functions*
1 a $\{4, 7, 10, 13\}$ **b** $\{0, 1, 2, 3\}$ **c** $\{2, 11\}$ **d** $\{y\in R: -1 \leqslant y \leqslant 1\}$
2 a $\{x\in R: x \leqslant 2\}$ **b** $\{x\in R: x \neq -1\}$
c $\{x\in R: x \leqslant -2 \text{ or } x \geqslant 2\}$ **d** $\{x\in R: x \neq 0 \text{ and } x \neq 2\}$

3 a

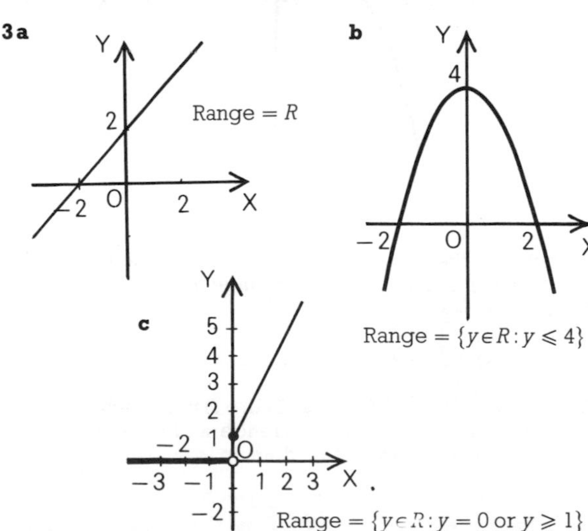

4 a (i) $9x^2+30x+25$ (ii) $3x^2+5$ (iii) $9x+20$ **b** (i) $\cos(\sin x)$ (ii) $\sin(\cos x)$ (iii) $\sin(\sin x)$

5 a

(i)

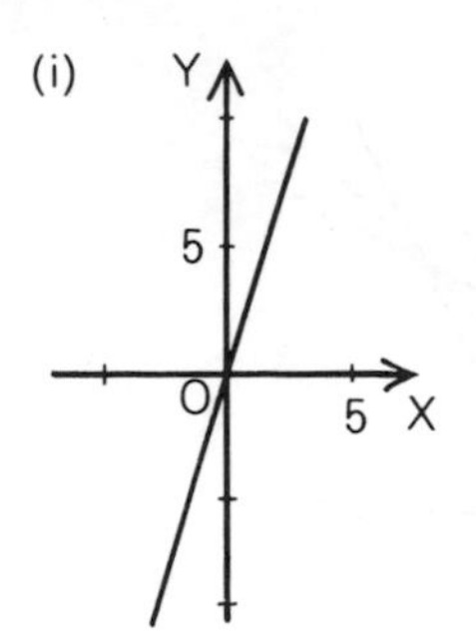

(ii)

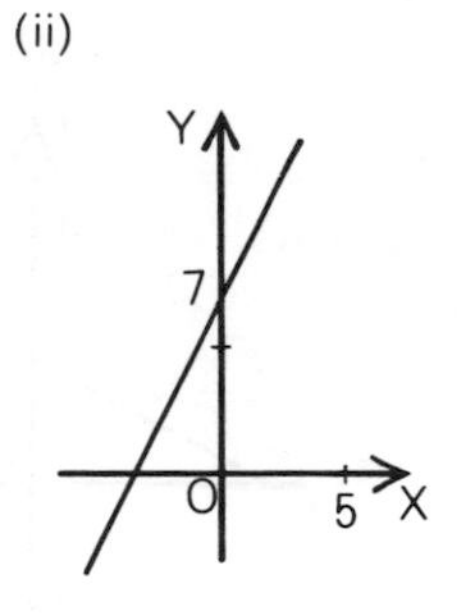

(iii)

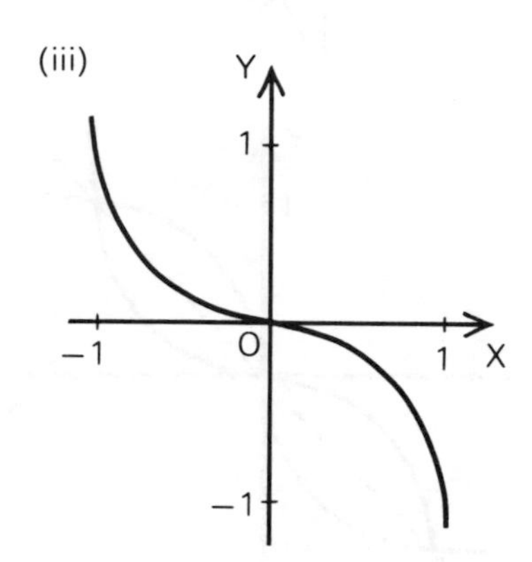

(iv)

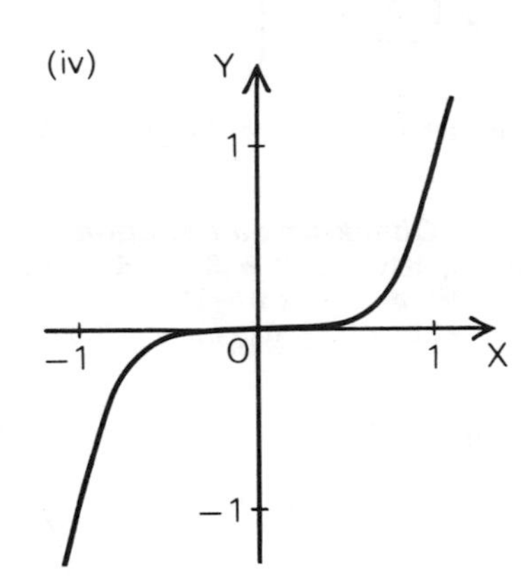

In each case there is a 'one-to-one correspondence' between the domain (R) and the range (R).

b (i) $f^{-1}(x) = \frac{1}{3}x$
(ii) $g^{-1}(x) = \frac{1}{2}(x-7)$
(iii) $h^{-1}(x) = \sqrt[3]{(-x)}$
(iv) $h^{-1}(x) = \sqrt[5]{x}$

6 In each case the restrictions ensure that for each $x \in$ domain there is only one $y \in$ range, and for each $y \in$ range there is only one $x \in$ domain.

a

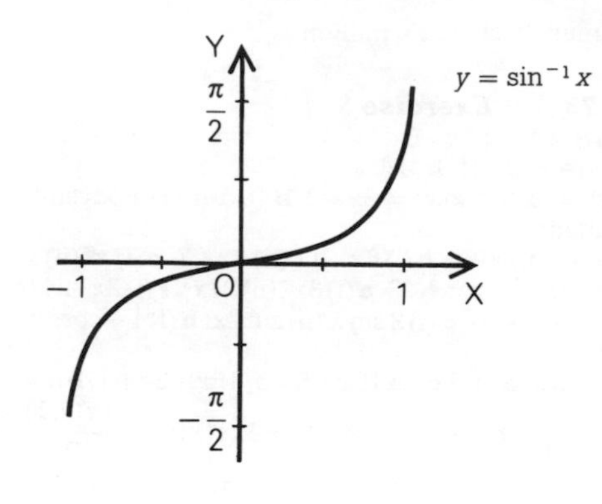

b

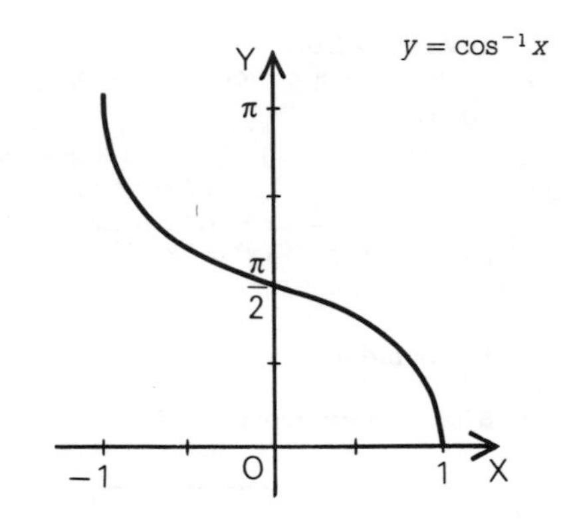

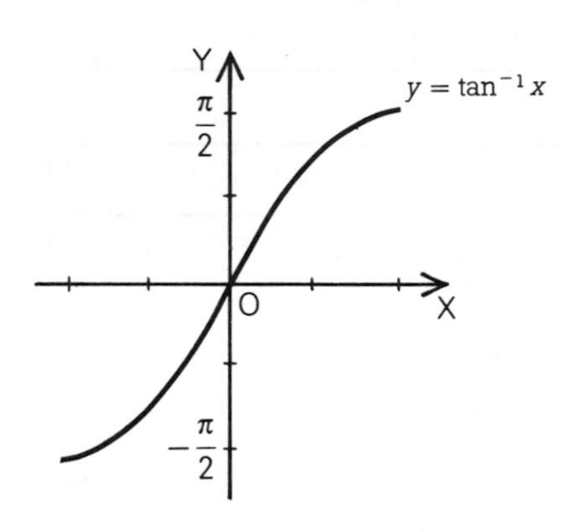

Page 79 ***Exercise 1***

1 a $\sin x°$ **b** $\sin y°$ **c** $-\sin t°$ **d** $\cos n°$
2 a $\tan 2u°$ **b** 1 **c** $\cos^2 4w$
3 a $\sin 80°$ **b** $-\cos 20°$ **c** $\sin 85°$ **d** $-\cos 69°$ **e** $\sin 3°$
5 (i) The sines of supplementary angles are equal. The cosine of an angle is equal to the negative of the cosine of its supplement
(ii) The sine of an angle is equal to the cosine of its complement, and vice versa

Page 81 ***Exercise 2***

1 a $\sin X \cos Y + \cos X \sin Y$ **b** $\cos X \cos Y - \sin X \sin Y$
c $\cos C \cos D + \sin C \sin D$ **d** $\sin P \cos Q - \cos P \sin Q$
2 a

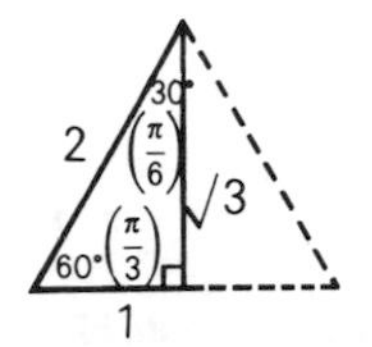

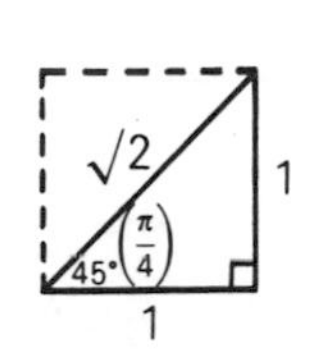

b

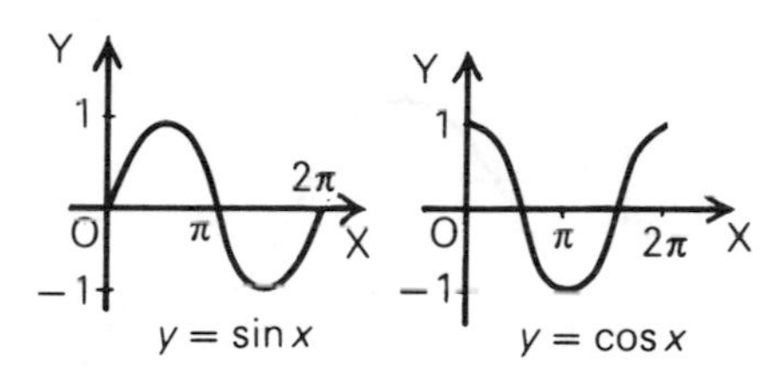

4 a $\sin(20°+70°) = 1$ **b** $\cos(170°-50°) = -\frac{1}{2}$
c $\cos(2B-B) = \cos B$ **d** $\sin(5y-y) = \sin 4y$
8 a False **b** True **c** False
13 a AB = 130, BC = 50 **b** $\angle ABC = 180° - (x°+y°)$
c $\cos ABC = -\frac{33}{65}$ **d** $AC = \sqrt{26\,000} \doteqdot 161.25$
14 a $F\cos(\alpha-\beta)$ and $F\cos(\alpha+\beta)$ **b** $2F\cos\alpha\cos\beta$
15 b (i) $\frac{336}{625}$ (ii) $-\frac{527}{625}$

Page 82 ***Brainstormer***

a $\tan(B-A) = \dfrac{\tan B - \tan A}{1+\tan B \tan A}$ **b** $\theta = 45$

Page 83 ***Exercise 3***

1 a $2\sin X\cos X$ **b** $2\cos^2 X - 1 = \cos^2 X - \sin^2 X = 1 - 2\sin^2 X$ **c** $2\sin Y\cos Y$ **d** $2\cos^2 Y - 1 = \cos^2 Y - \sin^2 Y = 1 - 2\sin^2 Y$
2 a $2\sin\frac{A}{2}\cos\frac{A}{2}$ **b** $2\cos^2\frac{A}{2} - 1 = \cos^2\frac{A}{2} - \sin^2\frac{A}{2} = 1 - 2\sin^2\frac{A}{2}$
4 a $\frac{1}{2}$ **b** $\frac{1}{2}$ **c** 1
5 a $\frac{2}{\sqrt{5}}$ **b** $\frac{4}{5}$ **c** $\frac{3}{5}$

Page 84 ***Investigation***

$\cos^4 x = \frac{3}{8} + \frac{1}{2}\cos 2x + \frac{1}{8}\cos 4x$
$\sin^4 x = \frac{3}{8} - \frac{1}{2}\cos 2x + \frac{1}{8}\cos 4x$

Page 84 ***Exercise 4***

1 a 30, 90, 150, 270 **b** 0, 180, 360 **c** 90, 210, 330
2 a 90, 120, 240, 270 **b** 30, 150 **c** 120, 180, 240
3 a $\frac{\pi}{2}, \frac{7\pi}{6}, \frac{11\pi}{6}$ **b** $\frac{\pi}{6}, \frac{5\pi}{6}, \frac{3\pi}{2}$ **c** $0, \frac{\pi}{3}, \pi, \frac{5\pi}{3}, 2\pi$
4 a π **b** $\frac{\pi}{2}, \frac{7\pi}{6}, \frac{3\pi}{2}, \frac{11\pi}{6}$ **c** $\frac{\pi}{3}, \pi, \frac{5\pi}{3}$

Page 85 ***Exercise 5***

5 a $\dfrac{\sin(i-\theta)°}{\sin R°} = \dfrac{\sin i°}{\sin r°}$

Page 86 ***Check-up on Compound Angle Formulae***

1 a $\sin u°$ **b** $\cos v°$ **c** $-\sin v°$ **d** $-\cos 80°$
3 $\cos(A+B) = \frac{16}{65}$ **5** $\cos 2A = \frac{7}{25}$ **7** $\sin^2 A = \frac{1}{2}(1-\cos 2A)$
9 a 90, 270 **b** 0, 180, 210, 330, 360 **c** 120, 240.

Page 90 ***Exercise 1***

1 a $3\cos x$ **b** $-4\sin x$ **c** $\cos x - \sin x$ **d** $1+\sin u$
e $-\sin t - \cos t$ **f** $5\cos t - 3\sin t$ **g** $3x^2 + 3\sin x$ **h** $u + 3\cos u$
2 a $10x - 2\cos x$ **b** $-\frac{1}{2x^2} + 3\sin x$ **c** $\frac{1}{2}x^{-1/2} - \frac{1}{2}\sin x$
d $\frac{1}{2}\cos x - \frac{1}{3}\sin x$ **e** $-10\sin x - 11\cos x$ **f** $-\frac{1}{2}x^{-3/2} + \sin x$
g $\frac{1}{2} - \frac{2}{x^0} - 4\cos x$ **h** $-\frac{1}{x^2} + \cos x$ **i** $-\frac{6}{x^3} - 2\sin x$
3 a $-\frac{1}{2}\sin x$ **b** $\frac{3}{2}\sin t$ **c** $\cos u$
4 a 1 **b** 0 **c** 1 **d** −1 **e** $\frac{1}{2}$ **f** $\frac{1}{\sqrt{2}}$ **g** $\frac{1}{2}$ **h** 0 **i** $\frac{\sqrt{3}}{2}$ **j** $-\frac{\sqrt{3}}{2}$
5 a 0 **b** $y = 2$
6 a $y = 3$ **b** $y = 3x - 1$ **c** $y + x = 1 + \frac{\pi}{4}$ **d** $y = x + 2$
e $y + \sqrt{2}x = \frac{3\sqrt{2}}{4}\pi$ **f** $2y - x = 1 - \frac{\pi}{2}$
7 a (i) 0·54 cm/sec (ii) −0·42 cm/sec **b** 1·57 secs
8 a 4 m **b** $2\cos t$ **c** Mean sea level
9 a $\frac{1}{2}l^2\sin\theta$ ft² **b** (i) $\frac{1}{2}l^2\cos\theta$ (ii) $\theta = 0$ (door closed)

Page 92 ***Investigation***

1

$f(x)$	$f'(x)$
$(x+4)^2 = x^2+8x+16$	$2x+8 = 2(x+4)$
$(x+4)^3 = x^3+12x^2+48x+64$	$3x^2+24x+48 = 3(x+4)^2$
$(x+4)^4 = x^4+16x^3+96x^2+256x+256$	$4x^3+48x^2+192x+256 = 4(x+4)^3$
⋮	⋮

2

$f(x)$	$f'(x)$
$(3x+1)^2 = 9x^2+6x+1$	$18x+6 = 6(3x+1)$
$(3x+1)^3 = 27x^3+27x^2+9x+1$	$81x^2+54x+9 = 9(3x+1)^2$
$(3x+1)^4 = 81x^4+108x^3+54x^2+12x+1$	$324x^3+324x^2+108x+12 = 12(3x+1)^3$
⋮	⋮

3

$f(x)$	$f'(x)$
$(x^2-1)^2 = x^4-2x^2+1$	$4x^3-4x = 4x(x^2-1)$
$(x^2-1)^3 = x^6-3x^4+3x^2-1$	$6x^5-12x^3+6x = 6x(x^2-1)^2$
$(x^2-1)^4 = x^8-4x^6+6x^4-4x^2+1$	$8x^7-24x^5+24x^3-8x = 8x(x^2-1)^3$
⋮	⋮

Results:
If $f(x) = (ax+b)^n$ then $f'(x) = na(ax+b)^{n-1}$
If $g(x) = (ax^2+b)^n$ then $g'(x) = 2nax(ax^2+b)^{n-1}$
Example
For $h(x) = (x^2+3x+4)^2$, $h'(x) = 2(2x+3)(x^2+3x+4)$

Page 94 ***Exercise 2***

1 a $8(x+1)^7$ **b** $4(2+u)^3$ **c** $6(2t-1)^2$ **d** $15(3v+1)^4$
e $-20(1-2x)^9$
2 a $8x(x^2+2)^3$ **b** $6u(1+u^2)^2$ **c** $-15t^2(4-t^3)^4$ **d** $24v^3(2v^4-1)^2$
e $2(2x+3)(x^2+3x+4)$
3 a $-6(3x-1)^{-3}$ **b** $-2x(x^2+2)^{-2}$ **c** $-12u(2u^2-5)^{-4}$
d $-\frac{2}{(2x+3)^2}$ **e** $-\frac{12}{(3t+4)^3}$
4 a $(2x+1)^{-1/2}$ **b** $(3u-2)^{-2/3}$ **c** $\frac{1}{2}(2x+1)(x^2+x)^{-1/2}$
d $\frac{4}{3}(3t^2+1)(t^3+t)^{1/3}$ **e** $(v+1)(2v^2+4v)^{-3/4}$
5 a $4(3x^2-4x)(x^3-2x^2)^3$ **b** $\frac{2}{\sqrt{(4u+3)}}$ **c** $-2(6t+1)^{-4/3}$
d $-\frac{24x}{(2x^2-1)^3}$ **e** $2\left(1-\frac{1}{u^2}\right)\left(u+\frac{1}{u}-1\right)$
6 $5y-4x = 9$ **7** Maximum of 18 (when $x = 5$)
8 a $3x^2-2\cos x$ **b** $\frac{\sin t}{2(\cos t)^{3/2}}$ **c** $-\frac{3}{(3u+1)^2}-3\sin u$
9 b $-\frac{1}{4\pi}\left[\frac{3}{4\pi}(1000-V)\right]^{-2/3}$; $\frac{dr}{dV}$ is negative as r decreases when V increases **10 a** 2

Page 95 ***Exercise 3***

1 a $2\cos 2x$ **b** $-2\sin 2x$ **c** $3\cos(3u+1)$ **d** $-4\sin(4v-1)$
e $a\cos(ax+b)$
2 a $5\sin^4 x\cos x$ **b** $-2\cos u\sin u$ **c** $4\sin^3 x\cos x$
d $-4\cos^3 x\sin x$ **e** $\frac{\cos x}{2\sqrt{\sin x}}$
3 a $-3(1+\cos x)^2\sin x$ **b** $4(\cos x+\sin x)(\sin x-\cos x)^3$
c $\frac{-\cos x}{\sin^2 x}$ **d** $\frac{2\sin t}{\cos^2 t}$ **e** $\frac{-6\cos x}{\sin^3 x}$
4 a $\cos^2(ax)+\sin^2(ax) = 1$, $\frac{d}{dx}(1) = 0$
5 $\frac{\pi}{180}\cos x°$, $-\frac{\pi}{180}\sin x°$
6 a $\left(\frac{\pi}{2}, 1\right)$, $\left(\frac{3\pi}{2}, -1\right)$
7 b $\frac{1}{\sqrt{2}}$ and $-\frac{1}{\sqrt{2}}$ **c** 70·6°

8 b $2y-2x = 1-\frac{\pi}{2}$ **c** (i) $0 < x < \frac{\pi}{2}$ (ii) $\frac{\pi}{2} < x < \pi$ **d** (0, 0) and $(\pi, 0)$ are minimum SPs. $\left(\frac{\pi}{2}, 1\right)$ is a maximum SP
e

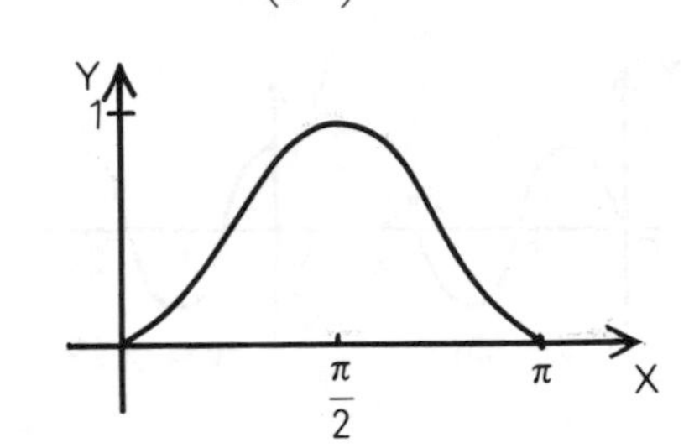

9 Since $f'(x) = 1+\cos x \geqslant 0$ then f is never decreasing on R. But $f'(x) = 0$ when $x = \pi$, etc.
10

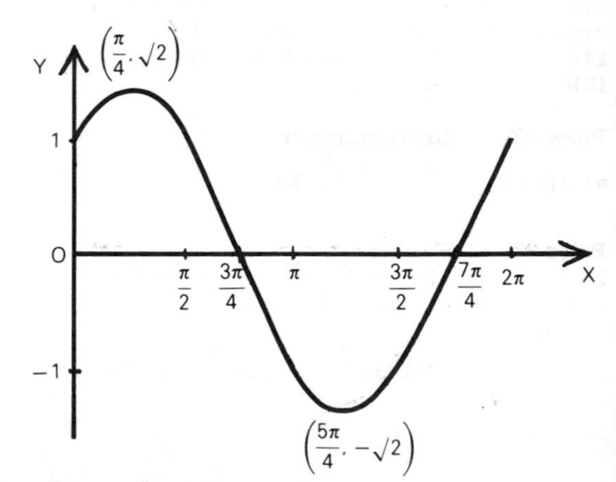

Page 97 ***Exercise 4***

1 a $f(-2)$ **b** $f(0)$ (maximum) and $f(2)$ (minimum)
c $f(-1)$ (maximum) and $f(1)$ (minimum)
2 a (i) $f(-1) = 3$, $f(2) = -3$ (ii) Maximum turning value 5 (iii) Maximum value 5, minimum value -3
b (i) $f(-2) = -2$, $f(3) = 18$ (ii) Maximum turning value 2, minimum turning value -2 (iii) Maximum value 18, minimum value -2 **c** (i) $f(-1) = 17$, $f(3) = 189$ (ii) min TV -64, P.I. (0, 0) (iii) Maximum value 189, minimum value -64. **d** (i) $f(0) = 1$, $f(\pi) = -1$ (ii) Maximum turning value $\sqrt{2}$ (iii) Maximum value $\sqrt{2}$, minimum value -1

e (i) $f(0) = 0$, $f(\pi) = \pi$ (ii) Maximum turning value $\frac{\pi}{3} + \frac{\sqrt{3}}{2}(1{\cdot}9)$, minimum turning value $\frac{2\pi}{3} - \frac{\sqrt{3}}{2}(1{\cdot}2)$ (iii) Maximum value π, minimum value 0 **f** (i) $f\left(-\frac{\pi}{2}\right) = -3$, $f\left(\frac{\pi}{2}\right) = -3$ (ii) Minimum turning value -3, maximum turning value 3 (iii) Maximum value 3, minimum value -3

Page 98 *Exercise 5*

1 Min. value 6 when $x = -1$; no max. value on R
2 Min. value 0 when $x = -1$; no max. value on R
3 Min. value -24 when $x = -3$; no max. value on R
4 Min. value 6 when $x = 3$; no max. value for $x > 0$
5 Min. value 2 when $x = 1$; no max. value for $x > 0$
6 Max. value 5 when $x = \frac{\pi}{2} + n\pi$; min. value 4 when $x = n\pi$ $(n \in Z)$

Page 100 *Exercise 6*

1 $S(6) = 12$ **2** 6 and -6, $P = -36$ **3** Max. $x^2y = 32\,000$ when $x = 40$, $y = 20$
4 $x = 2$ gives max. value of 128 cm³
5 3 cm × 3 cm × 1·5 cm **6** After 65 seconds
7 After 4 months **8** Diameter = 8 cm, height = 8 cm
9 8 staff (cost 77·5) [$x \doteqdot 7{\cdot}75$ but 7 staff give 77·9 for cost]
10 $\theta = \frac{\pi}{3}$ **11** $\frac{1}{24}$ square unit

Page 101 *Brainstormer*

a $AC = \sqrt{(x^2+s^2)}$, $DB = \sqrt{[(a-x)^2+t^2]}$, DC will be constant wherever it is placed. **d** If $\frac{x}{a-x} = \frac{s}{t}$ then by similar triangles AC || DB **e** At A

Page 102 *Exercise 7*

In questions **1–5**, $k > 0$; in **6, 7, 8** k is a constant.

1 $\frac{dN}{dt} = kN$ **2** $\frac{dQ}{dt} = -kQ$ **3** $\frac{dV}{dr} = kr^2$, $k = 4\pi$.
4 $\frac{dV}{dl} = kl^2$, $k = 3$ **5** $\frac{dV}{dA} = k\sqrt{A}$, $k = \frac{\sqrt{6}}{24}$
6 $\frac{d}{dx}(\cos x) = k \sin x$, $k = -1$ **7** $\frac{dx}{dt} = kt$
8 $\frac{dy}{dx} = \frac{k}{x^3}$ **9** $\frac{dV}{dt} = 400\pi$ cm³/s ($\doteqdot 1257$ cm³/s)

Page 102 *Check-up on Differentiation—2*

1 a $3\cos x + 4\sin x$ **b** $1 + 2\cos x$ **c** $-5\sin x - \frac{1}{2x^2}$
d $2\cos 2x - 3\sin 3x$
2 a $\frac{1}{3}(2x+3)(x^2+3x+1)^{-2/3}$ **b** $\frac{1-3x^2}{2\sqrt{(x-x^3)}}$ **c** $2\cos(2x+3)$
d $4\sin^3 x \cos x$ **e** $-\frac{3\sin 3x}{2\sqrt{\cos 3x}}$
3 b $4y + 2\sqrt{3}x = 5 + \frac{2\sqrt{3}}{3}\pi$ **c** Increasing for $\frac{\pi}{2} < x < \pi$ and decreasing for $0 < x < \frac{\pi}{2}$. **d** Max. SPs are $(0, 2)$ and $(\pi, 2)$; Min. SP is $\left(\frac{\pi}{2}, 1\right)$

e

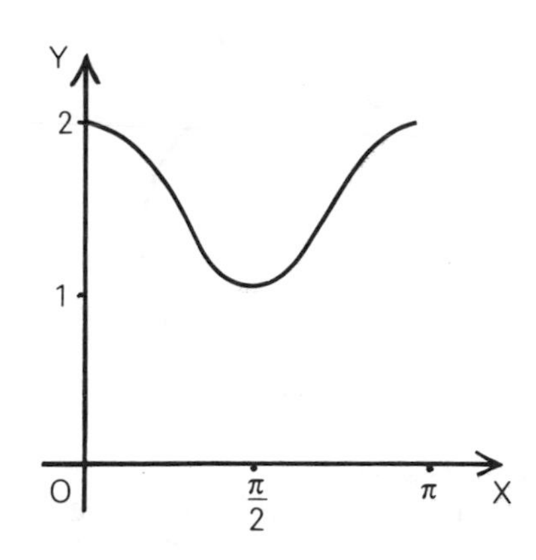

4 a Max. value 4 when $x = 1$, min. value -5 when $x = -2$.
b Max. value $\frac{\pi}{12} + \frac{1}{2}\sqrt{3}$ when $x = \frac{\pi}{12}$, min. value $\frac{5\pi}{12} - \frac{1}{2}\sqrt{3}$ when $x = \frac{5\pi}{12}$.
5 a Min. value of -13 when $x = -3$, no max. value on R
b Min. value of 12 when $x = 2$, no max. value on $x > 0$
c Max. value of 1 when $x = \frac{\pi}{2} + 2n\pi$, min. value of -1 when $x = \frac{3\pi}{2} + 2n\pi$ $(n \subset Z)$
6 125 m ends with 100 m opposite house, giving 12 500 m² maximum area.
7 a $\frac{dN}{dt} = k(A-N)$ **b** $\frac{dA}{dC} = kC$, $k = \frac{1}{2\pi}$.

Page 104 *Exercise 1*

1 a, **c**, **d**, and **e** **2** >
3 b $\{(x, y): 1 < x^2 + y^2 \leqslant 16\}$. Hit, 3 points
c $\{(x, y): 16 < x^2 + y^2 \leqslant 49\}$. Hit, 2 points
d $\{(x, y): 49 < x^2 + y^2 \leqslant 100\}$. Hit, 1 point
e $\{(x, y): x^2 + y^2 > 100\}$. Miss, 0 points
4 a (i) 2 points (ii) 3 points (iii) 1 point (iv) 0 points (v) 0 points **b** 4 points
5 The centres of all the circles are at the origin O. (Circles not drawn to scale.)

a Radius 1 **b** Radii 1 and 4 **c** Radii 4 and 5

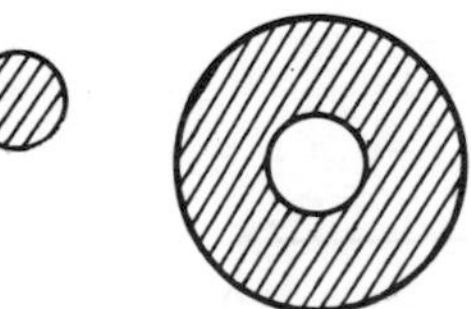

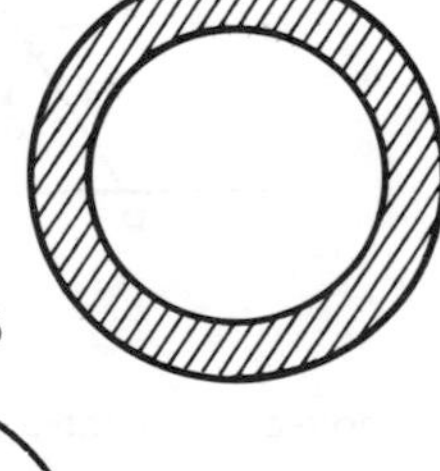

d Radius 1 **e** Radius 5

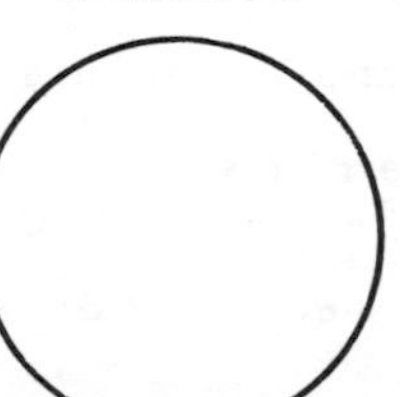

f Radius 10

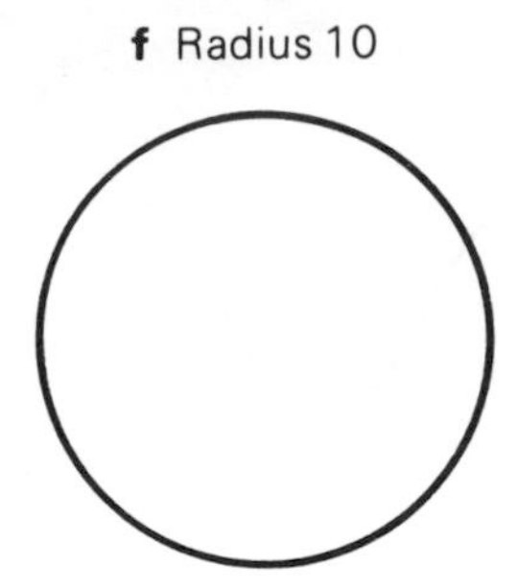

Page 105 *Exercise 2*

1 a $x^2+y^2=4$ **b** $x^2+y^2=49$ **c** $x^2+y^2=144$ **d** $x^2+y^2=625$
2 a 5, $x^2+y^2=25$ **b** 13, $x^2+y^2=169$ **c** $3\sqrt{2}$, $x^2+y^2=18$
d 7, $x^2+y^2=49$
3 a O(0, 0), 8 **b** O(0, 0), 13 **c** O(0, 0), 6 **d** O(0, 0), 2
4 a $x^2+y^2=4$ **b** $x^2+y^2=100$ **c** $x^2+y^2=196$ **d** $x^2+y^2=4t^2$

5 a Distance 5 units **b** $x^2+y^2=25$

c

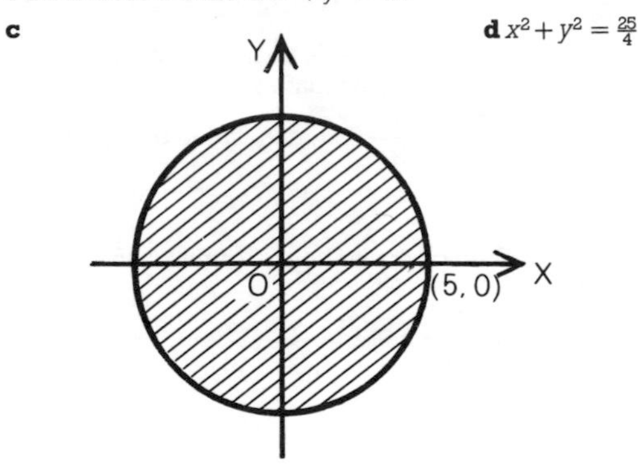

d $x^2+y^2=\frac{25}{4}$

6 a $x^2+y^2=64$ **b** $x^2+y^2=25$
7 a $x^2+y^2=12{\cdot}96$ **b** (i), (iii), (v); these have $x^2+y^2<12{\cdot}96$

Page 106 *Brainstormer*

a Denote length of strut by $2a$. A starts at point $(-a, 0)$ and ends at $(0, a)$. In any position of strut (PQ), A is centre of circle through O, P, Q, so OA = a. Locus of A is a quarter of a circle, centre O and radius a.

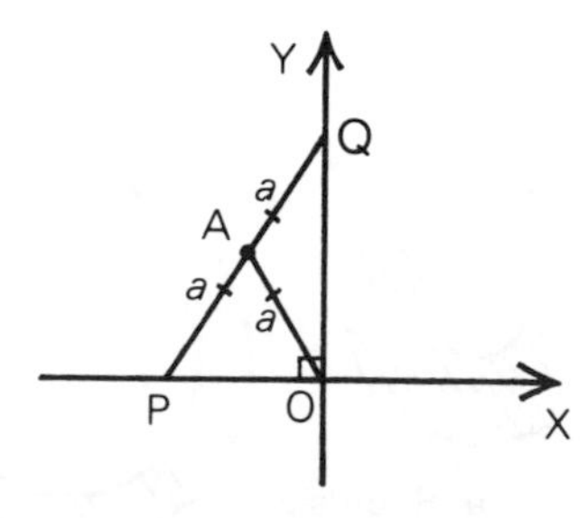

b Whole circle, centre O and radius a; $x^2+y^2=a^2$

Page 107 *Exercise 3*

1 a $(x-2)^2+(y-7)^2=16$ **b** $(x-3)^2+(y+1)^2=49$
c $(x+5)^2+(y+7)^2=6{\cdot}25$
2 a (3, 1), 5 **b** (4, 4), 1 **c** (−5, 2), 2 **d** (−2, −1), 1·2 **e** (1, 0), 4
f (0, −3), 6

3 a (i) $\sqrt{13}$ (ii) $(x-2)^2+(y-2)^2=13$ **b** (i) $2\sqrt{10}$
(ii) $(x-1)^2+(y+5)^2=40$ **c** (i) $4\sqrt{5}$
(ii) $(x+2)^2+(y+3)^2=80$ **d** (i) 6 (ii) $(x+3)^2+(y+2)^2=36$
e (i) 6 (ii) $(x+3)^2+y^2=36$ **f** (i) $\sqrt{\{(b_1-a_1)^2+(b_2-a_2)^2\}}$
(ii) $(x-a_1)^2+(y-a_2)^2=(b_1-a_1)^2+(b_2-a_2)^2$
4 a (5, 2) **b** 5 **c** $(x-5)^2+(y-2)^2=25$
5 $(x-6)^2+(y-8)^2=29$
6 a A(−7, −1), radius 10; B(3, −1), radius 10
b $(x+7)^2+(y+1)^2=100$, $(x-3)^2+(y+1)^2=100$
7 a $y=1$ **b** $(x-15)^2+(y-7)^2=36$
c (i) $(x-3)^2+(y-3)^2=4$ (ii) $(x-13)^2+(y-7)^2=36$
8 $(x-8)^2+(y-2)^2=16$, $(x-32)^2+(y-2)^2=16$

Page 108 *Brainstormer*

$(x-23)^2+y^2=9$

Page 109 *Exercise 4*

1 a yes, (2, 4), $\sqrt{3}$ **b** no **c** yes, (4, 1), 2 **d** yes, (1, −1), $\sqrt{7}$
e no, single point {(−3, 1)} **f** yes, (−2, −3), 3 **g** no
h yes, $(-\frac{3}{2}, -\frac{1}{2})$, 3.
2 a (vi) **b** (v) **c** (iv) **d** (ii) **e** (iii) **f** (i)
3 a $(-1)^2+5^2-6(-1)-4(5)-12=0$ **b** (7, −1)
4 a (1, 4), 5 **b** (i) A, B, D inside (ii) F on (iii) C, E outside
5 a (−4, −1), 5 **b** (i) $\sqrt{178}$ (ii) $\sqrt{153}$
c $\sqrt{(x_1^2+y_1^2+2gx_1+2fy_1+c)}$
6 a O. Point is on circle **b** $5\sqrt{2}$ **c** $2\sqrt{34}$
7 a $a=3$ **b** $b=-4, 2$ **c** $x=8$, a tangent

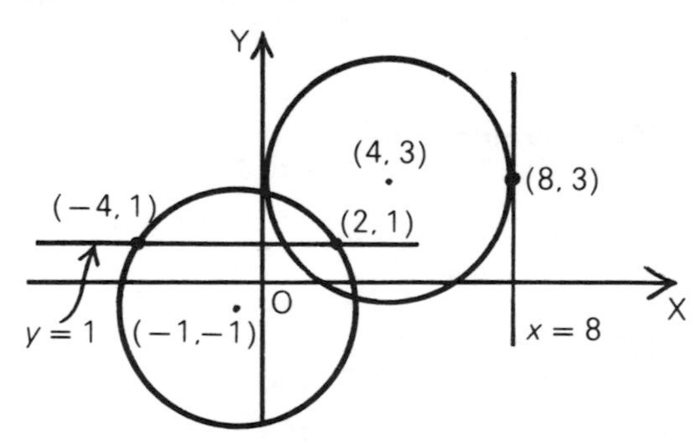

8 a $(-3)^2+4^2=25$ **b** (i) $-\frac{4}{3}$ (ii) $\frac{3}{4}$ **c** $3x-4y+25=0$
9 a $2x-3y+13=0$ **b** $x-3y+10=0$
c $y=3$ **d** $x-y-5=0$ **e** $7x-4y-25=0$ **f** $4x-y+6=0$
10 a $2g+5f=-18$, $2g-f=6$ **b** $g=1, f=-4$
c $x^2+y^2+2x-8y+7=0$, (−1, 4), $\sqrt{10}$
11 a $-2g+2f+c=-2(1)$, $6g+6f+c=-18(2)$,
$12g+4f+c=-40(3)$ **b** $2g+f=-4$ **c** $3g-f=-11$
d $g=-3, f=2$ **e** $c=-12$, $x^2+y^2-6x+4y-12=0$
12 a $x^2+y^2-2x-2y-43=0$ **b** $x^2+y^2+8x-6y=0$

Page 110 *Brainstormer*

A $(x-\frac{17}{8})^2+(y-\frac{7}{2})^2=\frac{49}{64}$
C $(x-\frac{7}{2})^2+(y-\frac{11}{4})^2=\frac{1}{16}$

Page 111 *Exercise 5*

1 a yes, (3, 1), (−3, 1) **b** yes, (0, 2) **c** no
d yes, (2, 2), (−2, −2) **e** yes, (1, 2), (−2, −1) **f** yes, (2, 1)
g yes, (−1, 0), (−2, −1) **h** yes, (4, 1), (2, 5)
i yes, (3, −1), (1, −7) **j** yes, (0, −4)
2 a (i) 6 (ii) 8 **b** (i) 24 (ii) 10 **c** (i) 1 (ii) 4 **d** (i) 0 (ii) 8. Circle touches the x-axis
3 a (3, 0) **b** (3, −1) **c** (−1, 3) **d** (3, 2) **e** (2, 0)
4 b (−2, 6), (10, 10). Length of tangent is $4\sqrt{10}$.

5a (4, 1) **b** $2x+y-19=0$
6a $x+y-5=0$ **b** $x=-1$
7a (−1, 4) **b** (1, 1), (−3, 1)
8a A(−3, 1), B(1, 3) **b** $(x+2)^2+(y-4)^2=10$
9a $x=2y-8$ **b** A(2, 5), B(6, 7) **c** 72° to 162°

Page 112 *Investigation*

1 For the position of P shown, the locus of centres of the cue ball is the shaded right-angled triangle whose vertices lie on the bisectors of the angles at A, B, C.
Investigate other positions of P.

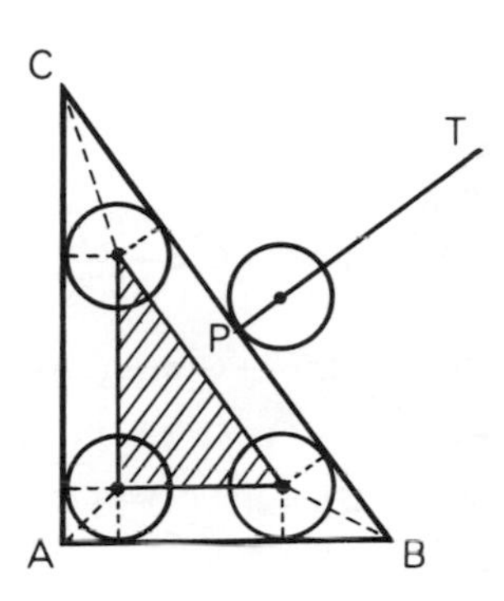

2a $c=0$ **b** (i) $f=0$ (ii) $g=0$ (iii) $g=f$ **c** (i) $c=f^2$ (ii) $c=g^2$ (iii) $c=f^2=g^2$ **d** $g^2+f^2-c=1$
3a $2g_1g_2+2f_1f_2=c_1+c_2$

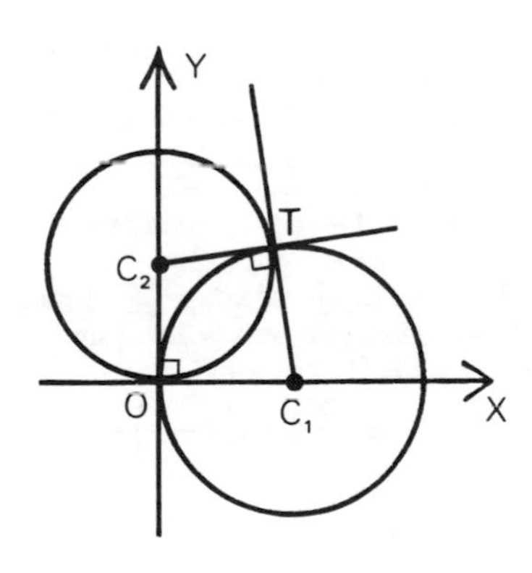

b Each of OC_1 and OC_2 is a tangent to one circle and a radius of the other, and $\angle C_1OC_2=\frac{\pi}{2}$. By symmetry about C_1C_2 the tangents at T are also perpendicular.
c There are intersecting and non-intersecting systems of circles. Applications in Physics are to equipotential lines and lines of force in Electrostatics.

Page 113 *Check-up on The Circle*

1a

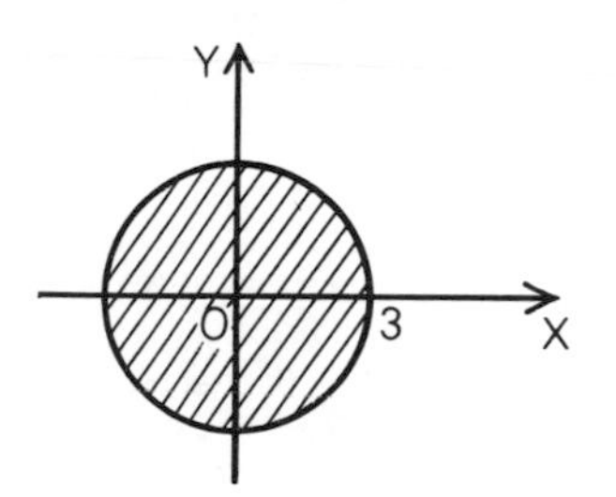

b

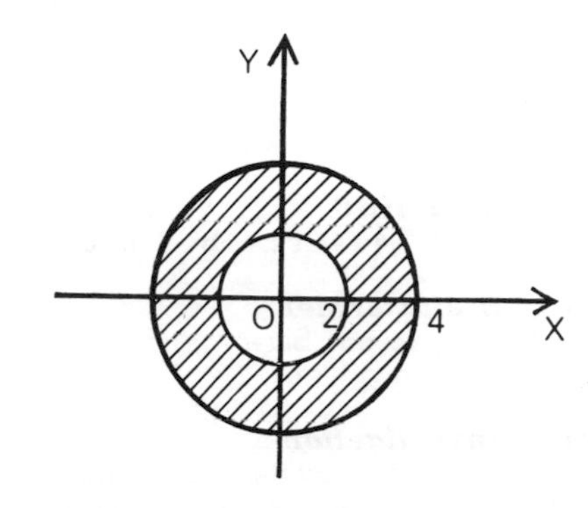

2a $x^2+y^2=25$ **b** Centre O (0, 0), radius $\frac{3}{2}$
3a (4, −3), 4 **b** $(x+5)^2+(y-7)^2=4$
4 $(x-2)^2+(y-2)^2=1$
5a (−3, −1), 5 **b** (1) (iii) (2) (i), (ii), (v) (3) (iv)
6 (0, 1), (5, 6)
7a (1, −7) **b** $3x+5y+5=0$
8 $x^2+y^2-4x+6y-4=0$.

Page 114 *Exercise 1*

1a 2, 1 **b** 3, −2 **c** 4, 7 **d** 5, 0 **e** 0, 0 **f** 2, −1 **g** 11, 0 **h** 10, 3
2a x^2-x **b** $-2x^2+5x+3$ **c** $4x^2-4x+1$
d $8x^3-12x^2+6x-1$ **e** $15x^4+5x^3-3x-1$
f $-x^5+4x^4+3x^3-13x^2+5x-4$ **g** $x^3-6x^2+11x-6$
3a $256x^4$, 4 **b** $-2x$, 1 **c** Not a polynomial **d** x^2-x, 2
e $1-\frac{1}{x^2}$, not a polynomial **f** Constant, 0 **g** $8x-4$, 1
h $x^6-6x^5+15x^4-20x^3+15x^2-6x+1$, 6 **i** $x-1$, 1
j $1+\frac{1}{x-2}$, not a polynomial **k** Constant, 0 **l** $4x^2-4x+1$, 2
4a $f(x)+g(x)=3x^3+x^2+1$, $f(x)+h(x)=3x^3-x$, $g(x)+h(x)=x^2+3x-1$ **b** $f(x)-g(x)=3x^3-x^2-4x+1$, $g(x)-f(x)=-3x^3+x^2+4x-1$, $f(x)-h(x)=3x^3-3x+2$, $h(x)-f(x)=-3x^3+3x-2$, $g(x)-h(x)=x^2+x+1$, $h(x)-g(x)=-x^2-x-1$
c $f(x)g(x)=3x^5+6x^4-2x^3-3x^2+2x$, $f(x)h(x)=3x^4-3x^3-2x^2+3x-1$, $g(x)h(x)=x^3+x^2-2x$

Page 116 *Exercise 2*

1a 29 **b** 13 **c** −3 **d** 39 **e** 21 **f** −5 **g** 618 **h** −1 **i** 178

Page 117 *Exercise 3*

1 $x+3$, 1 **2** $2x-5$, 7 **3** x^2+1, −2 **4** x^2+x+5, 6
5 $2t^2-2t+8$, 1 **6** $3u^2+2u+10$, 0 **7** $2x+2$, −1
8 $4x$, 3 **9** x^2-2x+4, 0 **10** $t^5+t^4+t^3+t^2+t+1$, 0
11a x, −3 **b** $x-2$, 9 **c** $2t^2+2t-2$, 1 **d** $\frac{3}{2}x^2-\frac{3}{4}x-\frac{5}{8}$, $\frac{45}{8}$

Page 118 *Exercise 4*

1 $(x-1)^2(x+1)$ **2** $(x+1)^2(x-2)$ **3** $(x-1)(x-2)(x+3)$
4 $(x-2)(x^2+4)$ **5** $(x+2)(x-3)(3x+1)$
6 $(x-3)(x+3)^2$ **7** $(x-1)(x+1)(x^2+1)$ **8** $(x+1)^3$
9 $(x+2)(2x+1)(3x-1)$ **10** $a=-4$, $(x+1)(x-2)(x+2)$
11 $b=-16$, $(2x-1)(x-4)^2$

Page 118 *Exercise 5*

1 1, −1 **2** 2, $\sqrt{2}$, $-\sqrt{2}$ **3** 2, −2 **4** 3, 4, −4 **5** −3
6 3, −3, $\frac{3}{2}$ **7** $k=-2$; 1, 2
8 Equation is $(x+1)(x^2+2)=0$; real root −1
9 c 1, −3; B(−3, −18) **10 a** $y=3x-2$ **b** (−2, −8)
11 a $x^3+x^2-36=0$ **b** Only real root 3, 3 m
12 a $y^3-6y^2+11y-6=0$ **b** years 1, 2, 3
13 3 seconds (the real root of $t^3-6t-9=0$)
14 a 1 million years **b** (i) 2 million years after end of first ice age (ii) a further 3 million years
15 20 days

Page 120 *Exercise 6*

1 1·3 **2** 1·2 **3** 0·5 **4 a** 4·9 cm **b** 4·7 cm **c** 6·1 cm

Page 121 *Brainstormer*

c m divides the product dn^3, but m has no factor (> 1) in common with n, so m divides d

Page 121 *Investigation*

a $f(10)=9q(10)+f(1)$, so 9 divides $f(10) \Leftrightarrow$ 9 divides $f(1)$. Let $f(x)=a_nx^n+a_{n-1}x^{n-1}+\ldots+a_1x+a_0$; then $f(10)$ is the decimal integer $a_na_{n-1}\ldots a_1a_0$ and $f(1)=a_n+a_{n-1}+\ldots+a_1+a_0$. The same method can now be extended to $f(x)$
b Same rule holds with 9 replaced by 3
c Use $f(x)=(x+1)q(x)+f(-1)$. In this case, $a_na_{n-1}\ldots a_1a_0$ is divisible by 11 $\Leftrightarrow a_n-a_{n-1}+a_{n-2}-a_{n-3}+\ldots$ is divisible by 11.

Page 121 *Check-up on Polynomials*

1 a 2, −3 **b** 0, 0 **c** 1, −5 **d** 4, −2
2 a $2x^2-x^3$ **b** $1-6x+9x^2$ **c** $12-3x-2x^2-x^3-2x^4$
3 a 252 **b** 64
4 a $3x^2+5x+12$, 19 **b** $2x^3-x^2+3x-8$, 31
c x^3+x^2-x+1, −3
5 a $(x+1)^2(x-3)$ **b** $(x-2)(x-5)(x+3)$
c $(x-2)(x+2)(2x^2+3)$ **d** $(2x-1)(x^2+x+1)$
6 a −3 **b** $\frac{3}{2}$, $-\frac{3}{2}$
7 c $x=1$, −2; B(−2, −10) **8** 1·2.

INTEGRATION

Page 123 *Exercise 1*

1 a $y=x^2+C$ **b** $y=2x^2+C$ **c** $y=3x^2+x+C$
d $y=3x-2x^2+C$
2 a $y=3x^2+3x+2$ **b** $y=6x^2+x-10$ **c** $y=5x-4x^2-3$
3 $y=x^2-x-1$ **4** $y=4x^2-8$ **5** $y=5x+5x^2$

Page 125 *Exercise 2*

1 a $\frac{1}{3}x^3+C$ **b** $\frac{1}{5}x^5+C$ **c** $-\frac{1}{2}x^{-2}+C$ **d** $-\frac{1}{5}x^{-5}+C$ **e** $\frac{2}{3}x^{3/2}+C$
f $\frac{3}{7}x^{7/3}+C$ **g** x^2+C **h** $2x^4+C$ **i** $5x+C$ **j** $-\frac{1}{x}+C$ **k** $2\sqrt{x}+C$
l $-\frac{6}{x^{1/2}}+C$
2 a $\frac{1}{6}x^6+C$ **b** $-\frac{1}{4}x^{-4}+C$ **c** $\frac{3}{4}x^{4/3}+C$ **d** $\frac{-1}{3x^3}+C$ **e** $8x^{1/4}+C$
3 a $x-x^2+C$ **b** $\frac{1}{3}x^3+\frac{1}{2}x^2+C$ **c** $\frac{1}{4}x^4-x+C$
d $2x^3-2x^2+2x+C$
4 a x^2-x^3+C **b** x^4-x+C **c** $\frac{1}{3}x^6-\frac{3}{2}x^2+x+C$
d $\frac{-2}{t}-\frac{1}{2}t^2+C$ **e** $\frac{2}{3}u^{3/2}-3u+C$ **f** $\frac{2}{5}v^{5/2}-4v^{1/2}+C$
5 a $y=x^4+x^2+C$ **b** $y=x^6-x^3+C$ **c** $y=x^3-x+C$
d $y=t^2-\frac{1}{t}+C$ **e** $v=t^5-\sqrt{t}+C$ **f** $p=\frac{1}{2}z^2+z^{1/4}+C$
6 a $y=\frac{1}{3}x^3-\frac{1}{x}+2$ **b** $y=2t^2-2t^3$ **c** $p=2\sqrt{u}+2u^{3/2}-8$
7 $y=x^3-3x^2+x+6$ **8** $y=x^4-2x^3+5x-1$
9 $f(x)=\frac{1}{3}x^{3/2}-2x^{5/2}+3$
10 a x^3-x^4+C **b** $\frac{4}{3}t^3+2t^2+t+C$ **c** $u-\frac{1}{2u^2}+C$
d $\frac{2}{3}v^3-\frac{1}{2}v^2+C$

Page 126 *Exercise 3*

1 a $\frac{1}{5}(x+1)^5+C$ **b** $\frac{1}{8}(2x+1)^4+C$ **c** $\frac{1}{3}(14+x)^3+C$
d $-\frac{1}{6}(1-x)^6+C$ **e** $\frac{1}{21}(3x-4)^7+C$
2 a $-(x-3)^{-1}+C$ **b** $\frac{2}{3}(x+4)^{3/2}+C$ **c** $-\frac{1}{6(2x+3)^3}+C$
d $\frac{1}{2}\sqrt{(4x-1)}+C$
3 a $\frac{1}{10}(2t-1)^5+C$ **b** $-(1-4x)^{7/4}+C$ **c** $-\frac{1}{6}(1-4u)^{3/2}+C$
d $(3v-1)^{2/3}+C$ **e** $x-\frac{1}{x+1}+C$
4 a $y=-\frac{1}{24}(3-4x)^6+C$ **b** $u=\frac{1}{3}(2t+1)^{3/2}+C$
c $y=\frac{1}{4(1-2v)^2}+C$
5 a $y=4-\frac{1}{x-1}$ **b** $y=\frac{2}{3}(x+1)^{3/2}+2$
6 Alison

Page 127 *Exercise 4*

1 a $-\cos x+C$ **b** $\sin x+C$ **c** $3\sin x+C$ **d** $-4\cos x+C$
e $-\frac{1}{2}\cos x+C$ **f** $\frac{1}{2}\sin 2x+C$ **g** $-\frac{1}{3}\cos 3x+C$ **h** $\sin 4x+C$
i $-3\cos 2x+C$ **j** $2\sin\frac{1}{2}x+C$
2 a $\sin(x+2)+C$ **b** $-\frac{1}{2}\cos(2x-1)+C$ **c** $\frac{1}{4}\cos(3-4x)+C$
d $\sin(ax+b)+C$
3 a $-5\cos x+C$ **b** $2\sin x-3\cos x+C$
c $\frac{1}{5}\sin 5\theta+\cos 3\theta+C$ **d** $-\frac{1}{2}\cos 2x+\sin 3x+C$
e $\frac{1}{3}t^3+\frac{1}{2}\sin 2t+C$ **f** $\frac{1}{2}\cos(3-4u)+C$
4 a $\cos^2 x=\frac{1}{2}(1+\cos 2x)$, $\int\cos^2 x\,dx=\frac{1}{2}x+\frac{1}{4}\sin 2x+C$
b $\sin^2 x=\frac{1}{2}(1-\cos 2x)$, $\int\sin^2 x\,dx=\frac{1}{2}x-\frac{1}{4}\sin 2x+C$ **c** $x+C$; $\int(\sin^2 x+\cos^2 x)dx=\int 1\,dx=x+C$
5 a $y=-\cos 2x+C$ **b** $y=3\sin\left(1+\frac{x}{3}\right)+C$
c $y=-2\cos\left(2x+\frac{\pi}{3}\right)+C$
6 a $y=1-\cos 3x$ **b** $y=-\cos\left(t-\frac{\pi}{4}\right)+\sin\left(t-\frac{\pi}{4}\right)+2$
c $y=\frac{3}{2}\sin(2t+\alpha)-\frac{1}{2}\sin\alpha$

Page 128 *Exercise 5*

1 a $P=5t+6t^{4/3}+C$ **b** $P=5t+6t^{4/3}+5000$ **c** 5136
2 $P(x)=70x-0\cdot2x^2-1200$
3 $V=1400t+30t^2$
4 a $\frac{dN}{dx}=7+6x$ **b** $N(x)=7x+3x^2+42$
5 $\frac{dw}{dt}=\frac{1}{30}(2+t)$, $w(t)=\frac{1}{15}t+\frac{1}{60}t^2+0\cdot3$
6 $x(t)=3-2\cos 2t$
7 a $\frac{ds}{dt}=30+6t$, $s(t)=30t+3t^2$ **b** 117 metres
8 $y=4x^2-\frac{1}{3}x^3$, $42\frac{2}{3}$ cm

Brainstormer

a $v = 400t - t^2$ **b** $\frac{dy}{dt} = 400t - t^2$ **c** $y = 200t^2 - \frac{1}{3}t^3$, 1 666 667 m

Page 132 ***Areas between x = a and x = b***

(i) For $y = x^4$, area $= \frac{1}{5}b^5 - \frac{1}{5}a^5$ (ii) For $y = x^5$, area $= \frac{1}{6}b^6 - \frac{1}{6}a^6$

Page 133 ***Exercise 6***

1 a 4 **b** 3 **c** 6 **d** $\frac{1}{4}$ **e** $4\frac{2}{3}$ **f** $2\frac{2}{3}$ **g** 4 **h** 4
2 a $24\frac{1}{5}$ **b** $\frac{1}{3}$ **c** $\frac{1}{6}$ **d** $\frac{2}{3}$ **e** $24\frac{1}{5}$ **f** $1\frac{1}{3}$
3 a $[\sin x]_0^{\pi/2} = 1$ **b** $[\sin t]_0^{\pi/2} = 1$ **c** $[\sin u]_0^{\pi/2} = 1$
The notation for the variable in a definite integral does not matter—it is a 'dummy variable'
4 a $\frac{1}{2}$ **b** 0 **c** 2 **d** $\frac{1}{2}$ **e** $\frac{7}{24} + \frac{1}{\pi}$ **f** $\frac{1}{2} - \frac{1}{4}\sqrt{3}$
5 a $1\frac{1}{9}$ **b** $1\frac{13}{15}$ **c** $38\frac{2}{3}$ **d** $1\frac{1}{12}$ **e** $\frac{1}{8}(\pi - 2)$
6 a $\int_1^3 \left(\frac{1}{2}x + 1\right) dx$ **b** $\int_{-1}^3 (3 + 2x - x^2) dx$ **c** $\int_2^6 \frac{1}{x} dx$

Page 134 ***Exercise 7***

1 0. $\int_{-1}^2 (x^2 - 1)dx = \int_{-1}^1 (x^2 - 1)dx + \int_1^2 (x^2 - 1)dx$
$= (-1\frac{1}{3}) + (1\frac{1}{3}) = 0$
The region involved has the same measure of area below the x-axis as above it

2 a $2\frac{1}{3}$ **b** 3

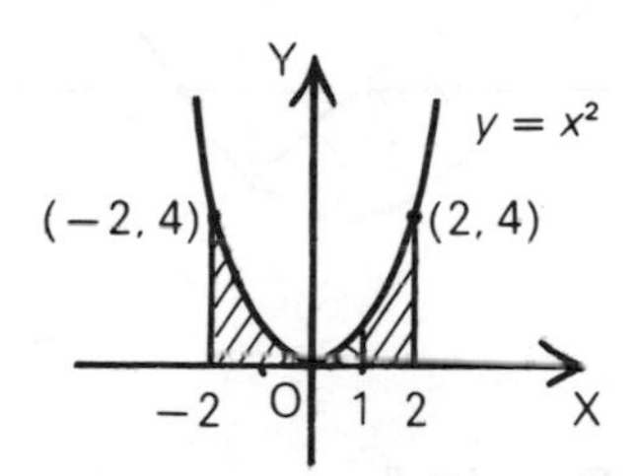

3 a 15 **b** 17

4 10

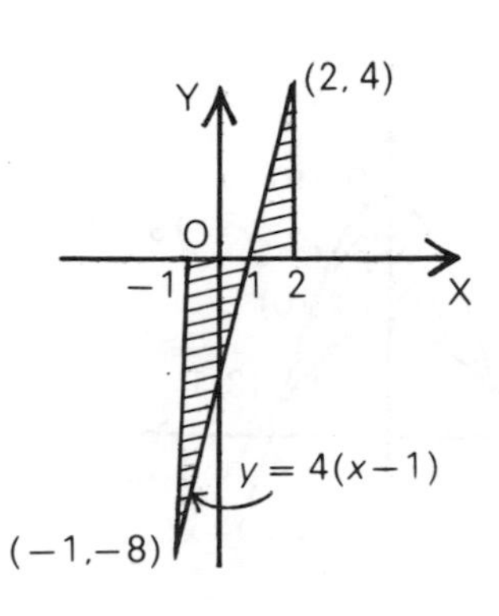

5

a $8\frac{2}{3}$ **b** $6\frac{1}{4}$

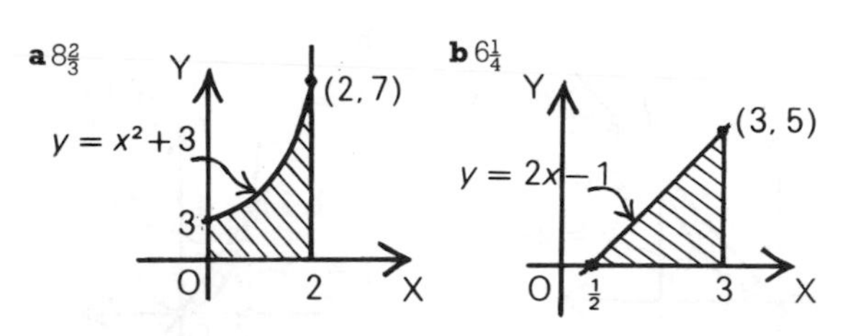

c $\frac{3}{8}$ **d** 2

6 a 40 **b** $10\frac{2}{3}$

7 $2 - \frac{1}{2}\sqrt{3}$

8 a $(\frac{2}{3}\sqrt{3} \doteqdot 1{\cdot}2, -\frac{16}{9}\sqrt{3} \doteqdot -3{\cdot}1)$ (min. TP) $(-\frac{2}{3}\sqrt{3}, \frac{16}{9}\sqrt{3})$ (max. TP) **b** $x = 0, 2, -2$

c

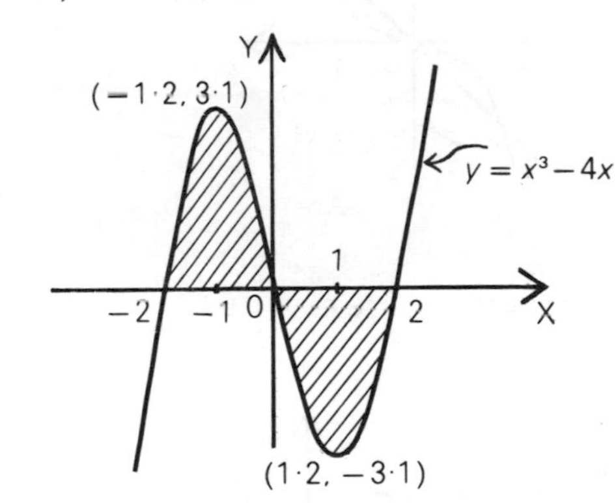

d 8 **e** Region has same area below the x-axis as above it

Page 135 ***Brainstormer***

a $\frac{\pi}{2}$ **b** π **c** $\sqrt{3} + \frac{2\pi}{3}$. For **c**

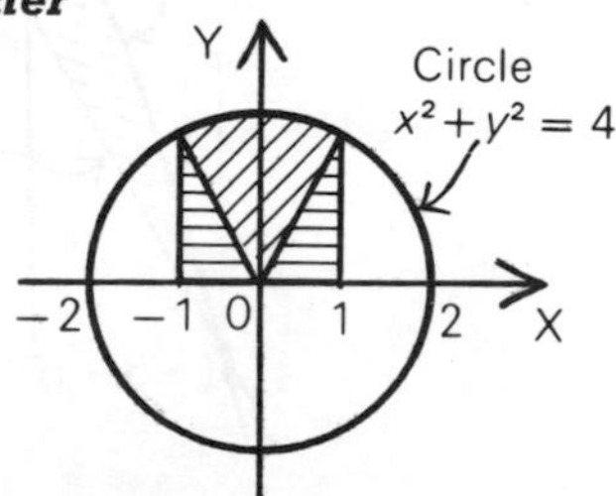

Page 136 *Exercise 8*

1 a

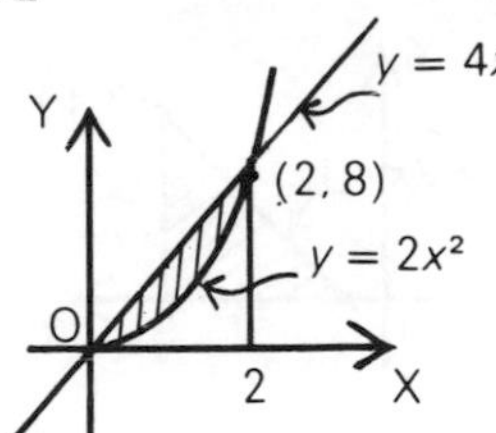

Area = $2\frac{2}{3}$

b

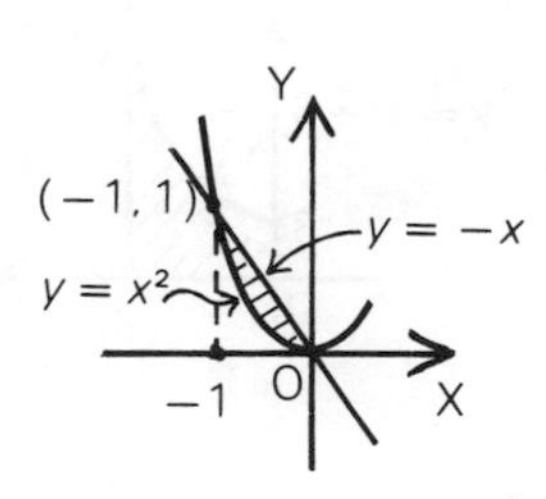

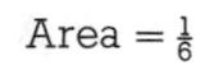

Area = $\frac{1}{6}$

c **d**

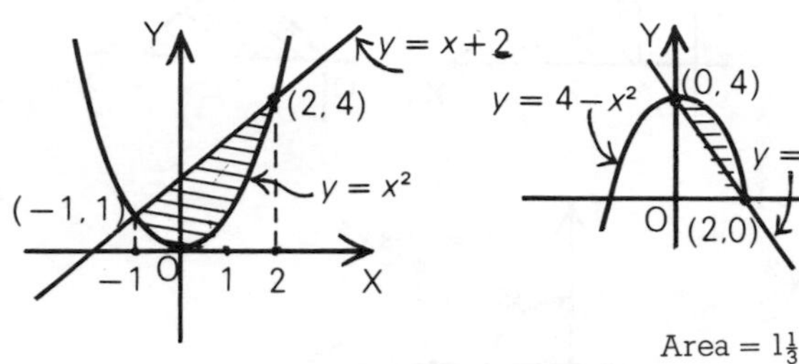

Area = $4\frac{1}{2}$

Area = $1\frac{1}{3}$

2

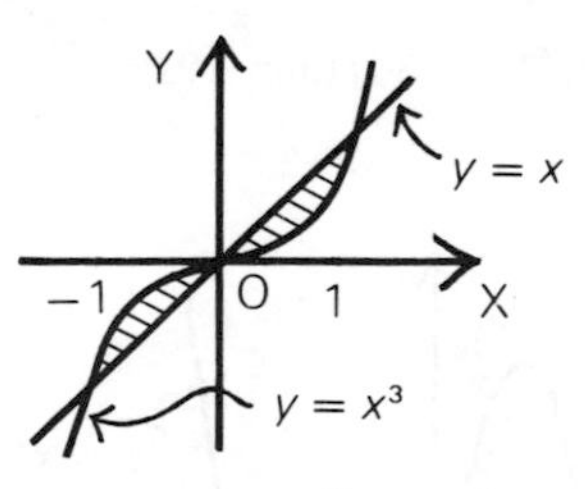

Area = $\frac{1}{2}$

3 a (1, 2), (−3, −10) **c** Area = $10\frac{2}{3}$

b

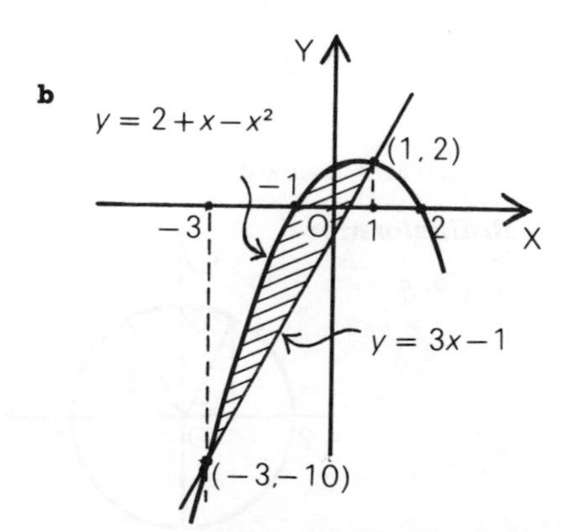

4 a

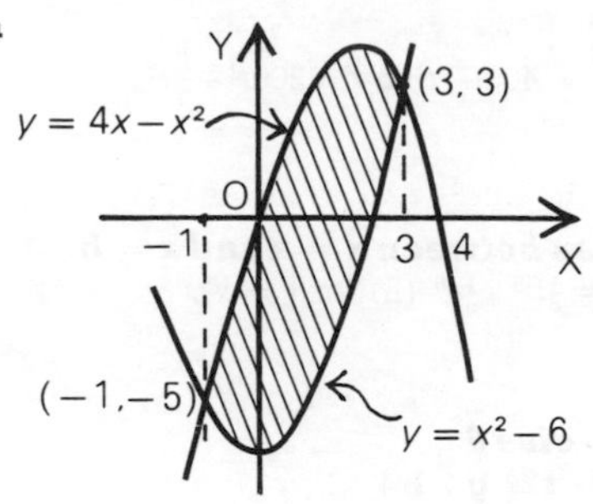

b Area = $21\frac{1}{3}$

5 Area = $1-\dfrac{\pi}{4}$

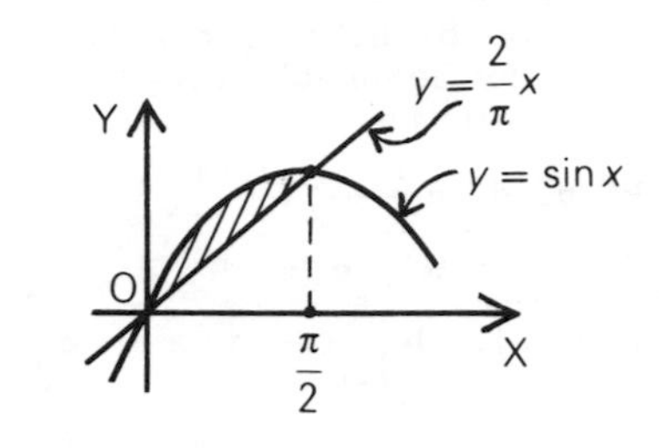

Page 136 *Brainstormer*

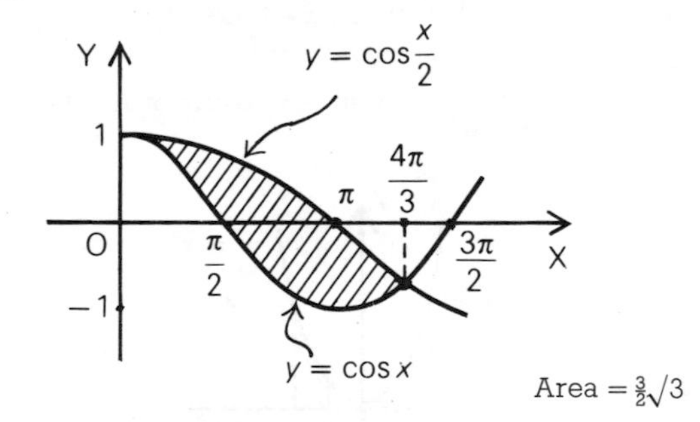

Area = $\frac{3}{2}\sqrt{3}$

Page 136 *Exercise 9*

1 $26\frac{2}{3}$ metres **2 a** 2000 **b** 375
3 £2000 **4 a** $5\sqrt{5}-5$ seconds **b** $5(\sqrt{61}-\sqrt{57})$ seconds
5 $\pi y^2 dx$ is the volume of a circular cylinder of radius y and height dx. Volume = $\frac{2}{3}\pi r^3$
6 $\dfrac{2\pi}{3}-1$

7 a

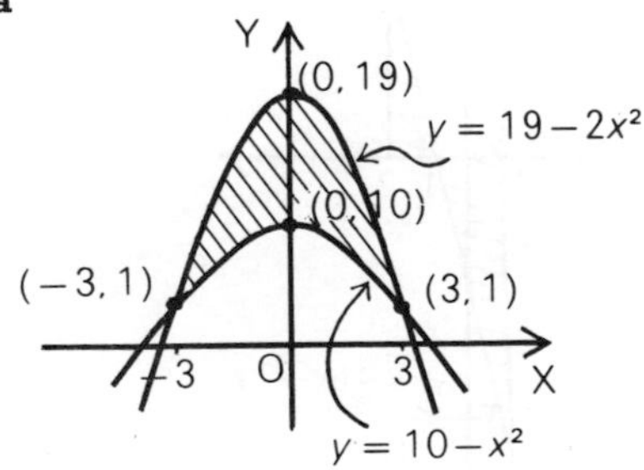

b 36

Page 138 *Check-up on Integration*

1 a $2x-\frac{1}{4}x^4+C$ **b** $\frac{1}{16}(4x+1)^4+C$ **c** $2\sqrt{t}-2t^3+C$
d $-\frac{1}{5}\cos 5u-\frac{1}{3}\sin 3u+C$ **e** $\frac{1}{3}x^3-2x-\frac{1}{x}+C$
2 a $3x+2x^2-x^3+C$ **b** $\frac{1}{3}\sin(3t+2)+C$ **c** $\frac{1}{3}(2u-3)^{3/2}+C$
d $\frac{1}{2}x^6-x^2+C$
3 a x^3-2x^2+C **b** $2x+\frac{1}{4}\sin 4x+C$ **c** $2x^3+\frac{1}{3}\cos 6x+C$
d $-\frac{2}{x}-\frac{1}{3}x^3+C$
4 a $y=3x-2x^3+C$ **b** $y=3x+\cos 2x+C$
c $y=4x+\frac{1}{2}\sin 4x+\frac{3}{2}\cos 2x+C$
5 a $y=3x-x^2+2x^3+1$ **b** $y=\frac{1}{2}\sin 4x+\frac{1}{2}\cos 6x$
6 $2+3\sin 2t$
7 a 5 **b** 2 **c** $\frac{1}{3}$ **d** $1\frac{7}{8}$
8 a

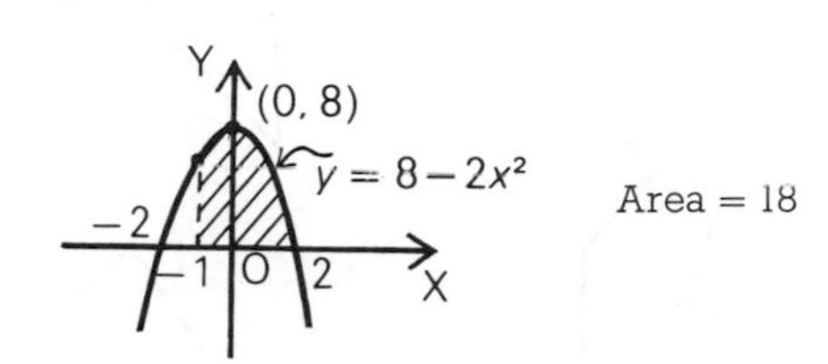

Area = 18

b

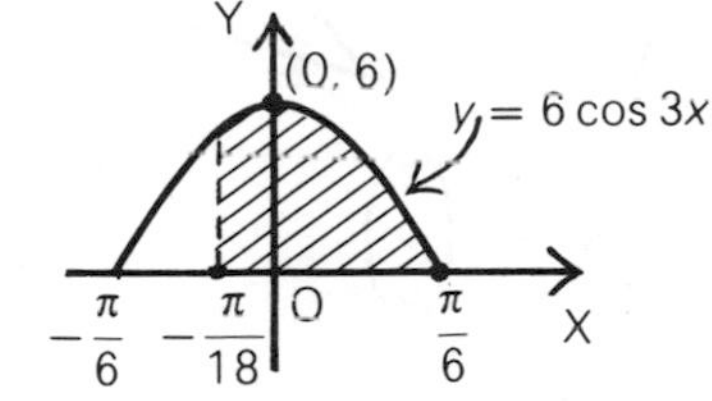

Area = 3

9 a

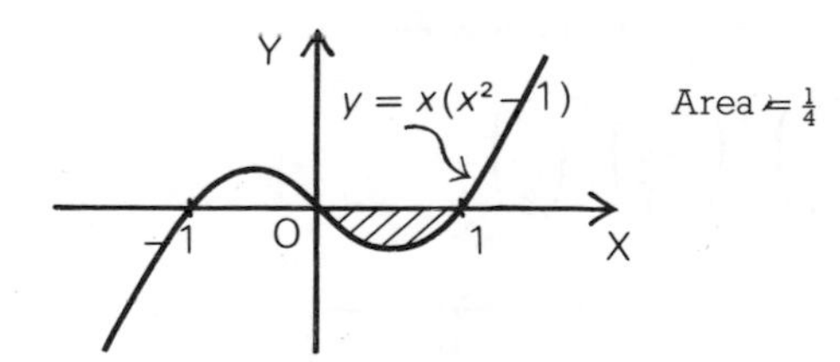

Area = $\frac{1}{4}$

b Area above x-axis = area below

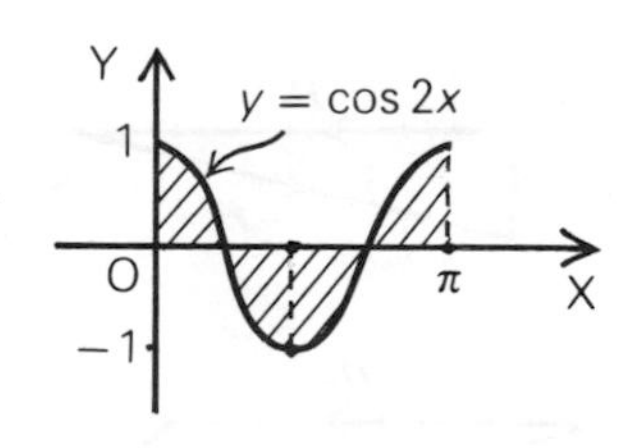

10 Area = $20\frac{5}{6}$

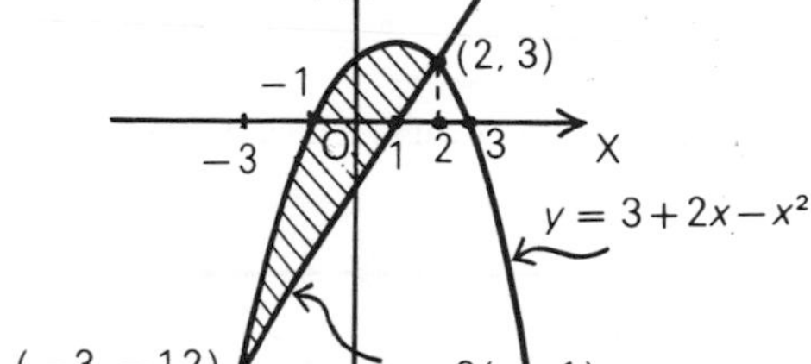

11 18·027 km

Page 139 *Exercise 1*

2

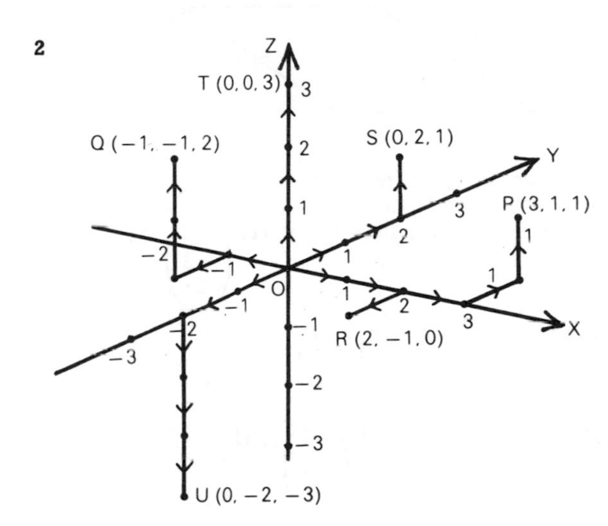

3 O(0, 0, 0), R(2, 0, 0), Q(2, 1, 0), V(0, 1, 0), S(2, 0, 3), T(0, 0, 3), U(0, 1, 3)
4 a O(0, 0, 0), A(2, 0, 0), B(0, 0, 3), C(0, −2, 0) **b** O(0, 0, 0), A(−2, 0, 0), B(0, 0, 3), C(0, 2, 0)
5 B(3, 2, 0), C(3, 4, 0), D(−1, 4, 0), E(−1, 2, 3), F(3, 2, 3), G(3, 4, 3), H(−1, 4, 3)
6 a (i) 5 (ii) 13 **b** $OC^2=x^2+y^2+z^2$
7 a (2, 2, −2) **b** (1, 5, 7) **c** (−1, −2, −4)

Page 142 *Exercise 2*

1 b, d are vectors; **a**, **c**, **e**, **f** are scalars

2 a $\begin{pmatrix}2\\0\\5\end{pmatrix}$ **b** $\begin{pmatrix}1\\3\\-6\end{pmatrix}$ **c** $\begin{pmatrix}0\\0\\2\end{pmatrix}$ **d** $\begin{pmatrix}-3\\-1\\-2\end{pmatrix}$ **e** $\begin{pmatrix}-5\\2\\-2\end{pmatrix}$ **f** $\begin{pmatrix}7\\-3\\-3\end{pmatrix}$

g $\begin{pmatrix}7\\7\\3\end{pmatrix}$ **h** $\begin{pmatrix}p-a\\q-b\\r-c\end{pmatrix}$

3 a $\sqrt{29}$ **b** $\sqrt{46}$ **c** 2 **d** $\sqrt{14}$

4 a (i) $\begin{pmatrix} x_B - x_A \\ y_B - y_A \\ z_B - z_A \end{pmatrix}$ (ii) $\begin{pmatrix} x_A - x_B \\ y_A - y_B \\ z_A - z_B \end{pmatrix}$ **b** $\begin{pmatrix} -u \\ -v \\ -w \end{pmatrix}$

5 $x = 3$, $y = 2$
6 PQ, RS are parallel, of same length and in same direction

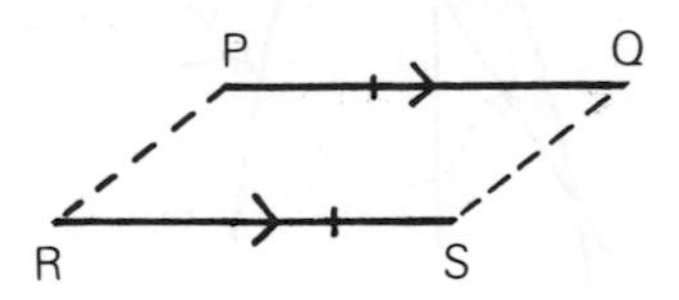

7 $\overrightarrow{OP} = \begin{pmatrix} 6 \\ 4 \\ 2 \end{pmatrix} = \overrightarrow{RQ}$, so OPQR is a parallelogram

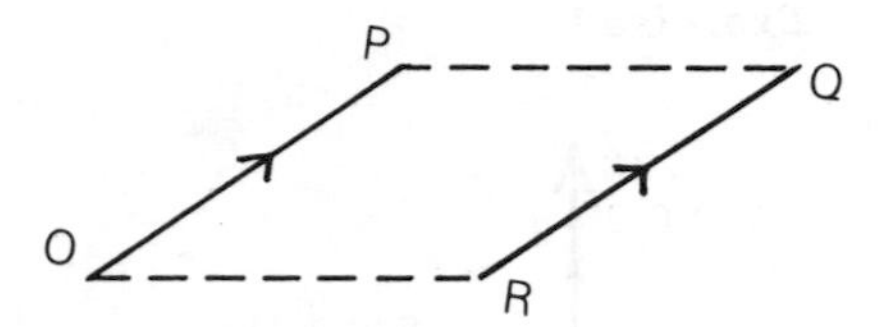

8 $AS^2 = BS^2 = CS^2 = 72$
9 (i) $LM^2 = 5 = LN^2$, so isosceles (ii) $LM^2 + LN^2 = 10 = MN^2$, so right-angled at L **10** B(10, 7, 2) **11 a** $\begin{pmatrix} 2 \\ 3 \\ -1 \end{pmatrix}$ **b** $\sqrt{14}$

c C(5, 2, 1) **d** B(3, 3, 4) **e** $\begin{pmatrix} 2 \\ -1 \\ -3 \end{pmatrix}$

f $\sqrt{14}$
12 a C(2, 0, 3), D(2, 2, 3) **b** $\begin{pmatrix} -1 \\ -1 \\ -3 \end{pmatrix}$, $\sqrt{11}$

c $7 + \sqrt{11}$ **d** $\begin{pmatrix} 1 \\ 1 \\ -6 \end{pmatrix}$, $\sqrt{38}$

Page 144 ***Exercise 3***
1 a (i) $\begin{pmatrix} 0 \\ 2 \\ 2 \end{pmatrix}$ (ii) $\begin{pmatrix} 3 \\ 0 \\ 10 \end{pmatrix}$ (iii) $\begin{pmatrix} 4 \\ -4 \\ -2 \end{pmatrix}$

b (i) $2\sqrt{2}$ (ii) $\sqrt{109}$ (iii) 6

2

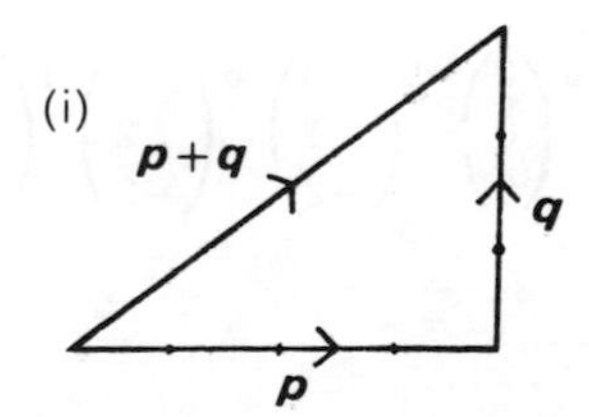

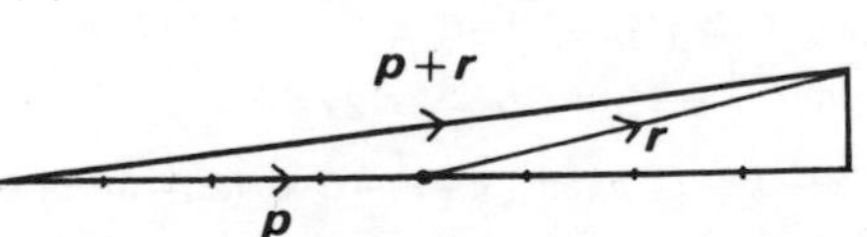

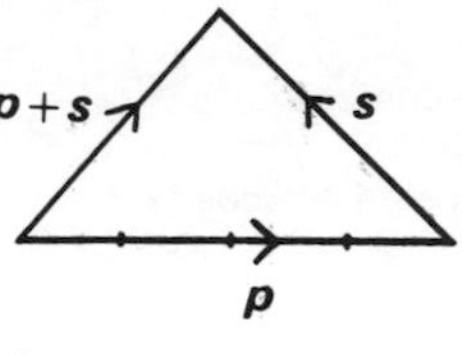

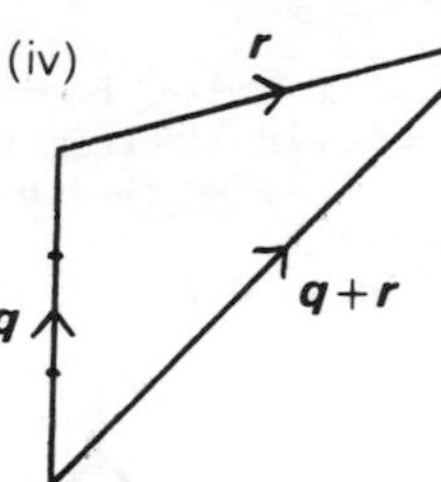

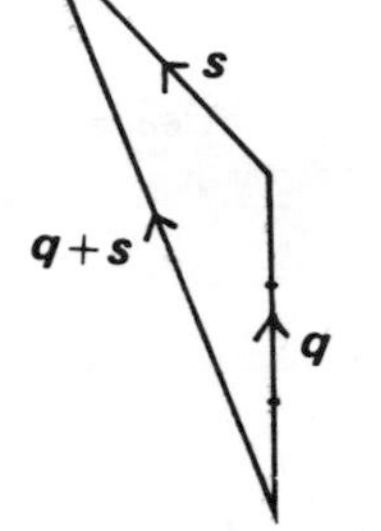

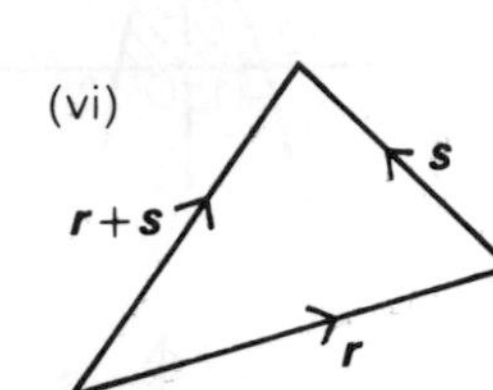

3 a (i), (ii), (iii) Each $\boldsymbol{u} + \boldsymbol{v} = \begin{pmatrix} 0 \\ 0 \\ 0 \end{pmatrix}$

b (i) $\begin{pmatrix} -3 \\ -2 \\ -1 \end{pmatrix}$ (ii) $\begin{pmatrix} 1 \\ 0 \\ -4 \end{pmatrix}$ (iii) $\begin{pmatrix} 2 \\ 3 \\ 4 \end{pmatrix}$ (iv) $\begin{pmatrix} -a \\ -b \\ -c \end{pmatrix}$

4 a Same length **b** Opposite direction

5 a $\begin{pmatrix} 6 \\ 5 \\ 1 \end{pmatrix}$ **b** $\begin{pmatrix} -3 \\ -2 \\ 1 \end{pmatrix}$ **c** $\begin{pmatrix} 3 \\ 3 \\ 2 \end{pmatrix}$

6

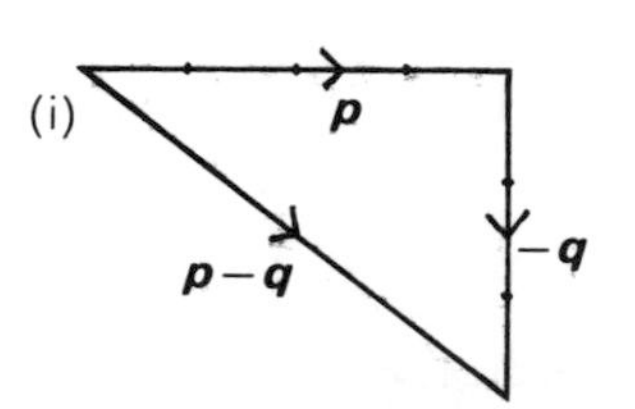

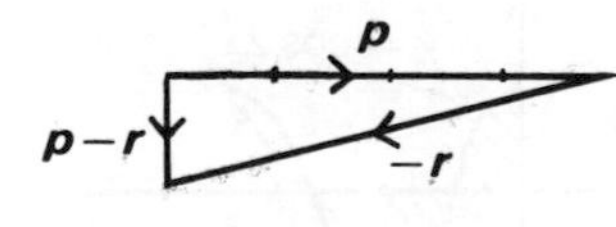

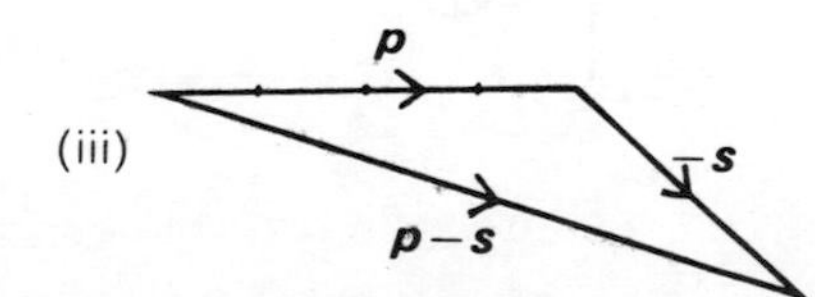

(iv) (v)

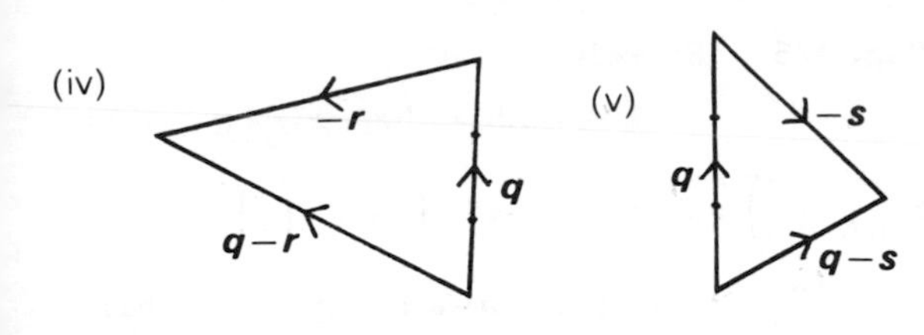

(vi)

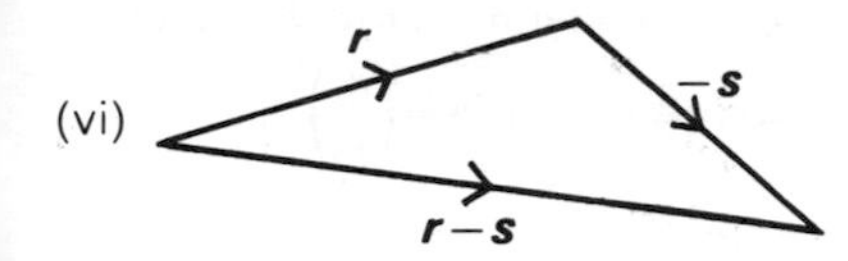

7 $\overrightarrow{BA} - \overrightarrow{BC} = \overrightarrow{BA} + \overrightarrow{CB} = \overrightarrow{CB} + \overrightarrow{BA} = \overrightarrow{CA}$

8 a $\overrightarrow{CB}$ **b** $\overrightarrow{CD}$

9 a $\boldsymbol{F}_1 + \boldsymbol{F}_2 + \boldsymbol{F}_3 = \begin{pmatrix} 8 \\ 0 \\ -15 \end{pmatrix}$ **b** 17

10 Speed is 13 mph. Direction makes 67° with the river bank.

11 Speed is 168 mph (to nearest unit). Direction—on a bearing of 015°.

12 a $\overrightarrow{DB}$ **b** $\overrightarrow{CD}$ **c** $\overrightarrow{CB}$ **d** $\overrightarrow{DC}$

13 a (i) $\overrightarrow{AB} = \overrightarrow{AC} + \overrightarrow{CB}$ (ii) $\overrightarrow{AB} = \overrightarrow{AC} + \overrightarrow{CF} + (-\overrightarrow{EF}) + (-\overrightarrow{BE}) = \overrightarrow{AD} + (-\overrightarrow{FD}) + (-\overrightarrow{EF}) + (-\overrightarrow{BE})$ **b** (i) $\overrightarrow{BE} = (-\overrightarrow{CB}) + \overrightarrow{CF} + (-\overrightarrow{EF}) = (-\overrightarrow{AB}) + \overrightarrow{AD} + \overrightarrow{DE}$ (ii) $\overrightarrow{BE} = (-\overrightarrow{AB}) + \overrightarrow{AC} + \overrightarrow{CF} + (-\overrightarrow{EF}) = (-\overrightarrow{AB}) + \overrightarrow{AD} + (-\overrightarrow{FD}) + (-\overrightarrow{EF}) = (-\overrightarrow{CB}) + \overrightarrow{CF} + \overrightarrow{FD} + \overrightarrow{DE} = (-\overrightarrow{CB}) + (-\overrightarrow{AC}) + \overrightarrow{AD} + \overrightarrow{DE}$

Page 147 *Exercise 4*

1 a (i) $\begin{pmatrix} 2 \\ -2 \\ 6 \end{pmatrix}$ (ii) $\begin{pmatrix} 6 \\ 12 \\ 6 \end{pmatrix}$ (iii) $\begin{pmatrix} 8 \\ 10 \\ 0 \end{pmatrix}$

b (i) $\begin{pmatrix} 0 \\ 2 \\ 4 \end{pmatrix}$ (ii) $\begin{pmatrix} 18 \\ -6 \\ 0 \end{pmatrix}$ (iii) $\begin{pmatrix} 18 \\ -4 \\ 4 \end{pmatrix}$

2 $\boldsymbol{p} - 2\boldsymbol{q} = \begin{pmatrix} 4 \\ 3 \\ -8 \end{pmatrix}$, $|\boldsymbol{p} - 2\boldsymbol{q}| = \sqrt{89}$

3 $\begin{pmatrix} -6 \\ 0 \\ 0 \end{pmatrix}$

4 a (i) $\overrightarrow{AB} = \begin{pmatrix} 1 \\ 3 \\ 3 \end{pmatrix}$ (ii) $\overrightarrow{CD} = \begin{pmatrix} 3 \\ 9 \\ 9 \end{pmatrix}$

b $\overrightarrow{CD}$, $\overrightarrow{AB}$ have the same direction, and CD = 3AB

5 a $\overrightarrow{RS} = \begin{pmatrix} 2 \\ 0 \\ -2 \end{pmatrix} = 2\overrightarrow{PQ}$, $k = 2$ **b** $\overrightarrow{PQ}$ is parallel to $\overrightarrow{RS}$, in the same direction and RS = 2PQ

6 $\overrightarrow{DF} = 4\overrightarrow{DE}$. $\overrightarrow{DF}$, $\overrightarrow{DE}$ are parallel with point D in common, so D, E, F are collinear

7 a $\overrightarrow{AC} = \frac{1}{3}\overrightarrow{CB}$; $\overrightarrow{AC}$, $\overrightarrow{CB}$ are parallel with C in common, so A, B, C are collinear; $\frac{AC}{CB} = \frac{1}{3}$

b $\overrightarrow{AC} = -\frac{3}{2}\overrightarrow{CB}$; A, B, C are collinear as in **a**; $\frac{AC}{CB} = -\frac{3}{2}$

8 B(0, 0, 3), C(4, 0, −9)

9 a $\overrightarrow{AB} = \frac{1}{2}\overrightarrow{BC} = \boldsymbol{u} + \boldsymbol{v}$, so A, B, C collinear (see **7a**)

b $\overrightarrow{DE} = \overrightarrow{EF} = \boldsymbol{u} + \boldsymbol{v}$, so D, E, F collinear

Page 148 *Brainstormer*

a (i) Both = $\overrightarrow{AC}$ (ii) Both = $\overrightarrow{AD}$

b

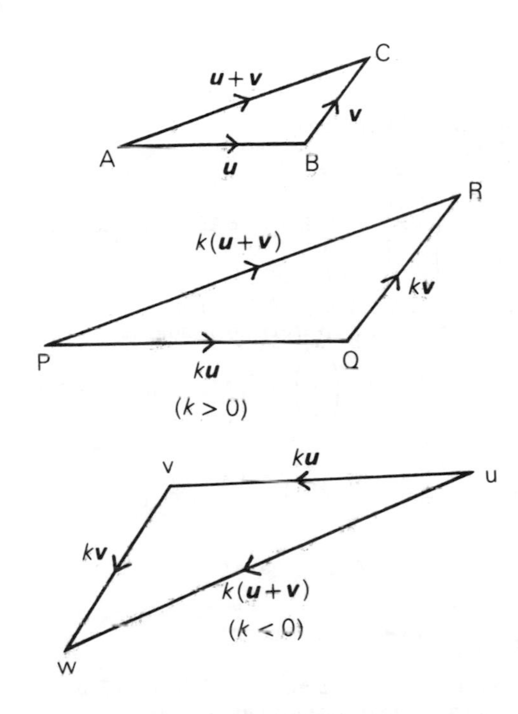

Triangles ABC, PQR, UVW have sides parallel in pairs. By similar triangles and addition of vectors, $k(\boldsymbol{u} + \boldsymbol{v}) = k\boldsymbol{u} + k\boldsymbol{v}$.

Page 150 *Exercise 5*

1 a (i) $\begin{pmatrix} 1 \\ 2 \\ 0 \end{pmatrix}$ (ii) $\boldsymbol{i} + 2\boldsymbol{j}$

b (i) $\begin{pmatrix} -1 \\ -5 \\ 2 \end{pmatrix}$ (ii) $-\boldsymbol{i} - 5\boldsymbol{k} + 2\boldsymbol{k}$

c (i) $\begin{pmatrix} 6 \\ 0 \\ -7 \end{pmatrix}$ (ii) $6\boldsymbol{i} - 7\boldsymbol{k}$ **d** (i) $\begin{pmatrix} 0 \\ 8 \\ -11 \end{pmatrix}$ (ii) $8\boldsymbol{j} - 11\boldsymbol{k}$

2 a (i) 7 (ii) 9 **b** $\pm\frac{1}{\sqrt{2}}\boldsymbol{i} + \frac{1}{2}\boldsymbol{j} - \frac{1}{2}\boldsymbol{k}$

3 a (5, 5, 2) **b** (−5, −2, 0)

4 a P(2, 1, 2) **b** Q(3, 5, 6)

5 a (i) P$(-1, -1, 5)$ (ii) Q$(2, 0, 8)$ **b** Both $= \begin{pmatrix} 3 \\ 1 \\ 3 \end{pmatrix}$

c PQ is parallel to AC, and PQ $= \frac{1}{2}$AC
6 a D$(0, -2, 0)$ **b** $BD^2 + DC^2 = BC^2$, so $\angle BDC = 90°$
7 a (i) $\overrightarrow{AB} = \begin{pmatrix} 0{\cdot}6 \\ 0{\cdot}8 \\ 0{\cdot}05 \end{pmatrix}$ (ii) $\overrightarrow{BC} = \begin{pmatrix} 1{\cdot}2 \\ 1{\cdot}6 \\ 0{\cdot}10 \end{pmatrix}$ **b** (i) 1·0 (ii) 2·0
c $\overrightarrow{AB} = \frac{1}{2}\overrightarrow{BC}$, so A, B, C are collinear (see question **7a** of *Exercise 4*)
8 a $-\boldsymbol{i} - 3\boldsymbol{j} + 5\boldsymbol{k}$, $\sqrt{35}$ **b** $\boldsymbol{i} + 3\boldsymbol{j} - 5\boldsymbol{k}$, $\sqrt{35}$ **c** $-2\boldsymbol{k}$, 2
9 a $\boldsymbol{e} + \boldsymbol{m} + \boldsymbol{s} = 6\boldsymbol{i} + 3\boldsymbol{j} + 2\boldsymbol{k}$ **b** 7

Page 152 *Brainstormer*

a $\overrightarrow{AP} = \frac{m}{n}\overrightarrow{PB}$

Page 153 *Exercise 6*

1 a $\sqrt{3}$ **b** $\frac{5}{2}\sqrt{2}$ **c** $-\frac{5}{2}\sqrt{2}$ **d** 0
2 a 8 **b** 1 **c** -5 **d** 5 **e** a **f** c
3 a 14 **b** -2 **c** 2 **d** 5
4 a (i) $\begin{pmatrix} 1 \\ -3 \\ -1 \end{pmatrix}$ (ii) $\begin{pmatrix} -5 \\ 3 \\ -4 \end{pmatrix}$ **b** -10 **c** AB $= \sqrt{11}$, AC $= 5\sqrt{2}$
d 115·2° **e** $\angle CBA \doteqdot 45{\cdot}3°$, $\angle ACB \doteqdot 19{\cdot}5°$
5 a 78·6° **b** $\boldsymbol{p} \cdot \boldsymbol{q} = 0$, so $\boldsymbol{p}$, $\boldsymbol{q}$ are perpendicular
6 $\overrightarrow{AB} = \begin{pmatrix} 3 \\ -5 \\ 9 \end{pmatrix}$, $\overrightarrow{AC} = \begin{pmatrix} 2 \\ 3 \\ 1 \end{pmatrix}$; $\overrightarrow{AB} \cdot \overrightarrow{AC} = 0$, so AB, AC are perpendicular
7 a T$(3, 5, 0)$ **b** $\overrightarrow{QT} = \frac{1}{2}\overrightarrow{TS} = \begin{pmatrix} 2 \\ -1 \\ 8 \end{pmatrix}$, so Q, T, S are collinear **c** 51·3° **8** 14·3° **9 a** (i) $\sqrt{6}$ (ii) $\sqrt{6}$ (iii) $\frac{\pi}{3}$
b Equilateral
10 $|\boldsymbol{F}|\cos\theta = \frac{\boldsymbol{F} \cdot \boldsymbol{a}}{|\boldsymbol{a}|} = 2\sqrt{3}$
11 a Both are $(5, -1, 3)$ **b** $\overrightarrow{AC} \cdot \overrightarrow{EF} = 0$, so AC, EF are perpendicular, so result using **a**
c $\overrightarrow{AB} = \overrightarrow{DC} = \begin{pmatrix} 5 \\ 2 \\ 2 \end{pmatrix}$, so ABCD is a parallelogram;

$\overrightarrow{AD} = \begin{pmatrix} 1 \\ 4 \\ -4 \end{pmatrix}$, AB $=$ AD $= \sqrt{33}$, so ABCD is a rhombus **d** $\frac{5}{33}$
12 a $M_1(1, 2, 2)$, $M_2(2, 2, 4)$, $M_3(6, 5, 5)$, $M_4(5, 5, 3)$
$\overrightarrow{M_1M_2} = \begin{pmatrix} 1 \\ 0 \\ 2 \end{pmatrix} = \overrightarrow{M_4M_3}$, so $M_1M_2M_3M_4$ is a parallelogram
b 58·2°

Page 155 *Brainstormer*

If $\boldsymbol{a} = \begin{pmatrix} a_1 \\ a_2 \\ a_3 \end{pmatrix}$, $\boldsymbol{b} = \begin{pmatrix} b_1 \\ b_2 \\ b_3 \end{pmatrix}$, $\boldsymbol{c} = \begin{pmatrix} c_1 \\ c_2 \\ c_3 \end{pmatrix}$, then
$\boldsymbol{a} \cdot (\boldsymbol{b} + \boldsymbol{c}) = a_1(b_1 + c_1) + a_2(b_2 + c_2) + a_3(b_3 + c_3) =$
$(a_1b_1 + a_2b_2 + a_3b_3) + (a_1c_1 + a_2c_2 + a_3c_3) = \boldsymbol{a} \cdot \boldsymbol{b} + \boldsymbol{a} \cdot \boldsymbol{c}$

Page 155 *Exercise 7*

1 a $\overrightarrow{AB} = \begin{pmatrix} x_B - x_A \\ y_B - y_A \end{pmatrix}$ **b** $|\boldsymbol{u}| = AB = \sqrt{[(x_B - x_A)^2 + (y_B - y_A)^2]}$
c $\begin{pmatrix} a \\ b \end{pmatrix} = \begin{pmatrix} c \\ d \end{pmatrix} \Rightarrow a = c, b = d$ **e** $\boldsymbol{i} = \begin{pmatrix} 1 \\ 0 \end{pmatrix}$, $\boldsymbol{j} = \begin{pmatrix} 0 \\ 1 \end{pmatrix}$
f $\overrightarrow{OP} = \begin{pmatrix} x \\ y \end{pmatrix} = x\boldsymbol{i} + y\boldsymbol{j}$ **g** $\overrightarrow{AB} = \boldsymbol{b} - \boldsymbol{a}$ **h** Midpoint of AB has position vector $\frac{1}{2}(\boldsymbol{a} + \boldsymbol{b})$ **i** $\boldsymbol{a} \cdot \boldsymbol{b} = |\boldsymbol{a}||\boldsymbol{b}|\cos\theta$ ($\boldsymbol{a}$, $\boldsymbol{b}$ non zero)
j $\boldsymbol{a} \cdot \boldsymbol{b} = x_1x_2 + y_1y_2$, where $\boldsymbol{a} = \begin{pmatrix} x_1 \\ y_1 \end{pmatrix}$, $\boldsymbol{b} = \begin{pmatrix} x_2 \\ y_2 \end{pmatrix}$.
2 a $\begin{pmatrix} 1 \\ 2 \end{pmatrix}$ **b** $\begin{pmatrix} -8 \\ -3 \end{pmatrix}$ **c** $\begin{pmatrix} 6 \\ -7 \end{pmatrix}$ **d** $\begin{pmatrix} 0 \\ -12 \end{pmatrix}$
3 a 5 **b** 6 **c** 13 **d** 4 **e** 2·5
4 a $a = -4, b = -1$ **b** $a = -7, b = -4$ **c** $a = -2, b = 2$
d $a = 1, b = -4$
5 a $\overrightarrow{OA} = \boldsymbol{p}$ $\overrightarrow{OB} = 3\boldsymbol{p}$ $\overrightarrow{OC} = 4\boldsymbol{p}$

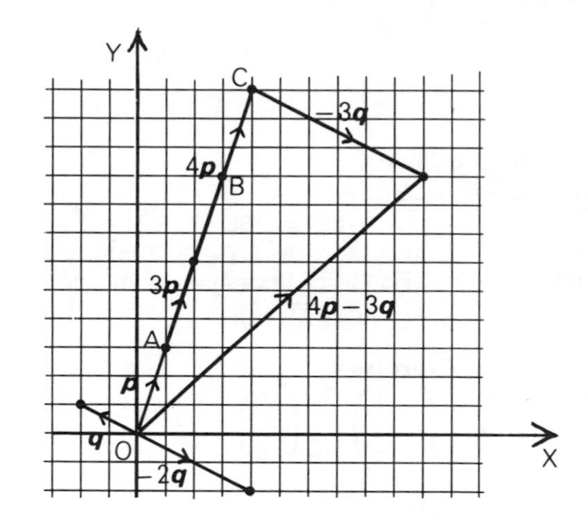

b $x = 3, y = 10$
6 a (i) $\begin{pmatrix} -3 \\ 5 \end{pmatrix}$ (ii) $\begin{pmatrix} 5 \\ 3 \end{pmatrix}$ (iii) $\begin{pmatrix} 8 \\ -2 \end{pmatrix}$ **b** $AB^2 + AC^2 = BC^2$, so $\angle A = 90°$
7 a $\begin{pmatrix} -3 \\ 2 \end{pmatrix}$, $\begin{pmatrix} 3 \\ 6 \end{pmatrix}$, $\begin{pmatrix} 0 \\ 4 \end{pmatrix}$ **b** $\begin{pmatrix} \frac{1}{3} \\ 2 \end{pmatrix}$ **c** $\overrightarrow{NG} = \frac{1}{2}\overrightarrow{GA} = \begin{pmatrix} -\frac{5}{3} \\ 0 \end{pmatrix}$, so N, G, A are collinear, and $\frac{NG}{GA} = \frac{1}{2}$ **d** $\overrightarrow{LG} = \frac{1}{2}\overrightarrow{GB}$, so L, G, B are collinear, and $\frac{LG}{GB} = \frac{1}{2}$ **e** G lies on all three medians AN, BL, CM, so they are concurrent at G
8 a $-\frac{63}{65}$ **b** $\boldsymbol{F}_1 + \boldsymbol{F}_2 = \begin{pmatrix} -2 \\ 8 \end{pmatrix}$, magnitude $2\sqrt{17}$, $\angle 104{\cdot}0°$

9 a (i) $\frac{7}{10}\sqrt{2}$ (ii) $\frac{7}{10}\sqrt{2}$ **b** Equation of OP is $y = \frac{3}{4}x$, of OQ is $y = \frac{4}{3}x$ and of OR is $y = x$, hence result

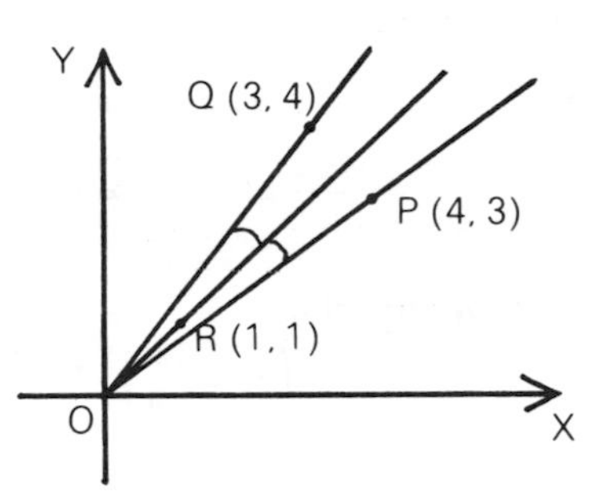

Page 157 *Investigation*

a For triangle with vertices A(***a***), B(***b***), C(***c***), check that G($\frac{1}{3}$(***a*** + ***b*** + ***c***)) lies on all three medians. **b** Simplify the problem by taking BC as the x-axis and the altitude through A as the y-axis. **c** Simplify the problem by taking BC as the x-axis and the perpendicular bisector of BC as y-axis.

d Put $\alpha = BC = |\boldsymbol{b} - \boldsymbol{c}|$, $\beta = |\boldsymbol{c} - \boldsymbol{a}|$, $\gamma = |\boldsymbol{a} - \boldsymbol{b}|$, and let I be point $\frac{1}{\alpha + \beta + \gamma}(\alpha\boldsymbol{a} + \beta\boldsymbol{b} + \gamma\boldsymbol{c})$. Show cos BAI = cos IAC, so IA bisects ∠BAC. Similarly for IB, IC and so result.

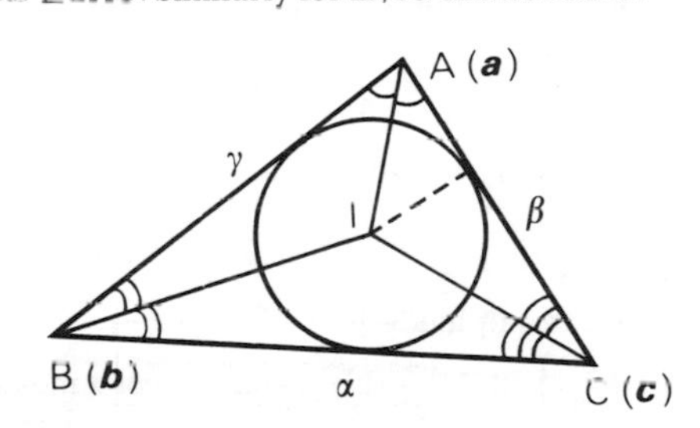

Page 157 *Check-up on Vectors*

1 a (i) $\begin{pmatrix} 4 \\ -4 \\ 8 \end{pmatrix}$ (ii) $\begin{pmatrix} 1 \\ -6 \\ -1 \end{pmatrix}$ **b** (i) $4\sqrt{6}$ (ii) $\sqrt{38}$

2 $a = -3$, $b = -6$

3 a (i) $\begin{pmatrix} 5 \\ 1 \\ -8 \end{pmatrix}$ (ii) $\begin{pmatrix} 6 \\ 3 \\ -9 \end{pmatrix}$ (iii) $\begin{pmatrix} -6 \\ 0 \\ 10 \end{pmatrix}$ (iv) $\begin{pmatrix} 0 \\ 3 \\ 1 \end{pmatrix}$

b (i) $\begin{pmatrix} -2 \\ -3 \\ 0 \end{pmatrix}$ (ii) $\begin{pmatrix} -12 \\ 0 \\ 3 \end{pmatrix}$ (iii) $\begin{pmatrix} -4 \\ 6 \\ 2 \end{pmatrix}$ (iv) $\begin{pmatrix} -16 \\ 6 \\ 5 \end{pmatrix}$

4 B(6, 1, 4), **5 a** $\begin{pmatrix} 4 \\ 4 \\ 2 \end{pmatrix}$ **b** 6 **c** (i) $\frac{2}{3}$ (ii) $\frac{2}{3}$ (iii) $\frac{1}{3}$

6 $\overrightarrow{AB} = \frac{1}{2}\overrightarrow{BC}$, so A, B, C are collinear and $\frac{AB}{BC} = \frac{1}{2}$

7 $\pm\frac{1}{3}\sqrt{7}\boldsymbol{i} + \frac{1}{3}\boldsymbol{j} - \frac{1}{3}\boldsymbol{k}$ **8** P(8, 6, 3)

9 a (i) $(-6, -7, -4)$ (ii) $(2, 1, 4)$ (iii) $(4, 3, 6)$ **b** $\overrightarrow{PQ} = 4\overrightarrow{QR}$, so P, Q, R collinear and $\frac{PQ}{QR} = \frac{4}{1}$

10 a -13 **b** 8 **11 a** 74·2° **b** $\boldsymbol{p}.\boldsymbol{q} = 0$, so ***p***, ***q*** are perpendicular

12 $\cos A = \frac{\overrightarrow{AB}.\overrightarrow{AC}}{|\overrightarrow{AB}||\overrightarrow{AC}|} = -\frac{8}{\sqrt{14}\sqrt{29}}$, so A is obtuse since cos A is negative, 113·4°

13 a $a = 3$, $b = 8$ **b** $a = -1$, $b = 1$

14 a $\frac{24}{25}$ **b** $\boldsymbol{F}_1 + \boldsymbol{F}_2 = 7\boldsymbol{i} + 7\boldsymbol{j}$; magnitude $7\sqrt{2}$, 45°.

Page 160 *Exercise 1*

a 1, 2π **b** 2, 2π

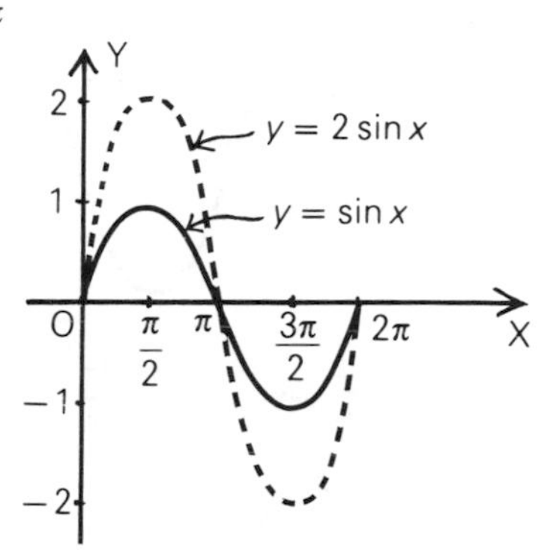

c 1, π **d** 2, π

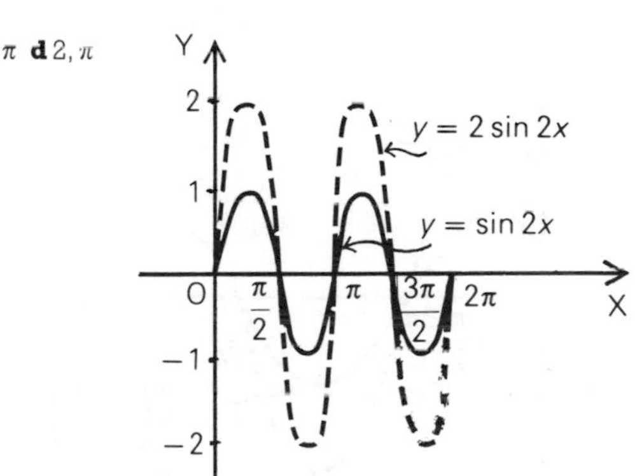

2 Intersections with x-axis: $\frac{\pi}{2}, \frac{3\pi}{2}$. Intersections with y-axis: $-1, -3$. Maxima at $x = \pi$, minima at $x = 0, 2\pi$. Amplitudes 1, 3. Periods 2π

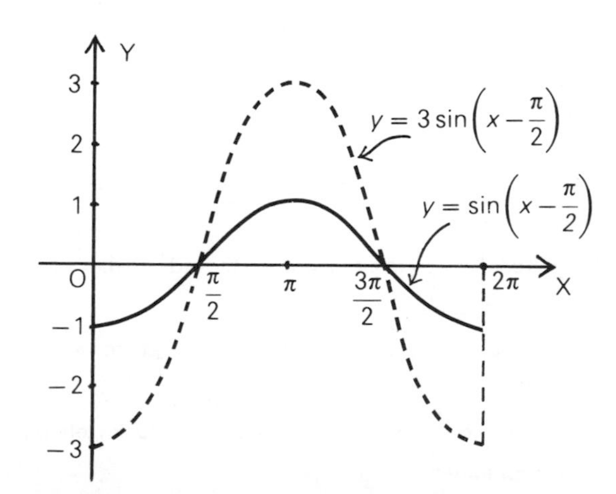

THE WAVE FUNCTION acos x + bsin x

3 a 1, 2π **b** 4, 2π **c** 3, π **d** 2, $\frac{2\pi}{3}$

4 $y = 3\sin x$: amplitude 3, period 2π

$y = \sin 3x$: amplitude 1, period $\frac{2\pi}{3}$

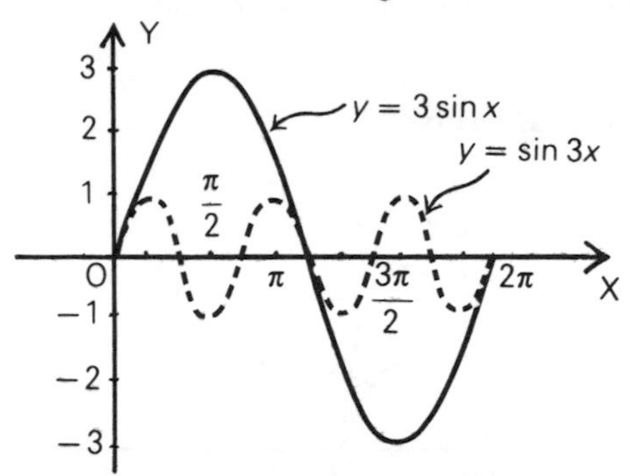

5 a Intersections with x-axis: $x = \frac{3\pi}{4}, \frac{7\pi}{4}$

Intersections with y-axis: $\frac{1}{\sqrt{2}}$, $\sqrt{2}$

Maxima at $x = \frac{\pi}{4}$ Minima at $x = \frac{5\pi}{4}$

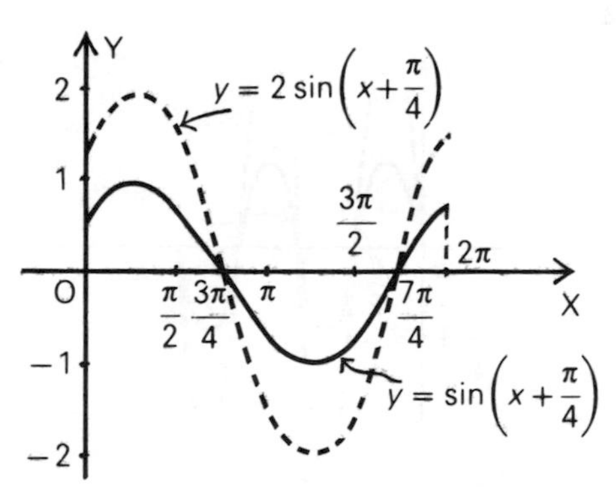

b $y = \sin\left(x+\frac{\pi}{4}\right)$ is same as $y = \sin x$, but moved $\frac{\pi}{4}$ to the left.

$y = 2\sin\left(x+\frac{\pi}{4}\right)$ is like $y = \sin\left(x+\frac{\pi}{4}\right)$, but with amplitude 2.

6 a $y = \cos 4x$ **b** $y = 5\sin 2x$ **c** $y = \sin\left(x-\frac{\pi}{3}\right)$

7 a $y = 2\sin\left(x+\frac{\pi}{2}\right)$ is like $y = \sin x$ moved $\frac{\pi}{2}$ to the left, but with amplitude 2

b $y = 3\cos\left(x-\frac{\pi}{6}\right)$ is like $y = \cos x$ moved $\frac{\pi}{6}$ to the right, but with amplitude 3

c $y = 4\cos\left(x+\frac{\pi}{6}\right)$ is like $y = \cos x$ moved $\frac{\pi}{6}$ to the left, but with amplitude 4

d $y = 5\sin\left(x+\frac{\pi}{3}\right)$ is like $y = \sin x$ moved $\frac{\pi}{3}$ to the left, but with amplitude 5

8 a First is $y = R\sin(x-\alpha)$, second $y = R\cos(x-\alpha)$

b (i) First: x-axis: α, $\pi+\alpha$, y-axis: $-R\sin\alpha$; Second: x-axis: $\frac{\pi}{2}+\alpha$, $\frac{3\pi}{2}+\alpha$; y-axis: $R\cos\alpha$

(ii) Maximum value R at $x = \frac{\pi}{2}+\alpha$ for first, $x = \alpha$ for second.

Minimum value $-R$ at $x = \frac{3\pi}{2}+\alpha$ for first, $x = \pi+\alpha$ for second.

Page 162 *Investigation*

a $y = \sqrt{2}\cos\left(x+\frac{\pi}{4}\right) = \sqrt{2}\cos\left(x-\left(-\frac{\pi}{4}\right)\right)$

or $= \sqrt{2}\cos\left(x-\frac{7\pi}{4}\right)$

Graph is like $y = \cos x$ moved $\frac{7\pi}{4}$ to the right $\left(\text{or } \frac{\pi}{4} \text{ to the left}\right)$ but with amplitude $\sqrt{2}$.

b $y = 5\cos(x-\alpha)$ where $\alpha = 53{\cdot}1°$ correct to 1 decimal place

Graph is like $y = \cos x$ moved 53·1° to the right, but with amplitude 5

Page 164 *Exercise 2*

1 a (i) $\sqrt{2}\cos\left(x-\frac{\pi}{4}\right)$ (ii) $\sqrt{2}$, $\frac{\pi}{4}$ (iii) graph $y = \cos x$ moved $\frac{\pi}{4}$ to the right, but with amplitude $\sqrt{2}$

b (i) $\sqrt{2}\cos\left(x-\frac{7\pi}{4}\right)$ (ii) $\sqrt{2}$, $\frac{7\pi}{4}$ (iii) graph $y = \cos x$ is moved $\frac{7\pi}{4}$ to the right with amplitude $\sqrt{2}$

c (i) $2\cos\left(x-\frac{\pi}{6}\right)$ (ii) 2, $\frac{\pi}{6}$ (iii) graph $y = \cos x$ moved $\frac{\pi}{6}$ to the right, but with amplitude 2

d (i) $2\cos\left(x-\frac{2\pi}{3}\right)$ (ii) 2, $\frac{2\pi}{3}$ (iii) graph $y = \cos x$ moved $\frac{2\pi}{3}$ to the right, but with amplitude 2

2 a (i) $5\cos(x-53)°$ (ii) 5, 53° (iii) graph $y = \cos x°$ moved 53° to the right, but with amplitude 5

b (i) $10\cos(x-323)°$ (ii) 10,323° (iii) graph $y = \cos x°$ moved 323° to the right, but with amplitude 10

c (i) $2\cos(x-300)°$ (ii) 2, 300° (iii) graph $y = \cos x°$ moved 300° to the right, but with amplitude 2

d (i) $\sqrt{5}\cos(x-63)°$ (ii) $\sqrt{5}$, 63° (iii) graph $y = \cos x°$ moved 63° to the right, but with amplitude $\sqrt{5}$

3 $R\sin(x+\alpha) = R(\sin x\cos\alpha+\cos x\sin\alpha) = (R\sin\alpha)\cos x + (R\cos\alpha)\sin x = a\cos x+b\sin x$ where $a = R\sin\alpha$, $b = R\cos\alpha$; $R = \sqrt{(a^2+b^2)}$, $\tan\alpha = \frac{a}{b}$.

4 a $\sqrt{2}\sin\left(x+\frac{\pi}{4}\right)$ **b** $\sqrt{2}\sin\left(x+\frac{3\pi}{4}\right)$ **c** $2\sin\left(x+\frac{\pi}{3}\right)$

d $2\sin\left(x-\frac{\pi}{6}\right)$ or $2\sin\left(x+\frac{11\pi}{6}\right)$

5 a $\sqrt{2}\cos\left(2x-\frac{\pi}{4}\right)$ **b** $\sqrt{2}\cos\left(3x-\frac{3\pi}{4}\right)$ **c** $5\cos(\omega t+0{\cdot}9)$

6 $2\cos(300t-60)°$

7 a $31{\cdot}62\sin(20t+18{\cdot}43)°$

b Cuts t-axis where $t = 8{\cdot}1$, $17{\cdot}1$

maximum 31·62 when $t = 3{\cdot}6$

minimum −31·62 when $t = 12{\cdot}6$

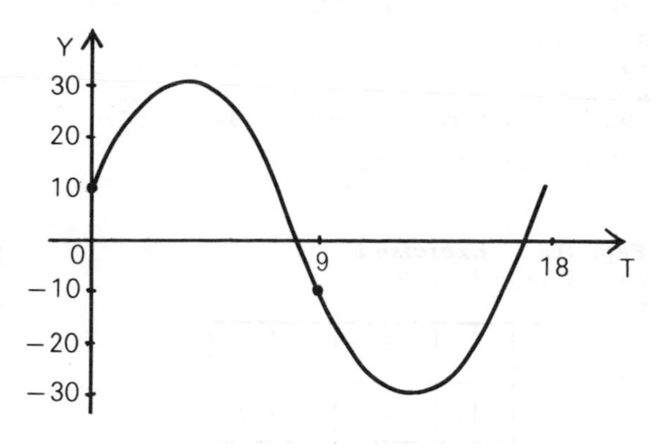

Page 165 ***Exercise 3***

1 a Max. $\sqrt{2}$, $x = 45$; min. $-\sqrt{2}$, $x = 225$
b Max. 25, $x = 286$; min. -25, $x = 106$
c Max. $\sqrt{13}$, $x = 146$; min. $-\sqrt{13}$, $x = 326$

2 a (i) Max. 4, $x = \frac{\pi}{3}$; min. 0, $x = \frac{4\pi}{3}$ (ii) max. $-2+\sqrt{2}$, $x = \frac{7\pi}{4}$; min. $-2-\sqrt{2}$, $x = \frac{3\pi}{4}$ (iii) Max. 3, $x = \frac{11\pi}{6}$; min. -1, $x = \frac{5\pi}{6}$

b

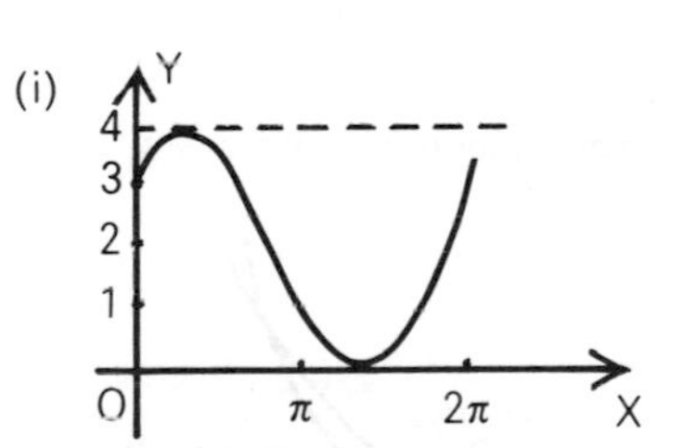

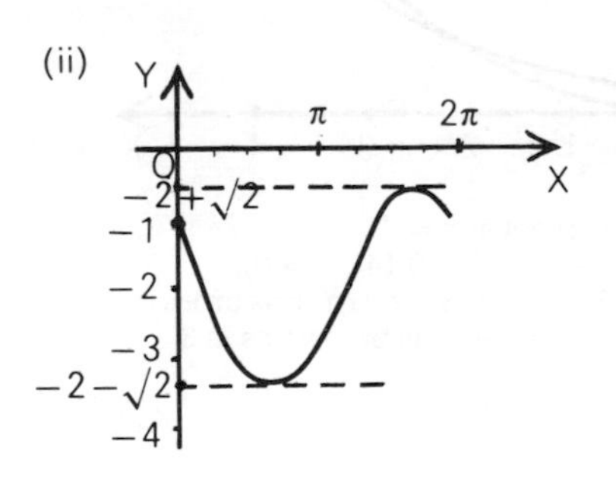

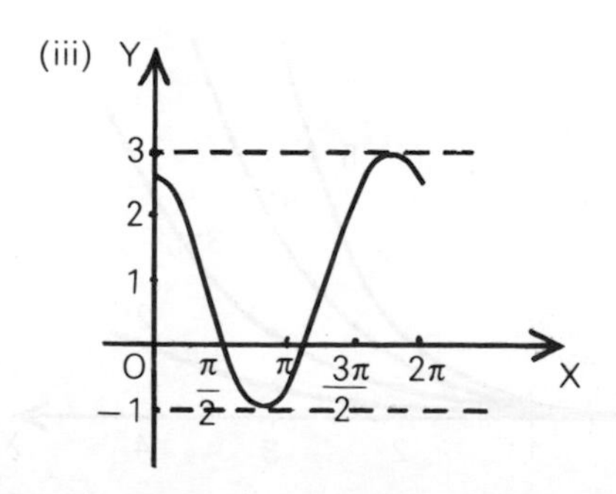

3 a Max. 10, $x = 18$; min. 0, $x = 108$
b Max. $8+\sqrt{5}$, $x = 148$; min. $8-\sqrt{5}$, $x = 58$

4 a Max. $2+2\sqrt{2}$, $x = \frac{\pi}{8}, \frac{9\pi}{8}$; min. $2-2\sqrt{2}$, $x = \frac{5\pi}{8}, \frac{13\pi}{8}$
b Max. $1+\sqrt{5}$, $x = \pi-0{\cdot}6, 2\pi-0{\cdot}6$;
min. $1-\sqrt{5}$, $x = \frac{\pi}{2}-0{\cdot}6, \frac{3\pi}{2}-0{\cdot}6$

5 b $1+\sqrt{2}, \frac{\pi}{8}$ **6 a** $2\cos\theta + 2\sin\theta$

b Max. $2\sqrt{2}$ metres at $\theta = \frac{\pi}{4}$

7 a Use $OA = 50\sin\theta$. $AD = 50\cos\theta$
b $1250 + 1250\sqrt{2}\cos\left(2\theta - \frac{\pi}{4}\right)$ **c** $1250(1+\sqrt{2}), \frac{\pi}{8}$

8 a Use $OM = \cos\theta$, $MP = \sin\theta$ **b** $\frac{1}{2} + \frac{\sqrt{5}}{4}\cos(2\theta - 2{\cdot}7)$

c $\frac{1}{2} + \frac{\sqrt{5}}{4}$, $\theta \doteqdot 1{\cdot}3$

Page 166 ***Exercise 4***

1 a 53 **b** 95, 191 **c** 162, 310
2 a $\frac{3\pi}{2}, 2\pi$ **b** $\frac{2\pi}{3}, 2\pi$ **c** $\frac{7\pi}{12}, \frac{23\pi}{12}$
3 a 0°, 135°, 180° **b** 24·2°, 102·7°

4 a Equation can be expressed as $\cos(x-\alpha)° = \frac{c}{\sqrt{(a^2+b^2)}}$,

so, for real roots, $\frac{|c|}{\sqrt{(a^2+b^2)}} \leqslant 1$, i.e. $c^2 \leqslant a^2 + b^2$

b 288·2, 329·1

Page 167 ***Brainstormer***

b 26

Page 167 ***Check-up on The Wave Function*** **a** ***cos*** **x** + **b** ***sin*** **x**

1 x-axis: $x = \frac{2\pi}{3}, \frac{5\pi}{3}$
y-axis: $y = \sqrt{3} \doteqdot 1{\cdot}7$
Max. 2, $x = \frac{\pi}{6}$
Min. -2, $x = \frac{7\pi}{6}$
Amplitude 2, period 2π

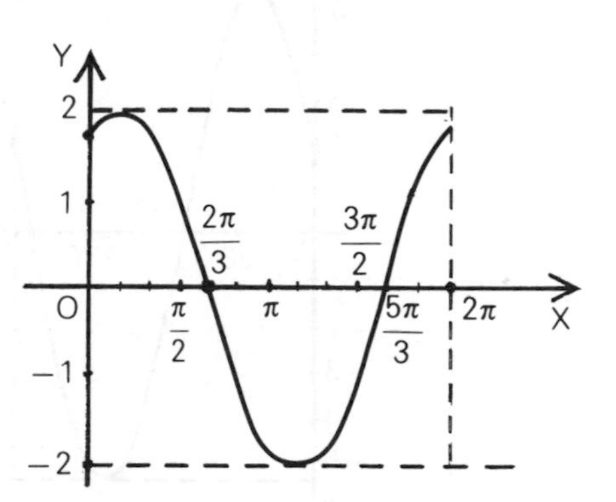

2 a (i) $2\cos\left(x-\frac{11\pi}{6}\right)$
Amplitude 2
Phase angle $\frac{11\pi}{6}$

(ii) $\sqrt{2}\cos\left(x-\frac{5\pi}{4}\right)$
Amplitude $\sqrt{2}$
Phase angle $\frac{5\pi}{4}$

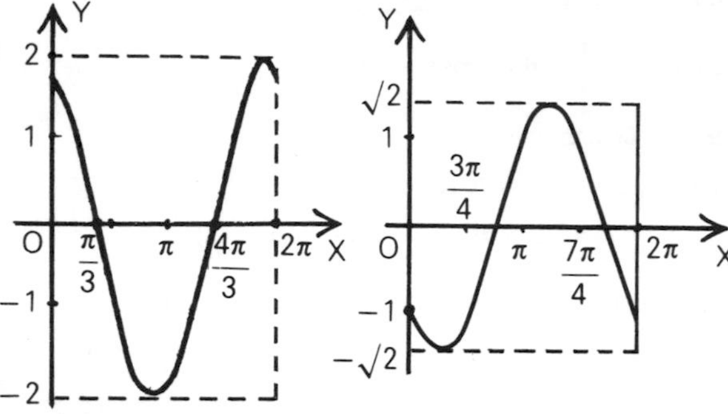

b (i) $5\cos(x°-127°)$
Amplitude 5
Phase angle 127°

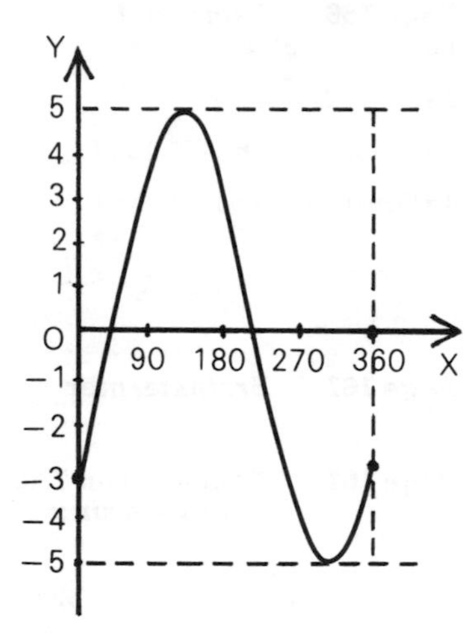

(ii) $\sqrt{29}\cos(x°-68°)$
Amplitude $\sqrt{29}$
Phase angle 68°
($\sqrt{29} \doteqdot 5{\cdot}4$)

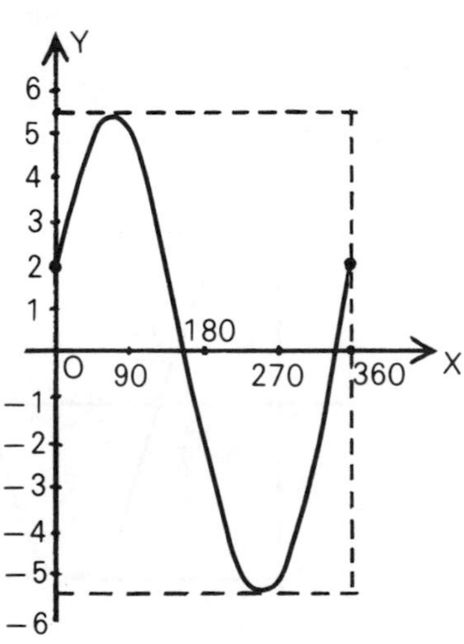

3 $3\sqrt{5}\sin(250t+63)°$, $3\sqrt{5}$, 63°
4 a Max. $3+\sqrt{2}$, $x = 45$; min. $3-\sqrt{2}$, $x = 225$
b Max. $-3+\sqrt{29}$, $x = 292$; min. $-3-\sqrt{29}$, $x = 112$
5 b 300 cm, 53° (to nearest degree) **6 a** 32·4, 290·8
b 45, 161·6, 225, 341·6 **7 b** $\frac{40}{1+\sqrt{2}}$, 45.

Page 168 ***Exercise 1***
1 a 2^{63} **b** $f(x) = 2^{x-1} (x \geqslant 1)$
2 a

x	1	2	3	4	5	6
No.	3	9	27	81	243	729

b $f(x) = 3^x$ $(x \geqslant 0)$
c

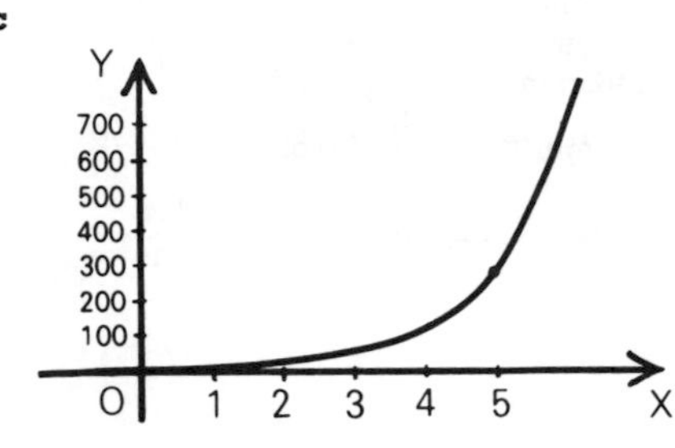

3 a Where $x = 2, 4$
b $y = 2^x$; it grows much more rapidly for $x > 4$

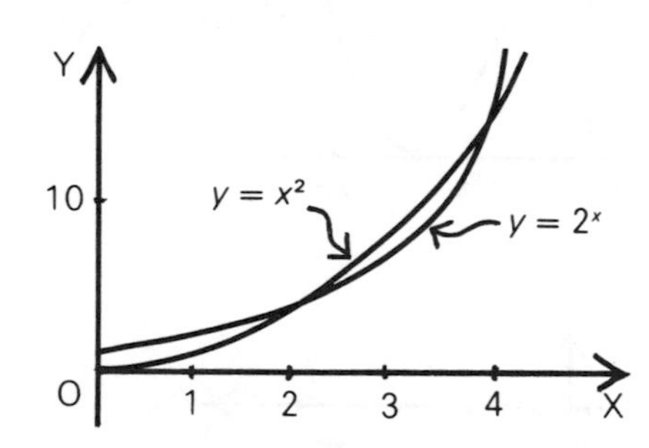

4 $y = 3^x$, $y = x^3$ meet at $x = 3$
Let $f(x) = 3^x - x^3$; $f(2{\cdot}4) = 0{\cdot}14\ldots (> 0)$,
$f(2{\cdot}5) = -0{\cdot}03\ldots (< 0)$, so the graphs cross between $x = 2{\cdot}4$ and $x = 2{\cdot}5$. Graphs behave as in **3**
5 a

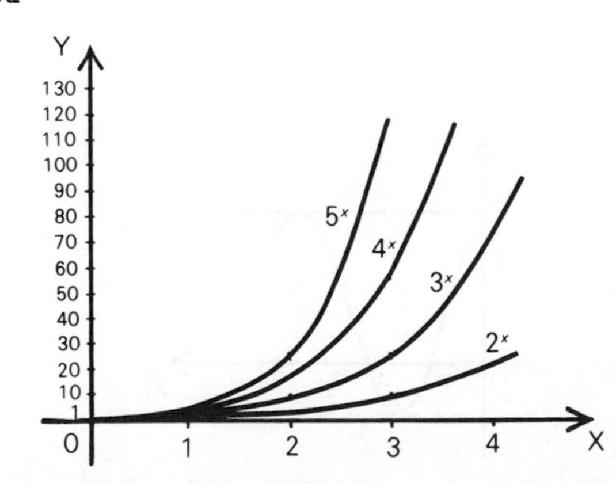

b $a^{\circ} = 1$ for all $a \neq 0$
6 a 1000.2^x $(x \geqslant 0)$ **b** 200.3^x $(x \geqslant 0)$ **c** 500.2^x $(x \geqslant 0)$

Page 169 *Exercise 2*

1

n	0	10	20	30	40	50
A(n)	100	311	965	2996	9305	28 900

2 a $100(1{\cdot}15)^n$
b

n	0	10	20	30	40	50
A(n)	100	405	1637	6621	26 786	108 366

3 a $P(t) = 100(1{\cdot}06)^t$ $(t \geqslant 0)$ **b** 429 million
4 a $400(1{\cdot}3)^t$ **b** (i) 5514 (ii) 9319
5 a $(1{\cdot}5)^x$ **b** (i) 25·6 units (ii) 291·9 units
6 a $500(1{\cdot}8)^t$ **b** (i) 17 006 (ii) 669 129 423

Page 170 *Brainstormer*

a $50.2^{t/30}$ **b** (i) 70·7 million (ii) 141·4 million

Page 171 *Exercise 3*

1 a $10(0{\cdot}4)^t = D(t)$
b

t	0	1	2	3	4	5	6
D(t)	10	4	1·6	0·6	0·3	0·1	0·04

c

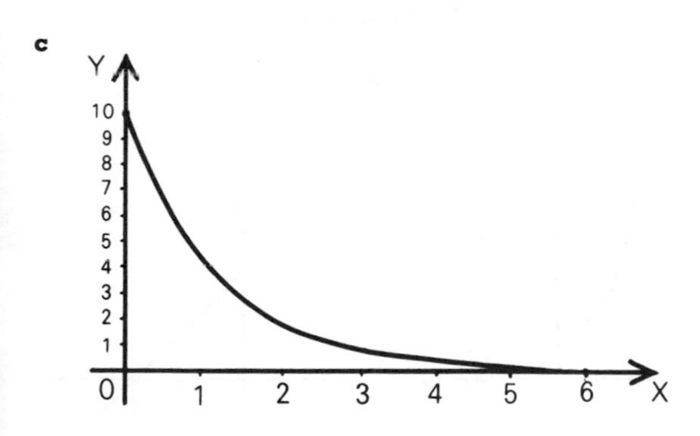

d After 10 weeks only 0·001 litre is left
2 a $4000(0{\cdot}85)^t$ **b** £1509
3 a Amount is halved every 1600 years **b** 3·86 **c** 0·00086 g left. Yes
4 a $5(0{\cdot}5)^{t/20}$ **b** 246 years
5 a $(0{\cdot}5)^{t/24\,400}$ **b** (i) 0·24 kg (ii) 0·06 kg
6 a $(0{\cdot}5)^{t/5720}$ **b** $(0{\cdot}5)^2 = 0{\cdot}25$, so $\dfrac{t}{5720} = 2$,
i.e. $t = 11\,440$ years

Page 174 *Exercise 4*

1 a 148·413 **b** 3·320 **c** 9·974 **d** 0·135 **e** 0·223 **f** 0·607
2 a 1, 0·607, 0·368, 0·223, 0·135 **b** (i) 0 (ii) 1 (iii) 5
3 a 1, 0·135, 0·018, 0·002 **b** (i) 1 (ii) 0

4 a

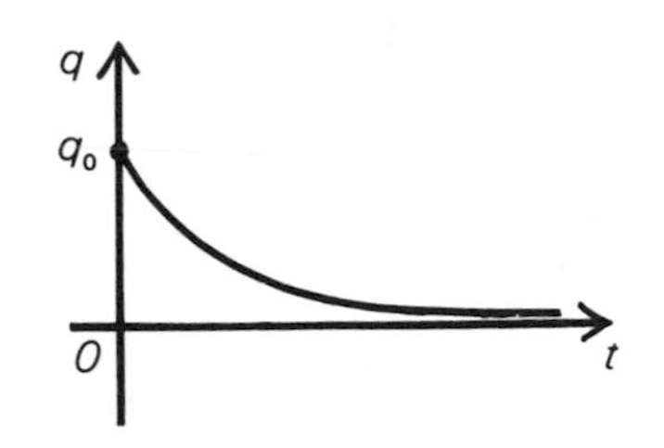

b (i) 0·7RC (ii) 2·3RC
5 a $i(0) = \dfrac{E}{R} + \left(i_0 - \dfrac{E}{R}\right) = i_0$ **b** $i(t) \to \dfrac{E}{R}$ as $t \to \infty$

c

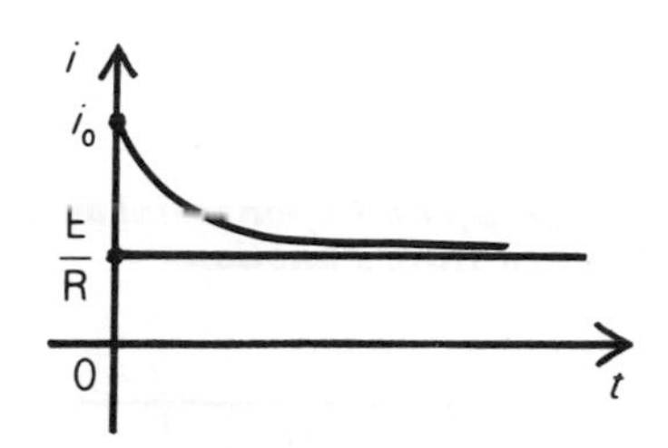

Page 175 *Investigation*

x	0	0·5	1	1·25	1·5	1·75
e^x	1	1·65	2·72	3·49	4·48	5·75

x	1	2	3	4	5	6
ln$x$	0	0·69	1·10	1·39	1·61	1·79

Axis of symmetry $y = x$
e^x and $\ln x$ are inverses of each other

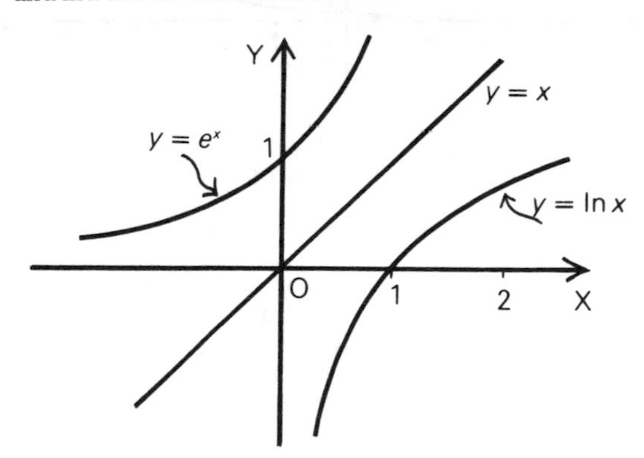

Page 176 *Exercise 5*

1 a $\log_3 81 = 4$ **b** $\log_{10} 1000 = 3$ **c** $\log_9 3 = \frac{1}{2}$ **d** $\log_5 \frac{1}{25} = -2$
e $\log_7 1 = 0$
2 a 2 **b** 4 **c** 2 **d** 3 **e** 1 **f** 2 **g** $\frac{1}{3}$ **h** -1 **i** -2 **j** $-\frac{1}{2}$
3 a $\log_{10} y = 2$ **b** $\log_e x = 2$ **c** $\log_e v = -1$ **d** $\log_{10} u = -10$
e $\log_x V = 3$ **f** $\log_q p = \frac{1}{2}$ **g** $\log_a c = b$ **h** $\log_y x = -4$

4 a $y = e^x$ **b** $p = 10^x$ **c** $w = 5^t$ **d** $v = a^{2u}$

5 1995 **6** 16 **7** $\frac{I}{I_0} = 10^{10}$

Page 178 ***Exercise 6***

1 a 2 **b** 1 **c** 0 **d** 1 **e** 2 **f** 1
2 a 2 **b** 2 **c** 2 **d** 1 **e** 4 **f** $\frac{3}{2}$
3 a 5 **b** 100 **c** $\frac{1}{3}$
4 $\log \frac{1}{2} < 0$ (for base $a > 1$), so last line should be '$\Rightarrow 4 > 3$'
5 18 **6 a** 50 **b** 74 **7** 8 **8** 19

Page 180 ***Exercise 7***

1 a $Y = 6X$ **b** $Y = 4X + \log_{10} 3$ **c** $Y = \frac{1}{2}X + \log_{10} 2$
d $Y = -4X + \log_{10} 1{\cdot}2$, where $X = \log_{10} x$, $Y = \log_{10} y$
2 a $Y = x + \log_e 2$ **b** $Y = x \log_e 10 + \log_e 3$ or
$Y(\log_{10} y) = x + \log_{10} 3$ **c** $Y = -x + \log_e 1{\cdot}3$
d $Y = 2x + \log_e k$, where $Y = \log_e y$
3 $y = x^{1{\cdot}5}$ **4** $y = (2{\cdot}1)^x$ **5** $y = 3x^{0{\cdot}5}$ **6** $y = 2{\cdot}3(5{\cdot}3)^x$

Page 180 ***Check-up on Exponential and Logarithmic Functions***

1 a

x	−5	−4	−3	−2	−1	0	1	2	3	4	5
3^x	0·004	0·012	0·037	0·111	0·333	1	3	9	27	81	243

b (Shape only)

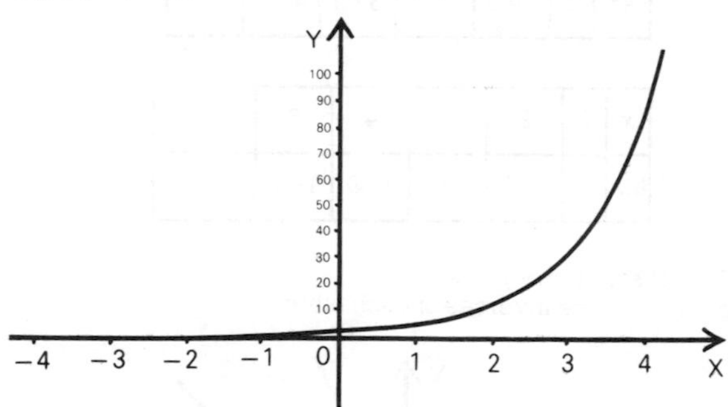

c (i) 12·5 (ii) 1·7 (iii) 0·7 (to 1 decimal place, by calculator)
d (i) $3^x \to \infty$ as $x \to \infty$ (ii) $3^x \to 0$ as $x \to -\infty$
2 a 7·1 billion **b** year 2009
3 a $40\,000(0{\cdot}75)^t$ **b** 9 years (whole number)
4 a 4 **b** 3 **c** $\frac{1}{2}$ **d** $-\frac{1}{2}$
5 a $\log_5 y = 3$ **b** $\log_e u = -2$ **c** $\log_v w = \frac{3}{4}$ **d** $\log_{10} V = -t^2$
6 a $y = e^2$ **b** $p = 10^{3v}$ **c** $2w = \alpha^{1/2}$ **d** $y = a^{x^3}$
7 a 1 **b** 2 **c** 2 **8 a** $y = \frac{1}{3}a^x$ **b** $y = ex^2$ **9** 32 **10 a** 20·4
b 40·8 **c** 67·7 (thousand feet) **11** $y = 2{\cdot}7(3{\cdot}6)^x$.

Page 182 ***Revision Exercise 1***

1 a $(1, -2)$, $\sqrt{5}$ **b** $(0, 0)$, $(2, 0)$, $(0, -4)$, **c** **d** $(2, 0)$ $(3, -1)$

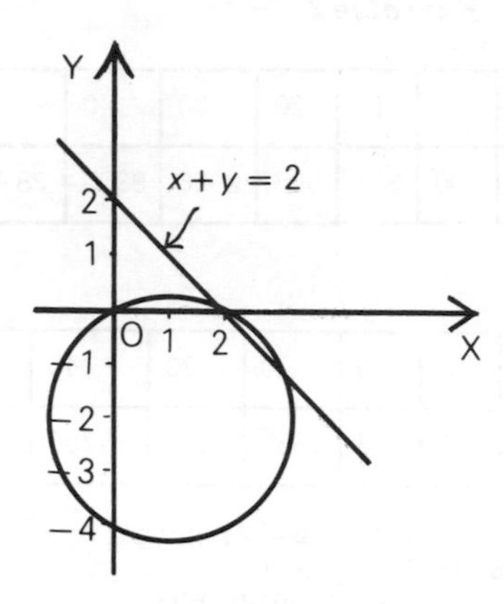

2 $(x-2)(2x^2+x+3)$
3 a $6(0{\cdot}9)^3 = 4{\cdot}4$ metres
b $6[1+2(0{\cdot}9)+2(0{\cdot}9)^2+(0{\cdot}9)^3]$
4 $y(x) = 5x - 2x^3 + 7$ **5** $7x - 8y + 12 = 0$
6 $y = \sqrt{(1-x^2)}$ is upper half of circle $x^2 + y^2 = 1$
Integral $= \frac{1}{4}$ area of circle $= \frac{\pi}{4}$

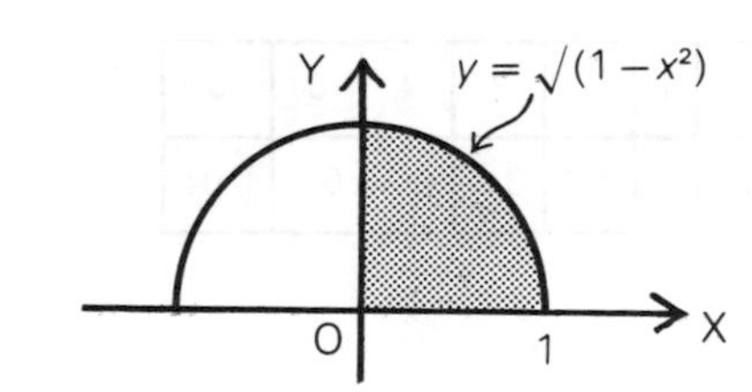

7 a

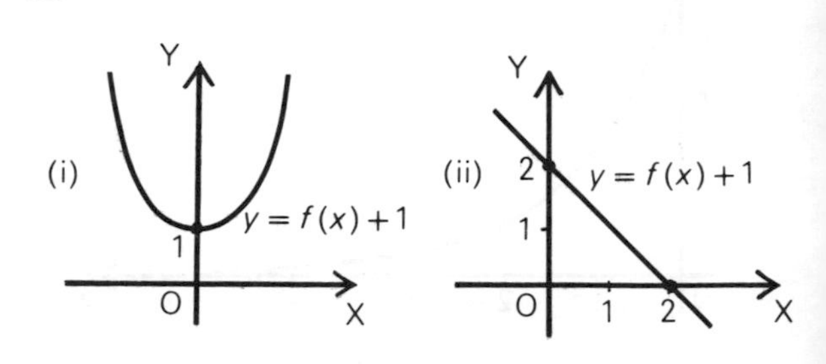

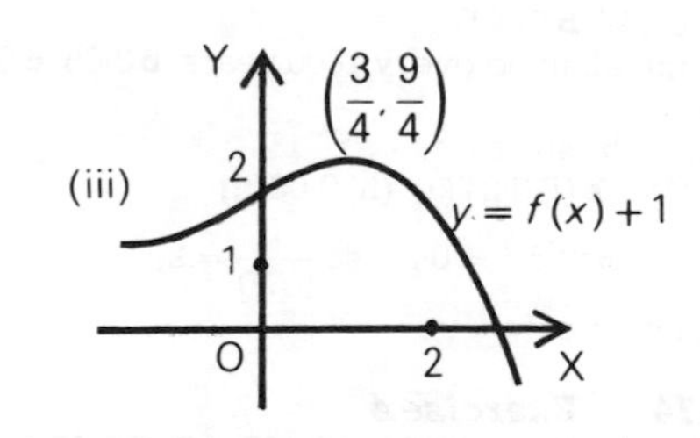

b

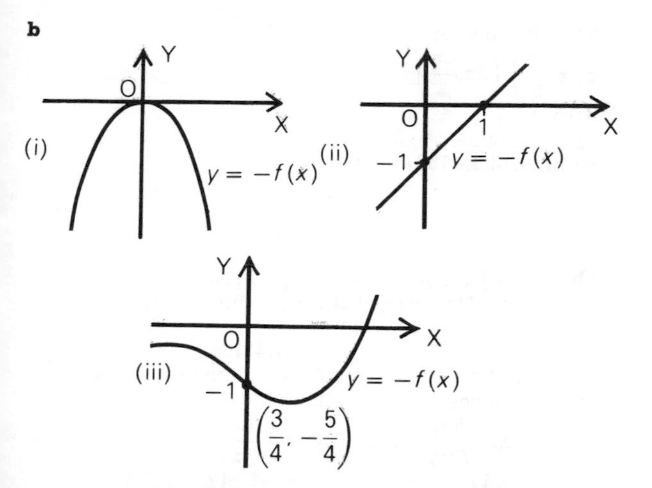

8 a $(x+1)^2+4$, min 4, -1 **b** $2(x-2)^2-7$ min, $-7, 2$
c $13-(x+3)^2$ max 13, -3
9 Step $\log\left(\sin\frac{\pi}{4}\right) = \log\left(\sin\frac{\pi}{4}\right) \Rightarrow 2\log\left(\sin\frac{\pi}{4}\right)$
$> \log\left(\sin\frac{\pi}{4}\right)$ is false since $\log\left(\sin\frac{\pi}{4}\right) = \log\left(\frac{1}{\sqrt{2}}\right)$
$= -\frac{1}{2}\log 2$ is negative for base $a > 1$
10 $f\circ g(x) = 18x^2-60x+51$, $g\circ f(x) = 2-6x^2$
11 Sin 3A = sin 2A cos A + cos 2A sin A =
(2 sin A cos A) cos A + (cos² A − sin² A) sin A, and now use
cos² A = 1 − sin² A
12 a $\frac{1}{\sqrt{3}}$ **b** $\sqrt{\left(\frac{2}{3}\right)}$ **13** $b = 2a$
14 a $-2, 1$ **b** $x > 1, x < -2$ **c** $-2 < x < 1$
15 $5\cos(200t° - 143{\cdot}13°)$
16 Perpendicular bisectors: AB: $3x+4y = -15$,
BC: $3x-y = 0$, CA: $2x+y = -5$; concurrent at $(-1, -3)$
17 $3\cos(3x+1)-(2-3x)^{-2/3}$
18 $0{\cdot}25, 2{\cdot}89, \frac{3\pi}{2}$ **19** $\pm\frac{1}{6}\sqrt{23}$
20 a cos (A + B) = cos A cos B − sin A sin B,
cos (A − B) = cos A cos B + sin A sin B
b Integral $= \frac{1}{2}\int_0^{\pi/4}(\cos 2x - \cos 4x)\,dx = \frac{1}{4}$
21 $k = 1{\cdot}6, n = -0{\cdot}1$
22 Domain $\{x\in R: x \geqslant 0\}$, range $\{y\in R: y \geqslant 3\}$
23 a 14×10^6 **b** 35 years **24** 32
25 -16 at $x = 2$ **26 a** $2\cos\left(x-\frac{11\pi}{6}\right)$ **b** $\frac{\pi}{2}, \frac{7\pi}{6}$
27 a $f^{-1}(y) = \frac{1}{3}(y+4)$ **b** $f^{-1}(y) = (y-1)^{1/3}$ **c** $f^{-1}(y) = \frac{1-y}{y}$
d $f^{-1}(y) = \frac{5y-3}{4y-2}$
28 ± 2 **29** cos 2A $= -\frac{3}{4}$, sin 2A $= \frac{1}{4}\sqrt{7}$
30 400π cm³/s

Page 184 *Revision Exercise 2*

1 a $(2, -3)$ **b** $3^2+4^2-4(3)+6(4)-37 = 0$, $x+7y-31 = 0$
c $x+7y+69 = 0$
2 a $f(-3) = -27$, minimum; $f(0) = 0$, horizontal point of inflexion
b $(0, 0), (-4, 0)$
c To the right, rising sharply. To the left, rising sharply

d

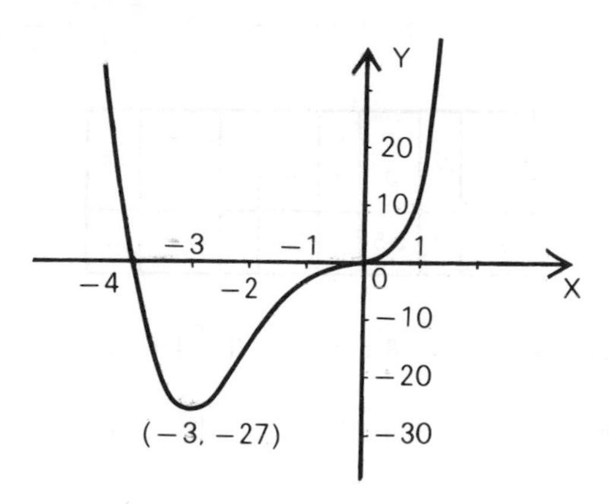

e $a < -27$
3 a $\cos B = -\frac{1}{\sqrt{57}} < 0$, so ∠B is obtuse
b P(−1, 1, 3), Q(5, −5, −9), R(0, 0, 1)
c $\overrightarrow{PQ} = -\frac{6}{5}\overrightarrow{QR}$, so $\overrightarrow{PQ}$, $\overrightarrow{QR}$ are parallel with Q in common, and therefore, P, Q, R are collinear
4 a $\frac{1}{2}S_n = 1-\frac{1}{2^n}$, so $S_2 = 2-\frac{1}{2^{n-1}}$
b (i) $3+\frac{1}{2}$, $3\left(1+\frac{1}{2}\right)+\frac{1}{2^2}$, $3\left(1+\frac{1}{2}+\frac{1}{2^2}\right)+\frac{1}{2^3}$
(ii) By extension of (i):
$u_n = 3\left(1+\frac{1}{2}+\frac{1}{2^2}+\ldots+\frac{1}{2^{n-2}}\right)+\frac{1}{2^{n-1}} (n \geqslant 2)$
(iii) By **a**, $u_n = 3\left(2-\frac{1}{2^{n-2}}\right)+\frac{1}{2^{n-1}} = 6-\frac{5}{2^{n-1}}$, which is true for $n \geqslant 1$
(iv) 6
5 b $\frac{5}{8}a^2\cos(2\theta-2{\cdot}5)+\frac{1}{2}a^2$ **c** $\frac{9}{8}a^2$, 1·2 radians
6 a $S = 2x^2+\frac{32}{x}$ **b** 2 **c** S(1) = 34 cm²
7 a $\frac{1}{\sqrt{2}}$ **b** 1 **c** 2·3
8 a $2+0+(-1) = 1$, so B ∈ plane; similarly C, D ∈ plane
b $M_1(3, 1, 2)$, $M_3(3, -3, 1)$, midpoint of M_1M_3 is $K(3, -1, \frac{3}{2})$; result follows since K is also the midpoint of M_2M_4 and M_5M_6 **c** $\frac{1}{17}$
9 a $M_1 - M_2 = 2$ **b** 89·1 **c** 1·5
10 a $f(x) = (x-1)^2(x+1)(x+2)$, roots $1, 1, -1, -2$
b $f'(x) = 4x^3+3x^2-6x-1$ (i) $f'(1) = 0$ (ii) $\frac{1}{8}(-7\pm\sqrt{33}) \doteqdot -0{\cdot}16, -1{\cdot}59$

c

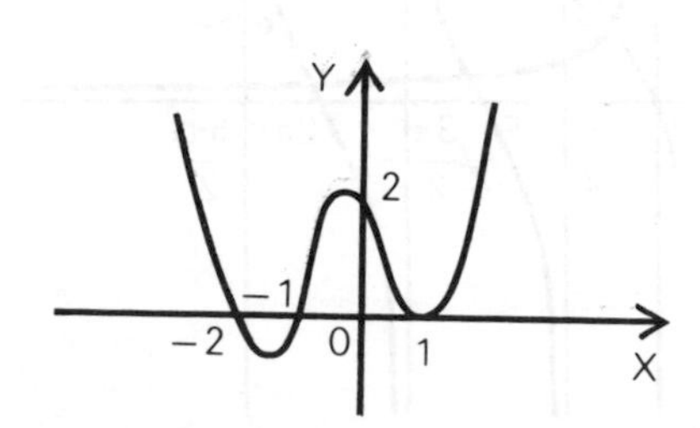

11 a $f'(\theta) = 24\sin^2\theta\cos\theta - 6\sin 2\theta$, $\theta = 0, \frac{\pi}{6}, \frac{\pi}{2}, \frac{5\pi}{6}, \pi, \frac{3\pi}{2}, 2\pi$

b

θ	0	$\frac{\pi}{6}$	$\frac{\pi}{2}$	$\frac{5\pi}{6}$	π	$\frac{3\pi}{2}$	2π
$f(\theta)$	3	2·5	5	2·5	3	−11	3

c Max. value $f\left(\frac{\pi}{2}\right) = 5$, min. value $f\left(\frac{3\pi}{2}\right) = -11$

12 a Both pass through points $(0, 0), \left(\frac{\pi}{4}, \frac{1}{2}\right), \left(\frac{\pi}{2}, 1\right)$

b

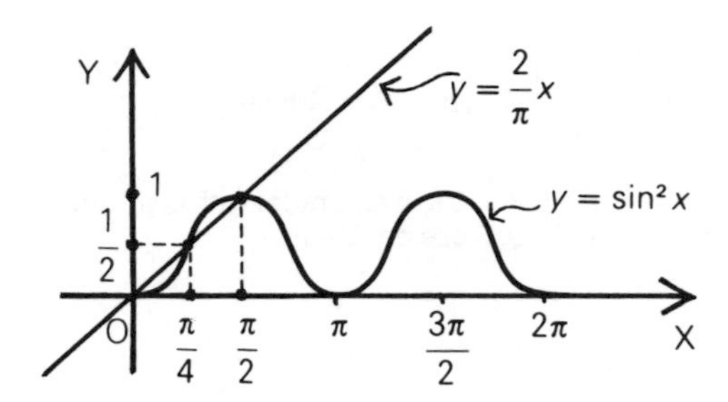

c Area $= \int_0^{\pi/4}\left(\frac{2}{\pi}x - \sin^2 x\right)dx + \int_{\pi/4}^{\pi/2}\left(\sin^2 x - \frac{2}{\pi}x\right)dx = \frac{1}{8}(4-\pi)$

13 a $(1-r)S_n = 1-r^n$, etc.

b (i) Flea is at $\frac{1}{2} - \frac{1}{2^2} + \frac{1}{2^3} - \ldots - \frac{1}{2^{n-1}} + \frac{1}{2^n}$ (n odd)

$= \frac{1}{2}\left[1 + r + r^2 + \ldots + r^{n-1}\right]\left(r = -\frac{1}{2}\right) = \frac{1}{3}\left(1 + \frac{1}{2^n}\right)$ (using **a**)

(ii) Similarly for n even

(iii) $\frac{1}{3}$

14 a

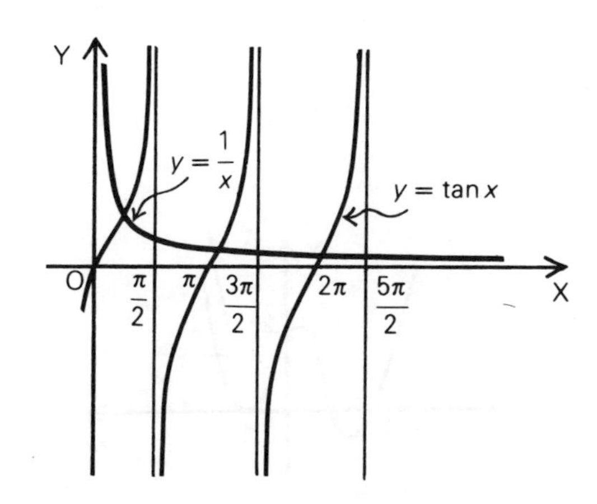

b Infinite number

c $\left(\pi, \frac{3\pi}{2}\right), \left(2\pi, \frac{5\pi}{2}\right)$

d 0·86

15 a $P = 260p - p^2 - 2600$

b £130 **c** Make 150 articles.

16 INVESTIGATIONS

1 b (i) $\frac{3}{2}$ (ii) $\frac{17}{12}$ (iii) $\frac{577}{408}$ (iv) $x_1, x_2, x_3, \ldots$ are, from the geometry, all to the right of $x = \sqrt{2}$ and getting closer and closer.

$x_4 = \frac{577}{408} = 1{\cdot}4142156\ldots$, $\sqrt{2} = 1{\cdot}4142135\ldots$, so x_4 gives $\sqrt{2}$ correct to 4 decimal places (v) $x_1 = 1$ gives $x_2 = \frac{3}{2}$, and the same sequence as above follows

c (i) $x_3 = x_2 - \frac{f(x_2)}{f'(x_2)}$ (ii) $x_{n+1} = x_n - \frac{f(x_n)}{f'(x_n)}$

(iii) $x_{n+1} = \frac{1}{2}\left(x_n + \frac{2}{x_n}\right)$ (iv) Use $x_{n+1} = \frac{1}{2}\left(x_n + \frac{21}{x_n}\right)$

(v) $x_1 = \frac{1}{\sqrt{5}}$ gives $x_2 = -\frac{1}{\sqrt{5}}$, which gives $x_3 = \frac{1}{\sqrt{5}}$ and this cycle continues

2 a $y = -2x^2 + x + 4$

b $\int_{-h}^{h}(kx^3 + ax^2 + bx + c)\,dx = \int_{-h}^{h}(ax^2 + bx + c)\,dx$ since $\int_{-h}^{h}x^3\,dx = 0$ (x^3 is an odd function), so the parabolic approximation is exact for every polynomial of degree $\leqslant 3$

3 a (i) $2, p, p+(p-2)$ has $2p-2$ even and so not a prime (p a prime)

(iii) No; $p, p+d, p+2d, \ldots, p+pd, \ldots$ has $p+pd = p(1+d)$ not a prime

b (i) Any even d can be used as a common difference.

(ii) $1, 2^r, 2^s$ $(s > r)$ has $2^r - 1$ odd and $2^s - 2^r$ even

4 a $n = 1 \Rightarrow n-1 = 0$; $n = 2 \Rightarrow$ sides 1, 2, 3, and so not a triangle

b $n = 4$ only **c** $n = 3$ only

d (i) $\cos B = \frac{n-4}{2(n-1)}$, so $n \geqslant 5$ for the largest angle B to be acute

(ii) $\cos B = \frac{1}{2}\frac{1-\frac{4}{n}}{1-\frac{1}{n}} \to \frac{1}{2}$ as $n \to \infty$, so $\angle B \to \frac{\pi}{3}$

(iv) $AD^2 = (n-1)^2 - y^2 = (n+1)^2 - x^2$, so $x^2 - y^2 = 4n$; but $x + y = n$, so $x - y = 4$

(v) $\Delta = \frac{\sqrt{3}}{4}n\sqrt{(n^2-4)} = \frac{\sqrt{3}}{2}n\sqrt{\left(\left(\frac{n}{2}\right)^2 - 1\right)}$

(vi) n must be even and $\left(\frac{n}{2}\right)^2 - 1$ of the form $3t^2$ for some integer t

5 a Number is $\frac{1}{2}n(n-1)$ **b** $V + R - E = 1$

c (i) 2, 4, 7, 11 (ii) $P_5 = 16$

(iii) (1) $2n$ (2) $\frac{1}{2}n(n-1)$ (3) $V_n = 2n + \frac{1}{2}n(n-1)$

(iv) (1) $2E$, since each edge counts twice

(2) $E_n = 3n + n(n-1)$

(v) $P_n = 1 + \frac{1}{2}(n + n^2)$

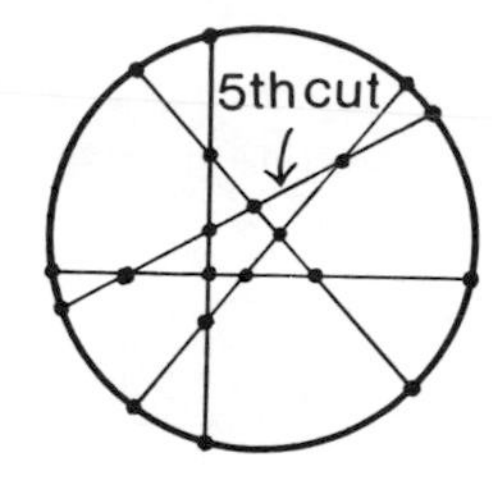

6 a $(m^2-n^2)^2+(2mn)^2=(m^2+n^2)^2$
c $m=dm_1, n=dn_1 \Rightarrow m^2-n^2=d^2(m_1^2-n_1^2)$, $2mn=d^2 2m_1n_1$, $m^2+n^2=d^2(m_1^2+n_1^2)$
d m, n with no common factor (>1) and one even, one odd
e $m-n=1$
7 a (ii) $S_n=\frac{1}{6}n(n+1)(2n+1)$
b (i) (2) $T_n=\frac{1}{6}n(n+1)(n+2)$
(ii) (2) Note that $1+2+3+\ldots+(n-1)=\frac{1}{2}n(n-1)$
(3) $U_n=\frac{1}{2}n(n-1)+\frac{1}{2}(n-2)(n-3)+\ldots+\frac{1}{2}.3.2$ (n *odd*)
8 a $N_{n+1}=4N_n$, $N_n=3.4^n$ $(n\geqslant 0)$
b $P_n=3,(\frac{4}{3})^n$ $(n\geqslant 0)$ **c** $P_n\to\infty$ as $n\to\infty$
d Area enclosed is $\Delta_n=$
$\Delta\left(1+\frac{1}{3}+4.\frac{1}{3^3}+4^2.\frac{1}{3^5}+\ldots+4^{n-1}.\frac{1}{3^{2n-1}}\right)$ $(n\geqslant 1)$, where Δ is the area of S_0

$$\Delta_n=\Delta+\frac{1}{3}\Delta\left[1+\left(\frac{2}{3}\right)^2+\left(\frac{2}{3}\right)^4+\ldots+\left(\frac{2}{3}\right)^{2n-2}\right]\to\Delta+\frac{1}{3}\Delta\frac{1}{1-(\frac{2}{3})^2}$$

as $n\to\infty$, i.e. $\Delta_n\to\frac{8}{5}\Delta$ as $n\to\infty$
9 a Let T_n denote minimum distance to deposit 1 tank at D_n:
$T_1=2$, $T_2=3T_1$, $T_n=2(T_1+T_2+\ldots+T_{n-2})+3T_{n-1}$ $(n\geqslant 3)$
Let R_n denote minimum distance to D_n without leaving a tank there, but returning to base:
$R_1=2$, $R_n=2(T_1+T_2+\ldots+T_{n-1})+2n$ $(n\geqslant 2)$
b Let N_n denote minimum distance to D_n (new base), with no return.
$N_1=1$, $N_2=2$, $N_3=T_1+3$, $N_4=2T_1+4$,
$N_n=2T_1+T_2+T_3+\ldots+T_{n-3}+n$ $(n\geqslant 5)$
10 a (i) $\frac{dx}{dt}=50$, $x=0$; $\frac{dy}{dt}=50\sqrt{3}$, $y=0$
(ii) $x=50t$, $y=50\sqrt{3}t-5t^2$
(iii) max. height 375 m at $t=5\sqrt{3}$ seconds; impact at $x=500\sqrt{3}$ m; $y=x\sqrt{3}-\frac{1}{500}x^2$, etc.
b Similarly, but now $x=25\sqrt{2}t$, $y=25\sqrt{2}t-5t^2+10$
c Here $x=(V\cos\alpha)t$, $y=(V\sin\alpha)t-\frac{1}{2}gt^2+h$
d $a=V\cos\alpha t$, $0=V\sin\alpha t-\frac{1}{2}gt^2+h$ at impact.
Case $h=0$ leads to $\sin 2\alpha=\frac{ga}{V^2}$.
Equation in case $h\neq 0$ is more complicated; try solving for $\tan\alpha$.
e $a=\frac{V}{\sqrt{2}}t$, $0=\frac{V}{\sqrt{2}}t-\frac{1}{2}gt^2+h$ at impact, so $V=a\sqrt{\left(\frac{g}{a+h}\right)}$.